How to prepare for College Board Achievement Tests

BIOLOGY

Including **Modern Biology in Review**

REVISED EDITION

by **MAURICE BLEIFELD**

Principal, Martin Van Buren High School, Queens Village, New York

BARRON'S EDUCATIONAL SERIES, INC

Woodbury, New York

About the Author

Maurice Bleifeld was formerly principal, Benjamin Franklin High School, New York City, principal of Samuel Huntington Junor High School, Jamaica, N.Y., chairman of Department of Biological Sciences at Newtown High School, Elmhurst, N.Y., and teacher of Biology at the Bronx High School of Science. He is the recipient of several awards from the National Science Teachers Association. Among his former positions: Lecturer on Science Education, New York University; Harvard-Newton Summer Program; City College of New York, Intensive Teacher Training Program. Teacher, Children's Science School, Woods Hole, Mass. Teaching guide writer, Scholastic Magazine *Science World*. President, New York Biology Teachers Association.

All inquiries should be addressed to:
Barron's Educational Series, Inc.
113 Crossways Park Drive
Woodbury, New York 11797

Library of Congress Catalog Card No. 61-18354
International Standard Book No. 0-8120-0374-8

PRINTED IN THE UNITED STATES OF AMERICA

Contents

Preface

A high school student preparing to apply for college admission may be asked to take one or more college entrance examinations. Among these are the College Board Achievement Tests. This book has been written to help students prepare for the College Board Achievement Test in Biology.

The book is organized in two parts as follows:

PART ONE — MODERN BIOLOGY IN REVIEW. This part has been revised and enlarged to include fundamentals in the following new areas:

Cellular structure and function, based on modern concepts of the cell.

Biochemistry, including the structural makeup of organic compounds, the role of enzymes, details of cellular respiration, and stages of photosynthesis.

Modern genetics, dealing with DNA, replication, RNA, protein synthesis and gene action.

Modern concepts of evolution, the heterotroph hypothesis, including population genetics, and the development of new species.

Modern ecology, including abiotic factors, ecological succession, energy flow, and world biomes.

Modern system of classification, based on an organization of three kingdoms, protist, animal and plant.

In addition to these new areas, there are other revisions on modern understandings of synthesis, transport, human physiology, hormones, mitosis, meiosis, heredity of man, and numerous other topics. Many additional illustrations have been added to illuminate the new concepts. Throughout the book there are key definitions, charts, tables, drawings and problems which present the essential concepts in clear fashion.

PART TWO — TEN PRACTICE ACHIEVEMENT TESTS which are similar to the College Board Achievement Tests in content, type of question and degree of difficulty. This part has been revised to include questions on modern biology. Each practice test has also been brought up to date with a series of challenging questions based on a reading selection taken from a scientific journal and dealing with such modern topics as: biochemistry, DNA, protein synthesis, nerve conduction, photosynthesis, replication, photoperiodism, imprinting and the species concept.

There is an answer key at the end of each test, and an explanation of all the answers. The tests were constructed specifically for this book, and do not contain questions actually used on any of the college board tests. They will help the student become familiar with the many types of questions, and give him experience in dealing with them.

A student who is preparing to take the College Board Achievement Test in Biology should profit from the information presented in PARTS ONE and Two. It would be advisable to study portions, or all, of the part reviewing modern biology, so as to be thoroughly familiar with the subject. Careful thought will be needed for understanding the questions and answering the Practice Achievement Tests.

The up-to-date revision of this book takes into account the recent adoption of a revised Biology syllabus in New York State. This new course of study evolved from an experimental syllabus that was tried out in a number of selected schools throughout the state over a four-year period. The book deals comprehensively with each of the units in the new syllabus. Consequently, it may be used to good advantage by students who are studying the new course, and who are preparing for the New York State Regents Examination in Biology.

Because of its comprehensive treatment of modern concepts in biology, the book will also be useful to those preparing for College Board tests, scholarship examinations, standardized tests, college biology courses, teaching examinations, and other types of challenging tests in biology. Despite the technical and advanced nature of some of the material, the terminology and vocabulary have been kept as meaningful as possible.

Grateful acknowledgment is made to the students, teachers and scientists who were kind enough to offer a number of constructive comments and suggestions which have been incorporated in this edition, and to the following publications for the use of various selections:

Scientific American. September 1968 ("How Light Interacts with Living Matter"), September 1961 ("The Living Cell"; "How Cells Transform Energy"; "How Cells Make Molecules"; "How Cells Communicate"; "How Cells Divide"), December 1958 ("The Amateur Scientist"), March 1958 ("Imprinting in Animals"), May 1952 ("The Control of Flowering"; "Inherited Sense Defects").

Ideas in Modern Biology, ed. John A. Moore, The Natural History Press, Garden City, N. Y. 1965.

Adventures in Biology, published by the New York Association of Biology Teachers.

Natural History. October 1960, "Young Scientist."

Humane Biology Projects. 1960, "Ecology and Conservation."

A special debt of gratitude is due my wife, Belle K. Bleifeld, for her valuable help in typing the manuscript.

MAURICE BLEIFELD

Information about the College Board Achievement Test in Biology

The College Board Achievement Test in Biology is a one-hour examination which consists of a varying number of multiple-choice questions of different types. It is different from most comprehensive examinations taken by high school students, in that it not only tests for a knowledge of the subject matter, but also tests ability to apply and reason with this knowledge.

Since courses in biology vary from school to school, the questions are selected for their general suitability to a wide variety of schools. It is not expected that a student will know all of the answers on any one test. However, a student with a good background in biology, who can think correctly about what he has learned, should be able to achieve a good score on the test.

There are no passing or failing scores on the test. The range of scores is from a low of 200 to a high of 800. The results are reported both to the high school and the college selected by the applicant. If an applicant's score is very low, a college admissions officer may doubt whether he is ready for college; if it is high, an applicant will usually be considered as being ready for the challenging work of college. It must always be remembered that a score on the achievement test is only one aspect of a student's total record—which also includes the other College Board scores, the scholastic record, the school's evaluation, and other pertinent information.

HOW TO PREPARE FOR THE TESTS

The first step in preparing for the tests is to review the subject matter of biology carefully and thoroughly. In doing this, it would be well to acquire both broad understandings and specific details about each of the units. The questions at the end of each chapter will help evaluate your progress. If a particular subject appears to present difficulty, it should be reviewed more intensively.

The Practice Achievement Tests will probably present a great challenge at first, because of the different types of questions. Some of these types may be new to you. It would be advisable to read the directions carefully, and to think through each question before attempting to answer it. The answer key at the end of each of the Practice Achievement Tests should be consulted only after you have tried to answer the entire test. Check on each incorrect answer. Use the review section to brush up on a particular area.

The subject matter background will be helpful on these tests as on other tests you have taken. However, you will need to think about and reason with this background. As you continue to answer the various Practice Achievement Tests, you will gain an understanding of the different types of questions. This type of familiarity will help you develop confidence and mastery. Above all, do not leave your review and practice until the last minute. Cramming will not help much on this test. Pace yourself so that you can work thoroughly, giving yourself as much time as you need.

TAKING THE TESTS

Since the test lasts an hour, you may not be able to answer all of the questions. It is advisable, therefore, to answer as many as you are sure about, first, and then to go back and work on the others. It would not be to your advantage to spend a great deal of time working out one answer, if you do not leave yourself enough time to answer the other questions that you do know.

The College Board tests do not encourage haphazard guessing. However, in answering questions about which you have only partial knowledge, you may be able to eliminate some of the wrong choices, and to improve your chances of selecting the correct answer.

You are advised not to worry if you are unable to answer all of the questions in a test, or if you do not have time to finish. No one is expected to know the answers to all of the questions. It may be of some comfort to realize that the other students taking the test undoubtedly find it just as challenging.

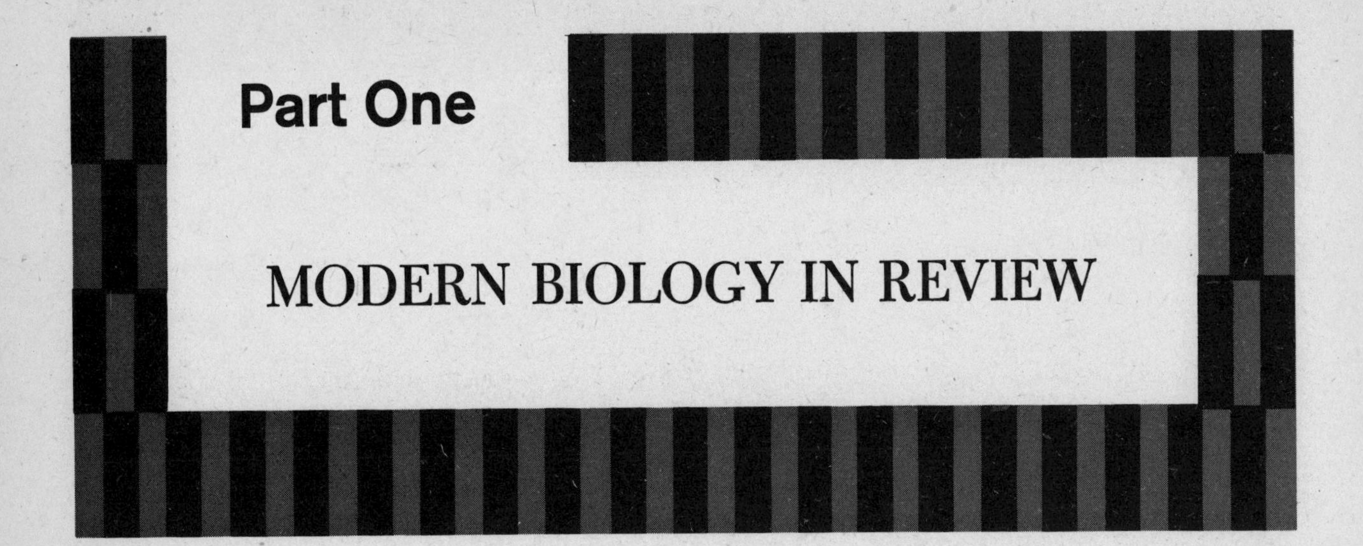

Part One

MODERN BIOLOGY IN REVIEW

HOW LIVING THINGS ARE CONSTRUCTED

1

1 Cells and protoplasm

USING THE MICROSCOPE

The average school microscope usually has an *eye-piece* (ocular) magnification of 10; the *low power objective* also has a magnification of 10. Objects viewed under low power will therefore be magnified 10 × 10, or 100 times. The *high power objective* generally has a magnification of 43; therefore objects studied under high power will be magnified 430 times (10 × 43). Not only is the image of an object enlarged by the lenses of the microscope, but it is also reversed and inverted. After first focusing on the image with the *coarse adjustment,* a sharper view can be obtained with the *fine adjustment.* The depth of the object can also be studied by careful use of the fine adjustment under high power. It is important for the *diaphragm* under the stage to be adjusted properly in order to admit the best amount of light for the clearest view. An even, north light is best; direct sunlight is to be avoided because of its glare and uneven qualities.

With the best equipment employing oil immersion lenses, the compound microscope is capable of magnifying up to about 1,800 times. The *electron microscope* permits magnifications of over 100,000 times. Viruses, heretofore, invisible, have now been photographed, as well as specific cell structures. A stream of electrons is focused on the object by means of a magnetic field, and the image is formed either on a fluorescent screen or on a photographic film.

Additional knowledge about the structure of cells has been obtained, through the use of modern techniques involving the: *phase microscope* for examining living cell structures without the use of stains which may distort or kill them; *time-lapse cinematography,* for observing the reactions of living cells treated with beams of X-rays, ultra-violet light or streams of atomic nuclei; selective stains used in combination with *laser* beams for studying the organelles of the cell; *ultracentrifuge,* for whirling cell fragments in a tube and isolating them in layers.

CELLS. The first discoveries of cells were made by such early microscopists as Antony van Leeuwenhoek (1632-1723), who made his own simple lens microscopes, and saw bacteria and protozoa; and Robert Hooke (1653-1703) who had several lenses in his *compound* microscope, and who first named cells, after studying the tiny boxlike structures in cork. Since then, it has been shown by other biologists such as Dutrochet, Schleiden and Schwann, that all living things are made of cells, whether they be whale, mouse, oak tree, or man.

Size of Cells. Cells are so tiny that they are measured in terms of microns (μ). There are 1,000 microns in a millimeter ($1\mu = 0.001$ mm.). Bacteria are the smallest cells that can be seen with the light microscope. They average 1-3 microns in size (0.001-0.003 mm.). Most other cells are about 10 microns in size. However, a nerve cell, which is microscopic, may extend more than a yard in length. And the yolk of a bird's egg, which is a single cell, may measure several inches across.

Cell Structure. Until quite recently, the structure of plant and animal cells was considered to be relatively simple, as shown in the following diagrams:

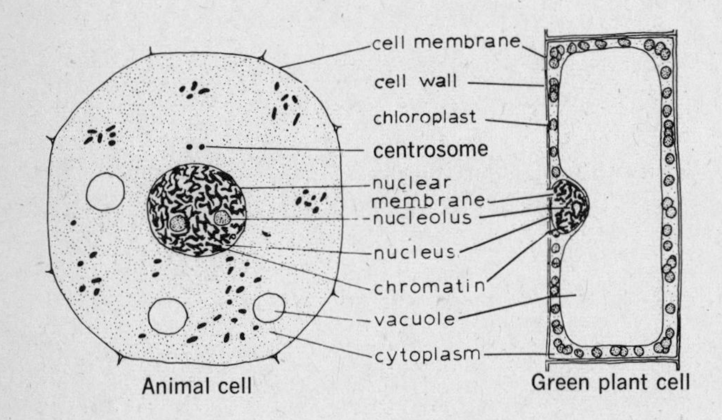

With the aid of the electron microscope, many details of the structure of the cell are being revealed, as shown in the modern version of a generalized cell:

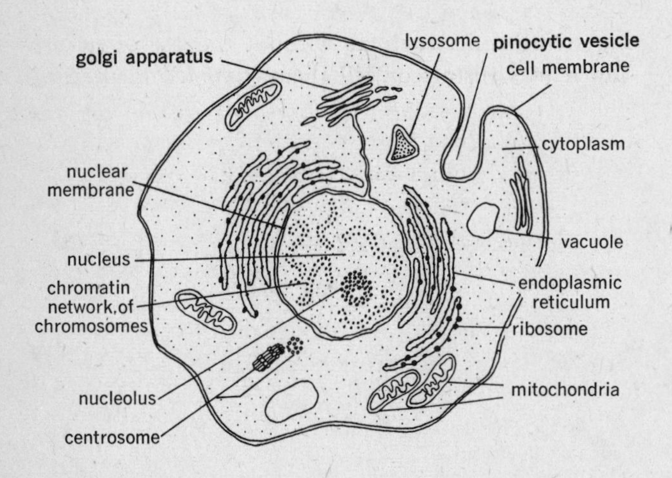

Protoplasm. A cell is a tiny unit of living material called protoplasm. It contains specialized areas called *organelles,* that perform certain functions:

1. Nucleus. The nucleus is a small spherical dense body. It controls the activities of the cell, cell reproduction, and heredity. It is made of a *chromatin* network of *chromosomes,* which contain the *genes,* the hereditary units. The genes have recently been shown to consist of molecules of DNA (deoxyribonucleic acid). There is usually at least one *nucleolus,* a tiny dot-like structure, inside the nucleus. It has been found to contain RNA (ribonucleic acid), which helps direct the synthesis of proteins. The nucleus is surrounded by the *nuclear membrane* which controls transport of materials into and out of the nucleus.

2. Cytoplasm, a granular, thick, grayish liquid which occupies most of the cell. It appears to to be somewhat like the white of a raw egg. It carries on most of the other activities of the cell. It contains:

a. Endoplasmic reticulum — a network of channels, or tubes, that extends throughout the cytoplasm. Its membranes connect with the nuclear membrane and the cell membrane. It is thought to function in the transport of materials throughout the cell.

b. Ribosomes — tiny granules that are distributed along the endoplasmic reticulum. They contain RNA, and function as sites for the synthesis of protein.

c. Mitochondria — rod-like structures with inner, folded surfaces called *cristae.* They contain enzymes associated with cellular respiration, leading to the release of energy from food. Most of the ATP molecules are formed here, serving as centers of energy storage. Mitochondria are called the "powerhouses" of the cell.

d. Lysosomes — contain digestive enzyzmes that break down large organic molecules, and worn-out organelles within the cell.

e. Vacuoles — are especially large in plant cells. They act as reservoirs for water and dissolved materials.

f. Golgi apparatus — produces lysosomes; concentrates enzyme secretions; secretes carbohydrates for plant cell wall formation.

g. Centrosome — present in animal cells, outside of the nucleus. It contains a pair of *centrioles,* which are composed of a bundle of small filaments, and which are active during nuclear division.

3. Cell Membrane. The cell, or plasma membrane, is the outer living layer of cytoplasm. It has a complex porous structure, composed mainly of protein and fatty material. It is semipermeable, and controls the passage of dissolved substances into and out of the cell. Some dissolved materials pass through the cell membrane by diffusion; this is *passive transport* in which the cell does not contribute energy for the movement. In *active transport,* the cell uses energy to move molecules across membranes, from a region of low concentration to a region of higher concentration. This is against the concentration gradient.

The process of diffusion can be explained in the following demonstration:

1. Fill a cellophane bag (or some dialyzing tubing) with a light starch suspension made by boiling

some cornstarch in water and allowing it to cool.

2. Seal the bag tightly with a rubber band, and place it in a jar of iodine solution. The effect of iodine on starch may be reviewed at this point by placing a drop of iodine directly on a small amount of starch suspension; a blue-black color results.

3. After a few minutes, it will be observed that the starch contents of the cellophane bag have turned a blue-black color. The iodine solution still retains its original color.

From this demonstration, it can be seen that iodine, which is in solution (is *soluble*) diffused through the membrane into the bag to stain the starch. The starch, which is not in solution (is *insoluble*), did not pass through the membrane. Other soluble materials, such as glucose, dissolved oxygen, and dissolved carbon dioxide can also diffuse through membrane. The diffusion of water through a membrane is known as *osmosis*.

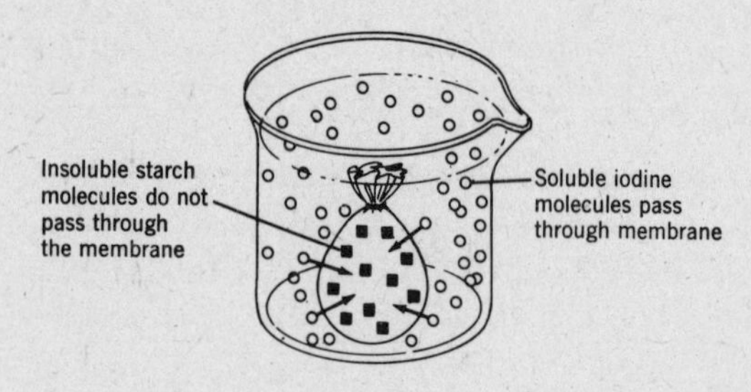

Demonstration: Diffusion

Explanation of diffusion. By referring to the *molecular theory* of matter, it is possible to explain the passage of these materials through a membrane. According to this theory, matter is made up of molecules, which are the smallest particles of a substance that still possess its properties. Molecules are in a constant state of movement, and tend to *diffuse*, or move from one place to another, especially in a gas or a liquid. Thus, you can smell ammonia from a distant bottle of ammonia water, because the ammonia molecules diffused through the air. Also, if a lump of sugar is added to a cup of tea, after a while without stirring you can taste the sugar in the first sip of tea, because the sugar molecules diffused through the tea.

In the above demonstration, the iodine solution contained a high concentration of iodine molecules; on the inside of the membrane, there were no iodine molecules. As they moved about constantly, many of the molecules of iodine passed through the membrane into the cellophane bag and stained some of the starch. As more and more iodine molecules continued to pass through the membrane, the entire starch suspension took on the blue-black color. The insoluble starch molecules, on the other hand, could not pass through the membrane, and so could not react with the iodine solution in the jar, which therefore retained its original light brown color.

In other words during diffusion through a membrane, there is a flow of dissolved molecules from a region of high concentration (the iodine solution) to a region of low concentration (the inside of the cellophane bag).

A *pinocytic vesicle* may form as a pocket of the cell membrane and engulf relatively large particles. These pockets are brought into the cell as vacuoles, which break down and release the particles into the cytoplasm. This process is known as pinocytosis, and also requires energy.

Protoplasm is a very complex substance composed chiefly of the elements carbon, oxygen, hydrogen and nitrogen, with smaller amounts of other elements such as phosphorus, sulphur, calcium, iron, etc. It also contains trace elements which are present in relatively minute amounts, such as copper, zinc, cobalt and flourine. It contains molecules of water (from 65% to 90% of the makeup of cells), protein, carbohydrates, fats, and minerals. Its content varies somewhat in the various living things. Even the protoplasm of the same organism may differ in specific parts, i.e., the protoplasm of skin cells is different from that of muscle cells.

Life Activities. Living things are able to carry on the following basic life functions:

Ingestion: The taking in of food.

Digestion: The breaking down of food to simpler soluble form, with the aid of enzymes.

Secretion: The formation of useful substances, such as enzymes.

Absorption: The passage of dissolved materials through the cell membrane, into and out of the cell.

Respiration: Release of energy from food.

Excretion: The waste products are passed out of the cell through the cell membrane.

Transport: Circulation of materials throughout the organism.

Regulation: Maintaining the stability of the organism's chemical makeup under a constantly changing internal and external environment (homeostasis).

Synthesis: The chemical formation of complex molecules from simple compounds.

Assimilation: The nonliving materials in food are changed into more protoplasm, resulting in growth and repair.

Reproduction: More living things are produced.

Irritability: The ability to respond to stimuli.

Bioluminescence: The production of light by a limited number of organisms (firefly; some bacteria, protozoa and fungi; deep-sea fish).

Movement: Most living things have the ability to change their position.

Living things perform the basic life functions with the help of specific structures or organs. In addition many of them have the ability to move from place to place. This function is largely restricted to animals, although a few plants such as the sensitive plant (mimosa), and the insectivorous plants (venus fly trap, sundew) have moving parts. Plant growth may also result in a change of position.

Viruses have represented an enigma to biologists. Are they living or nonliving? These ultramicroscopic materials are best known because of diseases they cause in living things (smallpox, polio, rabies, tobacco mosaic disease, etc.). They spread and invade living tissue. Once there, they reproduce. They thus show properties of protoplasm. On the other hand, they have been purified from infected tissue, and have appeared as dry protein crystals. In this condition, they are as nonliving as crystals of salt. However, once dissolved and placed in contact with living tissue, they become active and reproduce. Perhaps viruses are on the borderline between the living and nonliving. Recent investigations have shown that a virus generally contains a core of nucleic acid consisting of DNA or RNA, and a covering made of protein material.

Plant Cells. In addition to the nucleus, cytoplasm and cell membrane which are present in all living things, plant cells have certain other structures. If the cells of an onion skin are stained with a dye such as Lugol's iodine or methylene blue, the parts of the protoplasm can be readily seen.

The following nonliving structures of plant cells can also be observed:

1. A *cell wall* containing cellulose, surrounding the cell.
2. One or more clear vacuoles distributed throughout the cytoplasm, and containing *cell sap,* which is largely water with dissolved minerals.
3. Study of a green plant cell such as that from Elodea, a water plant, reveals a third type of structure present only in plant cells—*chloroplasts.* Chloroplasts are small oval bodies containing *chlorophyll* which is necessary for food-making (photosynthesis).

Ameba. One of the simplest of the protozoa is the one-celled ameba, which may be found in pond water. Although it consists of only one cell containing a tiny drop of streaming protoplasm, it is able to carry on all the life functions. It has no specific shape. As it crawls along, it extends projections called *pseudopods* (false feet), and flows into them. Ingestion is accomplished when the ameba comes in contact with a food particle which the pseudopods surround and engulf. Once inside the cell, the food particle becomes located within a *food vacuole.* Here, digestion occurs with the aid of enzymes which are secreted by the surrounding protoplasm into the food vacuole. As the food is digested and becomes soluble, it diffuses out of the food vacuole, into the protoplasm.

Some of the food is changed into more protoplasm during the process of assimilation. Respiration is also carried on: that is, oxygen is absorbed into the cell through the cell membrane; it is united with some of the food by special enzymes, to produce energy; carbon dioxide and water are formed as waste products. These waste products diffuse out through the cell membrane and are excreted. The *contractile vacuole,* which may be seen contracting and expanding, help eliminate excess water. After all the food in the food vacuole has been digested, the indigestible remains are eliminated in a simple manner—the cell membrane opens momentarily and these wastes are left behind as the ameba flows along.

When the ameba has grown to its maximum size, the chromatin material in the nucleus divides in half, forming two nuclei. Half of the cytoplasm collects around each nucleus, and the cell divides in two. By this simple process of reproduction known as *binary fission,* two identical daughter cells are produced. Each will move about independently, carry on all the life activities, grow and reproduce again. The ameba thus is potentially immortal, continuing to live on in its descendants.

The protoplasm of this simple cell also shows the property of irritability, or sensitivity to stimuli about it. If it comes in contact with a piece of plant material, it will engulf it and use if as food. On the other hand, it will stop moving in the direction of a grain of salt when it comes in contact with it, will reverse the flow of its protoplasm, and move away from it. A beam of strong light at one end of the ameba will cause it to flow in the opposite direction.

If conditions become unfavorable, such as drying of the pond, the ameba forms a hard protective wall around itself called a *cyst,* and its protoplasm becomes dormant. Under suitable conditions, the cyst breaks open and the ameba emerges to resume its usual activities. This ability to respond to the environment helps the ameba survive.

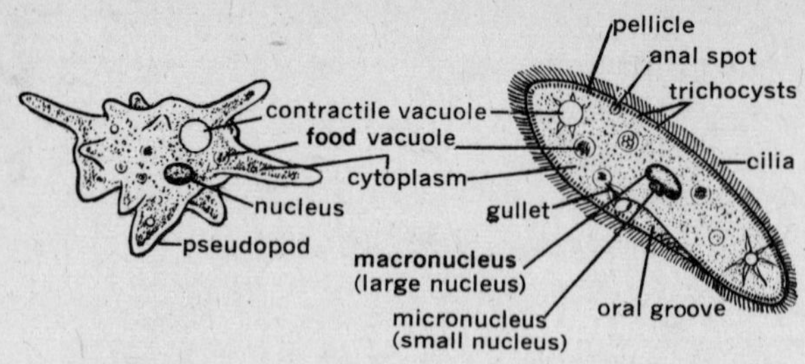

Paramecium. Another microscopic one-celled inhabitant of pond water is the well-known paramecium. It is a slipper-shaped organism that has a somewhat higher level of organization than the ameba for the performance of its life functions. It has a permanent shape maintained by the presence of a flexible exterior *pellicle*. The outside of the cell is covered with tiny hairlike extensions of protoplasm called *cilia*. These beat rapidly and move it through the water. There is an opening on one side called the *oral groove,* or mouth. It is also lined with cilia, which create a current of water that leads inward to the *gullet,* carrying with it tiny particles of food. At the end of the gullet, food vacuoles are formed which are carried slowly within the paramecium by the streaming protoplasm. As in the ameba, this is where digestion takes place. After the food within it has been digested and absorbed, a food vacuole will move close to a weak spot in the cell membrane called the *anal spot,* which will open suddenly and the indigestible solid material will be eliminated.

There are two contractile vacuoles, one at each end, surrounded by radiating canals, that alternately contract and expand to eliminate excess water. Since the water is being removed into an environment of fresh water, it occurs against a concentration gradient or level. Active transport is used to eliminate the water, in addition to the pumping action of the contractile vacuoles. The energy for active transport is derived from ATP molecules in the cell. The maintenance of the water balance is an example of *homeostasis*.

There are two nuclei, a large *macronucleus,* and a small *micronucleus*. During fission, both of them divide. The paramecium also carries on a type of sexual reproduction called *conjugation,* in which two cells lie close to each other and exchange parts of their micronuclei.

COMPARISON OF AMEBA AND PARAMECIUM

CHARACTERISTICS	AMEBA	PARAMECIUM
Shape	No specific shape	Slipper-shaped
Locomotion	By pseudopods—slow	By cilia—rapid
Ingestion	Engulfs food anywhere along cell membrane	Food enters through oral groove
Digestion	In food vacuoles	In food vacuoles
Elimination of solid wastes	Through temporary opening anywhere along cell membrane	Through definite anal spot
Excretion of wastes	By diffusion through cell membrane;	By diffusion through cell
Elimination of excess water	One contractile vacuole	Two contractile vacuoles having canals
Behavior	Moves toward favorable stimulus, away from unfavorable stimulus; forms cysts during unfavorable conditions.	Same; also expels trichocysts as protection or for anchoring
Reproduction	Binary fission	Binary fission; conjugation
Nucleus	One	Two (macronucleus and micronucleus)

The paramecium responds to stimuli in the environment by moving toward favorable factors and away from unfavorable ones. If it bumps into an object, the cilia reverse to carry it backward and then it goes forward again in another direction. There are small bodies called *trichocysts* lining the inside of the cell membrane, that discharge into long fine threads under certain conditions, i.e., when the paramecium is attacked by an enemy, or to anchor it while feeding on bacteria.

Euglena is another one-celled organism found in ponds, that is of interest to biologists. It is considered to be on the borderline between animals and plants, having characteristics of both. It moves by means of a whiplike *flagellum,* at the base of which is a contractile vacuole. Nearby is an *eye spot* that is sensitive to light. It does not seem to have a mouth for feeding, but instead, makes its own food by photosynthesis, by means of its numerous chloroplasts. Puddles of fresh water containing numerous euglenas will appear a greenish color.

Spirogyra is one of the green algae that lives in ponds. Its cells are arranged lengthwise in a long thread or *filament.* Each cell has a spiral chloroplast; in some species there may be two or more chloroplasts in each cell. Spirogyra manufactures its own food by carrying on photosynthesis. Starch is stored in structures on the chloroplast called *pyrenoids.* The individual cells reproduce by binary fission. At times, two filaments lying close to each other will send out projections from the cells, that touch and form a bridge; the protoplasm of each cell in one of the filaments streams across into the cells of the adjoining filament and combines with it. Hard-walled *zygospores* result which eventually form new filaments. This is a form of sexual reproduction called conjugation.

CHAPTER REVIEW

Select the correct choice for each of the following statements.

1. The average microscope having a high power objective marked 43x, and an ocular marked 10x gives a magnification of (A) 33x (B) 43x (C) 53x (D) 100x (E) 430x

2. In order to admit the proper amount of light, the part of the microscope that should be adjusted is the (A) coarse adjustment (B) fine adjustment (C) diaphragm (D) eyepiece (E) stage

3. All of the following scientists studied cells except (A) Hooke (B) Leeuwenhoek (C) Schleiden (D) Schwann (E) Linnaeus

4. The nucleus contains all of the following structures except (A) mitochondria (B) chromatin (C) genes (D) nucleolus (E) nuclear membrane

5. The most abundant substance in protoplasm is (A) protein (B) fat (C) carbohydrate (D) water (E) minerals

6. The conversion of nonliving material into living protoplasm is known as (A) assimilation (B) respiration (C) reproduction (D) absorption (E) digestion

7. A cell obtains energy during the process of (A) ingestion (B) respiration (C) irritability (D) excretion (E) secretion

8. Viruses resemble other living things because they (A) circulate (B) move (C) reproduce (D) are crystalline (E) are able to respond to stimuli in the environment

9. Animal cells do not possess (A) cell membrane (B) cell wall (C) cytoplasm (D) nucleus (E) nuclear membrane

10. Ameba moves by means of (A) cilia (B) flagella (C) pseudopods (D) pseudonyms (E) microscopic hairs

11. Both ameba and paramecium possess (A) contractile vacuole (B) anal spot (C) oral groove (D) trichocysts (E) gullet

12. Digestion in one-celled animals takes place in the (A) cyst (B) contractile vacuole (C) food vacuole (D) pellicle (E) gullet

13. Paramecium may reproduce by (A) binary fission only (B) conjugation only (C) both binary fission and conjugation (D) budding only (E) binary fission and budding

14. Dissolved gases pass in and out of paramecium through the (A) nuclear membrane (B) cell membrane (C) food vacuole (D) micronucleus (E) macronucleus

15. Plant and animal cells are alike in possessing (A) chlorophyll (B) chloroplast (C) cell wall (D) cellulose (E) cell membrane

16. Euglena is different from ameba and paramecium in possessing (A) cytoplasm (B) nucleus (C) cell membrane (D) chloroplasts (E) contractile vacuole

17. The spiral structure in spirogyra is the (A) spireme (B) chloroplast (C) chromatin (D) nucleus (E) filament

18. The only one of the following structures found in spirogyra is (A) flagellum (B) eyespot (C) pyrenoid (D) food vacuole (E) cilia

19. ATP is a chemical that is essential for (A) digestion (B) appetite (C) absorption (D) oxidation (E) assimilation

20. Enzymes are useful (A) only during digestion (B) only in respiration (C) in both digestion and respiration (D) only during ingestion (E) in both digestion and ingestion

21. All of the following are organelles except (A) endoplasmic reticulum (B) mitochondria (C) ribosome (D) Golgi bodies (E) ultracentrifuge

22. The life activity concerned with the taking in of food is known as (A) ingestion (B) digestion (C) secretion (D) excretion (E) assimilation

23. The life activity dealing with the stability of the organism's chemical makeup under its constantly changing environment is (A) respiration (B) irritability (C) reproduction (D) regulation (E) photosynthesis

24. DNA is found in the cell's (A) vacuole (B) nucleolus (C) nucleus (D) ribosomes (E) cell membrane

25. All of the following have recently given us greater knowledge about the cell except: (A) electron microscope (B) phase microscope (C) laser beams (D) time-lapse cinematography (E) single-lens microscope

ANSWER KEY

1-E	6-A	11-A	16-D	21-E
2-C	7-B	12-C	17-B	22-A
3-E	8-C	13-C	18-C	23-D
4-A	9-B	14-B	19-D	24-C
5-D	10-C	15-E	20-C	25-E

2 The Chemistry Of Life

Although it is only microscopic in size, the cell is a veritable chemical factory in which there is a constant interplay of biochemical processes and reactions. *Biochemistry* is the science that deals with the chemical compounds and processes in living things.

BASIC CHEMISTRY

Elements. All matter is composed of one or more elements. An element is the simplest form of a substance that cannot be broken down any further by ordinary chemical means, i.e., hydrogen, oxygen, carbon, nitrogen. There are 92 elements that occur naturally in nature. A number of additional elements, such as neptunium, plutonium and curium, have been produced artificially in recent years, bringing the number up to 103.

Compounds. Elements are combined to form compounds in a definite proportion by weight. A compound has different properties from the elements that compose it. Thus, water, H_2O, is a compound composed of two parts of hydrogen and one part of oxygen. Table salt, NaCl, is composed of sodium combined with chlorine.

Molecules. The smallest part of a substance that still has the properties of the substance, and is capable of stable independent existence, is called a molecule. Water can be broken down into smaller and smaller droplets until the smallest unit is reached, consisting of a molecule, H_2O. If it is broken down still further, it releases hydrogen and oxygen, two different substances.

Atoms. Elements are made of invisible building blocks called atoms. Each atom has a central nucleus surrounded by a definite number of moving negatively charged *electrons.* The electrons move so rapidly that they are thought of as a cloud surrounding the nucleus. Although the nucleus is only five-thousandth part of the atom in size, it is so dense that it contains nearly all of the atom's weight. The nucleus contains two kinds of particles, positively charged *protons,* and *neutrons* which have no charge. It has also been recently found to contain additional particles such as: positrons, anti-protons, mesons, neutrinos, and hyperons.

Atoms of different elements differ in their number of electrons, protons and neutrons. The simplest atom is that of hydrogen, the lightest element. It contains only 1 proton in its nucleus, and has 1 electron on the outside. Its atomic weight is 1. An atom of helium has 2 protons and 2 neutrons in its nucleus, giving it an atomic weight of 4. A carbon atom has 6 protons and 6 neutrons in its nucleus for an atomic weight of 12. The following table summarizes these facts for a number of elements:

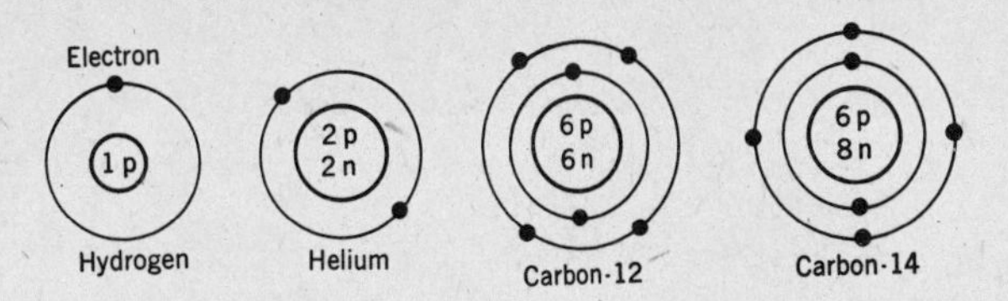

Atomic Structure

ELEMENT	SYMBOL	NUCLEUS PROTONS	NUCLEUS NEUTRONS	ATOMIC WEIGHT
Hydrogen	H	1	0	1
Helium	He	2	2	4
Carbon	C	6	6	12
Nitrogen	N	7	7	14
Oxygen	O	8	8	16
Aluminum	Al	13	14	27
Potassium	K	19	20	39
Iron	Fe	26	30	56
Cobalt	Co	27	32	59
Iodine	I	53	74	127
Lead	Pb	82	126	208
Uranium	U	92	146	238

For each proton in the nucleus, there is one electron flying around it, thus making the atom electrically balanced. Uranium, which until recently was the heaviest known element, has 92 protons and 92 planetary electrons. The number of protons makes one element different from another. If one proton could be removed from an atom of oxygen, an atom of nitrogen would remain.

Isotopes are different forms of the same element, whose nuclei contain the same number of protons but have a different number of neutrons. Thus, uranium-238 has 92 protons and 146 neutrons, for a total atomic weight of 238. Uranium-235 also has 92 protons, but only 143 neutrons, giving a total atomic weight of 235. There are three isotopes of hydrogen: the common type of hydrogen, with one proton and no neutrons; heavy hydrogen, or deuterium, with one proton and one neutron; and a heavier hydrogen, or tritium, with one proton and two neutrons. Some isotopes are found to occur naturally, while others have been made artificially. Some isotopes are stable, and remain the same at all times; others are radioactive, and are constantly undergoing change, either to another isotope or to another element.

Chemical Reactions. In the diagram of atomic structure, the electrons are shown arranged in rings or shells around the nucleus. Each shell holds a certain number of electrons. The innermost shell, designated as K, has two; the next shell, L, can hold eight; the third shell, M, also eight, etc. When an atom has a shell that does not have a complete set of electrons, it is said to be structurally unbalanced, and will tend to interact with other atoms. The chemical activity of an atom, then, arises from the number of electrons in its outer shell. Hydrogen has one electron in its K shell; it can hold two electrons. The number of electrons present in the outermost shell of some other atoms, (and the number of electrons the shell can hold) : Sodium — 1 (8) ; carbon — 4 (8) ; oxygen — 6 (8) ; chlorine — 7 (8) .

Each of these orbits represents a different energy level. The electrons in the inner shell are bound tightly to the nucleus. Those in the outermost shell are not held as tightly, and may react with the outermost shell of other atoms to form compounds. When chemical reactions normally occur, electrons are involved, not the nucleus. However, the nucleus may be changed under certain conditions involving atomic reactions.

Chemical Bonds. Atoms are combined to form molecules. The forces of attraction between them are referred to as chemical bonds. When bonds are made or broken, chemical reactions occur, involving energy changes.

A chemical bond may be made when two atoms

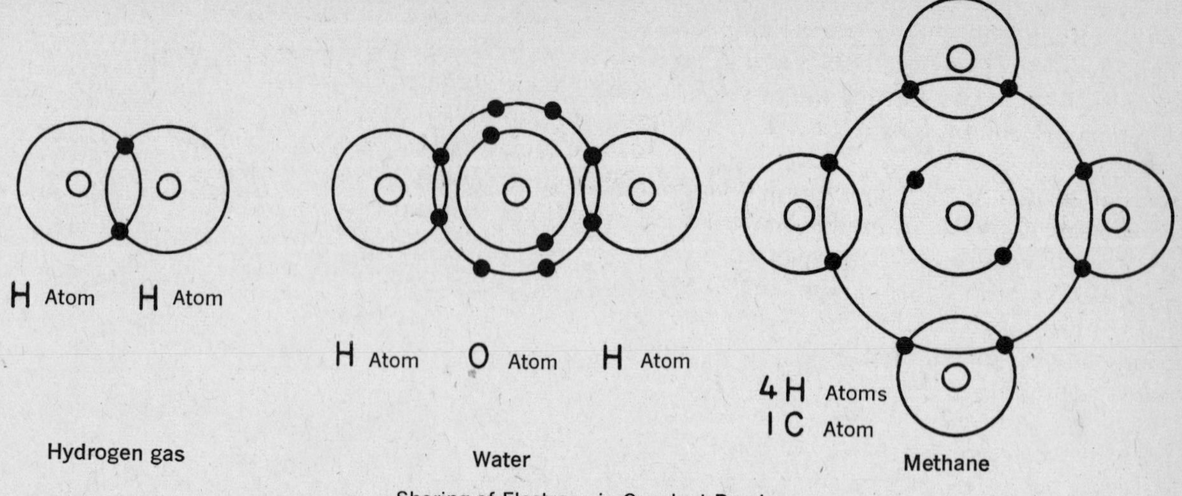

Sharing of Electrons in Covalent Bonds

share electrons. Thus, the hydrogen atom has one electron. Since its shell can hold two electrons, two atoms of hydrogen can share their electrons to form a molecule of hydrogen gas (H_2). In forming a molecule of water, H_2O, two hydrogen atoms share electrons with an atom of oxygen, which has six electrons in its outermost shell (eight is the stable number of electrons in its outermost L shell). Similarly, four atoms of hydrogen can share their electrons with an atom of carbon, which has only four electrons in its outermost shell, to form methane, CH_4. Such bonds formed when atoms share electrons are called *covalent bonds.*

In another type of bond, the *ionic bond,* electrons are transferred from one atom to another. This is the case in table salt or sodium chloride, NaCl. Sodium has only one electron in its outermost shell, which can hold eight electrons; chlorine has seven electrons in its outermost shell, and needs one more electron to complete this shell. When sodium and chlorine are united, an electron passes from the sodium atom to the chlorine atom. The chlorine atom is referred to as an electron acceptor.

When the sodium atom loses an electron, it loses its neutral status. Up to this time, it had an equal number of electrons and protons. Now, after losing an electron, it has an extra positive charge, and becomes an ion of sodium Na^+. Similarly, the chlorine atom gains an electron and has an extra negative charge, becoming a chloride ion Cl^-. An *ion* is thus an electrically charged atom with either a positive or negative charge.

Ionization. In ionization, the bonds of a substance are broken, and it separates into its ions. Sodium chloride ionizes when it is dissolved in water, to form ions of Na^+ and Cl^- in solution. Water also ionizes, with some of the H_2O molecules breaking down into hydrogen ions, H^+ and hydroxide ions, OH^-.

pH Scale. The number of hydrogen ions in a solution is the basis of pH. (Strictly speaking, this unit is the negative logarithm of the mass of hydrogen ions in a liter of solution. Thus, if a liter of a certain solution contains 10^{-7} grams H^+ ions, the $pH = -7$. The negative log is then equal to 7.)

Using various methods, the relative number of free H^+ and OH^- ions in solution can be determined. The pH scale extends from 0 to 14. A pH of 7 is neutral. Water, which has an equal number of H^+ and OH^- ions has a pH of 7, and is neutral. Below the pH of 7, the number of H^+ ions increases and the solution becomes more and more acid. The lower the pH number, the stronger the acid; the nearer the pH number approaches 7, the weaker the acid. Some common pH values for acids: Boric acid - 5.2; acetic acid - 2.9; hydrochloric acid - 1.1.

Similarly, above pH 7, the relative number of H^+ ions decreases, and the concentration of OH^- ions increases, to make the solution more *base,* or alkaline. The greater the number above 7, the stronger the base; the nearer the pH number approaches 7, the weaker the base. Some common pH values for bases: Ammonium hydroxide - 11.1; sodium hydroxide - 14.0; potassium hydroxide - 14.0.

The range of pH in the human body generally extends from 3 to 8.5. Blood has a pH of about 7.3. Living matter is sensitive to a change in pH, and does not tolerate wide variations in it.

There are special indicators that show the acidity or alkalinity of a solution. Litmus paper turns from blue to red in acid; from red to blue in base. A more specific range of pH can be shown with Hydrion paper, in which a color chart is used to identify the pH of a solution after a drop of it gives a certain color when placed on the paper. There are other methods of obtaining a more accurate calculation of pH, i.e., the use of special color indicators and the electric pH meter.

Energy of Bonds. When atoms unite to form molecules, energy is involved in the establishment of their bonds. Activation energy may be of several types, such as heat, light or electricity. There are varying amounts of bond energy in different compounds. During chemical reactions, such bonds are created or broken.

Structural Formulas. The chemical bond between the atoms of hydrogen that form a molecule of hydrogen gas, H_2, can be shown as H—H. In the case of oxygen gas, O_2, the bond can be shown as O=O; here, two pairs of electrons are shared by the two atoms. Water, H_2O, can be designated as H—O—H, where oxygen shares two electrons with the two hydrogen atoms. In the gas methane, CH_4, a carbon atom shares electrons with four hydrogen atoms, and can be represented as

```
      H
      |
  H — C — H
      |
      H
```

ORGANIC COMPOUNDS

Many of the compounds found in living things are organic, that is they contain carbon, hydrogen and other elements. Organic compounds were originally thought to be associated only with living things. Their molecules are comparatively large and complex. By comparison, inorganic compounds rarely contain carbon, but are composed of combinations of the other elements. Their molecules are generally smaller.

Because it has four electrons in its outer shell, carbon can form covalent bonds and share electrons with other carbon atoms, as well as with atoms of other elements, such as hydrogen, oxygen and nitrogen. Each of these bonds contains energy which is released when the atoms are separated in the life activities of cells. The principle organic compounds of living things include carbohydrates, lipids, proteins and nucleic acids.

Carbohydrates. The elements carbon, hydrogen and oxygen are found in carbohydrates. Hydrogen and oxygen are present in the same ratio as in water, 2:1. Some important carbohydrates are glucose, maltose, sucrose and starch. Simple sugars, such as glucose, fructose and galactose, have the same formula $C_6H_{12}O_6$, and are called monosaccharides. Although they contain the same atoms, their characteristics are different because of their different arrangement within the molecule. Thus, their structural formulas differ to a slight degree. Glucose is the source of energy for most organisms. Its structural formula is:

Glucose

Glucose molecules may be linked together to form the disaccharide, maltose, by a process called *dehydration synthesis*. This refers to the building up of complex molecules from simpler molecules, with the release of water. Other disaccharides are sucrose, which is ordinary table sugar, and lactose or milk sugar. Their formula is $C_{12}H_{22}O_{11}$. The structural formula of maltose is:

Maltose

Hundreds of glucose units bonded together make up more complex carbohydrates called polysaccharides. Such a collection of many similar, repeating units to form a large molecule is referred to as a *polymer*. Examples are starch, cellulose and glycogen. Starch is a storage form of sugar. Glycogen is known as animal starch. Cellulose is the supporting material found in the cell walls of plant cells.

Lipids. Lipids include fats, oils and waxes. Like carbohydrates, lipids also contain carbon, hydrogen and oxygen. However, there is proportionately much less oxygen in relation to hydrogen. Their molecules are relatively small, and are not arranged as polymers. Typically, a lipid molecule consists of a glycerol molecule bonded to three fatty acid molecules. A glycerol molecule can be represented as:

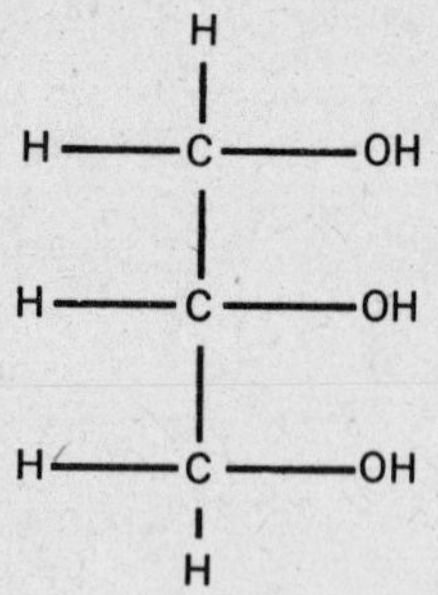

A fatty acid molecule consists of a long chain of carbon and hydrogen (hydrocarbon) atoms with a carboxyl (–COOH) group. The number of carbon atoms may vary from 4 to 24, although most of the common fatty acid molecules contain 16-18 carbon atoms. The general structural formula of a fatty acid is:

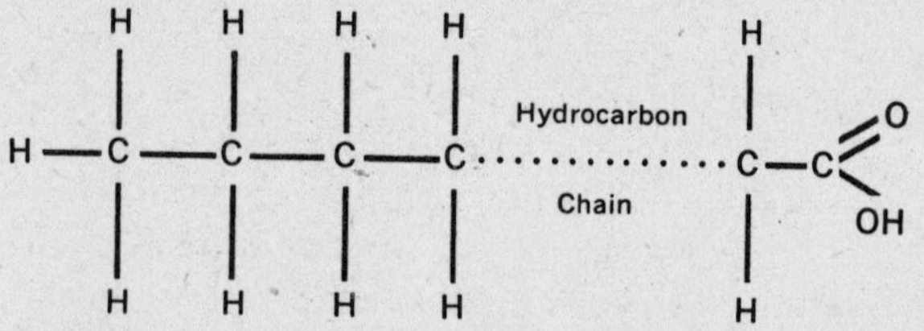

When the three fatty acid molecules are linked to a glycerol molecule, a lipid molecule is formed. Each of the glycerol's three OH groups becomes attached to one of the three fatty acid molecules, with the splitting off of three water molecules at the sites of linkage. This is another example of dehydration synthesis. The structural formula of a lipid can be represented as:

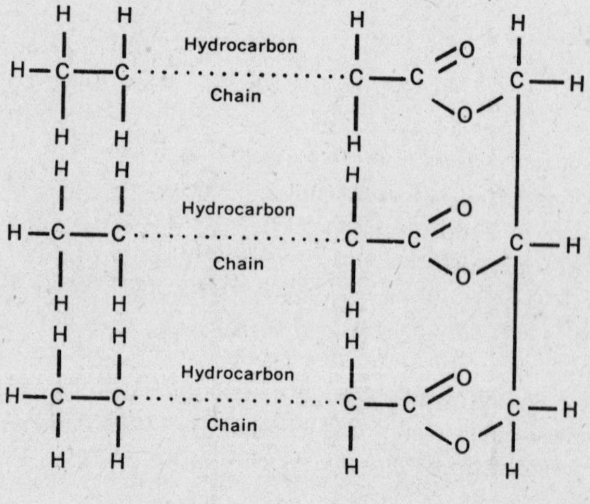

Fats and oils have the same molecular structure, although fats are solid at room temperature and oils are liquid. A common fatty acid is: stearic acid, $C_{17}H_{35}COOH$. When combined with glycerol, it forms the fat, stearin. Other fatty acids are palmitic acid ($C_{15}H_{31}COOH$), present in the fat palmitin, and oleic acid ($C_{17}H_{33}COOH$), in the oil, olein.

Proteins. Like carbohydrates and lipids, proteins also contain carbon, hydrogen and oxygen; in addition, they contain nitrogen, and in many instances, sulfur. Proteins are large polymers of many repeating amino acid units. There are more than 20 different types of amino acids. These can be combined in various ways to form many types of proteins.

The general structure of an amino acid shows a carbon atom bonded in four places with (1) a hydrogen atom, (2) an amino group (NH_2), (3) a carboxyl acid group (COOH), and (4) a side group (R) which is variable and is the basis for the variety of different amino acids. The structure of an amino acid can be shown as follows:

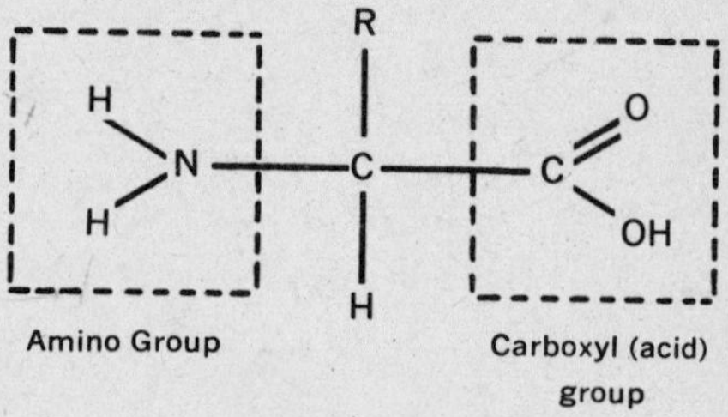

Amino Group — Carboxyl (acid) group

The simplest amino acid, glycine, has a hydrogen atom (H) as the R group. When the side group, R, is a methyl group (CH_3), the amino acid alanine is formed, which can be represented in this way:

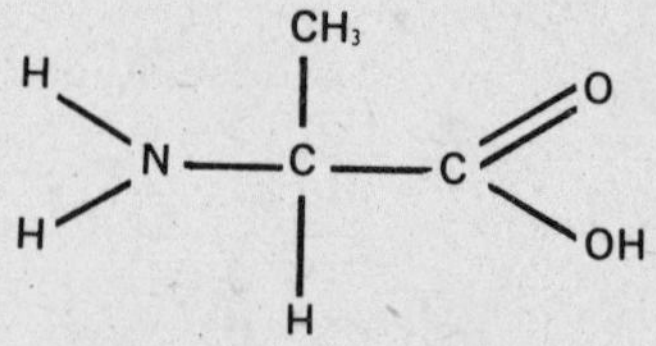

Alanine

Proteins are built up by the bonding of many amino acids. There may be more than 3,000 amino acids in a protein. In the bonding of amino acids, a C-N bond forms between the carboxyl group of one amino acid and the amino group of the next amino acid. The C-N bond is known as a peptide bond. A chain of amino acids bonded in this way is called a *polypeptide*. The shape of the protein molecule itself depends on the nature of the attraction between the different parts of the polypeptide chain, and it may be in the form of a straight chain, a globule, a twisted chain (helix), etc. The building up of the protein molecule is another example of dehydration synthesis; a molecule of water is given off when two amino acids become bonded together.

Each protein has its particular arrangement of amino acids. This gives the protein its properties. Since the structure of proteins is so complex, it has been difficult to analyze them. In 1954, the structural arrangement of the amino acids in a protein, insulin, was determined for the first time, by the English biochemist, Frederick Sanger.

Nucleic Acids. Nucleic acids contain carbon, oxygen, hydrogen, nitrogen and phosphorus. They are the largest organic molecules known. In structure, they are high-molecular-weight polymers that are made up of thousands of repeating units known as *nucleotides.* Nucleotides themselves are relatively complex molecules. They consist basically of three types of molecular units, (1) a phosphate, (phosphoric acid, H_3PO_4) (2) a five-carbon sugar and (3) a nitrogen base. Their arrangement may be shown as follows:

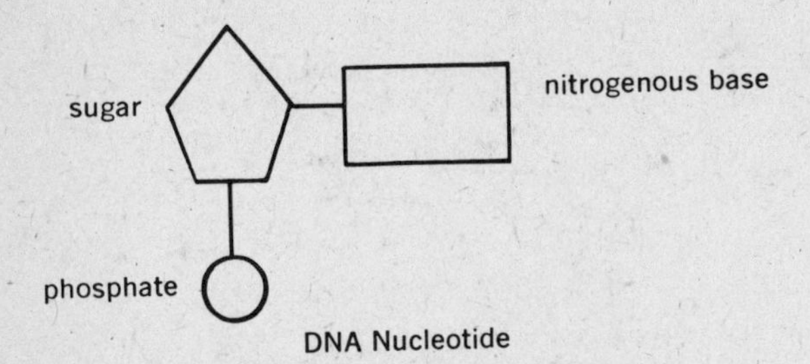

DNA Nucleotide

The sugar is either ribose, or deoxyribose. Deoxyribose, as the name indicates, has one oxygen less than ribose. The nitrogen base may be a purine, either adenine or guanine; or it may be a pyrimidine, either cytosine, thymine or uracil. The significant nucleic acids are DNA, deoxyribonucleic acid, and RNA, ribonucleic acid. DNA plays a key role in the determination of heredity. RNA is important in the synthesis of protein. Further details about the nucleic acids will be considered in the chapter on Modern Genetics.

ENZYMES. Living matter is constantly in a state of dynamic chemical activity. The reactions that take place are made possible by organic catalysts called enzymes. Each cell contains over a thousand enzymes. Enzymes help provide energy for the cell, assist in the building of new cell structures, digest food, and play a part in almost every activity that takes place.

Enzymes are considered organic catalysts, because they affect the rate of a chemical reaction without being changed. They can be used over and over again. They are protein in nature, and specific in their action. They often work together with coenzymes in reactions. Coenzymes have smaller molecules than enzymes, and are not protein. They are active only with enzymes. Such B-complex vitamins as thiamin, riboflavin and niacin, have been found to act as coenzymes in cellular respiration, leading to energy release.

How Enzymes Function. Enzyme molecules are huge compared to the molecules with which they interact. Apparently, only a small portion of the enzyme functions when it is active. The localized region is called the active site of the enzyme. The action of enzymes has been explained on the basis of the "lock and key" model. This analogy arises from the fact that a particular enzyme interacts only with a single type of substrate molecule. The substrate is the substance the enzyme acts on. The name of an enzyme usually has the ending – *ase,* added to the stem of the word which is taken from the substrate. Examples

ENZYME	SUBSTRATE
maltase	maltose
lipase	lipids
protease	protein

The association between enzyme and substrate is thought to be a close physical one, but does not lead to the formation of bonds between them. It is sometimes referred to as an enzyme-substrate complex. Enzyme action takes place while the enzyme-substrate is formed.

The following illustration shows how an enzyme may cause a complex molecule, designated as A-B, to separate into two smaller molecules, A and B.

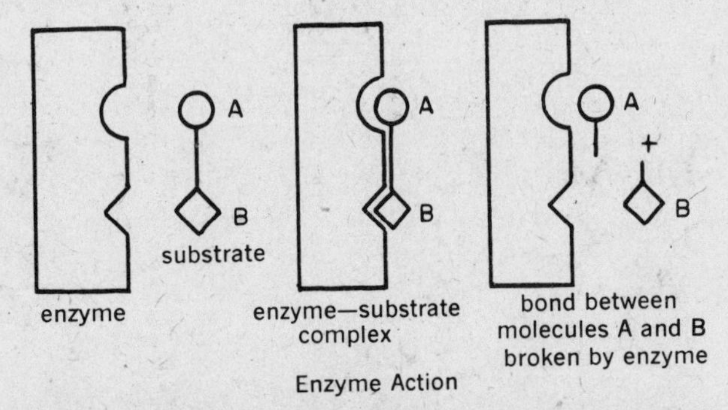

Enzyme Action

Factors affecting Enzyme Action. The rate of enzyme action is not fixed, but varies with such conditions as pH, temperature and relative amounts of enzyme and substrate.

1. pH. Some enzyme actions go on at a pH which is in the vicinity of 7, i.e., the enzyme maltase. If the solution became acid or base, this enzyme's action would be slowed down. The following graph illustrates the effect of varying the pH:

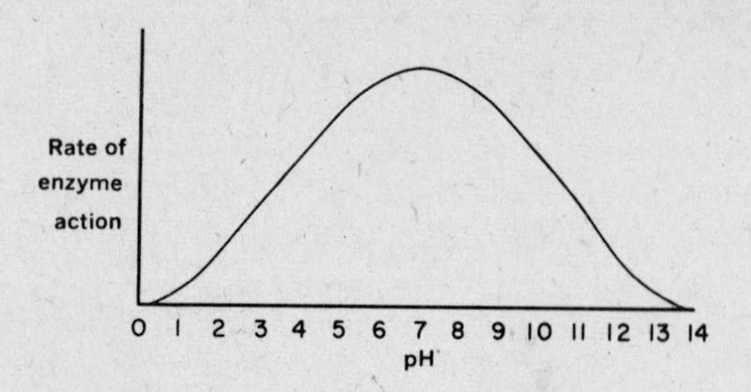

However, many enzymes function best at other pH values. Pepsin, found in the stomach, has its best activity at a pH of 1.5-2.2. Trypsin, in the small intestine, acts best at pH 7.9-9.0.

2. Temperature. Most enzymes function best at body temperature, 37°C. As the temperature is lowered, their rate of activity, decreases. As their temperature is raised, their activity increases, until a maximum is reached at about 40°C. Beyond this point, the shape of the enzyme molecule becomes distorted, and enzyme deactivation occurs. The effects of varying temperatures on the action of many enzymes is shown in the following graph:

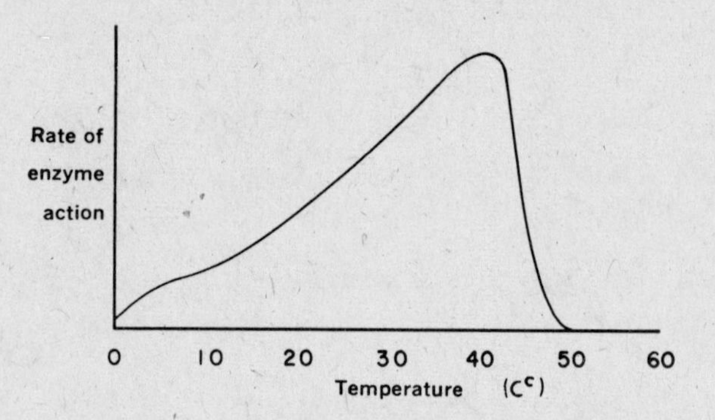

3. Relative Amounts of Enzymes and Substrate. If the amount of enzyme is increased, and the substrate remains constant, the rate of the reaction is increased. This increase continues up to a point and then remains at that level. The following graph illustrates the effect of adding an excess of enzyme when the amount of substrate remains fixed:

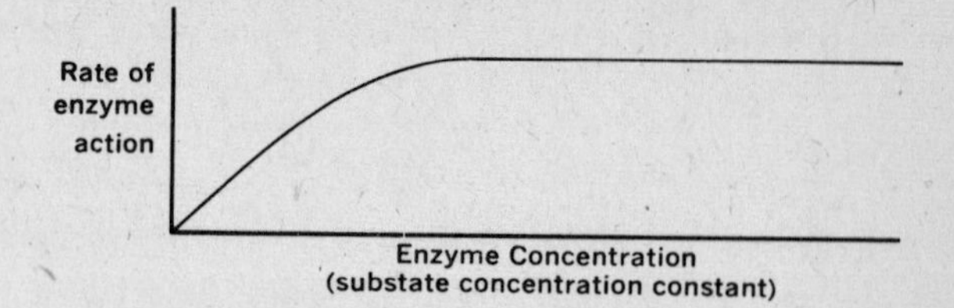

Similarly, if the amount of substrate is increased while the concentration of enzyme remains the same, the rate of the reaction will increase. This will continue up to the point where every available enzyme molecule is actively involved in the reaction.

Dehydration Synthesis. Among the reactions in which enzymes are involved is dehydration synthesis. In this process, large organic molecules are built up from their building blocks, with the release of water molecules. Thus, two glucose molecules are linked to form the disaccharide, maltose. A hydrogen atom (H) is removed from one glucose, and a hydroxyl group (OH) is removed from the other, to form the molecule of water. Dehydration synthesis also takes place when a lipid molecule is built up from fatty acids and glycerol; also, when a protein molecule is being synthesized from amino acids.

Hydrolysis. In hydrolysis, the reverse process takes place. Large molecules are broken down to their building blocks, with the addition of water. The enzyme maltase, for example, acts on maltose and breaks the—O—bond between the two glucose subunits. A hydrogen atom (H) from the water molecule combines with the O to form the OH part of the glucose molecule. The OH from the water molecule is added to make the other glucose molecule complete.

The reversible reactions, involving enzymes in dehydration synthesis and hydrolysis, are shown in the accompanying illustration:

Maltase

Water is "split off."

Dehydration synthesis

Hydrolysis

CELLULAR RESPIRATION

Energy of the Cell. Energy is needed for such activities as the contraction of a muscle cell, conduction by a nerve cell, synthesis of organic compounds by a cell, movement by a paramecium, active transport by a root cell, division of a cell, and so on. The energy to carry on these activities is derived from food, just as a candle flame burns fuel. However, the similarity ends there. In a fire, the fuel is burned all at once. In the cell, the organic bonds of food are broken down in a series of many small steps involving dozens of enzymes which are located in the mitochondria. Each enzyme is specific in performing only one step of the entire process. The release of the chemical bond energy of food into energy that can be used for the life activities of the cell is known as *respiration.*

Every cell contains molecules of the nucleotide, ATP, adenosine triphosphate, which is the actual source of energy. Each ATP molecule stores energy in its phosphate bonds. When one of these bonds is broken, energy is released to the cell for its various activities and ATP is converted to ADP, adenosine diphosphate. It is this change of ATP into ADP which supplies the cell with energy.

Without respiration taking place, the ATP supply of the cell would soon be used up. To produce more ATP, high energy phosphate bonds must be added to ADP. The energy for changing ADP into ATP comes from the chemical bonds of glucose. In short, the breakdown of glucose leads to the storage of energy in ATP. The breakdown of ATP provides the cell with energy. When the whole process takes place in the presence of oxygen, it is referred to as aerobic respiration; without oxygen, it is anaerobic respiration.

Aerobic Respiration. There are two phases in respiration, (a) glycolysis, and (b) the Krebs cycle.

a. Glycolysis. Through a series of steps, glucose, which is a six-carbon compound ($C_6H_{12}O_6$) is converted to two three-carbon molecules of pyruvic acid ($C_3H_4O_3$). Hydrogen is removed; this is oxidation. Energy is released in this stage and is stored in 4 molecules of ATP. However, in order for some of these processes to take place, the energy from 2 ATP's was used up, As a result, the total gain of energy during glycolysis is 2 ATP molecules. In this stage, oxygen has not yet been involved and it is considered to be an anaerobic phase.

b. The Krebs cycle. The pyruvic acid molecules are dehydrogenated (H removed) and decarboxylated (CO_2 removed) in a series of reactions. Some of the intermediate compounds formed are oxalacetate, citric acid, succinic acid and ketaglutaric acid. Hydrogen is passed on to a series of compounds which are referred to as hydrogen acceptors: NAD (nicotinamide-adenine dinucleotide), riboflavin coenzyme and cytochrome. The final hydrogen acceptor is oxygen, and the combination forms water. As the electrons of the hydrogen atoms are passed along to these acceptors, and the carbon bonds are broken, energy is released in small units, and stored in ATP molecules along the way.

During the succession of steps involved in the Krebs cycle, energy is released and stored in 36 ATP molecules. Added to the 2 ATP's resulting from glycolysis, the total yield of energy from one molecule of glucose is 38 ATP molecules. Aerobic respiration of glucose may be summarized in the following equation:

$$\text{glucose} + \text{oxygen} \xrightarrow[\text{enzymes}]{\text{numerous}} \text{water} + \text{carbon dioxide} + 38\,\text{ATP}$$

The detailed steps may be summarized as follows:

|— ANAEROBIC PHASE —||— AEROBIC PHASE —|

$$\text{glucose} \xrightarrow[\text{4ATP}]{\text{2ATP, enzymes}} 2\ \text{pyruvic acid} \xrightarrow[\text{36ATP}]{\text{Oxygen, enzymes}} \text{carbon dioxide} + \text{water}$$

Anaerobic Respiration. Most cells use oxygen and carry on aerobic respiration. However, in some cases, as in yeast, certain bacteria and fungi, and in the work of muscle cells, oxygen is not used and anaerobic respiration occurs. It is also referred to as fermentation. As in aerobic respiration, glycolysis first takes place, resulting in the formation of pyruvic acid. The net gain of energy is 2 ATP's. Following this, either lactic acid and carbon dioxide, or ethyl alcohol and carbon dioxide, are formed without any additional yield of energy. There is still a great deal of energy contained in the lactic acid or alcohol molecules, where the C-C and C-H bonds remain intact.

Lactic acid fermentation may be summarized in the following equation:

$$\text{glucose} \xrightarrow{\text{enzymes}} \text{lactic acid} + CO_2 + 2\text{ATP}$$

Lactic acid is produced in muscles that are very active. Under such conditions, anaerobic respiration takes place. The accumulation of lactic acid results in muscle fatigue. Alcohol fermentation may be summarized as follows:

$$\text{glucose} \xrightarrow{\text{enzymes}} \text{ethyl alcohol} + CO_2 + 2\text{ATP}$$

The net yield of energy in aerobic respiration is 38 ATP's; in anaerobic respiration, it is only 2 ATP molecules.

Although glucose is the chief source of energy, proteins and lipids may also be used for energy.

CHAPTER REVIEW

Select the correct choice for each of the following statements:

1. An atomic particle located outside of the nucleus is the (A) proton (B) positron (C) neutron (D) electron (E) meson

2. Of the following, the element with the greatest atomic weight is (A) lead (B) uranium (C) hydrogen (D) oxygen (E) carbon

3. Uranium-238 is different from uranium-235 in having (A) more neutrons (B) more protons (C) more electrons (D) fewer neutrons (E) fewer protons

4. The number of atoms in the compound H_2SO_4 (sulfuric acid) is (A) 3 (B) 5 (C) 7 (D) 8 (E) 9

5. All of the following are examples of elements except (A) hydrogen (B) oxygen (C) carbon (D) plutonium (E) water

6. The K and L electron shells are full when they hold the following number of electrons respectively (A) 2 and 2 (B) 2 and 4 (C) 2 and 6 (D) 2 and 7 (E) 2 and 8

7. Chlorine has 7 electrons in its outer shell. The number of electrons it needs to complete the shell is (A) 1 (B) 2 (C) 3 (D) 4 (E) 5

8. In a covalent bond, electrons are (A) shared (B) split (C) lost (D) added (E) transferred

9. In an ionic bond, electrons are (A) hydrolyzed (B) transferred (C) fused (D) neutralized (E) shared

10. The strongest acid has a pH of (A) 1 (B) 5 (C) 7 (D) 9 (E) 11

11. An organic compound in which H and O are present in the same ratio as in water is (A) fat (B) protein (C) amino acid (D) nucleic acid (E) carbohydrate

12. A substance composed of a large molecule that is made of many smaller similar repeating units is called (A) glycerol (B) polymer (C) fatty acid (D) monosaccharide (E) disaccharide

13. A lipid molecule is composed of glycerol and fatty acid molecules on a ratio of (A) 1:1 (B) 1:2 (C) 1:3 (D) 1:4 (E) 1:5

14. An amino acid molecule contains all of the following except (A) hydrogen atom (B) phosphate (C) NH_2 group (D) COOH group (E) R group

15. A combination of a phosphate, ribose and a nitrogen base is characteristic of a (A) carbohydrate (B) lipid (C) protein (D) nucleotide (E) glycerol

16. An enzyme is a (A) deoxyribose (B) lipid (C) protein (D) ribose (E) polysaccharide

17. An enzyme-substrate is the place where enzymes (A) are formed (B) are deactivated (C) are active (D) are reduced (E) are diluted

18. Most enzyme actions take place best at a temperature of about (A) 0°C (B) 10°C (C) 17°C (D) 27°C (E) 37°C

19. The most favorable pH for maltase action is (A) 1 (B) 3 (C) 5 (D) 7 (E) 9

20. In dehydration synthesis, a water molecule is (A) taken in (B) liberated (C) absorbed (D) bonded (E) electrolyzed

21. The storehouse of energy in the cell is (A) ATP (B) nucleus (C) cell membrane (D) DNA (E) RNA

22. The release of the chemical bond energy of food is called (A) digestion (B) irritability (C) respiration (D) ingestion (E) diffusion

23. The amounts of ATP formed in aerobic and anaerobic respiration respectively are (A) 30 and 2 (B) 32 and 2 (C) 34 and 2 (D) 36 and 2 (E) 38 and 2

24. As the result of glycolysis, the three-carbon molecule formed is (A) DPN (B) oxalacetate (C) citric acid (D) succinic acid (E) pyruvic acid

25. During fermentation, the final products formed are CO_2, and either lactic acid or (A) water (B) glucose (C) alcohol (D) riboflavin coenzyme (E) cytochrome

ANSWER KEY

1-D	6-E	11-E	16-C	21-A
2-B	7-A	12-B	17-C	22-C
3-A	8-A	13-C	18-E	23-E
4-C	9-B	14-B	19-D	24-E
5-E	10-A	15-D	20-B	25-C

3 Tissues and their functions

Single-celled organisms such as the ameba, paramecium and euglena carry on the basic life activities through special structures contained within the cell. Higher organisms consisting of millions of cells carry on similar activities by means of tissues, organs and systems which are specialized to perform the specific activities. They have a differentiation, or division of labor of their parts and functions in which all the different activities are coordinated for the mutual benefit of the individual cells and the total organism.

TISSUES. A tissue is a group of similar cells performing the same function. Mucous membrane cells from the tissue lining the inside of the cheek can readily be examined by scraping it lightly with a toothpick and depositing the material gently on a slide in a drop of stain. The arrangement of the cells next to each other can be seen, as well as the five or six-sided shape of each cell. The typical structures of protoplasm are noticeable, i.e., nucleus, cytoplasm and cell membrane. By special techniques, tissues from other parts of the body can also be examined. The study of tissues is known as *histology*. Tissues are classified as epithelial, muscle, nerve, connective and supporting, blood, and reproductive.

Epithelial tissue consists of a continuous layer of cells covering the body surfaces and lining the cavities within the body. It is found on the outer layer of skin, and in the lining of the digestive tract, the respiratory system, and the blood vessels; also in the glands. In these various parts of the body, it may have one or more of the following functions; protection, absorption, secretion and sensation. Epithelial cells may also differ in appearance, some being flat, some boxlike or cuboidal, and others tall and rectangular (columnar).

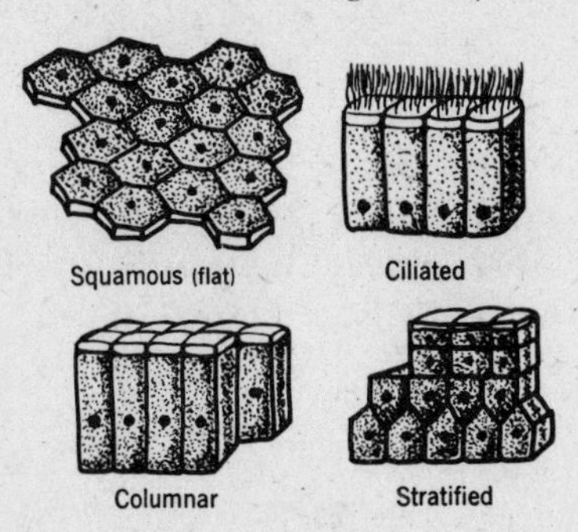

1. **Epidermis of the skin.** The outer layers of the skin protect the underlying cells from injury, bacteria, and from drying out. There are layers of these cells which are constantly being worn away and replaced from underneath.

2. **Flat (squamous) epithelium.** These cells are broad and flat, like tiles on a floor. They form a protective membrane that lines the inside of the mouth (mucous membrane) and the esophagus. They have special cells that secrete *mucus* to lubricate the passages and to trap dust and bacteria. They also line the smallest blood vessels (capillaries), where they allow dissolved materials to pass through them (absorption).

3. **Ciliated epithelium.** The epithelial cells lining the nasal cavities and the trachea (windpipe) have cilia which are constantly in motion, beating foreign particles upward and outward.

4. **Columnar epithelium.** These cells line the small intestines, where they absorb digested food. They contain special cells called goblet cells which secrete mucus, a sticky fluid that lubricates the interior of the intestine.

5. **Glandular epithelium.** The complex glands of the digestive system (i.e., salivary glands), the endocrine glands which secrete hormones (i.e., thyroxin), and the tear glands are specialized to secrete special liquids used by the body.

6. **Sensory epithelium.** Sensations from the outside of the body are received by certain other epithelial cells which are specialized to receive stimuli. The olfactory cells lining the inside of the nose help us to smell. The cells on the retina of the eye permit us to see.

Muscle tissue. This tissue has the ability to contract and produce movement. There are three types of muscle tissue:

1. **Voluntary or striated muscles** contract according to our will. Muscles of our arms and legs are examples of this type of muscle: they are usually attached to the bones of the skeleton, and so may also be referred to as skeletal muscles.

When the *biceps* muscles of the arm contract, the arm bends at the elbow; such muscles are called *flexor*. In straightening the arm out, the *triceps* muscles on the back of the arm contract; since they extend the arm, they are called *extensor*. The flexor muscles relax when the extensor muscles contract, and vice versa. Striated muscles consists of *fibrils* which have a striped or striated appearance, with alternating dark and light bands. Under the microscope, a small piece of beef which has been teased apart and stained shows the striated structure and many nuclei.

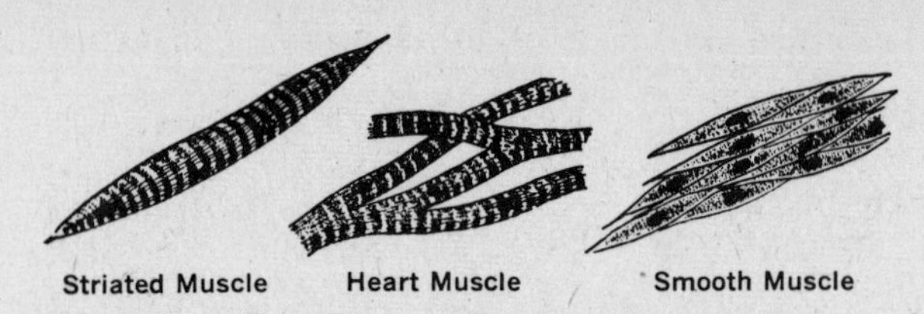

2. **Smooth muscle** is involuntary, and is found in the walls of the alimentary canal and the blood vessels. We are rarely aware of the activities of these muscles. Food is moved along and churned in the stomach and intestines by the wavelike contractions and expansions of the smooth muscles in the walls of these organs; this muscular activity is known as *peristalsis.* The relaxation of the muscles of the blood vessels in the face enlarges their diameter and increases the amount of blood they contain; such a sudden increased supply of blood in the skin results in blushing.

3. **Cardiac or heart muscle** is also involuntary, and is found only in the heart. Its cells have a striated appearance.

Nerve Tissue. Nerve cells are specialized for transmitting messages or impulses through the body. They appear different from other cells in having branched projections which are in close contact with each other. A *neuron* (nerve cell) has a cell body containing the nucleus and most of the cytoplasm of the cell. Extending out from the cell body are branched threads called *dendrites.* Many neurons also have a very long extension, or *axon,* which may be up to several feet long, and which is covered with a fatty sheath. Bundles of axons make up a nerve.

There are three types of neurons. (1) Sensory neuron—brings impulses from the sense organs (eye, ear, skin, etc.) to the brain or spinal cord. (2) Motor neuron—carries impulse from the brain or spinal cord to a muscle to make it move, or to a gland, causing it to secrete. (3) Associative neuron—connecting neuron between other neurons.

Connective and supporting tissue. This consists of several types of cells quite different from each other, whose function is either to support the body or to connect its parts. The cells secrete a nonliving substance or *matrix* around them.

1. **Bone.** "Solid bone" is not as solid as it seems. It consists of cells separated from each other by nonliving intercellular deposits of calcium and phosphorus which are secreted by the cells themselves. These cells are connected with each other by fine rootlike projections of cytoplasm. Blood vessels supply the cells with food and oxygen and remove the wastes. Bone tissue makes up the skeleton, which supports the body and protects vital organs such as the brain, and lungs. The bones of the skeleton also serve as levers for body movement, and as anchor sites for muscle action.

2. **Cartilage.** The cells of cartilage are contained in a nonliving matrix which is smooth, firm, and flexible, but not as hard as bone. Cartilage can be felt in the ears, nose and the windpipe. It also occurs in the joints at the ends of bones, where it allows for smooth action.

3. **White fibrous** connective tissue consists of a very tough matrix of white fibers containing the living cells. It is found in *tendons,* which connect muscles to bone, and *ligaments,* which connect bones together forming joints, e.g. knee, elbow.

4. **Yellow elastic** connective tissue contains fibers which stretch when necessary, and then return to their normal size. The arteries expand as each beat of the heart sends spurts of blood through them and then relax before the next beat or "pulse." The elastic fibers in the walls of the arteries permit them to expand in this way. "Hardening of the arteries" results when the elastic fibers lose their elasticity. This tissue is also found in the bronchial tubes of the lungs, and between the vertebrae of the backbone.

5. **Fat (adipose) tissue** has cells that store oil and fat. It is located beneath the skin, and around the heart and kidneys. It helps to retain body heat. Animals that hibernate in winter gradually use up this stored fat. Overweight people have a large amount of adipose tissue.

Blood tissue. The cells of this tissue are carried in a liquid called *plasma.* Plasma is a straw-colored fluid consisting of water and dissolved proteins, salts, nutrients, antibodies, hormones and wastes. There are three kinds of cells in it:

1. **Red blood corpuscles**—round disc cells which lose their nucleus after being manufactured in the marrow of the bones. They contain hemoglobin, which unites freely with oxygen. Their function is to bring oxygen to all the cells of the body. Blood appears red because of the great number of these red corpuscles in the plasma.

2. **White blood corpuscles**—ameba-like cells that move through the circulatory system and the body tissues. They engulf and destroy bacteria that may have entered the body. They are also called *phagocytes.*

3. **Platelets**—very small groups of cells that play a role in blood clotting.

Reproductive tissue. Special tissues of the body are set aside for the formation of the reproductive cells. *Egg* cells (female cells) are produced in the *ovaries,* and *sperm* cells (male cells) are formed in the *testes.* Sperm cells are extremely small, and have a tail of cytoplasm for movement. The nucleus is contained in the head end. Egg cells are much larger, and round. Both types of sex cell are formed by a special type of nuclear division in which they each receive half the amount of chromatin material normally found in cells. When these two types of cells unite during *fertilization,* the fertilized egg will then have the full amount of chromatin material.

Organs. Tissues are grouped together to form *organs.* For example, the stomach is an organ that contains: epithelial tissue covering its outer and inner surfaces: smooth muscle tissue by which it carries on peristalsis; nerve tissue which controls its muscular and glandular activity; connective tissue which holds it together; and blood tissue which supplies its cells with food and oxygen and takes away wastes. All of these tissues work together as the stomach performs its function of digestion. Similarly, other organs such as the heart, kidney and brain are also made of tissues working together.

Systems. Besides the stomach, there are other organs concerned with digestion, such as the mouth, esophagus, small intestine, large intestine, liver and pancreas. All together, they make up the digestive *system.* There are seven other systems composed of groups of organs working together to perform their particular function; circulatory system; respiratory system; excretory system; nervous system; reproductive system; skeletal system; muscular system.

Organism. In many-celled living things, cells are grouped together to form tissues; tissues work together to form organs; organs are organized into systems. All of the systems working together make up an *organism.* Thus, the basic life functions are performed by all organisms, plant or animal. Organisms may be one-celled, too, as in the case of ameba or euglena, which carry on all the life activities within the single cell.

SUMMARY OF ANIMAL TISSUES

TISSUE	TYPES	LOCATION	FUNCTIONS
Epithelial (covering tissue)	1. Epidermis	Outer layer of skin	Protection
	2. Flat (Squamous)	Lines digestive and respiratory systems; also blood vessels	Protection; lubrication; absorption
	3. Ciliated	Lines trachea and nasal cavities	Protection by sweeping out bacteria and dust
	4. Columnar	Lines small intestines	Absorption; lubrication
	5. Glandular	(1) Digestive system; (2) endocrine glands	Secretion of: (1) enzymes; (2) hormones
	6. Sensory	Sense organs; i.e., (1) nose, (2) retina of eye	Receive sensations (1) smell; (2) sight
Muscles	1. Striated	Attached to bones	Voluntary movement
	2. Smooth	(1) Digestive organs; (2) blood vessels	Involuntary; (1) peristalsis; (2) control of diameter of blood vessels
	3. Cardiac	Heart	Heart beat

SUMMARY OF ANIMAL TISSUES (Continued)

TISSUE	TYPES	LOCATION	FUNCTIONS
Nerve	1. Sensory	Connects sense organs with brain or spinal cord	Transmits impulses from sense organs to brain or spinal cord
	2. Motor	Connects brain or spinal cord with muscles or glands	Makes muscles contract or glands secrete
	3. Associative	Brain, spinal cord	Makes connections between neurons
Connective or Supporting	1. Bone	Makes up the skeleton	Supports body; protects organs
	2. Cartilage	Ear, nose, trachea, ends of bones	Support; frictionless movement of joints
	3. White fibrous	(1) Tendons; (2) ligaments	Connects (1) muscle to bone; (2) bone to bone
	4. Yellow elastic	Walls of blood vessels; between vertebrae	Elasticity
	5. Fat (adipose)	Under skin; around internal organs	Insulation against heat loss; stores fat
Blood	1. Red blood corpuscles	Plasma	Carry oxygen
	2. White blood corpuscles	Plasma	Destroy bacteria
	3. Platelets	Plasma	Clotting of blood
Reproductive	1. Male reproductive tissue	Testes	Produces sperm
	2. Female reproductive tissue	Ovaries	Produces eggs

PLANT TISSUES. The tissues of higher plants may be studied in their leaves, stems, and roots.

Outer covering. The epidermis is the outside layer of cells in a plant.

1. **Leaf.** In a leaf the epidermis is waterproof, containing a waxy substance called *cutin.* It covers the upper and lower sides of the leaf. In it are numerous tiny openings, *stomates,* on either or both surfaces. Stomates permit the passage of oxygen and carbon dioxide in and out of the leaf, and the liberation of water.

2. **Stem.** The epidermis of a stem has thickened outer cell walls. It prevents the loss of water, and protects the underlying tissues from injury. On woody stems, the outside layer consists of a *bark.* Stems have openings called *lenticels* for the exchange of gases.

3. **Roots.** The epidermis of a root is soft and thin, permitting the absorption of water and minerals. *Root hairs* occur as single-celled outgrowths of root epidermis. They absorb water and minerals and also help anchor the plant in the ground.

Conducting tissue. Special tissue located in ducts called *fibrovascular bundles* conduct materials throughout the plant. The ducts start in the roots, and continue up the stem into the leaves. Here, they branch out into the midrib and the veins. They are composed of two kinds of conducting tissue, xylem and phloem.

1. **Xylem**—consists of long, thick-walled, dead, woody cells. They are specialized for conducting water up from the roots to the stem and leaves. They also serve to support the plant.

2. **Phloem**—consists of thin-walled living cells located close to the xylem ducts. These cells include *sieve tubes* arranged longitudinally, end to end, with tiny holes in the adjacent walls. These openings permit the passage of food from cell to cell. It is through the phloem tissue that manufactured food is transported from the leaves to other parts of the plant.

In the stem of dicots, the fibrovascular bundles are arranged in the form of a concentric ring. In the stem of monocots, these bundles are scattered about.

Growing tissue. The growing tip of a stem contains actively dividing cells known as *meristem*. These cells also give rise to the different types of tissue in the stem (differentiation). At the other end of the plant, the tips of roots are similar in having a growing region also composed of meristem cells.

Another part of the plant that consists of actively dividing cells is the *cambium*. This important growing region is found in the fibrovascular bundles, between the phloem and the xylem. As cambium cells divide, they form more cambium, as well as more xylem and phloem, enlarging the stem, and causing the trunk of a tree to grow in diameter. Trees in the temperate zone show such growth only during the spring and summer of each year. A cross-section of a tree shows the presence of *annual rings* which are formed during these growing periods. Annual rings consist of large cells of xylem formed in the spring and smaller woody cells in the summer. The age of a tree can be determined by counting these rings. They also give a clue as to the climate conditions that prevailed, with wide rings indicating favorable seasons.

Food-making tissue. Between the upper and lower epidermis of a leaf, there are cells containing chloroplasts arranged in two regions, a *palisade* layer and a *spongy* layer. The palisade cells are elongated and are arranged in column-like fashion under the upper epidermis. Here, they receive sunlight and carry on their chief function, food-making or photosynthesis. The spongy layer is located below the palisade cells. These cells are more irregular, with large air spaces among them that are adjacent to the stomates. These cells are thus well suited for the exchange of gases with the surrounding atmosphere, and for the passage of carbon dioxide to the palisade layer above them. Cells of the spongy layer have fewer chloroplasts and so are not as important from the viewpoint of photosynthesis as the palisade cells. The green cells of young stems also carry on photosynthesis.

Supporting tissue. Plants are generally strengthened and supported in four ways.

1. The fibrovascular bundles have woody xylem cells that give support to the stem. These are long, pointed dead cells with greatly thickened walls which make up a considerable proportion of the wood of trees.

2. There are also groups of living cells with greatly thickened walls that lend support to various parts of a plant, i.e., in the leaf, just above and below the midrib; similar cells in stems, next to the epidermis.

3. In addition, there are scattered thick-walled dead cells in stems, located just outside of the phloem that help support the plant.

4. In parts of the leaf, especially between the midrib and the epidermis, there are large, thin-walled cells that strengthen the leaf by their turgidity. That is they are stretched full of water and are quite rigid in the same way that a hose full of water is rigid. When there is a loss of water from these cells, the leaf becomes softer and *wilted*. Similar groups of cells perform the same function of support by turgidity in succulent stems and in the young stems of woody plants.

Storage tissue. There are large, thin-walled cells in stems known as *pith*, that store food. They may be located either in the center of the stem, or scattered through the stem as rays of cells (medullary rays). These pith rays also serve to conduct food and water within the stem. Roots may become thickened and serve to store food, as in the case of the sweet potato, carrot, and radish.

Reproductive tissue. The reproductive organs of a plant are located in the flower. Male nuclei (sperm nuclei) are formed in the *pollen* grains, produced by the stamens. Female nuclei (egg nuclei) are formed in the *ovules* of flowers located within the ovary. When pollination occurs, pollen is deposited on the stigma; a pollen tube develops from each pollen grain, and grows down into an ovule, carrying the sperm nuclei within itself. The sperm and egg nuclei fuse (*fertilization*), and the ovule develops into a seed. Part of the seed contains the embryo of a new plant; stored food for the embryo occupies most of the remaining part of the seed.

SUMMARY OF PLANT TISSUES

TISSUE	TYPES	LOCATION	FUNCTIONS
Covering	1. Epidermis 2. Bark	Leaf, stem, root Woody stem	Protection; prevent moisture loss; form root hairs on roots
Conducting	1. Xylem	Fibrovascular bundles in leaf, stem, root	Conduct water up from root into stem, leaves
	2. Phloem	Fibrovascular bundles in leaf, stem, root	Conduct food from leaves to rest of plant
Growing	1. Meristem	Tips of stems, roots	Growth; differentiation
	2. Cambium	Fibrovascular bundles, between xylem, phloem	Growth in thickness; formation of new xylem, phloem
Food-making	1. Palisade cells	Upper layer of leaf	Photosynthesis
	2. Spongy layer	Lower layer of leaf	Some photosynthesis; help circulate gases for pallisade layer
	3. Green cells	Stems	Photosynthesis
Supporting	1. Wood	Xylem	Mechanical support
	2. Thick-walled living cells	Leaf, stem (under epidermis)	Mechanical support
	3. Thick-walled dead cells	Stem, outside of phloem	Mechanical support
	4. Thin-walled, large cells	Leaf, stem	Support by turgidity
Storage	1. Pith cells	Stems	Store food
	2. Root tissue	Enlarged roots	Store food
Reproductive	1. Pollen	Stamen of flower	Produces sperm nuclei
	2. Ovule	Ovary of flower	Produces egg nuclei

CHAPTER REVIEW

Select the correct choice for each of the following statements

1. A group of similar cells performing the same function is known as a (n) (A) organ (B) system (C) tissue (D) layer (E) membrane

2. Epithelial tissue may have all of the following functions except (A) protection (B) absorption (C) contraction (D) secretion (E) sensation

3. The chief tissue of which glands are composed is (A) epithelial (B) muscle (C) nerve (D) connective (E) supporting

4. The muscles attached to the bones are (A) voluntary and smooth (B) involuntary and smooth (C) voluntary and striated (D) involuntary and striated (E) smooth and striated

5. An axon is part of a (n) (A) muscle cell (B) nerve cell (C) epithelial cell (D) bone cell (E) cartilage cell

6. A tissue containing deposits of intercellular material is (A) nerve (B) muscle (C) epithelial (D) striated (E) connective

7. Of the following, the one that contains the largest number of different types of cells is the (A) mucous membrane (B) cardiac muscle (C) nerve (D) small intestine (E) blood

8. A tissue specialized for contraction is (A) nerve (B) muscle (C) adipose (D) blood (E) reproductive

9. Of the following, the one that includes all the others is (A) blood (B) connective (C) epithelial (D) kidney (E) nerve

10. The correct order of arrangement of the following (1-organ, 2-cell, 3-tissue, 4-organism, 5-system) is (A) 2–1–3–4–5 (B) 2–3–4–1–5 (C) 2–3–1–4–5 (D) 2–3–4–5–1 (E) 2–3–1–5–4

11. Stomates are largely found in a (A) leaf (B) stem (C) root (D) root hair (E) leaf hair

12. Structures that conduct materials throughout a plant are called (A) lenticels (B) meristem (C) fibrovascular bundles (D) cambium (E) epidermis

13. Food-making cells in plants are located in the (A) xylem (B) phloem (C) pith (D) palisade layer (E) ducts

14. The growing region of a plant stem is called the (A) bark (B) meristem (C) wood (D) spongy layer (E) phloem

15. Cambium is located next to the (A) stomates (B) guard cells (C) epidermis (D) root hairs (E) xylem and phloem

ANSWER KEY

1-C	4-C	7-D	10-E	13-D
2-C	5-B	8-B	11-A	14-B
3-A	6-E	9-D	12-C	15-E

GREEN PLANTS - BASIS OF ALL LIFE

2

4 How plants make food

In the fall of 1966, using radioactive carbon and phosphorus on chloroplasts which had been separated from freshly-ground spinach leaves, Dr. James A. Bassham and Dr. R. G. Jensen duplicated photosynthesis outside of a cell, and studied its step-by-step stages. Thus, another step was taken in our understanding of photosynthesis.

Autotrophs are organisms that can make their own food from inorganic raw materials. A few types of bacteria carry on *chemosynthesis,* by which they produce organic molecules from inorganic compounds of sulfur, iron and nitrogen. However, practically all other food is made by green plants during *photosynthesis.*

Since green plants can make their own food, they serve as the basis of practically all life on earth. Some animals eat plants as food. Some animals eat other animals, which in turn obtained their food from plants. However, all animals depend on green plants for their food, either directly or indirectly.

Photosynthesis is the manufacture of food (carbohydrates) from carbon dioxide and water in the presence of light. It takes place in the cells of green plants containing chlorophyll. During this process, the raw materials, carbon dioxide and water, are combined by the chlorophyll to form simple sugar, glucose. Oxygen is given off as a by-product. The following simplified equation summarizes the process:

$$\text{water} + \text{carbon dioxide} \xrightarrow[\text{chlorophyll enzymes}]{\text{energy}} \text{glucose} + \text{water} + \text{oxygen}$$

This equation can be restated with greater chemical detail as:

$$12H_2O + 6CO_2 \xrightarrow[\text{chlorophyll}]{\text{energy}} \underset{\text{glucose}}{C_6H_{12}O_6} + 6H_2O + 6O_2$$

Much information about the various steps in photosynthesis has been obtained recently through the use of isotypes of oxygen and carbon, and the application of refined techniques of chromatography. With the aid of heavy oxygen, oxygen-18, it has been shown that the oxygen that is given off comes from the water. Radioactive carbon, carbon-14, has been traced as a "tagged" molecule through the complex stages of carbon fixation with the aid of instruments such as the Geiger counter. In chromatography, various compounds are separated out at different levels as they are absorbed on a length of filter paper.

Photosynthesis is now known to take place in two sets of reactions, (a) a light reaction, in which oxygen is liberated, and light energy is converted into chemical energy, and (b) a dark reaction, in which the chemical energy is used to convert carbon dioxide into organic compounds, without the necessity of light.

(a) Light Reaction. In the presence of light energy, chlorophyll becomes activated, and releases high-energy electrons from its molecules to a protein compound named ferredoxin. Ferredoxin releases the electrons to form ATP (adenosine triphosphate), and to change the coenzyme NADP (nicotinamide-adenine dinucleotide phosphate) to $NADPH_2$. The H atoms for the latter, come from water molecules which are split. Oxygen atoms are left over, and are given off as O_2 gas. The energy stored in ATP and $NADPH_2$ is used to convert CO_2 into carbohydrates in the dark reaction.

(b) Dark Reaction (Carbon fixation). During this phase, a series of intermediate compounds are formed from the carbon dioxide molecules and hydrogen atoms, eventually giving rise to glucose. A number of enzymes are active along the way. Some of the intermediate compounds are: PGA (phosphoglycerate) formed from CO_2, water and ribose diphosphate (RDP); PGAL, formed when hydrogen is added to PGA from the hydrogen carrier, $NADPH_2$, fructose, formed by the combination of PGAL molecules. Fructose is converted into glucose.

The compound PGAL has several values, in addition to forming glucose; some of it is recycled to continue carbon fixation, with the formation of various intermediate products; it may be used in the synthesis of fats and proteins; it may also be used as a fuel to provide more energy for carbon fixation.

When photosynthesis is active and sugar is being formed faster than it can be used by the leaf, it is converted into starch for storage. Starch can be seen under the microscope as small grains. At night, starch is changed back into sugar and is conducted away from the leaf. In some plants such as sugar cane and the beet, reserve food is stored as *sucrose* ($C_{12}H_{22}O_{11}$). This is the form of sugar familiar to us as ordinary granulated table sugar. It is known as a *disaccharide,* having twice as many carbon atoms as the simple sugars such as glucose (grape sugar) and fructose (fruit sugar), which are *monosaccharides,* $C_6H_{12}O_6$.

The leaf factory. The leaf may be considered as the food factory of the plant, since it is here that photosynthesis is largely carried on. Leaves of different species of plants differ from each other; in fact, it is usually possible to identify most trees by the shape, size, and form of their leaves. In general, most leaves have an expanded broad portion, the *blade,* and the thin stalk, the *petiole,* by which the blade is attached to the stem. The veins, which contain the fibrovascular bundles, are arranged as a network in leaves of dicotyledonous plants (oak, maple, elm), but are parallel to each other in the leaves of monocotyledonous plants (grasses, wheat, corn).

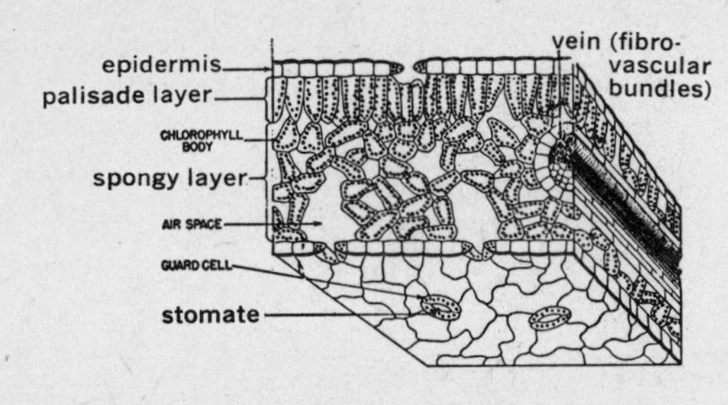

Cross Section of a Leaf

A microscopic study of the cross-section of a leaf reveals several different layers of cells and tissues which are adapted for photosynthesis:

1. The **upper** and **lower epidermis** make up the upper and lower surfaces of the leaf. They are composed of a layer of cells without chloroplasts whose outer walls are thickened and contain a waxy substance called *cutin.* They are thus well adapted to prevent the loss of water. Numerous tiny openings, *stomates,* are usually found only on the lower epidermis. The leaves of some plants have hundreds of stomates in an area of the leaf only a quarter of an inch square. Each stomate is surrounded by two kidney-shaped cells called *guard cells,* which contain chloroplasts. These guard cells move in such a way as to open or close the stomates. The stomates allow an exchange of oxygen and carbon dioxide between the moist surface of the cells lining the air spaces on the inside of the leaf and the external atmosphere; they also permit the passage of water vapor from the inside of the leaf to the outside.

2. The **palisade layer** is present below the upper epidermis as a row of column-like long cells, each containing many chloroplasts. Being near the top of the leaf, they are in a favorable position to receive much sunlight; their chief function is to carry on photosynthesis. Between the palisade cells are small air spaces which are connected with larger air spaces in the lower part of the leaf, which in turn are connected with the stomates. These air spaces make it possible for carbon dioxide to diffuse to the palisade cells and for oxygen to diffuse out.

3. The **spongy layer,** below the palisade layer, consists of irregular cells surrounded by large air spaces that are connected with the stomates. There is a free circulation of air around them and with the air spaces between the palisade cells. The spongy cells contain fewer chloroplasts, and are not in as favorable a position to carry on photosynthesis as the palisade cells.

4. **The veins** run through the spongy layer, and contain the *fibrovascular bundles* which help support the leaf, and distribute materials through the xylem and phloem. Water is brought up from the roots through the stem into the petiole and through

the blade by means of the *xylem*. Water diffuses from cell to cell and is combined with the carbon dioxide to form glucose. The glucose diffuses out through the cells to the *phloem* which carries it out of the blade, through the petiole, into the stem and roots.

Chlorophyll is found in tiny oval bodies called chloroplasts. It is formed only in the presence of sunlight. This explains why plants that are germinated from seeds in the dark are white. When it is analyzed, chlorophyll is found to be a mixture of pigments, including two green ones, chlorophyll A and chlorophyll B, and the yellow pigments, carotene and xanthophyll, which are hidden by the green color. The latter colors show up especially in the fall when the leaves change to yellow, orange and red colors before dropping from the trees. Chlorophyll contains the elements carbon, hydrogen, oxygen, nitrogen and magnesium. In order for chlorophyll to be formed, minute quantities of magnesium and iron must be available to the plant.

The chloroplasts have been found to reproduce when the cell divides. Under the electron microscope, each chloroplast is seen to be composed of disk-shaped units called *grana*. Each of these grana, has layers, or lamellae, of protein, fatty materials and the chlorophyll pigments. Chlorophyll is similar to hemoglobin in chemical structure. In chlorophyll, a magnesium atom is surrounded by four nitrogen atoms in the same way that the iron in hemoglobin is surrounded by four nitrogen atoms.

The value of chlorophyll in food making can be shown by the following demonstration. Choose a plant such as the silver-leaf geranium, in which there is a white fringe around the outer border of each leaf, or a coleus whose variegated leaves have colored areas which are not green. There is no chlorophyll present in these parts of the leaf, and therefore no food should be formed there.

1. Extract the chlorophyll from the leaf by boiling it gently in alcohol. If the leaf is first plunged into hot water before this is done, the waxy epidermal covering is broken down and the chlorophyll can be dissolved out more readily.

2. When the alcohol has turned green, remove the leaf and rinse it in water. It is now completely white.

3. Spread the leaf out in a flat dish and cover it with iodine solution. Iodine reacts with starch and turns it a blue-black color.

4. The parts of the leaf which contained chlorophyll, and stored starch as the result of photosynthesis, are now deeply stained. The parts of the leaf which did not contain chlorophyll, and therefore did not store starch, will merely have the light brown color of the iodine solution.

Demonstration: Value of Chlorophyll

Light. Chlorophyll is active in making carbohydrates only if light is present as the source of energy. In nature, sunlight is used, but artificial light can also be used. When there is no light, photosynthesis stops. Chlorophyll absorbs certain wavelengths of light energy more effectively than others. Most absorption takes place at the blue-violet, and orange-red ends of the spectrum. Here, the rate of photosynthesis is highest. Green light is reflected by chlorophyll, and is least effective in photosynthesis.

The following demonstration can be conducted to show the necessity of light:

1. Cover the upper and lower epidermis of part of a leaf on a geranium plant with two thin round pieces of cork. Place the plant in sunlight. The covered part of the leaf is now in the dark, and cannot make carbohydrates. The rest of the leaf is in the light, and can make carbohydrates.

2. Extract the chlorophyll as was done in the previous demonstration, and cover the leaf with iodine solution.

3. The leaf now shows two areas of color: the part that was exposed to the light, and was able to store starch, is deeply stained by the iodine. The covered part, which did not make food, will have only the light brown color of the iodine.

Carbon dioxide. The air around us normally contains about 0.03% of carbon dioxide. This gas is constantly being excreted by living things as they carry on respiration. Volcanoes, fires, automobiles, and industrial plants also add carbon dioxide to the air. Green plants use up some of this carbon dioxide as they carry on photosynthesis. The following demonstration shows the necessity of carbon dioxide for food-making:

1. Place a geranium plant under a bell jar (A) in which there is a container of dry potassium hydroxide. Seal the edges of the bell jar with vaseline to make it airtight. The potassium hydroxide absorbs carbon dioxide so that there should not be any of it present under the bell jar. Set up another bell jar (B), with a plant in the same way, but substitute sand for the potassium hydroxide. This is the control.

2 After a day in sunlight, remove a leaf from each plant, extract the chlorophyll, and cover with iodine solution.

3. Leaf A will only have the light brown color of the iodine showing that no starch was formed, because of the absence of carbon dioxide. Leaf B will be stained blue black, showing the presence of starch resulting from the availability of carbon dioxide for photosynthesis.

Oxygen. Photosynthesis is an important source of the oxygen in the air. An aquarium is stocked with green water plants to furnish fish with oxygen; on the other hand the fish give off carbon dioxide which is used by the plant in photosynthesis. This *oxygen-carbon dioxide cycle* shows the interrelationship that exists between animals and green plants. The giving off of oxygen by green plants during photosynthesis can be demonstrated as follows:

1. Place a quantity of the green water plant, Elodea, under a funnel in a wide-mouthed jar full of water. Cover the stem of the funnel with an inverted test tube filled with water. Keep in sunlight.

2. In a short time, bubbles of gas can be observed streaming from the plants up the funnel into the test tube. After some time, the test tube will be filled with the gas, all of the water having been displaced.

3. Insert a glowing splint, into the test tube, and it will glow brighter or even burst into flame, because of the presence of pure oxygen in the test tube.

FORMATION OF OTHER TYPES OF FOOD

Some of the glucose produced during photosynthesis is used for energy; some is stored as starch; and some is converted into other types of food materials, including fats and oils, and proteins.

1. **Fats and Oils.** The natural fats and oils are composed of carbon, hydrogen and oxygen, with oxygen occurring in very small amounts. Thus, the formula of one fat, olein of olive oil, is $C_{57}H_{104}O_6$. The oxidation of fats produces large amounts of energy. Fats and oils are liberally stored in the seeds of the coconut, olive, and peanut, and are extracted for their commercial value. The formation of fats from carbohydrates is known as *fat synthesis.*

2. **Proteins.** Glucose is combined with such elements as nitrogen, sulfur and phosphorus to form basic compounds known as *amino acids.* From these, complex proteins are built up. The formula of the protein in corn, zein, is $C_{736}H_{1161}N_{184}O_{208}S_3$. Proteins are the active constituents of protoplasm, and are used for forming new protoplasm, by the process of assimilation, for growth and repair. The production of proteins from carbohydrates is known as *protein synthesis.* The minerals used in this process are transported in the water up from the roots through the stem in the xylem ducts.

3. **Vitamins.** Plants are excellent sources of the vitamins. Vitamin A is made from carotene, one of the pigments of chlorophyll. It naturally follows that green leafy vegetables are good sources of this vitamin. The yellow vegetables, such as sweet potato and carrot, are also good sources of vitamin A. The vitamin B complex is found in the seeds of many plants, including the grains, and peas. Vitamin C is abundant in fresh vegetables and fruits.

CHAPTER REVIEW

Select the correct choice for each of the following statements.

1. Practicaly all living things depend for their food on the activity of (A) chromatin (B) chlorotone (C) chlorophyll (D) colchicine (E) cholesterol
2. For the process of photosynthesis, green plants require (A) carbon monoxide and water (B) carbon dioxide and carbon monoxide (C) carbon dioxide and water (D) oxygen and water (E) oxygen and carbon dioxide
3. A by-product of photosynthesis is (A) carbon dioxide (B) water (C) carbon monoxide (D) oxygen (E) nitrogen
4. The substance produced as the direct result of photosynthesis is (A) glucose (B) protein (C) fat (D) iodine (E) vitamin A
5. All of the following are carbohydrates except (A) sucrose (B) starch (C) fructose (D) fruit sugar (E) carotene
6. The passage of gases into and out of a leaf is controlled by the (A) petiole (B) guard cells (C) palisade layer (D) spongy layer (E) veins
7. Chlorophyll can be extracted from a leaf by (A) soaking it in water (B) boiling it in alcohol (C) electrolysis (D) adding iodine to it (E) adding glucose to it

8. When a leaf that was kept in the dark is treated with iodine after its chlorophyll was extracted, it will turn (A) blue-black (B) brick-red (C) green (D) the color of the iodine (E) white

9. The food made during photosynthesis may be used for all of the following purposes except (A) protein synthesis (B) fat synthesis (C) storage (D) energy (E) transpiration

10. In a balanced aquarium, the fish supply the green plants with (A) oxygen (B) water (C) carbon dioxide (D) glucose (E) starch

11. An autotroph makes food from (A) organic molecules (B) inorganic molecules (C) glucose (D) fats (E) proteins

12. The source of oxygen in photosynthesis is (A) CO_2 (B) glucose (C) ATP (D) H_2O (E) TPN

13. Chloroplasts are made of (A) grana (B) stomata (C) cutin (D) guard cells (E) xylem

14. All of the following are intermediate products of carbon fixation *except* (A) PGA (B) DNA (C) PGAL (D) RDP (E) fructose

15. The part of the spectrum in which photosynthesis is most active is (A) green-yellow (B) yellow-orange (C) blue-violet (D) blue-green (E) green-orange

ANSWER KEY

1-C	4-A	7-B	10-C	13-A
2-C	5-E	8-D	11-B	14-B
3-D	6-B	9-E	12-D	15-C

5 How plants live

In addition to carbon dioxide and water which are used in photosynthesis, plants need other materials for their activities, including oxygen and minerals. They also use water for a variety of other purposes.

Aerobic Respiration. Green plants, like animals, carry on respiration. That is, they take in oxygen, and combine it with food to release energy for their various activities, such as food-making and growth. In the process, carbon dioxide and water are excreted as wastes. Thus, respiration is the reverse process of photosynthesis. As a matter ot fact, during the day, both processes go on side by side, in different parts of the green cell. In photosynthesis, an excess amount of oxygen is produced by the chloroplasts. Some of it is used directly by the protoplasm of the cell for oxidizing food, while the rest of it diffuses out and is given off through the stomates. At the same time, while respiration is being carried on by the protoplasm, carbon dioxide and water are given off. These waste products are used up almost at once by the chloroplasts which are carrying on photosynthesis. Thus two different parts of the cell carry on contrasting activities—the protoplasm takes in oxygen and gives off carbon dioxide and water, to release energy, during respiration; while the chloroplasts in the same cell take in carbon dioxide and water, and give off oxygen, storing energy, during photosynthesis.

At *night,* however, when photosynthesis does not go on, only respiration is carried on; oxygen is taken in from the air through the stomates, and carbon dioxide and water are given off.

A demonstration to show that plants give off carbon dioxide during respiration may be performed as follows:

1. Place a geranium plant under a bell jar (A) containing a beaker of fresh lime water. Seal the edges with vaseline to make it airtight.
2. Set up another bell jar (B) in the same way, but without a plant. Keep both setups in the dark overnight.
3. The next day, observe the lime water in both bell jars. In bell jar A, it will be cloudy because of the effect of the carbon dioxide given off by the plant. In bell jar B, it will be clear, since there was no carbon dioxide.

COMPARISON OF PHOTOSYNTHESIS AND RESPIRATION

	PHOTOSYNTHESIS	RESPIRATION
When it occurs	Day only	Day and night
What is taken in	CO_2 and H_2O	O_2
What is given off	O_2	CO_2 and H_2O
Where it takes place	In green cells only	In all cells
What happens to glucose	Built up	Broken down
What happens to energy	Used and stored up	Liberated

During respiration, some of the energy is released as very small amounts of *heat.* It is possible to measure this by performing the following demonstration with bean seeds that are germinating and actively carrying on respiration:

1. Fill a thermos bottle (A) that is lined with moist absorbent paper with bean seeds that have begun to germinate. Seal with a one-holed rubber stopper containing a thermometer whose bulb is present inside the bottle. Make a record of the thermometer reading.

2. As a control, set up another thermos bottle (B) in the same way, but without any seeds. Make a record of the thermometer reading.

3. After twenty-four hours, compare the thermometer readings. In thermos bottle A, there will be a rise, indicating the liberation of heat by the seedlings. Thermos bottle B will show no change.

Anaerobic Respiration. Some of the simpler plants, such as yeast and certain bacteria, carry on anaerobic respiration, in which oxygen is not needed. They produce as wastes carbon dioxide and ethyl alcohol. The process is known as *alcoholic fermentation.* It has commercial values in the brewing, baking and wine-making industries. Other types of bacteria and certain molds carry on *lactic acid fermentation,* in which carbon dioxide and lactic acid are produced. The amount of energy released during anaerobic respiration is relatively small compared to aerobic respiration, since most of the C-C and C-H bonds in the alcohol or the lactic acid have not been broken, and still retain their chemical energy.

Transport. Food, water and minerals are moved throughout the plant, from the roots to the leaves, through xylem and phloem tissues; also, from cell to cell by active transport and diffusion. The movement of these materials is referred to as *translocation.*

The roots. Minerals and water are absorbed from the soil by the roots of the plant. Roots also serve to anchor the plant to the soil. A study of a root shows several regions: The *tip* consists of a growing point made of rapidly dividing cells. There is a *root cap* over this growing tip which protects it as it pushes through the soil. Just back of the growing tip is a region which elongates considerably, and which is the most rapidly growing part of the root.

Farther back, the cells of the epidermis produce projections called *root hairs.* This part of the root appears fuzzy to the naked eye because of the large number of root hairs. Most of the water and minerals absorbed by a root are taken in by these root hairs. A root-hair cell is a single cell with a large vacuole, in which the nucleus is usually found

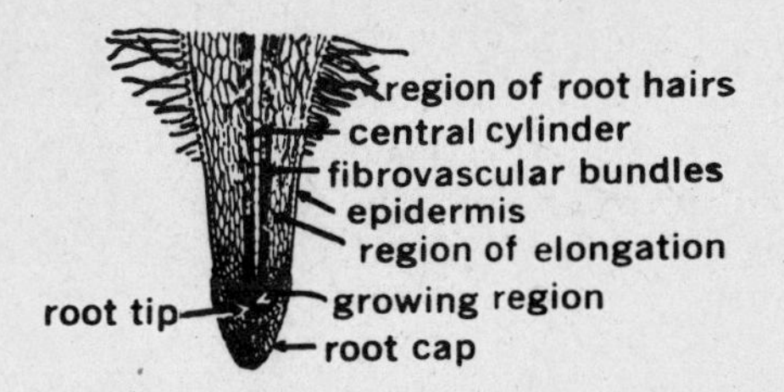

Structure of root tip

in the root-hair portion. The formation of root hairs increases the absorbing surface of the root; it also allows for closer contact, with the film of water around the soil particles. The contact with these soil particles is so close that the root hairs adhere to them. This results in two effects: the root binds the soil together; and the root is firmly anchored in the soil.

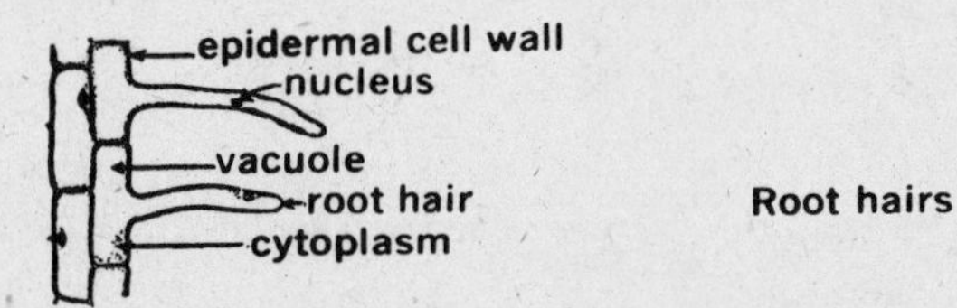

Root hairs

Diffusion of soil water. Water in the ground containing minerals is known as *soil water.* This soil water enters the root hair by active transport and diffusion. It diffuses through the cell membrane into the root-hair cell because (1) there is a higher concentration of minerals in it than there is inside the cell (2) the water outside is purer, and therefore in higher concentration than the water of the cell sap in the vacuole. After having entered the root hair, soil water then diffuses into the adjoining cells inside the root where there is a lower concentration of these molecules. As this process continues from cell to cell, the water and minerals then diffuse into the xylem of the fibrovascular bundle and are conducted up the stem to the leaves.

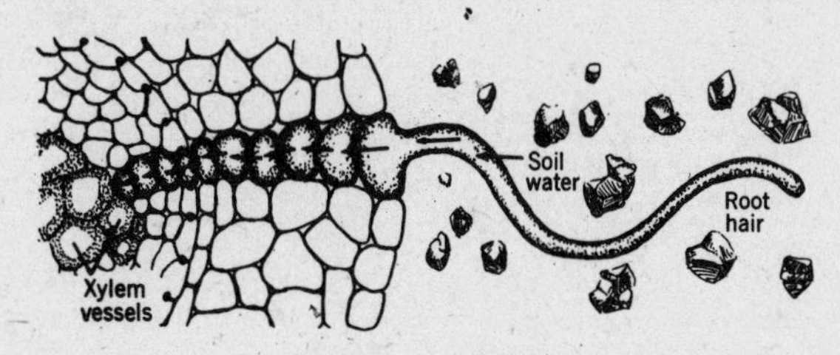

Diffusion of Soil Water

A cell membrane does not permit all materials to pass through it. Only certain types of dissolved substances can pass through it. Cell membranes are therefore said to be *semi-permeable.* In certain cells of the plant, the cell membrane will permit more of one substance to accumulate than in others: thus, in some roots (carrots), stored food accumulates as it enters the root cells; also, in leaf cells, the element magnesium will pass through the cell membranes to form more chlorophyll.

A demonstration to show the rise of dissolved substances in a stem to the leaves may be performed as follows:

1. Place a celery stalk containing leaves in a jar of red ink.

2. After a day, observe that the leaves are stained red, showing that the ink has risen up the stem.

3. To see the conducting tubes of xylem, hold the stalk against a bright light. They are stained red and can be seen through the light-colored stalk, running the length of the stalk. They can be seen more clearly if some of the tissue over them is cut away with a razor blade.

4. Cutting a thin cross-section of the stalk will reveal the fibers stained red, and surrounded by unstained cells.

Transpiration. After water is brought up to the leaves, it is passed through the veins to the spongy layer, and then evaporates into the air spaces, and diffuses out through the stomates. This evaporation of water from the leaves is known as *transpiration*.

By keeping the cells of the spongy layer around the air spaces of the leaf constantly moist, transpiration makes the absorption of carbon dioxide for photosynthesis possible. An average tree may transpire 50 gallons of water a day. Transpiration may be demonstrated in the following way:

1. Cover the pot and soil of a potted geranium plant completely with a rubber sheet. Place it under a bell jar whose edges are sealed with vaseline.

2. In the same way, cover completely another pot that does not contain a plant, and place it under a bell jar. This is the control.

3. After several hours, the inside of the first bell jar will be covered with moisture resulting from transpiration. The control bell jar will have no moisture in it.

4. If it is desired, the potted plant can be weighed at intervals of three hours, and a record kept of the weight of the water lost.

If water is lost in transpiration faster than it is replaced by the roots, the leaves will lose their turgidity and will wilt. During such periods, the stomates close, reducing the rate of transpiration. The cutinized wall of the epidermis also helps retard water loss. Plants that grow in dry regions have fewer stomates than those from moist areas; their stomates are also sunk in pits covered by hairs to reduce evaporation. Cactus plants in the desert have spines as modified leaves; their water loss is very small.

The rise of water in a tree several hundred feet high may be explained as resulting from several factors:

1. **Cohesion.** There is a long, fine continuous column of water in the ducts. As the water at the top end evaporates, the water column is pulled upward, with more water entering from the roots. The force holding the water particles together is known as cohesion. It gives the column of water considerable tensile strength. The resulting upward pull is known as *transpiration pull.*

2. **Root pressure.** If the stem of some plants is cut off near the root, water can be seen coming out of the cut end of the stump. The pressure in the root that causes this rise of water may be due to the turgidity of the root cells. This causes water to diffuse inwardly from the epidermal region of the root, and forces water up through the xylem.

3. **Capillarity.** Water rises in fine capillary tubes. This can be shown by placing a piece of glass tubing in a glass of water. The water in the tube will be higher than that in the glass.

These explanations probably tell only part of the story of how water rises in plants. Full agreement has not been reached by scientists on this subject.

Minerals. Minerals enter a plant when there is a greater concentration of them outside than inside the plant. There are fourteen elements which are regarded as essential for green plants. The three in carbohydrates, carbon, hydrogen and oxygen, are obtained from carbon dioxide and water during photosynthesis. They make up most of the plant. The others are obtained from the soil—nitrogen, phosphorus, potassium, calcium, magnesium, iron, sulfur, manganese, boron, zinc, and copper. The first four of these are needed in larger amounts. Nitrogen, sulfur and phosphorus are needed in protein synthesis. Iron and magnesium are necessary for the production of chlorophyll; although iron is not part of the chlorophyll molecule, chlorophyll cannot be formed without it; magnesium is part of the chlorophyll structure. For a complete healthy condition, however, plants need about thirty other elements such as chlorine, aluminum, molybdenum, and many others in minute amounts; these are known as *trace elements.*

Farmers and gardeners make use of our knowledge about the mineral requirements of plants by adding fertilizers to the soil containing specific amounts of such elements as nitrogen, phosphorus and potassium. Another way of increasing the amount of nitrogen compounds in the soil is to grow legume plants, which bear nodules on their roots containing nitrogen-fixing bacteria; these bacteria fix nitrogen into nitrates. One interesting method of growing plants is to do without soil and instead to use a liquid solution of the required minerals. This is known as *hydroponics.*

A demonstration to show the absorption of

phosphorus by a potted geranium plant may be performed as follows:

1. Water the soil with radioactive sodium phosphate containing the radioactive isotope of phosphorus, P^{32}.

2. At five-minute intervals, hold a geiger counter close to the lower and then the upper leaves of the plant. Record the time required for P^{32} to enter the lower leaves and finally all the leaves of the plant.

Radioactive isotopes that can be traced through a living organism by the geiger counter are known as *tagged atoms.* Radioactive carbon in carbon dioxide has added to our knowledge of photosynthesis.

Types of roots. There are several kinds of roots that vary in their structure and function.

1. **Taproot.** This is the long, strong primary root of a plant such as the dandelions and the oak tree.

2. **Fibrous roots.** These form a network of roots just below the stem, i.e., grasses.

3. **Fleshy root.** This is the large root that stores food, as in the carrot and radish.

4. **Brace roots.** In some plants as corn and the mangrove tree, large roots grow down from the main stem into the ground, and brace or support the stem.

5. **Aerial roots.** Vines such as ivy attach themselves to walls or trees by aerial roots.

Digestion of food. After the green plant has made food, it stores it as starch, fat or proteins. These stored nutrients are insoluble and remain in their area of storage until they are either to be used by the plant, or to be transported from one part of the plant to another. Then they must be made soluble. This change of nutrients from insoluble to soluble form is known as *digestion.* Certain compounds called *enzymes* bring about these changes: they are examples of catalysts, substances that accelerate chemical changes without being used up in the process. Since this digestion takes place within the cells, it is called *intracellular* digestion.

The enzyme *diastase* converts starch, which is insoluble, into glucose, which is soluble. Starch grains that are stored in the leaves during photosynthesis, are broken down into glucose at night, which is then transported to other parts of the plant. A demonstration to show digestion in plants may be performed with bean seeds as follows:

1. Germinate six bean seeds. Starch may be shown to be present by testing some bean seeds with iodine; the blue-black color indicates the presence of starch. Glucose may be shown to be absent by heating some crushed beans in Benedict's solution; the Benedict's solution remains blue.

2. In a few days when the bean seeds have started to germinate and are beginning to produce roots, crush them and heat in Benedict's solution. The change of the liquid to an orange color now indicates the presence of glucose.

Without digestion taking place, the growing embryo would be unable to make use of the insoluble stored starch. Glucose is soluble, and can move from the cotyledons to the various parts of the young plant when it is needed. Other enzymes produced by plants are: *Lipase,* which digests fat into fatty acids and glycerin. *Bromelin* splits protein into amino acids.

Extracellular Digestion. Heterotrophic protists such as fungi and bacteria, do not contain chlorophyll, and cannot make their own food. Although these organisms are classified as protists by many biologists, they exhibit plant-like characteristics and so are included here. They secrete enzymes which digest food outside of them, changing it into simpler form which is then absorbed. Because such digestion takes place outside of the organism's cells, it is known as *extracellular* digestion.

Methods of plant nutrition. There are various ways by which plants obtain their food:

1. **Independent plants.** Green plants make their own food, and so may be considered to be *independent,* or *autotrophs.* They are the basis of food for heterotrophs such as saprophytes and parasites.

2. **Saprophytes** are plants that do not contain chlorophyll and obtain their nourishment from organic matter, i.e., various fungi such as mushrooms, molds, and bacteria. Mushrooms live on soil containing the decayed remains of plants. Molds may grow on bread, fruits, and leather. Bacteria of decay in the soil help to break down plant and animal remains, thereby enriching the soil.

3. **Parasites** are plants that attach themselves to living things for their nourishment. Wheat rust is a fungus that lives on the wheat plant, destroying it. Part of the life history of this parasite is spent on another host, the barberry plant. Another fungus causes athlete's foot in man. Some bacteria are the cause of serious human diseases.

4. **Symbiotic** plants live together in an intimate relationship. The lichen has already been described as being made up of a fungus and algae cells. The algae make food by photosynthesis, while the fungus supplies water and minerals. Another example: Legume plants (peas, clover) have *nodules,* or swellings on their roots which contain nitrogen-fixing bacteria. The bacteria live on the plant material, and fix nitrogen converting it into nitrates which the plant uses.

5. **Insectivorous plants** have leaves that trap insects and use them as food. All of these plants also carry on photosynthesis. In some cases the trapped insects are digested by enzymes secreted by the leaves. In other cases, they are decomposed by bacteria; the soluble nitrogenous materials are then absorbed by the plants:

a. **Pitcher plant** has its leaves modified in the form of an enclosed pitcher, in which water collects. Small insects are drowned in the water, not being able to climb out because there are hairs on the inner walls of the pitcher that point downward. The insects are decomposed by bacteria, and the products of decay are absorbed by the plant.

b. **Sundew** has small leaves covered with glands borne at the ends of many thin stalks. Insects become stuck on the sticky fluid given off by these glands. The thin stalks bend over and enclose the insect which is then digested by enzymes secreted by the glands.

c. **Venus's-flytrap** has paired leaves, each with three bristles. If an insect alights on either leaf and touches the bristles, the leaf closes, trapping it. Enzymes are secreted from glands on the leaves, digesting the insect. After the digested material is absorbed, the leaf opens and the insect remains fall out.

d. **Bladderwort** is a plant that lives in fresh-water ponds, which has many tiny bladders the size of a pinhead along its stem. There is a trapdoor entrance to each bladder. This prevents tiny water animals from escaping, once they have touched the little bristles around it, causing them to be swept into the bladder by the sudden closing of the trapdoor. Enzymes digest them.

Photoperiodism. The production of flowers by a plant depends on the changing daily cycle of light and darkness. This was first demonstrated by Wrightman Garner and Harry Allard in 1918 in experiments with a tobacco mutant called Maryland Mammouth. This plant normally does not flower before the onset of frost in the fall. In early summer, these two plant physiologists reduced the hours of daylight for the plant by carrying potted plants into a dark room at 6 P.M. each evening, where they remained until 8 A.M. the next morning. Before the end of July, the experimental plants flowered, while the control group living under normal long July days did not flower.

Plants can be classified according to their photoperiodism into three main groups: (1) Short-day plants — flower when the hours of daylight are less than 12-13. Examples: asters, chrysanthemums, soybeans. (2) Long-day plants — flower when the day length is more than about 12 hours. Examples: hollyhock, black-eyed Susan, iris, sweet clover. (3) Day neutrals — flower over a wide range of periods. Examples: tomato, rose, nasturtium.

It has been demonstrated that the length of darkness is more important than the length of day.

Plant Growth Substances. In recent years scientists have discovered the existence of certain substances that have pronounced effects on the growth of plants.

Gibberellin. Among the newest of these substances is the group of compounds known as *gibberellin.* It was discovered in 1926 by a Japanese scientist on Formosa named E. Kurosawa, who was investigating a fungus disease of rice plants which caused them to grow very tall and spindly. He grew the fungus on a liquid culture medium, and was surprised to discover that this liquid would produce the symptoms of the disease if he rubbed it on healthy rice plants. The substance in the liquid was isolated and named gibberellin after the fungus which produced it, *Gibberella fujikuroi.* It was not until 1950 that gibberellin came to the attention of the western world. Research since then has shown it to have the following unusual effects on plants:

1. It causes corn, wheat, and many other plants to grow very rapidly, showing an increase in height that is three to five times the normal in a short period of time.
2. It makes dwarf plants that by heredity should always be stunted, such as dwarf pea or dwarf corn, grow to the size of normal plants.
3. Seeds that are soaked overnight in it germinate ahead of time.
4. Biennial plants such as foxglove and carrot, that flower in the second year of their life cycle, burst into bloom in only one year.
5. Tomatoes and cucumbers develop from flowers that are not pollinated if the flower buds are sprayed with it.
6. Garden and house plants, such as geranium and petunias, bloom ahead of time, and have larger flowers.

Auxin. The growing tips of leaves and stems produce growth-promoting substances or hormones called auxins. They cause the cell walls of young cells to elongate. Auxins also plays a part in the tropisms of plants, making them grow toward the light. Besides these two effects, auxins and other growth-promoting substances also produce the following results: they induce the formation of roots on cuttings; they speed up the germination of seeds; they increase the growth of weeds to such an extent that they interfere with their life functions and cause them to die; they prevent the premature dropping of apples from trees; they produce seedless tomatoes.

Cytokinins (kinetin). Coconut milk was originally found to contain factors that stimulated cell division in plant tissue cultures. These cytokinins have also been found in corn kernels and other plant sources. They have been used together with auxins to cause the differentiation of roots, stems, leaves and buds from small sections of plant tissue growing in tissue culture. They also have been found to break seed dormancy and to promote leaf expansion.

Phytochrome (chromophore). The bluish pigment in plants named phytochrome, seems to have an effect on photoperiodism. It is affected by light near the limit of vision in the far red end of the spectrum.

Florigen. This is believed to be a flowering hormone which brings about flowering in plants. Although it has not yet been isolated in pure form, there is much experimental evidence that it exists.

Dormin. Some growth factors inhibit or prevent growth and development. Dormin keeps buds, leaves, tubers, seeds and fruits from developing. Its effects have been found to be reversed by auxin, gibberellin and cytokinin.

CHAPTER REVIEW

Select the correct choice for each of the following statements.

1. Green plants carry on respiration (A) only during the day (B) only at night (C) during both day and night (D) only when photosynthesis is going on (E) only when photosynthesis is not going on

2. During respiration, green plants take in (A) water (B) carbon dioxide (C) oxygen (D) carbon dioxide (E) oxygen and water

3. In a green plant, respiration takes place (A) in all the cells (B) only in the green cells (C) only in the cells without chlorophyll (D) only in the guard cells (E) only in the stomates

4. A root may perform all of the following functions except (A) absorption of water (B) absorption of minerals (C) anchorage (D) storage (E) photosynthesis

5. All of the following materials can diffuse through a membrane except (A) soil water (B) dissolved minerals (C) dissolved iodine (D) starch (E) red ink

6. A cell membrane (A) stores food (B) is semipermeable (E) is not permeable (D) is soluble (E) keeps molecules out of a cell

7. The evaporation of water from a leaf is known as (A) conduction (B) circulation (C) translocation (D) transpiration (E) transformation

8. Mineral salts enter a plant largely through its (A) root hairs (B) root cap (C) root tip (D) stomates (E) guard cells

9. All of the following factors may be involved in transpiration except (A) evaporation (B) cohesion (C) root pressure (D) condensation (E) capillarity

10. Two elements needed for the production of chlorophyll are (A) phosphorus and sulfur (B) chlorine and copper (C) magnesium and iron (D) calcium and phosphorus (E) manganese and boron

11. Green plants need nitrates for making (A) carbohydrates (B) sucrose (C) starch (D) fats (E) proteins

12. A method of growing plants on a liquid solution of minerals is known as (A) hydrotropism (B) hydroponics (C) herpetology (D) symbiosis (E) commensalism

13. In the dark, a green plant excretes (A) hydrogen (B) carbon dioxide (C) oxygen (D) nitrogen (E) none of the above

14. A fibrous type of root is present in (A) grasses (B) carrot (C) radish (D) dandelion (E) oak tree

15. All of the following are used for digestion by plants except (A) enzymes (B) diastase (C) nodules (D) lipase (E) bromelin

16. Organisms that live together in a mutually beneficial relationship are called (A) saprophytic (B) symbiotic (C) parasitic (D) independent (E) insectivorous

17. A mushroom cannot manufacture its own food because it lacks (A) minerals (B) cytoplasm (C) cellulose (D) chlorophyll (E) root hairs

18. An example of a saprophyte is (A) clover (B) bread mold (C) wheat rust (D) athlete's foot fungus (E) legume

19. Pitcher plant, sundew, and Venus flytrap are alike in that they (A) are saprophytes (B) are symbiotic (C) trap insects (D) lack chlorophyll (E) do not carry on respiration

20. A plant growth substance produced by a fungus is (A) auxin (B) bromelin (C) called a trace element (D) gibberellin (E) molybdenum

ANSWER KEY

1-C	5-D	9-D	13-B	17-D
2-C	6-B	10-C	14-A	18-B
3-A	7-D	11-E	15-C	19-C
4-E	8-A	12-B	16-B	20-D

HOW WE LIVE

3

6 Value of good nutrition

"Man does not live by bread alone." All of the food we eat contains one or more substances useful to the body known as *nutrients*. There are six such groups of nutrients: carbohydrates, lipids, proteins, minerals, water, and vitamins. Some foods, such as sugar, consist of only one type of nutrient; most have several types. For good health, the body must receive a sufficient supply of them all. It therefore is essential that we eat a balanced diet. Our food is needed for several purposes: (1) for energy (2) growth (3) repair of body parts (4) maintenance of good bodily health (5) to supply the requirements of specific parts of the body, i.e., teeth, bones, blood, thyroid gland, etc. (6) synthesis of compounds needed for the maintenance of the body, such as enzymes, hormones, neurohumors, and other complex molecules. Plant foods are also useful in providing "roughage," which contributes indigestible material such as cellulose, and which stimulates the large intestine to function properly in excreting solid wastes.

FOOD AND ENERGY. Carbohydrates and fats are chiefly used for energy, although proteins may also be used for this purpose. One of the results of oxidation of food in warm-blooded animals is a constant body temperature—in humans, it is 98.6°F; in some birds, it may vary from 100°F to 112°F. Oxidation in animals is basically similar to that in plants. Certain respiratory enzymes in each cell are used to break the food down, with the release of energy, and formation of the waste products, carbon dioxide and water.

Recent research has revealed the true nature of this oxidation. It has been found that the food nutrients are not used directly for energy. Instead, the cells use a chemical compound called *adenosine triphosphate* (ATP), which is derived from the food. When a cell needs energy, part of ATP is broken away, and this liberates the needed energy. More ATP is then built up from the nutrients by a complex process involving numerous enzymes. The energy is used by the cells to carry on their activities and to grow.

We use energy all the time. Even when we are asleep, the heart is beating, we are breathing, and digestion is going on. The minimum amount of energy that keeps the body alive is known as *basal metabolism*. It can be determined by measuring the amount of oxygen consumed and the amount of carbon dioxide given off when the body is at rest. It is usually measured in the morning before a person has had any food. *Metabolism* is the sum total of all of the body's activities. It is possible to measure a person's rate of metabolism while he is engaged in various activities. Thus, a person who is reading a book has a lower rate of metabolism than one shoveling snow or swimming.

The energy value of food is measured in *Calories*. A Calorie is the amount of heat needed to raise the temperature of a kilogram of water (1,000 cc., or about a quart) one degree centigrade. To determine the number of Calories in a portion of food, it is burned in a *calorimeter*. The food is placed in a metal container surrounded by a water jacket, and is burned in pure oxygen. The heat that is released causes the temperature of the water to rise. This is measured by a thermometer and the number of

Calories is determined. In this way, it has been found that fat has twice the Calorie value of carbohydrates or protein. There are tables showing the Calorie values of different foods, i.e., a slice of white bread—60 Calories; an average size egg—100 Calories; a slice of pie—300 Calories; a portion of cabbage—40 Calories.

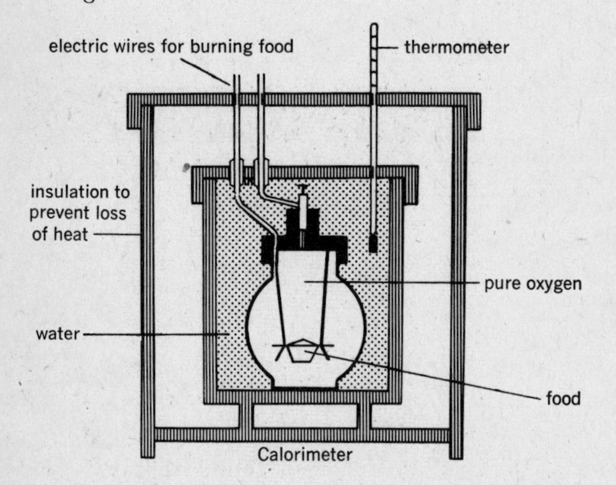

Calorimeter

The Calorie requirements of people vary with their age, sex, weight, activity, and climate. Babies and children have a higher basal metabolism than grownups; it is higher for boys than for girls; it is higher for heavy people; it is higher in a cold climate than in a warm one. An average-size man may have a basal metabolism of 1,800 Calories. A growing boy may need up to 4,000 Calories for his daily activities; a clerk, only about 2,500 Calories; a ditch digger, at least 5,000 Calories. When more food is eaten than can be used by the body, it is stored as fat under the skin and elsewhere in the body. Some of this fat is useful in insulating the body and preventing heat loss. Part of it serves as a reserve source of energy. The rest of it is excess and results in an overweight condition, being carried around as an added burden to the body's work. In middle-aged people it may cause a strain on the heart, since extra blood must be pumped to the added fat tissues. Overweight is usually a sign of overeating. It may also result from eating too many high-calorie foods, from lack of exercise, or a glandular disturbance, especially of the thyroid gland. An overweight person should consult a doctor as to the proper steps to be taken to modify his diet, and to reduce his Calorie intake.

Carbohydrates. The carbohydrate nutrients are present in foods as either *starch* or *sugar*. They are primarily used as a source of energy. Before starch can be utilized by the cells of the body, it must first be digested into glucose. Excess sugar is converted for storage either into fat, or into a form of animal starch known as *glycogen*. Glycogen is stored in the liver and the muscles. When required, glycogen is changed back again into glucose and used for energy. Foods rich in carbohydrates are: potatoes, bread, cake, spaghetti, corn, and candy.

Lipids. Like the carbohydrates, lipids consist of the elements carbon, hydrogen and oxygen. They have twice the energy value of carbohydrates and proteins. Excess fats are stored under the skin and in the internal body cavity. Foods containing large amounts of fats are: butter, mayonnaise, lard, bacon, ham, peanuts, and sweet cream.

Proteins. Like carbohydrates and fats, proteins may be used for energy. However, since they contain nitrogen, sulfur and sometimes phosphorus, in addition to carbon, hydrogen and oxygen, proteins are useful for making new protoplasm, and for repairing the body tissues (assimilation). Proteins are complex compounds composed of various *amino acids*. Some of these are more useful to the body than others. Among the most valuable amino acids are: lysine, and tryptophane. Growing children need more proteins than adults. Severe protein malnutrition in very young children probably results in retarded brain development; it may thus be responsible for low educational achievement among the poor. Proteins are not stored in the body and so should be part of the daily diet. On the other hand, a continued excess of proteins may be harmful to the kidneys, which have the function of removing nitrogenous wastes from the body. Foods rich in proteins are: meat, fish, eggs, bread, beans and milk.

Minerals. Minerals in small amounts are important for maintaining the health of various parts of the body. Some of the most important mineral compounds are those containing calcium, phosphorus, iron, copper, iodine, sodium, potassium, chlorine, and fluorine.

1. Calcium and phosphorus are needed for building bones and teeth. Calcium is also essential for the proper clotting of blood. They are found in milk and dairy products and leafy vegetables.

2. Iron is part of hemoglobin, the red protein found in red blood cells. If iron is lacking, not enough red blood cells are formed and *anemia* results. Minute amounts of copper also seem to be needed in the formation of red blood cells. Foods rich in iron are: liver, beef, raisins, and spinach.

3. Iodine is needed for a healthy condition of the thyroid gland, an endocrine gland located in the neck. If iodine is lacking, the gland enlarges, producing a condition known as simple goiter. Sea foods contain large amounts of this element.

4. Sodium and potassium help maintain the proper water balance in the body fluids. Sodium is found in table salt and both occur in vegetables.

5. **Chlorine** is part of hydrochloric acid made in the digestive fluid of the stomach, gastric juice. It is found in table salt and in vegetables.

6. **Fluorine** seems to be needed for proper tooth formation. In a number of cities and towns, small traces of it are being added to the water supply, as the result of scientific studies like those conducted in Newburgh and Kingston, N.Y. In the former city, fluorine was added to the drinking water, and the teeth of the children were observed over a ten-year period. They were found to show less tooth decay than in the comparable city of Kingston, which was used as a control, and which did not receive fluorine.

Water. Water makes up most of protoplasm. About 70 per cent of the body is made of water. It serves to carry dissolved nutrients, oxygen, and wastes for diffusion into and out of cells. Most of blood consists of water, which makes it a circulating medium of exchange. By evaporating from the skin, it helps regulate the body temperature. The body needs about eight glasses of water every day.

Vitamins. Today the value of vitamins is known and accepted. Fifty years ago, they were exciting, mysterious substances in food that were just beginning to be known as being essential for health. They were originally named vitamine by the Polish scientist Casimir Funk in 1911. Minute quantities of them are needed to help keep the body in vigorous health, and to prevent the *deficiency diseases.* Until the discovery of the vitamins, the causes of the age-old diseases of scurvy and beri-beri were a baffling mystery.

Vitamin A. This vitamin is formed in the intestinal wall from the yellow pigment *carotene,* found in yellow vegetables such as corn, carrots and sweet potatoes, and in green leafy vegetables such as spinach and lettuce. It is also obtained from liver, butter, and eggs. Vitamin A is needed for growth. It is also used by the body to make *visual purple,* a part of the retina needed for vision, especially in dim light. A deficiency of the vitamin results in night blindness, in which a person has difficulty in seeing at night. A healthy, moist condition of the eyes also depends on vitamin A; in its absence a dry, scaly condition, *xerophthalmia,* results. It also keeps the epithelial tissue of the respiratory system healthy, reducing susceptibility to colds. This vitamin is stored in various parts of the body, including the liver, and the fatty layer under the skin.

Vitamin B complex. Vitamin B is known to consist of a number of vitamins many of which have been isolated in crystalline form.

Thiamin or vitamin B_1, is used by the body to keep the nerves in healthy condition; it stimulates a good appetite and it is one of the substances needed in cell respiration. For many years, natives of oriental countries suffered from beri-beri, a disease characterized by weakness and inflamed nerves. It was discovered by Eijkmann and others that this condition was brought on by a diet consisting largely of polished rice; when rice is polished, the husks containing thiamin are discarded, leading to the disease. Thiamin is found in whole grains, pork liver, yeast, and fresh green vegetables.

Riboflavin or vitamin B_2, is one of the essential substances needed in oxidation and energy release in all cells. A deficiency of it leads to weakness and sores around the mouth. It is found in such foods as milk, liver, green vegetables, and yeast.

Niacin is also important in cell respiration. It is required in maintaining the good health of the skin and nervous system. Without it, a severe disease, *pellagra,* results. It had been common in underprivileged parts of the South until Goldberger discovered that the basic diet of corn, molasses and pork was deficient in the necessary vitamin, which at first was called vitamin G. Foods containing niacin, such as liver, lean meats, yeast and peanut butter prevent the disease.

Vitamin B_{12} is found in liver, kidney, meats, and milk. It is needed by the bone marrow to form red blood cells. Without it, a person develops pernicious anemia. It is also found in the mitochondria of cells where it plays a role in cell metabolism.

Other B complex vitamins whose functions in human beings are not entirely understood, include: *biotin,* needed for the growth of bacteria; *pantothenic acid,* needed for cellular respiration and for general body health; *pyridoxine* (vitamin B_6), needed for cellular respiration and good skin condition; *folic acid,* which helps relieve anemia.

Vitamin C, or *ascorbic acid,* helps in the formation of connective tissue. It also keeps the walls of the capillaries in good condition. It is found in citrus fruits (orange, lemon, etc.), and in other fresh fruits and vegetables. Its absence in the diet of sailors was for many years the cause of *scurvy,* until, during the eighteenth century, the British navy began to include limes in its menu. This disease leads to hemorrhages (bleeding), swollen gums, loose teeth, and general weakness.

Vitamin D, calciferol, is also known as the sunshine vitamin. It is formed from dehydro-cholesterol present in the skin, upon exposure to the direct rays of the sun. It is also formed in milk, bread, and yeast when they are irradiated with ultraviolet rays: in this process, the substance *ergosterol* is transformed into a form of vitamin D. It occurs naturally in cod-liver oil and other fish oils, egg yolk, and butter. Vitamin D is needed by the body in order for it to assimilate calcium and phosphorus in the proper formation of bones and teeth. Children lacking the vitamin have soft leg bones which become bowed, and bad teeth, resulting in *rickets.*

Vitamin E, tocopherol, seems to be needed for reproduction in rats. It also plays a part in the health of the muscles. Its value for humans has not been definitely established. It is found in vegetables and whole grains.

Vitamin K, naphthaquinone, plays a part in the clotting of blood. It seems to be necessary for the formation of prothrombin, one of the substances involved in the clotting process. Vitamin K is made in the liver from material present in such foods as green leafy vegetables, and liver. It is also formed by bacteria in the intestines and absorbed. In the absence of vitamin K, there is bleeding, because of the difficulty of the blood to clot. It has recently been found that an adequate supply of bile salts is needed for the absorption of vitamin K from the small intestine. Vitamin K deficiency may occur if the bacteria of the large intestine are destroyed by the administration of antibiotic drugs, preventing them from synthesizing the vitamin.

Conserving the Nutrients in Food. Some of our foods are *enriched,* by having vitamins and minerals added to them. When flour is being refined, valuable B complex vitamins and minerals are lost. In the baking of bread, fixed amounts of thiamin, riboflavin, niacin, iron, and calcium are added to the dough. Milk is usually enriched by the addition of Vitamin D. Cereals, spaghetti and other foods are also enriched, as can be determined by reading the label. Improper cooking may result in a loss of valuable vitamins and minerals. Thus, the water in which vegetables are cooked will contain a large amount of these dissolved nutrients; it should not be thrown away. Vitamin C is oxidized and destroyed at high temperatures; vegetables should therefore not be overcooked. Milk loses practically all of its vitamin C content during the pasteurization process; for this reason, young babies are given orange juice or ascorbic acid at an early age. Fruits and vegetables that were picked long before they are eaten lose part of their vitamin content. Quick-frozen vegetables retain a high percentage of their vitamins. Canned fruit juices keep their vitamin C value because they are heated in the absence of air.

IMPROVING THE DIET.

Since we are what we eat, it is obvious that our health depends to a large extent on our taking in the correct types of food. A vigorous, alert, disease-resistant condition can result from having the proper amounts of vitamins, water, minerals, proteins, fats, and carbohydrates in our diet.

Although we are the richest and best-fed nation on earth, many of our people are suffering from marginal malnutrition because of improper food habits. Nutrition experts say that girls in particular are prone to eat the wrong kinds of food in the mistaken belief that they are watching their weight. One of the most valuable foods is milk, the almost perfect food. It has large amounts of vitamin A, riboflavin, vitamin D, calcium, phosphorus, and the important amino acids.

The Basic Four is a balanced diet recommended by nutritionists, consisting of four groups of food. Other foods may be added, as desired.

Basic Group 1 — Milk, cheese, ice cream — two or more servings a day; provides proteins, fats, minerals, vitamin D.

Basic Group 2 — Meat, poultry, fish, eggs, nuts — two or more servings a day; provides proteins and other nutrients.

Basic Group 3 — Fruits and vegetables — four or more servings a day; provides minerals and vitamins A and C.

Basic Group 4 — Bread and cereals — four or more servings a day; provides carbohydrates, proteins, iron, and B-complex vitamins.

SUMMARY OF THE NUTRIENTS

NUTRIENT	USE	GOOD SOURCES
Carbohydrates	Energy	Potatoes, bread, spaghetti
Fats and Oils	Energy; storage; insulation	Butter, mayonnaise, bacon
Proteins	Energy; growth; repair	Meat, fish, eggs, beans, milk
Mineral Salts of:		
Calcium	Formation of bones and teeth; clotting of blood	Milk and dairy products, vegetables
Phosphorus	Bone and tooth formation	Whole wheat bread, cheese, meat

SUMMARY OF THE NUTRIENTS (Continued)

NUTRIENT	USE	GOOD SOURCES
Iron	Formation of hemoglobin for red blood cells	Liver, beef, spinach, raisins
Potassium	Maintains water balance in body fluids	Vegetables
Sodium	Maintains water balance in body fluids	Vegetables, table salt
Chlorine	Formation of hydrochloric acid in stomach	Table salt, vegetables
Fluorine	Tooth formation; prevents tooth decay	Dissolved in drinking water
Water	Carries dissolved nutrients, oxygen, wastes, for diffusion	Liquid and solid foods
Vitamins:		
A	Growth; formation of visual purple in retina; prevents night blindness, xerophthalmia; keeps epithelial tissue healthy	Yellow and green vegetables, liver, butter, eggs
B-Complex		
Thiamin (B_1)	Healthy condition of nerves; good appetite; cellular respiration coenzyme; prevents beriberi	Whole grains, liver, pork, fresh green vegetables
Riboflavin (B_2)	Cellular respiration coenzyme; prevents sores around mouth	Milk, liver, green vegetables; yeast
Niacin	Cellular respiration coenzyme; prevents pellagra; healthy condition of nerves and skin	Liver, lean meats, yeast, peanut butter
B_{12}	Prevents pernicious anemia; cell metabolism	Liver, kidney, meats, milk
Ascorbic Acid (C)	Formation of connective tissue; formation of capillary walls; prevents scurvy	Citrus and other fruits, fresh vegetables
D (Calciferol)	For assimilation of calcium and phosphorus in tooth and bone development; prevents rickets	Cod liver oil, other fish oils, made in skin in sunlight, irradiated foods
E (Tocopherol)	Reproduction in rats; healthy muscles; function in humans uncertain	Vegetables, whole grains
K (Naphthaquinone)	Clotting of blood; formation of prothrombin	Liver, green leafy vegetables, intestinal bacteria

CHAPTER REVIEW

Select the correct choice for each of the following statements.

1. A food rich in carbohydrates is (A) bacon (B) bread (C) lard (D) butter (E) ham
2. Of the following, the function that includes all the others is (A) digestion (B) excretion (C) metabolism (D) irritability (E) respiration
3. ATP is used directly in the process of (A) growth (B) repair (C) roughage (D) respiration (E) reproduction
4. A calorimeter is used in testing food for its value of (A) carbohydrate (B) fat (C) protein (D) vitamin (E) energy
5. An overweight condition may be caused by any of the following factors except (A) overeating (B) eating too many high calorie foods (C) lack of exercise (D) lack of iron in the diet (E) thyroid gland disturbance
6. A nutrient used to build protoplasm is (A) protein (B) fat (C) starch (D) sugar (E) iodine
7. All of the following are rich in proteins except (A) potatoes (B) meat (C) fish (D) eggs (E) milk
8. Two elements needed for building bones and teeth are (A) iron and calcium (B) calcium and phosphorus (C) phosphorus and iron (D) iron and sodium (E) sodium and potassium
9. Scientific evidence indicates that an element needed to prevent tooth decay is (A) chlorine (B) iodine (C) bromine (D) fluorine (E) iron
10. All of the following apply to vitamin A except (A) it prevents night-blindness (B) it is formed from carotene (C) it is used in making visual purple (D) it is used in making hemoglobin (E) it is found in green, leafy vegetables
11. A vitamin used in the formation of red blood cells is (A) vitamin B_1 (B) vitamin B_2 (C) vitamin B_{12} (D) calciferol (E) ascorbic acid
12. All of the following are members of the vitamin B complex except (A) thiamin (B) gibberellin (C) riboflavin (D) niacin (E) biotin
13. A vitamin formed in the skin upon exposure to ultra-violet rays is vitamin (A) A (B) B_2 (C) C (D) D (E) E
14. The proper clotting of blood depends on an adequate supply of vitamin (A) K (B) B_1 (C) B_6 (D) C (E) A
15. If equal quantities of the following are burned, the highest calorie value will be found in (A) fat (B) protein (C) glucose (D) starch (E) minerals

ANSWER KEY

1-B	4-E	7-A	10-D	13-D
2-C	5-D	8-B	11-C	14-A
3-D	6-A	9-D	12-B	15-A

7 Ingestion and digestion

Ingestion refers to the activities of animals in which they take in food. Before food can be absorbed into the cells of the body, it must first be made soluble, or *digested*. During digestion the complex molecules of food are reduced into smaller ones capable of passing through the membranes of the body. Digestion may occur within the cell (intracellular digestion), as in protozoa; or outside the cell, in a special cavity, or tube, as in most animals (extracellular digestion), after which the soluble molecules are absorbed into the cells.

THE DIGESTIVE SYSTEM. Food is passed along through the body in the tube called the *alimentary canal*. This consists of the mouth, esophagus or gullet, stomach, small intestine, large intestine, and rectum. The rectum leads to the outside through the anus. There are digestive glands leading into the alimentary canal from the mouth, stomach, and small intestine. Two large glands, the liver and the pancreas, are outside of the food tube, and send their digestive fluids into it through ducts. The sum total of the alimentary canal and the digestive glands makes up the *digestive system*.

Mouth. In the brief time that food is in the mouth, two main things happen to it: (1) It is chewed and broken down to smaller pieces by the teeth, aided by the tongue. This is known as *mechanical digestion*. (2) It is attacked by *enzymes* in the digestive liquid, the saliva; this is *chemical digestion.*

The teeth. Human beings and most mammals develop two sets of teeth during a lifetime. The first, or milk teeth, number twenty. In an adult, there are thirty-two teeth arranged in a semicircle, half on the upper jaw and half on the lower jaw. These include eight front chisel-like teeth, the *incisors,* which cut the food; four *canines,* or eye teeth which are pointed, and which help bite the food; eight *bicuspids* or premolars, for grinding; and twelve *molars* for grinding. The last set of molars, the wisdom teeth, usually appear late in adolescence. A tooth consists of the following parts:

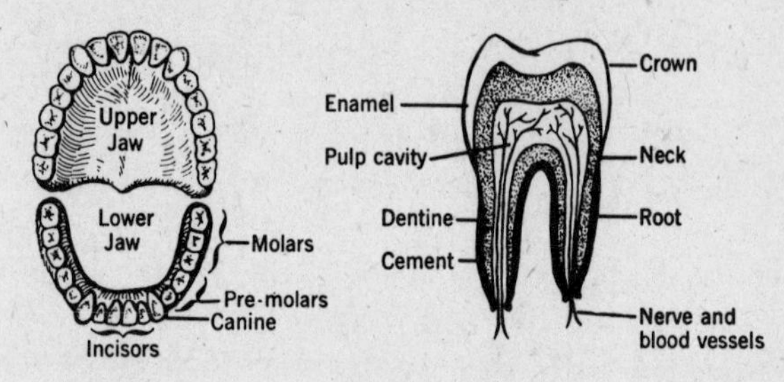

1. **The crown,** the exposed part, is covered with hard *enamel.* Enamel is the hardest substance in the body. When it is chipped or decayed by bacteria (*caries*), cavities develop which reach into inner parts of the tooth causing pain and sometimes infections. It is then that the dentist installs a filling to seal the cavity.

2. **The root,** the part of the tooth below the level of the gum, fits into sockets in the jaw bone. It is covered with a layer of cement.

3. **The dentine** makes up the main body of the tooth. It has a very strong bony structure, and is beneath the enamel.

4. **The pulp** cavity is a chamber in the center of the dentine, containing pulp, consisting of blood vessels, connective tissue and nerves. From this section, food and oxygen are brought into the tooth. The nerves spread into the dentine.

5. **The gum** is the living tissue which holds the teeth in place. Brushing the teeth not only removes particles of food that might decay and start cavities; it also massages the gums and keeps them in a firm, healthy condition, which is one of the requirements for a set of good teeth.

Salivary digestion. While the food is being chewed into smaller particles, it is being moistened by saliva, a digestive liquid produced by three pairs of salivary glands. These glands are located under the tongue, in the cheek near the ear, and in the lower jaw. They empty into the mouth through *ducts.* Saliva contains the enzyme *pytalin,* which digests starch into a sugar, *maltose.* A demonstration to show the chemical digestion of starch by saliva may be performed in the following way:

1. To some boiled 1 per cent starch suspension in a test tube, add a drop or two of iodine. After this has turned blue-black, add about an inch of saliva. Warm by holding it in your hand. Set up a control test tube in the same way, but without saliva.

2. After a few minutes, it will be observed that the blue-black color in the first test tube is fading, a sign that the amount of starch is being decreased. There is no change in the control.

3. When the liquid in the test tube has become clear, heat a small amount of it in a test tube of Benedict's solution. It will become orange, showing the presence of sugar.

4. If the control is heated with Benedict's solution, there will be no change. Also, if saliva is tested with Benedict's solution, it will be shown not to have any sugar, either.

5. The conclusion is reached, therefore, that the only way that sugar appeared was from the digestion of starch by saliva.

Esophagus. After food has been chewed, lubricated by saliva, and partially digested by ptyalin, a small mass of it (bolus) is pushed into the throat (pharynx), where it is swallowed. A flap of tissue, the *epiglottis,* closes over the adjacent windpipe and prevents the food from going into the wrong tube. The esophagus is lubricated by mucus, which is secreted by glands in its lining. *Peristalsis,* the alternate wavelike contractions and expansions of the involuntary muscles in the wall of the esophagus, moves the food along into the stomach.

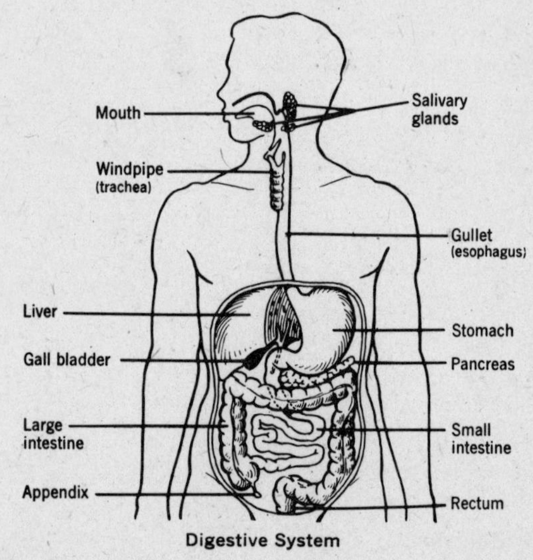

Digestive System

Stomach. Food remains in the stomach from three to four and a half hours. During this time, it is constantly being churned and mixed by peristaltic action with gastric juice, the secretion of the stomach glands (mechanical digestion). For a brief period, saliva continues to act on starch; this ceases when the contents of the stomach become acid. Gastric juice contains hydrochloric acid, water, and the enzymes, *pepsin* and *rennin* and a tiny amount of *lipase.* Pepsin partially digests proteins into *proteoses* and *peptones,* which are smaller combinations of amino acids and which are intermediate products of digestion. Rennin curdles the protein of milk, casein, and prepares it for digestion by pepsin. Complete digestion of proteins will take place in the small intestine. Pepsin is active only in an acid medium, so that advertisements for patent medicines that imply that "acid stomach" is undesirable, are misleading. Hydrochloric acid is further useful in dissolving minerals for absorption, and in killing bacteria that may have been swallowed. The very small amounts of the enzyme lipase digests only a tiny amount of fat into fatty acids and glycerol. The contents of the stomach eventually take the form of a semiliquid paste, *chyme,* and gradually pass out of the far, narrow end, the *pylorus,* into the small intestine, through a muscular valve, the *pyloric sphincter.*

A demonstration to show the digestion of protein may be shown by using artificial gastric juice:

1. Prepare artificial gastric juice by adding a small amount of pepsin powder to water and hydrochloric acid having a pH (acidity) of 1.5.
2. Place some chopped-up hard-boiled egg-white (consisting of the protein, albumen) in a test tube, and add the gastric juice. Set up a control with water.
3. Keep overnight in an incubator set at 37°C (98.6°F), the temperature of the body.
4. The next day, the egg-white will be seen to have been completely dissolved; in the control test tube, it will be unaffected.

Small Intestine. Digestion of food is completed in the small intestine, a narrow tube over twenty feet long. Mechanical digestion continues as peristalic action moves the food along, and constantly mixes it with the digestive liquids. There are three such liquids secreted by (1) numerous intestinal glands in the wall of the small intestine, (2) the pancreas, (3) the liver.

Intestinal juice contains the following enzymes:

Erepsin—digests proteins into amino acids.

Maltase—digests maltose into glucose.

Lactase—digests lactose (milk sugar) into glucose and galactose.

Sucrase—digests sucrose (cane sugar) into glucose and fructose.

Small quantities of *lipase*—digest fats into fatty acids and glycerol.

Pancreatic juice enters the upper end of the small intestine from the pancreas by means of the pancreatic duct. It contains the enzymes:

Trypsin—digests peptones and proteoses into amino acids

Amylopsin—changes starch into maltose

Lipase—changes fats into fatty acids and glycerol

The liver produces bile which is stored in the *gallbladder.* When food is being digested in the small intestine, the gallbladder contracts, and sends bile into it by means of the bile duct. Bile does not contain any enzymes. Instead, it prepares fat for digestion by *emulsifying* it, breaking it into very small particles. This is really a type of mechanical digestion, and permits lipase to digest the small fat globules more readily. The liver is the largest organ in the body. Besides forming bile, it also has a number of other functions, including: It changes excess glucose to glycogen and stores it; it destroys old, worn-out red blood cells; it converts amino acids into glucose, and a nitrogenous waste called urea which is removed from the blood in the kidneys; it produces prothrombin, one of the substances needed in the clotting of blood; it prepares fat for use by the body; it serves as a reservoir for storing blood.

Absorption of Digested Food. As the result of digestion in the mouth, stomach, and small intestine, the nutrients are converted into their *end products:* carbohydrates into glucose; lipids into fatty acids and glycerol; proteins into amino acids. These soluble end products are absorbed through the walls of the small intestine. Here, there are numerous microscopic, finger-like projections called *villi,* which increase the absorptive surface of the small intestine. Each villus has a single-celled layer of epithelial cells covering it. Inside, there is a network of *capillaries* and a central *lacteal,* a lymph vessel. The nutrients diffuse into the capillaries and become dissolved in the plasma of the blood. The capillaries of the villi lead into the *portal vein,* which goes into the liver. Here, most of the rich supply of glucose is removed and changed into glycogen, or animal starch, for storage. Some glycogen is also stored in the striated muscles. The blood that leaves the liver circulates the nutrients throughout the body. Several hours after a meal, when the glucose level in the blood has fallen, glycogen is changed back into glucose and is transferred out of the liver into the blood. The fatty acids and glycerol do not enter the blood directly, but diffuse into the lacteals. These lead into larger lymph vessels, and finally into the *thoracic duct* which empties into the blood stream in the neck region.

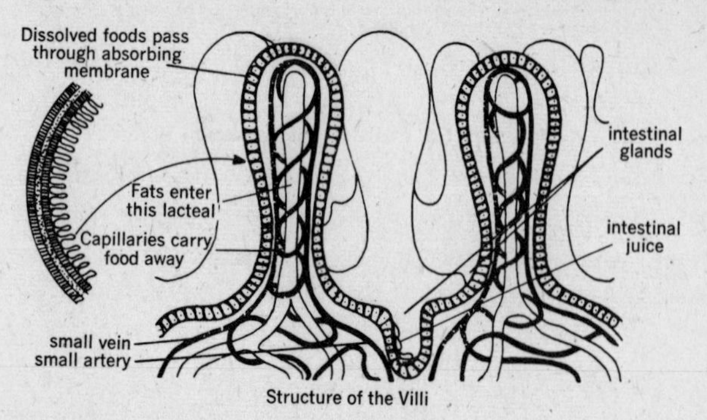

Structure of the Villi

Large intestine, or colon. After food has been digested and absorbed in the small intestine, the indigestible remains pass into the large intestine. A short distance from this point, there is a small outpocketing of the large intestine, about the length of the little finger, called the appendix. It has no apparent use in humans, although in herbivorous animals it serves for digestion. Like the esophagus, the large intestine has no digestive functions. The mucous cells in its lining secrete mucus that provide lubrication for the solid matter being moved along by peristalsis. Water is absorbed during this time and is used over again by the body. The solid wastes consisting of indigestible food, especially the cellulose remains of plant food, intestinal secretions, and bacteria, are known as *feces,* and are eliminated from the lower end of the large intestine (rectum) through the anus.

Ingestion and digestion in lower animals. All animals require food for energy, growth, and repair. Their survival depends on their ability to obtain food—in fact, their lives may be said to be dedicated to this purpose.

Protozoa. Although protozoa are classified by many biologists as protists, they are included here because they show animal-like characteristics. Many of them take in food through a special structure, the *oral groove,* as in the case of paramecium; or they may *engulf* their food, as is done by the ameba. Other protozoa have specific modifications for food getting. Food is digested in the *food* vacuoles. Digestion is intracellular.

Hydra, a commonly studied member of the coelenterate phylum, shows the adaptation of a simple multicellular animal for ingestion and digestion. Food is obtained with the aid of *tentacles* that surround the mouth. These have numerous stinging cells (*nematocysts*) that paralyze small animal prey. The tentacles then deposit this food into the mouth leading to the central body cavity. Digestion takes place with the aid of enzymes secreted by special gland cells located in the endoderm layer (extracellular digestion). The food is circulated by means of currents created by the beating of long flagella attached to some of the cells. Many of the broken-down food particles are engulfed by other cells lining the cavity, and digestion is completed within their food vacuoles (intracellular digestion). The indigestible remains left in the central cavity are eliminated through the mouth.

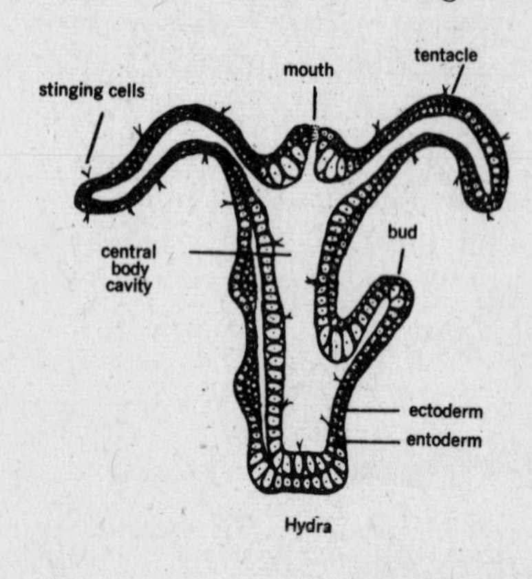

Hydra

Earthworm has a digestive tube with a mouth opening at one end, and an anus at the other. It swallows soil containing organic plant and animal remains, which passes from the mouth through a muscular *pharynx* into the *esophagus,* and then into an enlarged *crop,* where it is collected. Behind the crop, the food is throughly ground up in the *gizzard,* which has thick muscular walls, with the help of small stones. Then the food passes into the intestine where it is digested by enzymes secreted by gland cells of the lining epithelium. Soluble food is absorbed into the blood vessels of the intestinal wall. The lining of the intestine is folded, providing a greater area for absorption. The solid wastes are eliminated through the anus. Earthworms spend most of their time swallowing soil below the surface and depositing it on the surface as waste known as *castings.* Charles Darwin first pointed out the immense value of the lowly earthworm to agriculture because of its method of improving the soil as it burrows through it, aerating it, grinding it, enriching it, and bringing it to the surface in castings.

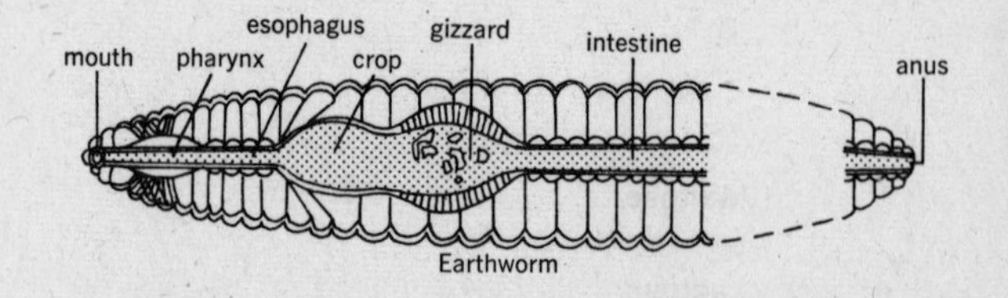

Earthworm

Insects. Some insects like the butterfly and the moth have sucking mouth parts. The grasshopper has chewing mouth parts consisting of toothed horny *mandibles* that work from side to side and *maxillae* that help to hold and cut the food; the other mouth parts consist of the *labrum* (the upper lip), and the *labium* (the lower lip). The digestive tract has three parts: fore-gut, mid-gut, and hind-gut. The fore-gut consists of the mouth, into which the salivary glands secrete saliva, the esophagus leading to the *crop* where food is stored, and the muscular *gizzard* containing chitinous teeth. The mid-gut consists of the stomach which is the main organ of digestion and absorption. As in the earthworm, there is an infolding of the digestive tube to increase the area for the absorption of soluble nutrients. There are six pairs of stomach *pouches* opening into the stomach, which secrete digestive juices. The intestine, or hind-gut, eliminates the indigestible food matter through the anus.

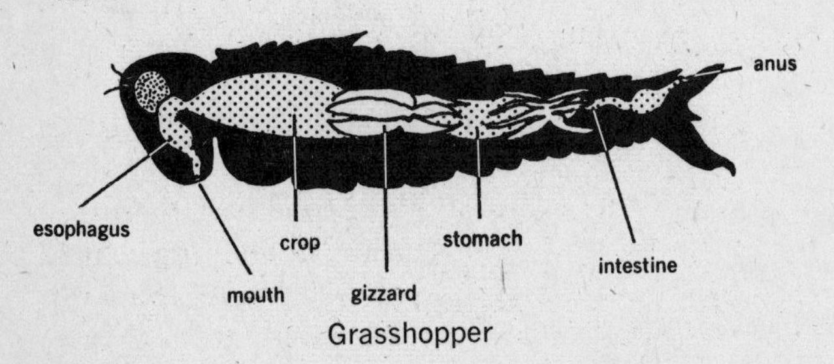

Grasshopper

Fish. The mouth of a fish contains small teeth that point backward and function mostly in holding large masses of food before they are swallowed whole. Tiny particles of food are strained by long projections of the gills called *gill rakers,* as they are carried in the water passing over the gills. After food is swallowed it enters the stomach where digestion is started by strong digestive juices. Just behind the stomach, where it joins the intestine, there are digestive glands, or *caeca,* which send secretions into the intestines. The *liver* forms bile which is stored in the *gallbladder* until it enters the intestine through a duct. Indigestible wastes are excreted through the anus.

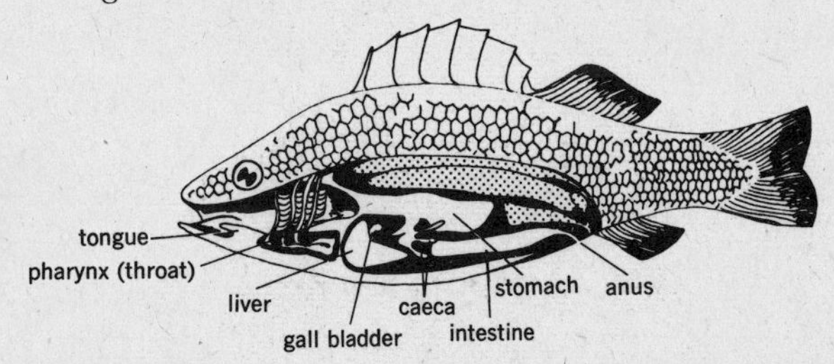

Frog. The *tongue* of a frog is well adapted for catching insects and worms. It is soft and sticky, and is attached at the front end of the lower jaw. It can flip out quite a distance to trap its food. There are small toothed structures in its mouth for holding the food. The *esophagus* leads to the *stomach* where digestion starts. It continues in the *small intestine* with the aid of secretions from the *pancreas,* and bile from the *gallbladder* and the *liver.* After soluble food has been absorbed, the indigestible remains are passed into the large intestine. This opens into the *cloaca,* which also collects wastes from the bladder and leads out of the body through the anus.

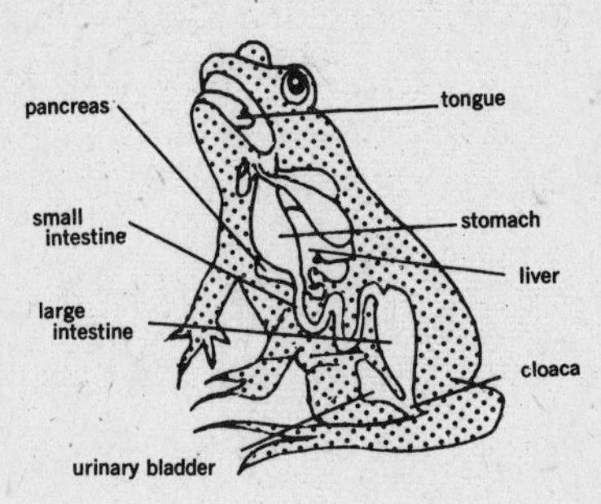

SUMMARY OF DIGESTION

NUTRIENT	WHERE DIGESTED	ENZYMES	PRODUCTS
Starch	1. Mouth (starts, and continues in stomach for short time)	Ptyalin	Maltose
	2. Small intestine	Amylopsin (pancreatic juice)	Glucose
Maltose	Small intestine	Maltase (intestinal juice)	Glucose
Lactose	Small intestine	Lactase (intestinal juice)	Glucose, galactose
Sucrose	Small intestine	Sucrase (intestinal juice)	Glucose, fructose
Lipids	Small intestine	Lipase (pancreatic juice)	Fatty acid and glycerol
	Small intestine	Bile (not an enzyme)	Emulsifies fats

SUMMARY OF DIGESTION (Continued)

NUTRIENT	WHERE DIGESTED	ENZYMES	PRODUCTS
Protein	Stomach	Pepsin	Proteoses, peptones
		Rennin	Curdles casein for pepsin digestion
	Small intestine	Erepsin (intestinal juice)	Amino acids
	Small intestine	Trypsin (pancreatic juice)	Amino acids

CHAPTER REVIEW

Select the correct choice for each of the following statements.

1. A part of the digestive system which is not in contact with food is (A) small intestine (B) stomach (C) liver (D) large intestine (E) trachea
2. Digestion of food starts in the (A) mouth (B) esophagus (C) stomach (D) pancreas (E) gullet
3. All of the following are parts of a tooth except (A) enamel (B) dentine (C) root (D) caries (E) pulp
4. A secretion that digests both carbohydrates and proteins is (A) ptyalin (B) saliva (C) pepsin (D) gastric juice (E) pancreatic juice
5. The end product of carbohydrate digestion is (A) sucrose (B) glucose (C) cellulose (D) peptone (E) trypsin
6. Pepsin is active only in the presence of (A) rennin (B) erepsin (C) amylopsin (D) hydrochloric acid (E) lipase
7. Villi are present in the (A) stomach (B) small intestine (C) large intestine (D) appendix (E) colon
8. Large amounts of carbohydrate are stored in the (A) liver (B) stomach (C) gallbladder (D) adipose tissue (E) portal vein
9. Bile is useful in the digestion of (A) starch (B) sucrose (C) fat (D) protein (E) amino acids
10. Fatty acids are absorbed by the (A) capillaries (B) lacteals (C) pylorus (D) thoracic duct (E) colon
11. Food is moved along the alimentary tract by the contractions known as (A) epiglottis (B) circulation (C) secretion (D) proteoses (E) peristalsis
12. Castings are produced by the (A) hydra (B) paramecium (C) crayfish (D) grasshopper (E) earthworm
13. The food tube in man is known as the (A) digestive system (B) trachea (C) alimentary canal (D) crop (E) gizzard
14. Greater surface for food absorption is provided in the earthworm by (A) food vacuoles (B) infolding of the intestinal wall (C) the pharynx (D) stomach pouches (E) nematocysts
15. Intracellular digestion takes place in both (A) protozoa and earthworm (B) protozoa and insects (C) protozoa and fish (D) protozoa and frog (E) protozoa and hydra

ANSWER KEY

1-C	4-E	7-B	10-B	13-C
2-A	5-B	8-A	11-E	14-B
3-D	6-D	9-C	12-E	15-E

8 Circulation

After it is digested, food is distributed throughout the body by means of the circulatory system. Circulation is the transport of materials throughout organisms, and within cells. Each of the billions of cells in a human being depends for its existence on the continuous delivery of food, oxygen and other materials to it, and the removal of wastes from it.

THE CIRCULATORY SYSTEM

The circulatory system of the body is arranged in a closed system of blood vessels, in which the heart serves as a pump that keeps the blood moving. This was first demonstrated by William Harvey (1578–1657) in the early part of the seventeenth century.

The heart is a four-chambered, muscular organ, about the size of a man's fist. It is located approximately in the center of the chest, pointing to the left. It has two smaller upper chambers, the *atria* (auricles), which receive the blood, and two thick-walled ventricles below them, which pump the blood away from the heart. Between the atria and ventricles are valves, which prevent the blood from flowing backward when the ventricles contract. Located in the right atrium is an area of specialized muscle tissue called the *pacemaker,* where the pulsations of the heart appear to originate, and from which they spread like a wave to the rest of the heart.

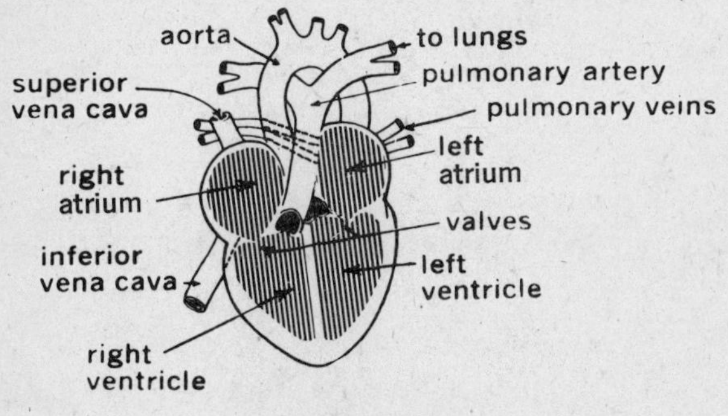

The Heart

A doctor listening to the heartbeat with a stethoscope, may hear a "murmur" if there is a defect in the *valves,* when some of the blood flows back into the atrium. The right and left sides of the heart are completely separated from each other by a thick wall. The cells of the heart itself are nourished by the coronary arteries; if any of them become blocked by a clot, as in coronary thrombosis, the heart cells are deprived of their food and oxygen, and death may result. The right and left sides of the heart really function as two pumps, since they pump blood to different parts of the body, i.e., the right to the lungs (pulmonary circulation), and the left to the rest of the body (systemic circulation). The rhythmic pumping of the heart takes place in two parts: the period when the ventricles are contracting is called the *systole,* and the phase immediately afterward, when they are relaxing is called the *diastole.* When a doctor measures a person's blood pressure with an instrument called a sphygmomanometer, he makes a record of both the systolic and the diastolic pressure. Thus, the average blood pressure for a young man of twenty years might be 120/80. In a baby, it might be 90/15. In an old man, 150/90. In high blood pressure, or hypertension, the pressure is higher than the average. Its cause is not well understood; it may arise from nervous tension or a kidney condition.

Arteries are large blood vessels that carry blood *away from* the heart. They are thick-walled, containing much elastic fiber tissue. There are also layers of involuntary muscle that control the size of the opening of the arteries. In older people, the walls of the arteries sometimes lose their elasticity, resulting in a condition known as arteriosclerosis, or hardening of the arteries. Each time the heart contracts, it forces blood into the arteries, making them stretch. Following a contraction, the heart rests momentarily and then the arteries relax. This alternate pulsation of the arteries can be felt as the *pulse* beat in parts of the body where they lie close to the surface, i.e., the wrist, the temple, and the neck. The average pulse rate varies with age, sex, activity, etc. It may vary from 60 to 85 per minute in different people. Exercise will cause it to increase. The largest artery of the body is the *aorta,* which leads away from the left ventricle to most of the body. All arteries with the exception of the pulmonary artery, carry oxygenated blood.

Veins are blood vessels that carry blood *to* the heart. They have thinner walls than the arteries, with less elastic fiber and muscular tissue. Blood passes through them in a steady flow, as compared with the spurting action of the arteries. Blood in the veins is kept moving toward the heart by both the pressure behind it, and by the movements of the voluntary muscles. Each time these muscles move they squeeze the veins, forcing the blood forward. There are *valves* along the length of the veins which prevent the blood from flowing backward. All veins carry deoxygenated blood (lacking in oxygen), with the exception of the pulmonary vein. Some of the veins lie close to the surface and can be seen through the skin as having a bluish color; deoxygenated blood actually has a dark red color. The largest vein is the *vena cava,* leading into the right auricle.

Capillaries connect the smallest arteries (arterioles) and the smallest veins (venules). They are microscopic, with walls only one cell thick. They are located everywhere in the body, close to the cells. When the skin is pricked with a needle, a drop of blood will appear, coming from broken capillary walls. Food, oxygen and other materials diffuse out of the blood through the capillary walls, into the intercellular fluid (ICF) surrounding the cells, and then into the cells. By the reverse process, wastes leave the cells, diffuse into the ICF, and then enter the blood through the capillary walls.

The circulatory route. Blood enters the *right atrium* of the heart through the (1) *superior (upper) vena cava,* carrying blood from the head and upper part of the body, and (2) *inferior* (*lower*) *vena cava,* carrying blood from the rest of the body. When the right atrium contracts, blood is forced into the *right ventricle.* It in turn contracts, and sends blood to the *pulmonary artery* leading to the *lungs.* Here, the blood enters capillaries surrounding the air sacs. Oxygen is absorbed through the capillary walls; carbon dioxide and water vapor are excreted. The oxygenated bright red blood then collects in the *pulmonary veins* which lead back to the heart, this time, into the left atrium. It contracts and forces the blood into the *left ventricle.* This chamber has the thickest walls of the heart; it pumps blood into the *aorta,* which carries blood to the entire body, except the lungs. The aorta divides into arteries that go into the head, the arms, legs, the heart itself (coronary arteries) and various parts of the body. The arteries branch into *arterioles,* which in turn lead to the *capillaries.* Oxygen and food enter the cells via the lymph; carbon dioxide and other wastes diffuse into the capillaries. They combine to form *venules,* which lead to the veins. The blood is now largely deoxygenated and dark red in color; it is carried once more into the heart through either *vena cava.* This circuit of blood takes less than 20 seconds, and includes:

1. **Pulmonary circulation.** Blood from the right side of the heart is pumped through the pulmonary artery to the lungs, where it absorbs oxygen and excretes carbon dioxide and water vapor; then back to the left side of the heart through the pulmonary veins. Deoxygenated blood is refreshed, or oxygenated in this part of the circulatory route. The pulmonary artery is the only artery containing deoxygenated blood. The pulmonary veins are the only veins with oxygenated blood.

2. **Systemic circulation.** In the "body" circulation, blood from the left ventricle goes through the aorta to the upper and lower parts of the body and then back to the right atrium. Arteries deliver blood containing food and oxygen to the internal organs. Veins leaving these organs collect the blood now containing carbon dioxide and nitrogenous wastes. The arteries entering the kidneys carry excretions of the cells, which are filtered out by the kidney tubules. The large kidney veins which return the blood to the circulatory route are relatively free of wastes.

3. **Portal circulation.** Digested food is absorbed in capillaries of the villi in the small intestine. These capillaries lead out of the small intestine by means of the *portal vein,* which then enters the liver. Here, the blood goes into capillaries which bring it to the liver cells. Excess glucose is removed and stored as glycogen, or animal starch. The blood is then carried out of the liver by means of the *hepatic vein* to the inferior vena cava. The portal circulation is essentially a part of the systemic circulation.

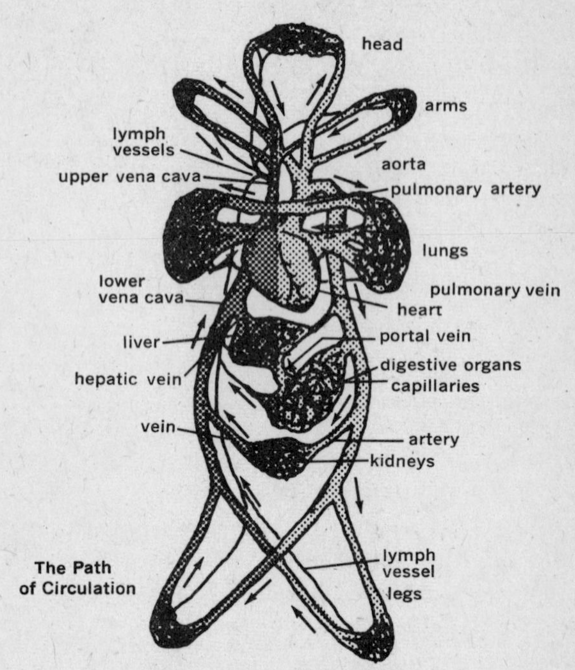

The Path of Circulation

Lymph and Lymph Circulation. All the cells of the body are surrounded by, and bathed by the watery intercellular fluid which originates in the blood plasma. It slowly oozes out through the capillary walls into the surrounding spaces. It consists of plasma without its large protein molecules. It is largely water, with small amounts of dissolved nutrients that will be used by the cells, wastes given off by them, and secretions. It also contains white blood cells that pass through the capillary walls. It serves as the intermediate fluid.

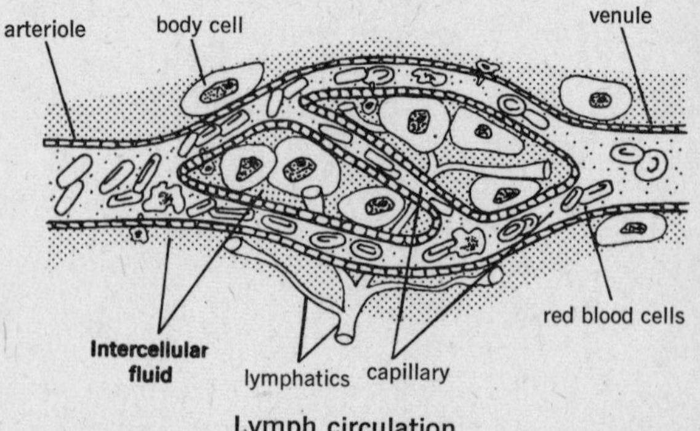

Lymph circulation

It collects in tiny lymph vessels called *lymphatics* which, like the capillaries, are present in every part of the body. It is called *lymph* when it is absorbed into the lymph vessels. It moves slowly along, under the gentle pressure of the fluid behind it, and as a result of the body's movements. There are valves along the lymphatics, which keep lymph moving in a forward direction. The lymphatics join into larger vessels and include the *lacteals* of the small intestine which carry digested *fats*. They finally form two large lymphatics, one from the upper part of the body and one from the lower. The latter is the *thoracic* duct, the largest lymph vessel. It joins the circulatory system in the neck region, returning lymph back into the blood stream. In various parts of the body such as the armpits and the groin, there are small *lymph glands,* or *nodes,* containing masses of lymphatic tissue. They manufacture certain white blood cells. They also serve to filter germs out of the body. During an infection, they may become swollen.

THE BLOOD. There are almost six quarts of blood in an average-sized person. Blood consists of a liquid, *plasma,* in which three types of cells are carried, the *red blood corpuscles,* the *white blood corpuscles,* and the *platelets:*

Plasma. The liquid part of the blood is straw-colored, consisting of 90 per cent water, and containing: dissolved nutrients (glucose, amino acids, fatty acids, salts, vitamins), wastes, antibodies, hormones, respiratory enzymes, and fibrinogen. When a person donates a pint of blood at a blood bank, the plasma is separated, dried, and later dissolved in distilled water, for use in operations and emergencies. Recently, specific parts of plasma have been fractionated (separated out), such as gamma globulin, an antibody used against measles; serum albumin, a protein used in transfusions to restore the blood volume.

Red blood corpuscles. Almost all the cells in the blood are the red blood cells. Their function is to transport oxygen. The red blood cells, or erythrocytes, are so small that a cubic millimeter of blood contains about 5,000,000 of them; and so numerous, that blood appears red—without them, blood is pale yellowish in color. They are disclike cells which are concave on both sides. They contain the protein *hemoglobin,* which gives them their color. When blood passes through the lungs, the oxygen of the air combines with the hemoglobin of the red blood corpuscles to form *oxyhemoglobin.* As these cells circulate through the body, oxygen diffuses from them into the cells. Red blood corpuscles are manufactured in the marrow of the bones. They lose their nucleus when they enter the bloodstream, and have recently been shown to remain active for about four months. Worn-out red cells are removed from the blood by the liver and the spleen. The spleen is also a storage place for active red blood cells, and contracts when extra amounts of red blood cells are needed in the bloodstream. The element iron is needed for the formation of hemoglobin. If it is lacking in the diet, *anemia* results, in which a person has lowered vitality, since there are not enough red blood cells to bring oxygen to the tissues. Vitamin B_{12} has been found useful in treating serious cases of pernicious anemia.

White blood corpuscles. Approximately one out of every 500 blood cells is a white blood cell. The white blood cells, or *leucocytes,* move like an ameba, and engulf bacteria that have entered the body, as well as other foreign matter. For this reason, they were named *phagocytes* (eating cells) by their discoverer, Eli Metchnikoff (1845–1916). They form pseudopodia and can crawl through the capillary walls into the tissues. They thus help protect the body against disease germs. Their action is known as phagocytosis. *Pus* is a thick, yellowish liquid that often results when there is an infection; is consists of bacteria, white blood corpuscles, both dead and alive, and broken-down tissue. The white blood cells are formed in the marrow of the bones, and in the lymph glands. There are several types of such cells. During a case of appendicitis, or other infection, there is a striking rise in the number of these cells above the average amount normally found in a cubic millimeter (about 7,000). In another disease, leukemia, which is believed to be a form of cancer, enormous numbers of white blood corpuscles are produced. Attempts have been made to treat it with radioactive phosphorus.

Platelets are even smaller in size than the red corpuscles. They are important in starting the set of reactions that take place when blood *clots.* When a wound occurs, the platelets break up and give off an enzyme. This causes the substance *prothrombin* to change into *thrombin.* Thrombin, in turn, acts on *fibrinogen,* one of the dissolved proteins in plasma, making it insoluble, and turning it into threads of *fibrin.* The red blood cells become entangled in this network, and a clot forms. This whole process takes from three to six minutes. Fresh blood which is allowed to stand in a jar will slowly separate into two parts—an upper clot, and a clear, yellowish liquid below called *serum.* Serum is like plasma except that it has no fibrinogen. Some people inherit a disease known as *hemophilia,* in which the blood does not clot. This is produced by a hereditary factor, or gene, and is not curable. In other cases, there may be poor clotting of a cut or wound, if a person is lacking in vitamin K, or the element calcium. At times, older people may suffer from the appearance of a clot in the circulatory system; if it occurs in the coronary arteries that bring blood to the cells of the heart, *coronary thrombosis* results, and the heart may be damaged, or death may result. Doctors inject *heparin,* a substance extracted from the liver and lungs, or *dicumarol,* which is obtained from clover, to reduce or eliminate such clots.

Blood types. Although all blood contains plasma and the three types of blood cells mentioned above, it is chemically different in various people. At the beginning of the century, Dr. Karl Landsteiner discovered why blood transfusions from one person to another sometimes resulted in death. Some red blood cells contain substances called *agglutinogens,* that react with antibodies known as *agglutinins* in the plasma of other people. On the basis of these differences, he classified blood into four types, now referred to as A, B, AB, and O. The agglutinogens of Type A are different from those of Type B blood. Therefore if blood from one is mixed with the other, they will agglutinate or clump, because of the action of the agglutinins in the opposite type blood. Type AB has both agglutinogens A and B, and Type O has neither. Therefore, it can be seen that a person with type AB blood can receive blood from A, B, AB, or O; it is known as the universal receiver. Since Type O blood has no agglutinogens, it may be transfused to any of the other three types, and is known as the "universal donor." A demonstration to show the blood type of a person may be performed as follows:

1. Divide a microscope slide in two halves with a glass marking pencil. Place a drop of serum from Type A on one side (it contains agglutinins against Type B) ; on the other side, place a drop of serum from Type B (it contains agglutinins against Type A) (Blood typing serum is obtainable from a medical supply company).

2. To each drop, add a drop of blood from the person's finger. Carefully mix the contents on each side of the slide with a separate toothpick.

3. After about three minutes, examine each side of the slide with the low power of the microscope.

4. If the red blood cells on side A have gathered together in clumps, but not on side B, the blood is Type B.

If the cells on side A have not clumped, but those on side B have clumped, the blood is Type A

If both side A and B have clumped, the blood Type is AB.

If neither side has clumped, the blood Type is O.

Other types of blood have also been described, including the M, N, Hr, and P factors, and the Rh factor. About 85 per cent of white Americans are *Rh positive,* indicating that they have this latter factor; the rest are called *Rh negative,* since they do not have it. All of the various blood types are inherited in definite ways. In the case of the Rh factor, if both parents are Rh positive, there is normally no difficulty in the development of their children. However, if the father is Rh positive and the mother is Rh negative, the red blood cells of the developing baby may be destroyed (erythroblastosis) because of the production of antibodies against them in the mother's body. This usually occurs after the birth of a first baby, after a large supply of antibodies has been formed. Doctors now use a method that is often successful in saving the life of such a newborn baby; they drain out all of the baby's blood as soon as it is born, and replace it with a fresh supply. In one case, this was done seven times before the baby was declared safe. Doctors now classify the Rh factor of expectant mothers ahead of time, so as to be ready with emergency transfusions if necessary.

SUMMARY OF THE BLOOD GROUPS

BLOOD TYPE	AGGLUTINOGENS IN RED BLOOD CELLS	AGGLUTININS IN PLASMA	DONOR TO	RECIPIENT FROM
A	A	Anti B	A, AB	A, O
B	B	Anti A	B, AB	B, O
AB (Universal recipient)	A and B	none	AB	A, B, AB, O
O (Universal donor)	none	Anti A and Anti B	A, B, AB, O	O

Circulation in lower animals. Food and oxygen are needed by all living cells. In single-celled animals, such as paramecium, the circulating protoplasm distributes these needed materials throughout the cell by its cytoplasmic streaming, known as cyclosis, and by diffusion.

Hydra has cells with long flagella lining its central cavity. The beating of these flagella creates currents which circulate the food throughout the cavity, where it diffuses into the cells, and where the larger particles are engulfed by special cells.

Earthworm has a closed transport system. It has a long large blood vessel on its upper (dorsal) side, just above the digestive system, and two long blood vessels below it. They are connected by side branches. Blood moves forward by the contractions of the dorsal vessel. There are five enlarged side vessels around the esophagus (aortic arches) that serve as hearts, and pump the blood back through the lower, or ventral, blood vessel. There are valves in the hearts and the dorsal vessel that prevent the blood from flowing backward. Muscular movements of the body also keep the blood moving along. Digested food enters the capillaries in the intestine. Oxygen is combined with hemoglobin which is dissolved in the blood fluid rather than being contained in blood cells. Capillaries branch out to the various parts of the earthworm, carrying food and oxygen to the cells, and removing the wastes.

Insects. The *heart* is a tubular structure that is located in the abdomen and pumps blood forward through the *aorta*. From this blood vessel, the blood flows through spaces among the head tissues and passes back through the thorax and the abdomen, bathing the various structures in them. It absorbs digested food from the stomach and re-enters the heart from the rear part of the aorta. Circulation is thus through an *open system,* rather than a closed one of specific blood vessels; there are no capillaries or veins. Oxygen is not carried in the circulatory system to any great extent because of the insect's branching system of air tubes.

Fish. Fish have a two-chambered tubular heart consisting of an *auricle* and a *ventricle*. It is located in the front part of the body, below the gills. Blood from the body enters the *auricle,* which contracts and pumps it into the thick-walled *ventricle*. It in turn pumps blood into the *ventral aorta* which sends the blood to the gills, where it branches into capillaries that absorb oxygen from the water and give off carbon dioxide. Then the capillaries combine to form the *dorsal aorta* which runs along the back and branches into arteries going throughout the body. The veins carry blood back to the heart. Digested food is absorbed as the blood passes through the intestine. The red blood cells are oval-shaped and contain a nucleus.

Frog. The heart of a frog has three chambers, consisting of *two auricles,* and *one ventricle*. Blood from the body enters the *right auricle,* from which it is pumped into the ventricle. Most of the blood is pumped out into the *pulmonary artery* and then to the lungs. Here, the blood flows in capillaries and absorbs oxygen and excretes carbon dioxide. The *pulmonary vein* carries blood back to the left auricles. It is pumped into the ventricle from which it is sent through arteries to the various parts of the body. There is thus a mixing to some extent of deoxygenated and oxygenated blood in the ventricle. Food is absorbed in the small intestine. The blood consists of plasma, oval red blood cells containing hemoglobin and having a nucleus, and white blood cells. There is a lymph circulation, in which four contracting lymphatics, or lymph hearts, pump lymph and eventually send it into the veins.

CHAPTER REVIEW

Select the best choice for each of the following statements.

1. A part of the heart that receives blood from the rest of the body is the (A) left ventricle (B) right ventricle (C) right auricle (D) valves (E) aorta
2. The cells of the heart are nourished by the (A) carotid arteries (B) coronary arteries (C) jugular vein (D) portal vein (E) pulmonary arteries
3. All of the following statements about arteries are true except (A) they are thick-walled (B) they pulsate (C) they contain much elastic fiber tissue (D) they carry blood away from the heart (E) they contain valves
4. All of the following veins carry deoxygenated blood except (A) superior vena cava (B) inferior vena cava (C) pulmonary vein (D) renal vein (E) hepatic vein
5. In the portal circulation, blood flows through the (A) liver (B) lungs (C) skin (D) kidneys (E) entire body
6. The liquid that bathes all the cells of the body is called (A) fibrinogen (B) lymph (C) plasma (D) blood (E) fibrin

7. The thoracic duct is part of the (A) skeletal system (B) digestive system (C) excretory system (D) lymphatic system (E) enzyme system

8. All of the following are found in plasma except (A) fibrinogen (B) antibodies (C) hormones (D) starch (E) amino acids

9. Oxygen is carried in the (A) phagocytes (B) platelets (C) white blood cells (D) red blood cells (E) gamma globulin

10. Iron is needed for (A) formation of white blood cells (B) formation of red blood cells (C) formation of platelets (D) prevention of goiter (E) prevention of diabetes

11. The circulation of blood was first demonstrated by (A) Metchnikoff (B) Pasteur (C) Cohn (D) Leeuwenhoek (E) Harvey

12. Red blood cells are formed in the (A) kidneys (B) skeletal muscles (C) marrow of the bones (D) cartilage (E) heart

13. All of the following statements about white blood cells are true except (A) they are formed in lymph glands (B) they are formed in bone marrow (C) they move like a paramecium (D) they destroy bacteria (E) they have a nucleus

14. The correct sequence of steps in clotting of blood (1-platelet; 2-fibrin; 3-fibrinogen; 4-thrombin; 5-prothrombin) is (A) 1–3–2–5–4 (B) 1–2–3–5–4 (C) 1–2–3–4–5 (D) 1–5–4–3–2 (E) 1–5–4–2–3

15. An inherited disease in which blood does not clot is (A) thrombosis (B) atherosclerosis (C) hemophilia (D) heparin (E) anemia

16. A person with Type A blood may safely receive a transfusion of (A) Type AB (B) Type A, and Type AB (C) Type AB, and Type O (D) Type A, and Type O (E) none of these

17. Worn out red blood cells are removed from the blood by the (A) lungs (B) spleen (C) small intestine (D) large intestine (E) pancreas

18. Most people (A) are Rh positive (B) are Rh negative (C) have anemia (D) have erythroblastosis (E) have hemophilia

19. Circulation of blood takes place in an open system in (A) fish (B) frog (C) paramecium (D) grasshopper (E) toad

20. Serum is like plasma except that it has no (A) water (B) fibrinogen (C) hemoglobin (D) fibrin (E) red blood cells

ANSWER KEY

1-C	5-A	9-D	13-C	17-B
2-B	6-B	10-B	14-D	18-A
3-E	7-D	11-E	15-C	19-D
4-C	8-D	12-C	16-D	20-B

9 Respiration

After food has been delivered to the cells, it must be combined with oxygen in order to release energy. When this occurs, carbon dioxide and water are given off as wastes. During this process of *respiration,* numerous respiratory enzymes participate in the series of reactions by which energy is produced in every cell of the body.

THE RESPIRATORY SYSTEM. The respiratory system includes the structures through which oxygen comes into the body to reach the bloodstream, and through which carbon dioxide and water vapor leave. Air first enters through the two nostrils in the nose, where large hairs hold back some of the dust. Farther along, the *nasal passages* are lined with cilia which beat bacteria and dust outward. The sticky mucus secreted by the epithelial cells also traps bacteria and dust. Thus, the air is filtered as it enters the body. It is also moistened and warmed as it passes over the thin lining layer of cells, and through the sinuses. The sinuses are cavities in the head connected with the nasal passages by narrow openings. During a cold, their membranes may become infected and swollen, blocking the openings and causing the troublesome condition known as sinusitis, or "sinus trouble."

Air continues into the *pharynx,* or throat, at the back of the mouth. It is obvious that breathing by mouth does not permit the air to be cleansed, moistened or warmed, as in nasal breathing. The *tonsils,* and the *adenoids,* lymph masses at the entrance to the throat and at the back of the nasal cavity, often have to be removed when they become diseased. The pharynx also has two small openings leading to each ear through the narrow *Eustachian tubes.*

The trachea, or windpipe, next receives the air. The upper end is covered with a flap of tissue, the *epiglottis,* which is open during breathing. However, when food is being swallowed into the nearby opening of the esophagus, the epiglottis closes and keeps food from entering the trachea. The *larynx,* or voice box, is located just below the opening of the trachea. It contains the vocal cords, and sometimes protrudes in the neck as the "Adam's apple." The wall of the trachea contains rings of cartilage which keep the passageway open. The inner surface is lined with ciliated epithelial cells, which beat dust and bacteria upward.

The trachea divides into two *bronchi,* tubes that carry air into the lungs. Each bronchus branches into smaller and smaller *bronchial tubes,* or bronchioles, which spread through the lungs. At the end of each of the millions of microscopic bronchial tubes there is an *alveolus* or air sac. The wall of the air sac is made up of one layer of cells surrounded by a network of capillaries. As the blood flows by in these capillaries, oxygen diffuses from the air through the walls of the air sacs and combines with hemoglobin in the red corpuscles to form oxyhemoglobin. Carbon dioxide and water vapor diffuse out of the blood into the air sacs, and are eliminated when the air is carried out.

Summary of route of air through respiratory system: (1) nostrils (2) nasal cavity and sinuses (3) pharynx (4) larynx (5) trachea (6) bronchi (7) bronchial tubes in the lungs (8) air sacs (alveoli).

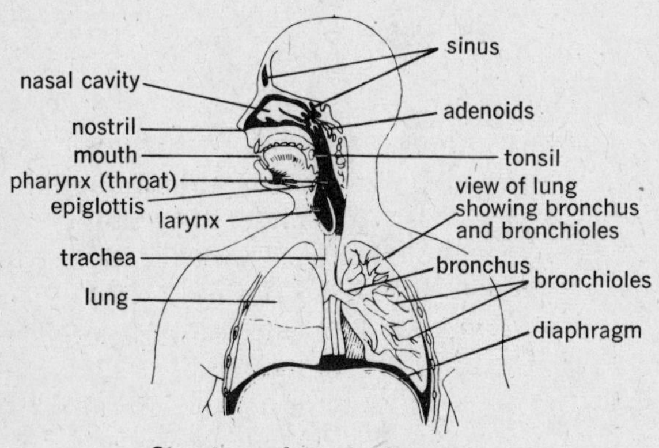

Structure of respiration system

When air is inhaled, it contains about 79 per cent nitrogen, 20.9 per cent oxygen, and 0.04 per cent carbon dioxide. After it has reached the air sacs, and diffusion has occurred, it is *exhaled* with 79 per cent nitrogen 16.3 per cent oxygen and 4.5 per cent carbon dioxide. The increased content of carbon dioxide may be demonstrated as follows:

1. Blow through a straw into a test tube of clear lime water. As a control, use an air pump to bubble air through another test tube of clear lime water.

2. In a short time, the first tube of lime water will become cloudy because of the accumulation of carbon dioxide. The second test tube will continue to remain clear since it is unaffected by the minute amount of carbon dioxide present in the air.

The presence of water vapor in exhaled air is easily demonstrated by breathing on a cold window pane, and observing the collection of moisture that results.

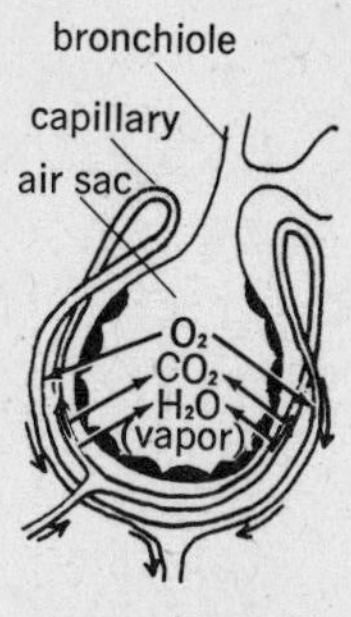

gas exchange in
an air sac (alveolus)

Breathing mechanism. The lungs are elastic, spongy organs located in the *thorax,* or chest cavity, which is a closed chamber. The outside of the lungs is covered with a moist, smooth membrane, the *pleura,* which also lines the thorax. A condition known as pleurisy results when the pleura becomes infected. The diaphragm, the muscular floor of the thorax, consists of voluntary striated muscle, and separates it from the abdominal cavity below. The walls of the thorax contain the ribs in a cagelike arrangement; most of them are attached to the breastbone in front, and all of them are attached to the spinal column in the back. Layers of voluntary muscle are attached to the ribs.

The act of breathing involves the movement of both the diaphragm and the ribs, to enlarge or decrease the size of the chest cavity. During this process, the lungs are either filled with air or are made to give up their air. They are thus passive, and cannot expand or contract by themselves. Specifically, breathing takes place as follows:

1. **Inspiration,** or inhaling. The diaphragm, which is curved upward when it is relaxed, contracts and flattens out. The rib muscles also contract, causing the ribs to be raised and spread apart. Both of these actions enlarge the chest cavity, and reduce the internal pressure on the lungs. Air from the outside, where the atmospheric pressure is higher, is then forced into the lungs, and they inflate. This brings air into the air sacs where the exchange of gases takes place.

2. **Expiration** or exhaling. The diaphragm relaxes, curving upward. The rib muscles relax, lowering the position of the ribs. The chest cavity is thus de-

creased in size, compressing the air in the lungs and forcing it out.

Normal breathing takes place about 15 to 18 times a minute. Exercise or emotional condition can cause it to increase. When breathing stops altogether, because of drowning, gas suffocation, or electric shock, *artificial respiration* must be applied in order to continue the supply of oxygen to the cells. There are several methods of artificial respiration, the latest of which includes mouth-to-mouth breathing. In this method, the rescuer blows into the mouth of the victim about 15 times a minute, thus inflating the lungs and enlarging the chest cavity. The exhaled air of the rescuer contains enough oxygen to supply the cells of the victim and revive him. In an attack of polio, the nerves that control the voluntary muscles in the chest may be damaged and the person prevented from breathing. He is placed in an "iron lung," a device which alternately expands and contracts the chest cavity, thus making him breathe artificially.

Control of breathing. Although breathing in a voluntary activity involving voluntary muscles, it is automatically controlled by a part of the brain, the *medulla*. It therefore goes on without our having to think about it. The rate of breathing is controlled by the amount of carbon dioxide in the blood. As the blood circulates, an increased amount of carbon dioxide in it will stimulate the cells of the medulla to send impulses to the chest muscles and diaphragm, causing them to contract and expand more rapidly; the breathing rate then increases. This is what happens when we are exercising and the cells are giving off large amounts of carbon dioxide. When we are resting, less carbon dioxide accumulates, and the cells of the medulla are not stimulated as much, so that the breathing rate is lower. The amount of air reaching the air sacs of the lungs is controlled to some extent by the size of opening of the bronchial tubes. Under certain conditions, the elastic tissue contained therein may contract, diminishing their diameter, and reducing the supply of air passing through. This happens in cases of asthma and other allergies, sometimes requiring the person to take adrenalin or antihistamines to relax the elastic tissue and to expand the bronchial tubes. On other occasions, during periods of excitement or strenuous exertion under the influence of extra adrenalin produced by the body, these openings may enlarge, permitting an added supply of air to reach the air sacs.

Effects of high altitudes. The coming of the air age has presented us with the problem of breathing at high altitudes. At sea level, the weight of the air produces a pressure of 15 pounds per square inch. Higher up, there is less pressure, and less oxygen in a given volume of air. Therefore, it is necessary to "pressurize" the cabins of airplanes flying above 10,000 feet, and to add oxygen to the air being breathed. Without this additional oxygen, a condition called *anoxia* results, which is characterized by a loss of co-ordination, vision, hearing and judgment, and finally results in unconsciousness.

Another problem of reduced pressure at high altitudes concerns the nitrogen which is normally dissolved in the body fluids. This nitrogen has the same concentration as in the air, about 78 per cent. As a plane climbs to higher altitudes where the air pressure is low, there is less pressure on the nitrogen, and it begins to come out of solution and leave the body. If the ascent is too fast, bubbles of nitrogen form in the tissues and block small blood vessels, producing a very painful case of the "bends." This is similar to the condition affecting deep sea divers who ascend too rapidly from underwater regions of high pressure to sea-level pressure. Besides the use of pressurized cabins, high-altitude fliers may also wear pressurized suits. These and other problems must be overcome before the astronauts can explore outer space in rockets.

Mountain climbers who ascend the tallest mountains must be equipped with oxygen tanks if they are to survive and supply their cells with sufficient oxygen. It has been found that the natives of Peru who live high up in the Andes Mountains have a much higher red blood count than the average person at sea level. This greater supply of hemoglobin enables them to absorb enough oxygen from the thin air to maintain the life of their cells.

Respiration in lower animals. Oxygen must be obtained by practically all living things in order for them to be able to oxidize their food and obtain energy. *Anaerobic* bacteria are an exception; they live without oxygen and are actually killed by it.

Protozoa absorb oxygen directly from the water in which they live; it diffuses in through the cell membrane, while carbon dioxide diffuses out into the water. Respiration in hydra is also carried on directly by diffusion.

Earthworm. The earthworm has a thin, moist skin through which oxygen diffuses into the blood. The hemoglobin dissolved in the blood then carries oxygen throughout the body. Carbon dioxide diffuses out of the body as the blood passes through the skin.

Insects. Insects such as the grasshopper have small openings on the side of the segments of the abdomen called *spiracles*. These lead to a system of branched air tubes, *tracheae*, spreading throughout the body to all the cells. Air is pumped in and out of the tracheae by the muscular contractions and expansions of the abdomen. In this way, oxygen is brought in, and carbon dioxide is expelled from the body.

Fish. Water is constantly entering a fish's mouth, and passing out through openings on both sides of the head. There are *gills* located in gill chambers on each side. The gills have numerous thin-walled, thread-like *gill filaments* which are well supplied with capillaries. As the water passes over the gill filaments, oxygen diffuses through the thin membrane into the blood, while carbon dioxide leaves the blood and diffuses out into the water. The oxygenated blood is circulated through the fish by means of the arteries. In order for a fish to survive in an aquarium, it is necessary to supply it with oxygen, either by stocking it with green plants that will give off oxygen during photosynthesis, or by pumping air into it.

CHAPTER REVIEW

Select the correct choice for each of the following statements.

1. Energy is obtained from food in the process of (A) assimilation (B) digestion (C) excretion (D) respiration (E) storage

2. Ciliated epithelial cells are located in the (A) esophagus (B) trachea (C) villi (D) aorta (E) cardiac muscle

3. The correct order of structures through which air passes (1-nasal cavity; 2-bronchi; 3-larynx; 4-air sacs; 5-trachea) is (A) 1–5–3–2–4 (B) 1–5–3–4–2 (C) 1–3–4–5–2 (D) 1–3–5–4–2 (E) 1–3–5–2–4

4. Oxyhemoglobin is formed in the (A) lungs (B) spleen (C) right auricle (D) left auricle (E) bone marrow

5. The percentages of nitrogen, oxygen, and carbon dioxide in exhaled air are about (A) 79–20–0.04 (B) 79–16–4 (C) 20–79–0.04 (D) 16–79–4 (E) 79–4–16

6. The passageway of the trachea is kept open by rings of (A) striated muscle (B) adenoids (C) cartilage (D) mucus (E) sinuses

7. During inhalation (A) the diaphragm flattens out, and the ribs are raised (B) the diaphragm is raised, and the ribs are lowered (C) the diaphragm flattens out, and the ribs are lowered (D) the diaphragm is raised and the ribs are raised (E) the diaphragm is raised and the ribs are stationary

8. The rate of breathing is controlled by the (A) lungs (B) bronchi (C) air sacs (D) medulla (E) diaphragm

9. The breathing rate is increased by an increase in the content of (A) oxygen (B) nitrogen (C) water vapor (D) carbon monoxide (E) carbon dioxide

10. Lack of oxygen at high altitudes produces (A) bends (B) anoxia (C) asthma (D) artificial respiration (E) antihistamines

ANSWER KEY

1-D	3-E	5-B	7-A	9-E
2-B	4-A	6-C	8-D	10-B

10 Excretion

A review of the activities carried on by the various parts of the body reveals that the cells and tissues are really members of a well-organized community. Each organ and system does its specific job for the good of all. The digestive system prepares food for use by the cells. The circulatory system distributes food, oxygen, and other materials to the cells, and removes carbon dioxide and other wastes. The respiratory system provides for the intake of oxygen and the excretion of carbon dioxide and water vapor. The excretory system completes the job of removing wastes from the body.

THE EXCRETORY SYSTEM. The wastes given off by the cells would poison them if they accumulated in the body. The lungs, kidneys, and skin serve as organs of excretion and remove such wastes as carbon dioxide, water, urea, and mineral salts from the blood. The large intestine does not excrete indigestible wastes; it merely eliminates them, since they were never really part of the body.

The lungs. When food is oxidized by the cells, carbon dioxide and water are given off as waste products. They diffuse out of the blood into the alveoli of the lungs, and are carried out of the body in the exhaled air.

The kidneys. The two kidneys are dark red, kidney-bean-shaped organs located in the rear of the abdominal cavity, above the small of the back. A large artery, the renal artery, carries blood containing a high percentage of wastes into each kidney. It branches into smaller blood vessels that become capillaries arranged in a dense network, or *glomerulus.* Each glomerulus is surrounded by a thin-walled cup, or *Bowman's capsule,* which extends into a long *nephron tubule.* The nephron is enclosed in a separate network of capillaries, and it connects with other tubules to form small tubes that lead into the *central cavity* of the kidney. As the blood flows through each glomerulus, a plasma-like liquid containing water, urea, salts, fatty acids, amino acids and glucose diffuses out into the capsule end of the nephron. As this liquid moves along the long tubule of the nephron, water, minerals, and the digestive end-products are reabsorbed into the capillaries by active transport. The remaining liquid collects as *urine* at the far end of the tubule and joins urine from the million other tubules in the central cavity of the kidney. From here, the urine passes out of each kidney through a long tube, the *ureter,* leading into the urinary *bladder.* Urine leaves the bladder at intervals through the tube called the *urethra.*

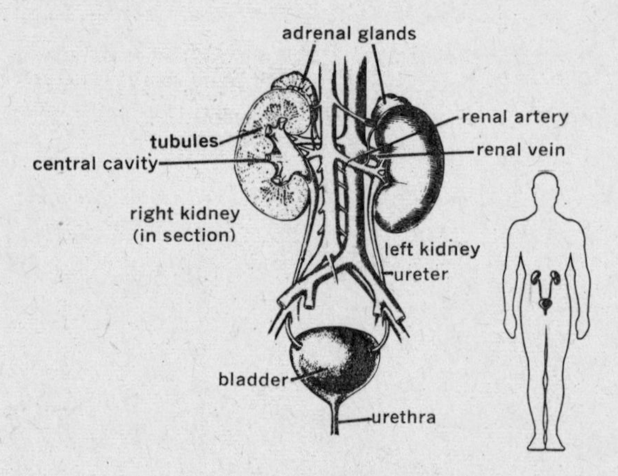

Kidney excretory system

Urine contains about 95 per cent water. Much of the remainder is the nitrogenous waste—urea—and mineral salts, uric acid and other materials. Urea is produced as a waste when the liver breaks down amino acids in a process called *deamination.* The amino group ($-NH_2$) is removed from the amino acid molecule through enzyme action, and rearranged into ammonia (NH_3). This is then combined with carbon dioxide, and through several reactions, becomes urea.

Under normal conditions, there are predictable amounts of each part of the urine, giving a clue to the healthy functioning of the various cells. If an abnormal condition is suspected, a doctor studies a *urinalysis.* Extra amounts of glucose may point to diabetes; high amounts of protein may indicate a disease of the kidney itself; pus is a symptom of an infection.

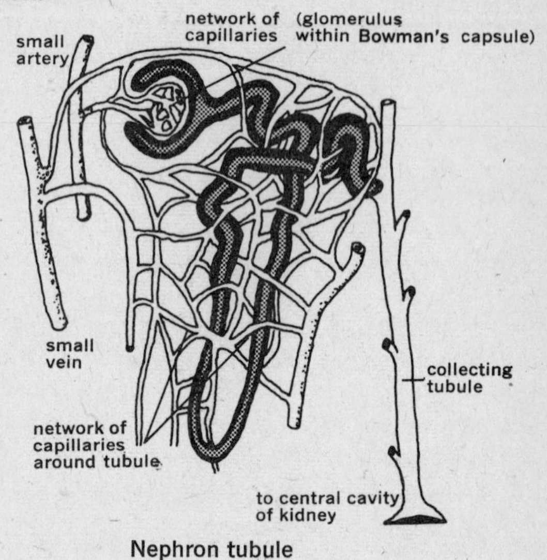

Nephron tubule

The blood that was flowing through the capillaries around the tubules collects into small veins, and finally leaves each kidney through a large vein, the renal vein. This leads into the inferior vena cava, and carries blood that has been purified of much of its liquid wastes. The kidney thus is a vital organ for removing the waste materials that would otherwise poison the body. It also helps to maintain the correct balance of water and mineral salts in the body.

The skin. The skin consists of two main layers of cells, (1) the thin, outer *epidermis,* and (2) the underlying, thicker *dermis.*

The **epidermis** consists of layers of flat squamous epithelial cells. The upper part is made of dead cells that are constantly being replaced by the cells below them.

The **dermis** contains sweat glands, connective tissue, blood vessels, nerves, nerve endings, oil glands, muscles, and hair follicles.

A **sweat gland** consists of a coiled section which leads up through a tube to an opening on the skin called a *pore.* A network of capillaries surrounds the coiled part of the sweat gland. As blood flows through these capillaries, the thin wall of the sweat glands absorbs a large amount of water as well as some salts. This liquid fills the tube, and comes out from the pore as *perspiration,* or sweat. The sweat glands excrete large quantities of water, to maintain the balance of various salts in the body.

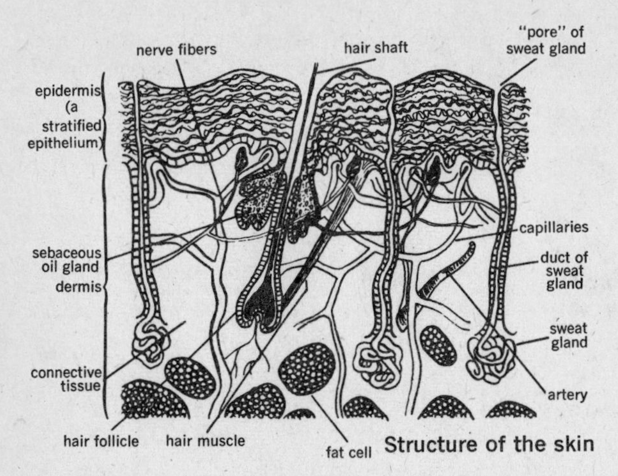

Structure of the skin

Regulation of body temperature. The excretion of perspiration helps to regulate the temperature of the body. As the perspiration evaporates from the surface of the skin, it cools it. The cooling effect of an evaporating liquid may be demonstrated in the following way:

1. Wrap some absorbent cotton around the bulbs of three thermometers. Wet one with a rapidly evaporating liquid, such as ether or alcohol; wet the second with water; leave the third dry, as a control.

2. After a few minutes, compare the readings of the three thermometers. It will be seen that the first will show the lowest temperature, since the rapid evaporation cooled it the most. The second will show a higher temperature than the first, since the water did not evaporate as quickly; however, it will show a lower temperature than the dry thermometer because evaporation was taking place, thereby cooling the theremometer.

When perspiration evaporates on the skin, it cools it, and lowers its temperature. This also lowers the temperature of the blood as it circulates through the skin. During vigorous exercise, when the cells of the body are releasing much heat energy, the blood vessels in the skin increase in diameter, thus carrying more blood to the skin, which may appear flushed. This results in more perspiration being excreted from the sweat glands. As it evaporates, it removes heat from the skin, cooling it. The reverse is true in cool weather. The blood vessels in the skin decrease in diameter, thus carrying less blood, and resulting in less excretion of perspiration. With less perspiration, there is less evaporation, and less cooling.

On a sunny, dry day, perspiration evaporates quickly, giving one a sensation of coolness and comfort. On a humid day, when the air contains a high percentage of moisture, there is less evaporation of perspiration, resulting in an uncomfortable, muggy feeling. An electric fan helps to provide a feeling of coolness by removing moist air which was directly in contact with the surface of the skin; this allows more rapid evaporation of the perspiration, with a consequent lowering of the temperature. In hot weather, prolonged physical activity may lead to so much perspiration that a large amount of salt may be lost, leading to a weakened condition known as *heat exhaustion*. For this reason, people engaging in heavy work or exercise are advised either to take salt tablets or to add salt to their food.

Other functions of the skin. Besides (1) excreting water, to maintain the balance of salts in the body; and (2) helping to regulate the temperature of the body, the skin also has the following other functions: (3) Protection against the entrance of bacteria into the body. (4) Protection of the underlying tissues against mechanical injury, and drying up. (5) Sensation of outside stimuli such as touch, heat, cold and pain, by means of the nerve endings located in it. (6) Formation of vitamin D upon exposure to ultraviolet rays.

Excretion in Lower Animals. When living cells carry on their activities, they give off wastes such as carbon dioxide, water, and nitrogenous compounds. In protozoa, these wastes diffuse directly out through the cell membrane. The nitrogenous wastes in protozoa are in the form of ammonia.

Hydra. Excretion also takes place by diffusion.

Earthworm. Each segment of the earthworm contains a pair of excretory organs called *nephridia.* These are tiny tubes which lead to the outside of the worm by means of openings located on the lower surface. Liquid wastes, including water, mineral salts, and nitrogenous wastes, are collected in the tube by means of a ciliated, funnel-like entrance inside the segment. Contractions of the body wall pass the wastes out. Carbon dioxide is excreted from the blood as it circulates through the thin, moist skin. The major nitrogenous wastes are ammonia and urea.

Insects. The excretory system consists of a number of small tubes called *Malpighian tubules,* which lie in the open blood spaces, and extract water, mineral salts and nitrogenous wastes from the blood. These wastes are excreted by the tubules into the hind gut, and are then eliminated through the anus with the indigestible wastes. The nitrogenous wastes are in the form of uric acid, which leaves in solid form. This is a water-conservation mechanism of great advantage in a dry environment. Carbon dioxide diffuses out of the blood into the air tubes (tracheae) and is excreted through the spiracles by the pumping action of the abdominal muscles.

Fish. There are two long kidneys which remove nitrogenous wastes from the blood and send them to the urinary bladder. From here, the wastes are expelled to the outside. Carbon dioxide diffuses out of the blood as it passes through the gill filaments, and is taken up in the water flowing past the gills and out of the openings on the sides of the head.

Frog. The frog's *kidneys* are deep red, flat oval structures located in the back of the body cavity. They remove nitrogenous compounds from the blood, and send them through *ureters* into the *cloaca* from which the urine passes into the urinary *bladder*. The cloaca is a common passageway for the liquid wastes of the bladder and the indigestible solid wastes of the large intestine, both of which leave the frog's body at the same time through the anus. Carbon dioxide is excreted through the lungs, as well as through the moist skin.

CHAPTER REVIEW

Select the correct choice for each of the following statements.

1. Two wastes resulting from the oxidation of food are (A) oxygen and water (B) carbon dioxide and water (C) oxygen and carbon dioxide (D) oxygen and ATP (E) carbon dioxide and ATP

2. Urea is removed from the blood by the (A) lungs (B) liver (C) kidneys (D) spleen (E) bladder

3. The microscopic structures in the kidneys which remove wastes from the blood are (A) renal arteries (B) renal veins (C) tubules (D) ureters (E) urethra

4. The dermis of the skin contains all of the following except (A) layer of dead cells (B) blood vessels (C) nerve endings (D) hair follicles (E) oil glands

5. The part of the skin that removes liquids from the blood is the (A) pores (B) epidermis (C) network of capillaries (D) sweat glands (E) flat squamous layer

6. All of the following are functions of the skin except (A) formation of vitamin D (B) sensation (C) protection against entrance of bacteria (D) excretion of carbon dioxide (E) regulation of body temperature

7. On a humid day, a person is uncomfortable because (A) he perspires more (B) there is less evaporation of perspiration (C) perspiration evaporates quickly (D) larger amounts of salt are excreted by the skin (E) the surface of the skin is too dry.

8. A structure through which protozoa excrete wastes is (A) food vacuole (B) cell membrane (C) trichocyst (D) oral groove (E) cilia

9. Urine leaves the body through the (A) glomerulus (B) epiglottis (C) alveoli (D) ureters (E) urethra

10. Malpighian tubules are used for excretion in the (A) earthworm (B) insect (C) hydra (D) fish (E) frog

ANSWER KEY

1-B	3-C	5-D	7-B	9-E
2-C	4-A	6-D	8-B	10-B

HOW LIVING THINGS ADJUST TO THEIR ENVIRONMENT

4

11 The Endocrine System

The body carries on many activities, with the help of its various organs and systems. These activities are all co-ordinated so that they go on at the proper rate, at the proper time, and in the proper amounts. For example, when food leaves the stomach and enters the small intestine, the pancreas and the liver send their secretions to join intestinal juice in digesting it. These digestive organs do not send their secretions until that time. How is this controlled?

The Glands of the Endocrine System

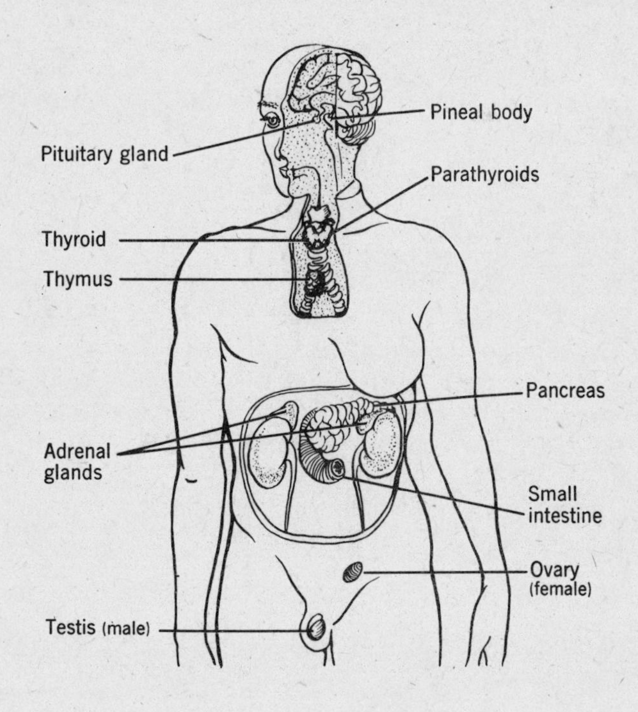

Secretin. Certain cells in the lining of the small intestine secrete a chemical substance called *secretin,* when food enters from the stomach. Secretin diffuses directly into the bloodstream and is carried through the body. When it reaches the pancreas, it stimulates the digestive glands there to produce pancreatic juice, which pours into the small intestine through the pancreatic duct. Secretin thus serves as a chemical messenger, or *hormone* (endocrine) . A hormone is a chemical produced in one part of the body, that is transported to other parts where it exerts specific effects.

Other hormones of the alimentary canal. The lining of the small intestine also secretes two other hormones when food enters from the stomach, that serve similar functions: one, *cholecystokinin,* stimulates the gall bladder to contract, and send bile into the small intestine; the other, *enterocrinin,* stimulates the digestive glands of the small intestine to secrete intestinal juice. Like secretin, these hormones are circulated in the bloodstream. Thus, the start of digestion in the small intestine is coordinated by three different hormones that are produced at the time when food is arriving from the stomach. The lining of the stomach itself also secretes a hormone, *gastrin,* when food enters it. Gastrin is carried in the bloodstream to the digestive glands in other parts of the stomach which are stimulated to produce gastric juice.

Endocrine glands. Endocrine glands are different from digestive glands, which have a duct leading to the digestive system, in that they are "ductless"; their secretions, the hormones, diffuse directly into the blood and are transported to other parts of the body where they are used for specific purposes. Hormones may be secreted by groups of individual cells, such as those that produce secretin and the other hormones of the alimentary canal, or by entire endocrine glands. These endocrine glands are: thyroid, parathyroid, pancreas (islands of Langerhans), adrenal, pituitary, and reproductive organs.

Thyroid gland. The thyroid gland is located at the base of the neck, below the larynx. It absorbs large amounts of iodine from the blood for making the hormone, *thyroxin.* When the diet is deficient in iodine, the thyroid gland enlarges and produces a swelling on the throat. This condition is known as simple, or *endemic goiter.* It used to be much more prevalent in the inland areas of the country where iodine is not available in the food. Along the seacoast, iodine is plentiful in sea food. Iodized salt is now used to prevent goiter.

Thyroxin is brought by the bloodstream to all the cells of the body, where it controls the rate at which oxidation takes place, and energy is released. The *basal metabolism test* is used by doctors to measure the rate of oxidation in a person who is at rest. If it is too low, it indicates an underactive thyroid with a low rate of oxidation; if it is too high, it is a sign of an overactive thyroid, with a high rate of oxidation.

Undersecretion of thyroxin (hypothyroidism). (1) In children: Infants whose thyroid gland is not producing enough thyroxin are known as *cretins.* They are underdeveloped, physically and mentally. Their bodies appear stunted. They become idiots. However, treatment with thyroxin helps correct the condition. (2) In adults: Grownups whose thyroxin production is below normal develop a condition known as *myxedema.* Their body becomes puffy, and they lose their alertness and even intelligence. They have a slow pulse, and a low rate of oxidation. Treatment with thyroxin makes the abnormal conditions disappear.

Oversecretion of thryoxin (hyperthyroidism). When the thyroid gland is overactive and is producing an excess of thyroxin, an individual has a higher rate of oxidation. He remains thin because his food is being oxidized rather than being stored. He is restless and sleeps less. His pulse beat is rapid. He becomes irritable. His eyes bulge, and the thyroid gland is usually enlarged. He is said to have *exophthalmic goiter.* Treatment may consist of removing part of the thyroid gland, as a method of reducing the production of thyroxin. Or, radioactive iodine is given to the individual; it is absorbed and concentrated in the thyroid gland, where its rays cause a reduction in the production of thyroxin. The chemical, *thiouracil,* also may be used to decrease the activity of the thyroid cells.

Frog metamorphosis from the tadpole stage depends on thyroxin. Tadpoles have been made to develop into frogs in a very short time by feeding them with thyroxin.

The parathyroid glands. There are four small parathyroid glands located next to the thyroid gland, They secrete the hormone, *parathormone,* which regulates the absorption and use of the element calcium by the body. It is also important in bone formation. If it is lacking, the muscles do not function properly, and the individual may develop a case of involuntary muscle contractions or convulsions known as *tetany.* This can be relieved by injecting either parathormone or calcium salts.

The pancreas. In addition to producing pancreatic juice, the pancreas has collections of cells called the *islands of Langerhans,* that secrete the hormones insulin and glycagon. The value of insulin to the body: (1) It helps the cells to utilize glucose for energy; (2) it also causes the liver to convert excess glucose in the blood into glycogen for storage. In these ways, the amount of glucose in the blood is always reduced to a constant level. Glycagon, a recent discovery, seems to counteract insulin. It increases the blood level of glucose by causing the liver and small intestine to release it. In the condition called *diabetes,* the production of insulin is affected. As a result, the cells cannot use glucose for energy, and the liver does not store it. The person becomes lacking in energy, and loses weight. There is an excess of glucose in the blood, and a urinalysis reveals the presence of glucose.

Diabetes cannot be cured, but a diabetic can live a normal and useful life span if he receives daily injections of insulin. In mild cases, the drug tolbutamide (or orinase) may be taken orally. He should eat a diet low in carbohydrates. Insulin was discovered and extracted from the pancreas of animals in 1922 by Dr. Frederick Banting and his associates in Canada. He received the Nobel Prize for this research.

Adrenal glands. The adrenal glands are located on top of the kidneys. They consist of two parts, (1) the *medulla,* or inner section, and (2) the *cortex,* or outer layer.

1. Medulla. The medulla produces the hormone *adrenin.* During periods of great emotional stress such as anger or fear, a large amount of adrenin is secreted into the bloodstream. Because of its effects in giving the body extra energy, the adrenal glands have been called glands of emergency, or glands of combat. Adrenin produces the following effects: (1) The liver is stimulated to convert glycogen to glucose. (2) The heart beats faster. (3) The smooth muscles of the blood vessels contract, causing an increase in blood pressure. (4) The breathing rate is increased and the diameter of the bronchioles enlarges. (5) The blood vessels in the skin and the digestive system contract, and digestion is reduced. (6) The clotting rate of blood is speeded up. (7) With extra amounts of glucose and oxygen being supplied more rapidly to the cells of the body, including the skeletal muscles and the brain, there is an increase in energy, strength and alertness.

Under these conditions, a boy will find that he can run as he had never run before. A man will be able to lift very heavy objects that he would normally be unable to budge. The synthetic hormone, *adrenalin,* is injected directly into the heart by doctors in order to stimulate it during an operation or during a heart attack. Adrenalin is also used in cases of asthma or other allergies, to enlarge the size of the bronchial tubes and relieve breathing. During a tooth extraction, a dentist may apply adrenalin to contract the size of the blood vessels, thus reducing bleeding and hastening clotting.

2. Cortex. At least thirty hormones are secreted by the cortex of the adrenal glands. In general, they seem to be important in maintaining the proper balance of liquids and mineral salts in the body; they also play a role in carbohydrate metabolism. One of them, *cortisone,* plays a role in the healthy condition of the cartilage in the joints between bones. It has been found to be important medically in successfully treating cases of rheumatoid arthritis, and in relieving the painful swellings of the joints. It has also been used in severe cases of asthma and skin conditions. However, its effects are not completely understood, at present; it sometimes produces undesirable side effects.

Cortin is another hormone of the cortex. It appears to control the use of water and salts by the cells. It also affects blood pressure. When it is lacking, an individual develops Addison's disease, in which there is marked weakness, a bronze coloration of the skin, and loss of weight. Injections of cortin relieve this condition.

Other hormones of the cortex appear to have an influence on the development of secondary sexual characteristics. The production of the cortex hormones is stimulated by one of the hormones of the pituitary gland called ACTH.

Pituitary gland. This gland is about the size of a pea, and is situated at the base of the brain. It has three main parts. Despite its small size, it secretes many hormones that interact with and control the other ductless glands. For this reason, it is referred to as the "master gland." Some of its hormones have the following effects:

1. **ACTH** (adrenocorticotropic hormone) stimulates the cortex of the adrenal gland to produce a number of hormones, including cortisone. In severe cases of rheumatoid arthritis, it has brought about dramatic relief from disability and pain.
2. **Growth hormone** promotes the growth of the body. In rare cases, too much or too little of it may be produced in children. If there is an excess of it, a child continues to grow to extraordinary height (*gigantism*) ; if there is too little, the child remains a midget (*dwarfism*) . If an adult begins to produce too much of it, the extremities of the body (hands, feet, face) enlarge to produce a condition known as *acromegaly;* this may be treated with x-rays.
3. **Prolactin** is a hormone from the anterior (front) part of the pituitary that stimulates the mammary glands to produce milk.
4. Another hormone stimulates the thyroid gland to produce thyroxin.
5. One of the hormones from the posterior pituitary, *pitocin,* stimulates contraction of the smooth muscles of the uterus during childbirth.
6. Other hormones interact with the reproductive organs—the testes and the ovaries (gonads) — and stimulate them to develop, and to produce additional hormones.

Reproductive organs (gonads). Various hormones are produced by (1) the male gonads, the *testes,* and (2) the female gonads, the *ovaries.*

1. **Testes** secrete male hormones, including *testosterone.* This hormone stimulates the formation

of such male secondary sexual characteristics as: appearance of hair on the face, deepening of the voice, increase in height. Other male animals show the effects of this hormone in their own development: the lion has a mane; the male deer has long antlers; the rooster has a comb and can crow; and male songbirds have bright plumage and a beautiful voice.

2. **Ovaries** produce female hormones, including *estrogen*. This hormone stimulates the formation of such female secondary sexual characteristics as: development of the breasts, retention of a high-pitched voice, and softening of the body features. Other female animals produce this hormone and develop accordingly; the lioness has no mane; the female deer has no antlers; the hen has no comb, and has a clucking voice; female songbirds have dull feathers and lack the ability to sing. Under the influence of the pituitary, the ovaries produce mature egg cells at monthly intervals. A great deal of research is presently being conducted on the value of female sex hormones in treating certain types of cancer.

Thymus gland. This is a large gland in children and is located in the chest. It decreases in size with age. It is thought to play a part in growth and development, but this has not been confirmed. Recent evidence indicates that the thymus plays a role in antibody production and immunity.

Pineal gland. Another little-understood gland is the pineal, which is located in the brain. Its function is not known.

Coordination of the body's activities. Very small amounts of the hormones are needed, yet they play an important part in regulating the various activities of the body. The ductless glands not only stimulate some of these activities, but they also interact with each other. Thus, in the oxidation of glucose: thyroxin regulates the rate of oxidation; insulin permits the cells to oxidize glucose, and stimulates the liver to store it as glycogen; adrenin stimulates the liver to convert glycogen back into glucose, and stimulates the rapid oxidation of glucose; the pituitary hormones control the thyroid and adrenal glands. The development of the reproductive organs is a complex process, involving hormones from the pituitary gland, and the reproductive organs themselves.

Insect hormones. It has recently been found that the life history of insects is regulated by at least three hormones: (a) A brain hormone, which is produced when the insect egg hatches. It stimulates the formation of growth and juvenile hormones in the larva. (b) A growth hormone (ecdysone), which is produced in the thorax, and which regulates growth and molting of the larva. (c) The juvenile hormone which is produced in the head of the larva. It also influences growth, but inhibits metamorphosis. When it is absent, the larva is induced to develop into the pupa stage. In experiments, it was injected into Cecropia moth caterpillars, causing their metamorphosis to be delayed, and resulting in giant adult moths.

Plant hormones. In animals, it has been shown that a hormone is a substance formed in one part of an organism which is transported to another part where it produces its effects. Plants have also been shown to produce hormones. Growth-promoting hormones called *auxins* are formed in the growing tips of stems. One of them, auxin A, has the formula $C_{18}H_{32}O_5$. Auxins cause the stem cells to elongate, resulting in their growth. It has also been determined that auxins initiate the formation of roots on stems. A number of chemical substances similar to auxins have been found that produce these and other effects. Thus one of them, called *2,4-D* for short, makes broad-leaved weeds grow so rapidly that the metabolism of the plants is upset and they die. Others make unpollinated flowers develop into fruits without seeds; seedless tomatoes have been produced in this way. They are also used to prevent fruits, such as apples, from dropping off a tree too soon. If auxin is distributed evenly on all sides of a stem, a plant grows straight up. If more accumulates on one side, growth proceeds more rapidly there, resulting in a bending toward the other side. This will be discussed more fully in the next chapter dealing with plant tropisms.

Another type of plant hormone, vitamin B, is unusual in that it is also useful to humans as a vitamin. The vitamin B complex are plant hormones produced in the leaves of plants and transported down the stem to the roots. Roots have been shown to require the vitamin B complex in order to grow.

SUMMARY OF THE ENDOCRINE GLANDS

GLAND	LOCATION	HORMONE	EFFECTS
Lining of Alimentary Canal	1. Stomach	Gastrin	Stimulates gastric glands to secrete gastric juice
	2. Small intestine	Secretin	Stimulates pancreas to secrete pancreatic juice
	3. Small intestine	Cholecystokinin	Stimulates gall bladder to send bile to small intestine
	4. Small intestine	Enterocrinin	Stimulates flow of intestinal juice
Adrenal Gland	Top of kidneys		
1. Medulla	Inner part of adrenal	Adrenin	Stimulates liver to change glycogen to glucose; stimulates heart beat, breathing rate, blood pressure, clotting rate
2. Cortex	Outer part of adrenal	1. Many hormones	Maintain balance of liquids and mineral salts; carbohydrate metabolism; secondary sexual differences
		2. Cortisone	Useful in treating rheumatoid arthritis, allergies, skin conditions
		3. Cortin	Useful in treating Addison's disease
Islands of Langerhans	Pancreas	1. Insulin	Regulates utilization of glucose by cells, and storage of glucose as glycogen by the liver; treatment for diabetes
		2. Glycagon	Stimulates liver and small intestine to release glucose.
Parathyroid	On thyroid gland	Parathormone	Regulates calcium metabolism; lack causes tetany
Pineal	Base of the brain	?	?
Pituitary	Base of the brain	1. ACTH	Stimulates production of adrenal cortex hormones, including cortisone
		2. Growth	Promotes growth; deficiency in children = dwarfism; oversecretion in children = gigantism; in adults, acromegaly
		3. Prolactin	Stimulates production of milk
		4. Thyroid-stimulating	Stimulates production of thyroxin
		5. Pitocin	Stimulates contraction of smooth muscles of uterus
		6. Gonad-stimulating	Stimulates development of gonads, and production of their hormones
Reproductive organs (Gonads)	1. Testes	Testosterone	Regulate male secondary sexual characteristics
	2. Ovaries	Estrogen	Regulate female secondary sexual characteristics
Thymus	Chest of children	?	? (Disappears during growth) Appears to be involved in immunity
Thyroid	Base of neck	Thyroxin	Regulates rate of oxidation of glucose. Undersecretion in children = cretinism; in adults = myxedema. Oversecretion = exophthalmic goiter. Lack of iodine = endemic goiter

CHAPTER REVIEW

Select the correct choice in each of the following statements.

1. Hormones are distributed throughout the body by (A) ducts (B) ductless glands (C) blood (D) nerves (E) connective tissue

2. All of the following hormones are produced in the alimentary canal except (A) secretin (B) cholecystokinin (C) enterocrinin (D) gastrin (E) insulin

3. The pancreas is stimulated to secrete its digestive enzymes by (A) the liver (B) bile (C) gallbladder (D) secretin (E) insulin

4. The adrenal glands are located next to the (A) kidneys (B) heart (C) neck (D) brain (E) stomach

5. An element required in large amounts by the thyroid gland is (A) iron (B) iodine (C) fluorine (D) calcium (E) phosphorus

6. The basal metabolism rate is regulated by the (A) parathyroid (B) thyroid (C) spleen (D) thymus (E) pineal

7. All of the following conditions are related to the thyroid gland except (A) goiter (B) cretinism (C) myxedema (D) high rate of metabolism (E) tetany

8. The absorption and use of calcium are regulated by (A) parathormone (B) adrenalin (C) thyroxin (D) thiamin (E) prolactin

9. A hormone that enables the cells to utilize glucose is (A) gastrin (B) insulin (C) testosterone (D) cortisone (E) pitocin

10. A gland known as the gland of emergency is the (A) pituitary (B) adrenal (C) thyroid (D) parathyroid (E) pancreas

11. All of the following effects are produced by adrenin except (A) increase in heart beat (B) conversion of glucose to glycogen (C) increase in breathing rate (D) clotting rate of blood is speeded up (E) digestion is decreased

12. A hormone that stimulates the production of cortisone is (A) adrenin (B) cortin (C) auxin (D) ATP (E) ACTH

13. The gland that is referred to as the master gland is the (A) thyroid (B) adrenal (E) islands of Langerhans (D) pituitary (E) ovary

14. All of the following are related to the pituitary gland except (A) gigantism (B) dwarfism (C) diabetes (D) acromegaly (E) rheumatoid arthritis

15. A hormone that affects secondary sexual characteristics is (A) insulin (B) adrenalin (C) estrogen (D) thyroxin (E) parathormone

16. A plant hormone that stimulates growth is (A) axon (B) auxin (C) adenoid (D) anaphase (E) annelida

17. A giant results from the overproduction of a hormone of the (A) pancreas (B) pituitary (C) parathyroid (D) thyroid (E) lining of esophagus

18. Radioactive iodine is sometimes used to treat an overactive (A) liver (B) pancreas (C) thyroid (D) adrenal cortex (E) adrenal medulla

19. A gland that serves as both a duct and a ductless gland is the (A) pituitary (B) thyroid (C) parathyroid (D) pancreas (E) adrenal

20. During a heart attack, a doctor may inject (A) thyroxin (B) prolactin (C) insulin (D) adrenalin (E) enterocrinin

ANSWER KEY

1-C	5-B	9-B	13-D	17-B
2-E	6-B	10-B	14-C	18-C
3-D	7-E	11-B	15-C	19-D
4-A	8-A	12-E	16-B	20-D

12 Reaction of plants and lower animals

One of the characteristics of protoplasm is its ability to react to the environment, known as *irritability*. Single-celled organisms, as well as multicellular plants and animals, show a *response* to *stimuli* about them. Indeed, their *behavior* may be said to form a basis for their survival. Without the ability to move away from unfavorable stimuli, or to react positively toward favorable ones, they would soon die out. Higher animals have a well-developed nervous system for receiving stimuli and responding to them. However, plants and lower animals are able to do this, too.

Plant behavior. Most plant behavior consists of a turning or growing away from or toward a particular stimulus. This type of behavior is known as a *tropism.* When the tropism is in the direction of the stimulus, it is said to be *positive;* when it is away from the stimulus, it is *negative.*

Phototropism. Plants grow toward the light. This is known as *positive phototropism.* The leaves of a geranium plant on a window sill are seen to face the window. If the plant is turned around, the leaves will grow toward the light again. A simple demonstration to show this positive phototropism:

1. Place a flower pot containing a bean seedling that has just begun to germinate, in the right-hand corner of a tall box that is entirely enclosed, except for a small opening in the upper left corner. Next to it, as a control, set up another plant in the same way but in a box that has no opening. Water both plants frequently.

2. After a few weeks, it will be observed that the first plant is growing in the direction of the light from the opening, and will grow out through it. The control plant will grow straight up.

Geotropism. Different parts of a plant may respond differently to the same stimulus. Thus the stem and leaves will grow upward, away from gravity; they show *negative geotropism.* On the other hand, the roots will grow downward, toward gravity; they show *positive geotropism.* A demonstration to show this can be performed as follows:

1. Germinate some bean seedlings in moist toweling paper. When the roots are about two inches long, arrange three of the seedlings on a strip of moist absorbent cotton which is then put between two panes of glass. Keep the set-up in place with two rubber bands, and stand it upright in a vertical position in a pan containing an inch of water, to keep the seedlings moist. Place it in the dark so that it will not be affected by the light.

2. The three seedlings should be arranged in this manner: (A) root pointing up and stem down; (B) root and stem in a horizontal position; (C) root pointing down, with stem up.

3. After two or more days it will be observed that: (A) the root has curved around and is growing downward, while the stem has curved around and is growing upward. (B) The root has begun to grow downward, and the stem has begun to grow upward; (C) The root is continuing to grow downward and the stem upward.

Hydrotropism. Roots show positive *hydrotropism,* and grow in the direction of water. This tropism sometimes causes complications in streets where roots of poplar trees grow into drain pipes and plug them up. A demonstration to show positive hydrotropism:

1. Germinate a dozen bean seedlings in moist toweling paper. When the roots are about an inch or two long, place them with the roots pointing downward, at intervals against the glass of a dry aquarium containing dry sawdust. Put a large sponge against the glass at the far end of the aquarium.

2. Water the sponge liberally so that it is the only source of moisture available to the plants.

3. Within a week, it will be seen that the roots of all the plants are growing toward the end of the aquarium containing the wet sponge.

Thigmotropism. The growing stem of a climbing plant such as a bean or grape shows a positive reaction to contact, as it coils around a supporting stick. This is known as *positive thigmotropism.*

Auxins and tropisms. The plant hormones known as *auxins* are formed at the growing tip and are transported downward in the stem. They stimulate the cells to elongate, resulting in growth. When auxins are distributed equally on all sides of the stem, the plant grows straight. It has been found that auxins accumulate in higher concentrations on the shaded side of a stem. The increased growth of the cells on this side makes the stem grow toward the light, resulting in positive phototropism.

Our knowledge of auxins also helps explain geotropism. When a young plant is blown over on its side so that it is in a horizontal position, its stem soon turns upward. The reason for this is that auxins accumulate on the lower side of the stem, under the influence of gravity. They cause the cells there to elongate, resulting in a bending upward. On the other hand, auxins have the reverse effect on root cells. When a root is in a horizontal position, auxins accumulate on the lower surface. Here, they inhibit the growth of the cells. The roots then bend downward, resulting in positive geotropism.

Other plant responses. The "sensitive plant" (*mimosa*) has many leaflets on its stems. If a part of it is touched, these leaflets can be seen to fold up along a stem. The petiole of the leaf will droop if the stimulus is a strong one. The heat of a lighted match held near the end of a leaf produces the same results. These movements are thought to result from rapid changes in the turgidity of the cells.

The carnivorous plants also show rather rapid responses. The leaves of the *Venus-flytrap* fold over to trap an insect that has touched any of the three hairs on one of the blades. The *sundew* has numerous slender stalks with sticky glands at their ends, that bend over to trap an insect that is located on the center of the leaf. In the case of the *bladder-*

wort, a group of sensitive filaments at the entrance to each bladder stimulates the nearby entrance to open suddenly, trapping the small water animal that touched them.

Behavior of protozoa. The ameba, which is a simple blob of protoplasm, shows the property of irritability when it reacts to a strong beam of light shining on one end; it moves in the opposite direction. If it comes in contact with a particle of food, it reacts positively, and engulfs it. If, on the other hand, it comes in contact with an irritating substance such as a grain of salt, it flows away from it.

Paramecium responds to a weak acid such as mild vinegar by moving toward it. It moves away from salt and towards sugar. It moves toward a warm region, but away from one that is hot. If it bumps into an object, it reverses its movement and then comes forward again at a different angle. If a drop of blue ink is placed on a slide containing paramecia, they are killed by it; but first they discharge their long, hairlike trichocysts. It is thought that trichocysts may normally help repel small animals that annoy the paramecium. A demonstration to show its response to a mild electric current may be performed as follows:

1. Place a culture of paramecia in a U-tube. Insert a wire into each arm of the tube and connect it with the terminals of a dry cell.
2. In a few seconds the paramecia will be seen by the naked eye moving as tiny white specks toward the arm of the tube containing the wire connected to the negative terminal. Soon, practically all of the organisms will be collected in that area.
3. If the terminals are now reversed so that the current flows in the opposite direction, the paramecia will be seen to migrate to the other arm of the tube where the negative terminal is located.

The form of behavior exhibited by paramecium and other animals, in which they move in a particular direction in response to a stimulus is known as *taxis.*

Vorticella is a bell-shaped protozoan on a slender stalk. The stalk contracts rapidly when another animal touches it. Euglena has a sensitive eyespot which attracts it to light. If a beam of light is directed to one end of a culture dish containing euglena, they will all move toward the light, where they can carry on photosynthesis with the aid of their chloroplasts.

Behavior of other lower animals. The positive phototaxis of a moth is well known. Other insects are similarly attracted to light, as can be seen on a summer evening. Fish in a stream of water align themselves facing into the current; this is an example of positive rheotaxis. Earthworms move away from the light into a shaded or dark place. They also move toward a moist area from a dry one. If *hydra* is touched, it contracts to a compact mass, little larger than a pinhead.

CHAPTER REVIEW

Select the correct choice in each of the following statements.

1. The characteristic of protoplasm that enables it to respond to stimuli is known as (A) absorption (B) ingestion (C) digestion (D) irritability (E) excretion

2. The response of a plant stem in growing upward in the dark is known as (A) positive geotropism (B) negative geotropism (C) positive phototropism (D) negative phototropism (E) negative photosynthesis

3. The response of a plant to contact is known as (A) chemotropism (B) geotropism (C) thigmotropism (D) rheotropism (E) hydrotropism

4. The action of the roots of a tree in growing into a drain pipe is an example of (A) positive hydrotropism (B) negative hydrotropism (C) positive geotropism (D) negative geotropism (E) negative rheotropism

5. The growth of a leaf toward the light results from (A) a greater concentration of auxin on the shaded side (B) a greater concentration of auxin on the lighted side (C) an equal concentration of auxin on all sides of a stem (D) a negative phototropism (E) a positive geotropism

6. All of the following plants show rather rapid movements except the (A) sensitive plant (B) Venus's flytrap (C) bladderwort (D) sunflower (E) sundew

7. An animal that shows positive phototropism is (A) earthworm (B) hydra (C) vorticella (D) ameba (E) moth

8. The growth of a bean vine around a stake is an example of (A) geotropism (B) galvanotropism (C) thigmotropism (D) chemotropism (E) phototropism

9. An earthworm will move from a dry area to a moist area; this is an example of (A) phototropism (B) geotropism (C) hydrotropism (D) rheotropism (E) thigmotropism

10. The upward growth of a stem may result from (A) positive phototropism and negative geotropism (B) positive phototropism and positive geotropism (C) negative phototropism and negative geotropism (D) negative phototropism and positive geotropism (E) positive phototropism and negative thigmotropism

ANSWER KEY

1-D	3-C	5-A	7-E	9-C
2-B	4-A	6-D	8-C	10-A

13 Nervous system

The nervous system of man is a complex communication system that branches throughout his billions of cells, receiving and sending messages, and directing his numerous voluntary and involuntary activities. In addition, man has the most highly developed brain in the animal kingdom, giving him the unique ability to think, reason and create. The nervous system is organized into several parts: the central nervous system, consisting of the brain, spinal cord, and nerves; the autonomic nervous system; and the sense organs.

Neurons. The nervous system is composed of neurons, or nerve cells, which are specialized to possess the property of irritability. There are three chief types of neurons:

1. Sensory (afferent) neuron—transmits impulses from the various sense organs (eye, ear, skin, etc.) to the brain or spinal cord.
2. Motor (efferent) neuron—transmits impulses from the brain or spinal cord to muscles, causing them to contract, or to glands, stimulating them to secrete.
3. Associative (connective) neuron—makes connections between sensory, motor, and other associative neurons; it is found in the brain and spinal cord.

Structure of a neuron. Neurons are quite different in appearance from the other cells of the body. A motor neuron has a cell body, or cyton, containing the nucleus and most of cytoplasm. Spreading out from it are many branched, threadlike *dendrites.* On one side there is a very long extension, the *axon,* which may be several feet long, even though it is microscopic. There is a white fatty covering, the *axon sheath* around the axon, which insulates it. The end of the axon branches into an end brush which lies in a muscle or a gland. Although the other two types of neurons are somewhat different in their appearance from a motor neuron, they are basically alike in their structure. Axons of thousands of neurons are arranged in bundles called *nerves,* as they leave the brain or spinal cord. At this point the nerves are quite thick. They become thinner as they branch out to the various parts of the body, very much as the wires in a thick telephone cable spread out in various directions.

Nerve impulse. Neurons have a high degree of irritability. They transmit messages, or impulses, along their length, and from one neuron to another, at the rate of over 300 feet a second. Their transmission is accompanied by the use of oxygen and the excretion of carbon dioxide. There is a change in the balance of sodium and potassium ions along the length of the neuron. Heat energy is released. In addition to this chemical reaction, there is also an electrical charge which can be measured. The record of electrical impulses in the brain is called an *electroencephalograph* (EEG); the electrical impulses in the heart can be studied in an *electrocardiogram* (ECG). A type of chemical stimulant (neurohumor) is secreted at the point where the end brushes transmit an impulse into a muscle, making it contract. One of these neurohumors is called *acetylcholine.* Acetylcholine is also produced along the length of the axon, and is thought to play a part in the transmissions of an impulse.

Synapse. Neurons are not connected with each other, like the wires in an electrical circuit. Instead, the end of an axon lies very close to the dendrites of another neuron, forming a branched network in which the fibers do not touch, but are almost connected. This area is called a *synapse.* The direction of flow of an impulse is always from the axon of one neuron across a synapse into the dendrites of another neuron. Acetylcholine is secreted by the end brush during the passage of an impulse across the synapse. When we are first learning a new skill, such as typewriting, impulses pass with difficulty across the many synapses involved in the new nerve pathways. As the habit becomes established, these impulses cross over with ease, with resulting facility and speed in typewriting.

Reflex arc. One of the pathways through the nervous system may be illustrated by tracing the impulses involved in a simple reflex such as suddenly pulling your finger back when it touches a flame. (1) First the sensation is received in the sense organ, the skin, where there are numerous nerve endings. (2) The impulse that is generated there travels along the sensory neuron to the spinal cord. (3) In the spinal cord, the impulse crosses a synapse and enters an associative neuron. (4) The impulse passes through this neuron, crosses a synapse, and goes onto the dendrites of a motor neuron. (5) It then travels out of the spinal cord along the axon to the end brush, where it passes on to the muscle fibers of the arm, making them contract, to pull the finger away from the flame. The pathway of the impulse involved in this reflex act is known as a *reflex arc*.

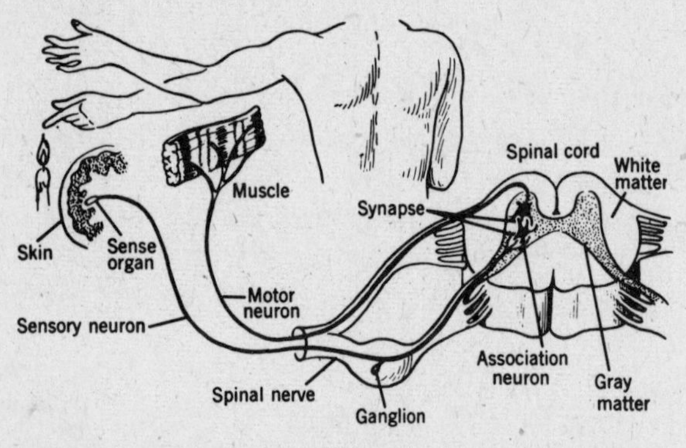

The Reflex Arc

During an actual reflex, many more neurons of all three types, as well as many muscle fibers are involved. It takes only a moment for the arm to be withdrawn quickly and the finger saved from the flame. There was no thought involved in this reflex act. It was centered in the spinal cord. The individual becomes aware of the incident when the associative neurons of the spinal cord send impulses along each other up to the brain. Many neurons now become involved, as there is thought about it, and a sensation of pain. It is interesting to note that although countless neurons are closely located next to each other, the impulses take just the right path.

Spinal cord. The spinal cord runs down the length of the back from the brain, and is enclosed in the vertebrae, the bones of the backbone. There are 31 pairs of nerves that leave the spinal cord and branch out to the body. One of them, the sciatic nerve which runs into the leg, is the largest in the body and is as thick as a pencil. The center of the spinal cord has a butterfly-shaped mass of *gray matter,* which contains the cytons of the neurons. Surrounding it is the *white matter* in which the axons are located. One of the functions of the spinal cord is to serve as the pathway for impulses from the brain to the various parts of the body and back. Another function, as was mentioned in the description of the reflex arc, is to serve as a center of reflex actions. A demonstration to show this latter function may be performed with a frog:

1. Pith the frog painlessly by applying a dissecting needle through the back of its neck into the brain, and moving it about to destroy the brain tissue. The spinal cord is left intact.

2. With the frog limp, place it in a flat position on a table. Pinch the large toe with a pair of forceps, and notice the kicking reflex.

3. Straighten the leg out. Apply a small square of paper that was dipped into acetic acid to the back of the thigh. Notice the reflex movements of the leg as though an attempt is being made to kick the irritating paper off.

4. These two reflexes take place without the presence of a brain. They can also be shown to occur in a living frog which is handled very gently.

The brain. The brain of man is the largest and most highly developed of all the animals. It is protected by the skull, or cranium, and lies enclosed by a layer of fluid. It is covered with three sets of membranes, the *meninges.* These also cover the spinal cord. When they become infected, the disease meningitis results. There are twelve pairs of nerves, the cranial nerves, that branch out from the brain. They include the optic nerve, which brings sight impulses to the brain from the retina of the eye; the auditory nerve which transmits sound sensations from the ears; and the olfactory nerve which carries sensations of smell from the nose.

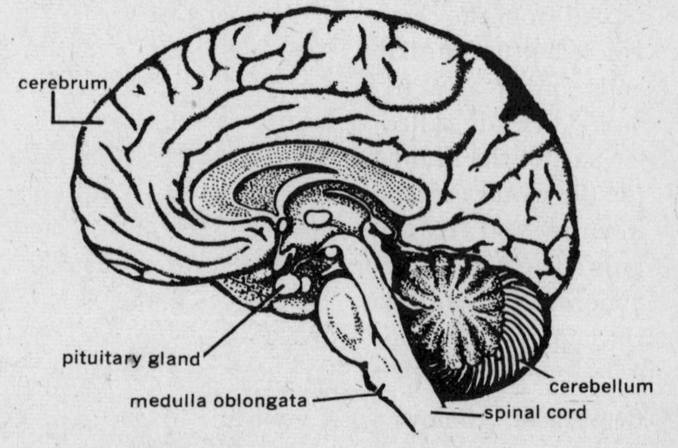

The number of neurons in the brain is fixed at birth, and does not change. However, there is an increase in the number of dendrites and neuron connections as an individual grows. The brain consists of three main areas: (1) cerebrum, (2) cerebellum, and (3) medulla. Their functions have been learned from experiments with animals in which parts of the brain have been removed, and from medical records of illness and injury to various areas of the brain.

Cerebrum. This is the largest part of the human brain. It has two halves, known as *cerebral hemispheres.* The surface has many folds, or *convolutions,* which increase the volume of the cerebrum. The outer part, the cortex, contains the *gray matter,* which is made up largely of cytons. The *white matter* is within this layer, and it contains the nerve fibers, or axons. This arrangement is the reverse of the spinal cord where the white matter is on the outside of the gray matter.

The cerebrum is the seat of intelligence and thought. In some way that is not understood, it gives us the ability to learn, reason and remember; it helps us create great music, literature, art and scientific theories. Various areas of the cerebrum have been mapped. The *sensory* areas are connected by the cranial nerves with the sense organs (eye, ear, tongue, etc.), and interpret the stimuli they receive from them. Thus, we "see" with the part of the cerebrum located in the back of the head. A sharp blow there can render a person temporarily or permanently blind, although the eyes themselves may be uninjured. There are also *motor areas* that control the voluntary muscles of the arms, legs, face, and the trunk. The nerves to these areas cross over in the brain, so that the left side of the cerebrum sends impulses to the right side of the body, and vice versa. When a person suffers a case of apoplexy, or a stroke, paralysis often occurs, because of damage to a part of the motor area resulting from either a blood clot or the bursting of a blood vessel.

Electrical impulses called *brain waves* can be recorded from the cerebral cortex. In a person who is awake but not thinking hard, continuous waves can be recorded at a frequency of about 10 per second. These are called *alpha* waves. When a particular part of the brain becomes very active, for example the motor area, additional *beta* waves of higher intensity and frequency are formed. During sleep, only a few straggly *delta* waves are recorded. Various abnormalities of the brain are being studied through their brain-wave patterns.

Cerebellum. Below the cerebrum, in the rear of the cranium, is the smaller cerebellum. It coordinates the movement of the voluntary muscles so that they work together. The act of picking up a piece of candy with the fingers and putting it in the mouth requires the coordination of many muscles, all working together. The control of the body's balance is also centered in the cerebellum.

Medulla. The part of the brain located at the top of the spinal cord is the medulla. It is the center for breathing and heartbeat. When it is stimulated by the presence of increased amounts of carbon dioxide in the blood, it sends impulses to the diaphragm and chest muscles which speed up the rate of breathing. The medulla also serves as the area through which impulses travel from the spinal cord into the brain, and back. Certain reflexes, such as blinking, sneezing, swallowing, and the secretion of salivary juice, are centered in the medulla.

Autonomic nervous system. We are usually not aware of the autonomic nervous system, because it controls the internal, involuntary functions. It consists of masses of cytons called *ganglia* which are located on either side of the backbone. This chain of ganglia is connected by nerves with the spinal cord on the one side and the internal organs on the other. There are also nerves connecting them with a cluster of ganglia called a *plexus* in several parts of the body, which serve as local centers of control. Thus, the *solar plexus* is located near the stomach; others are located at the heart, in the lower part of the abdomen, and the side of the head.

The involuntary actions controlled by this system of ganglia and plexuses include heartbeat, secretion of digestive juices, size of the opening of the blood vessels, sweating rate, peristalsis, and size of the pupil of the eye. Emotions such as anger, fear, and joy, which stimulate the production of adrenin, act on the autonomic nervous system first, which in turn sends impulses to the adrenal glands, causing them to secrete. There are two sets of nerves to the inner organs, one stimulating them to action, the other causing them to relax. Thus, peristalsis of the digestive tract is stimulated by certain nerves of the autonomic system. The end brushes of the motor neurons secrete acetylcholine which has a stimulating effect on the smooth muscles. There are also nerves whose end brushes secrete a substance like adrenin, which inhibits and slows down peristalsis. The secretions of the glands, contractions and relaxations of the heart, the size of the bronchioles in the lungs, the extraction of urine by the kidneys, and the other internal activities are similarly regulated by this balance between the nerves of the autonomic nervous system.

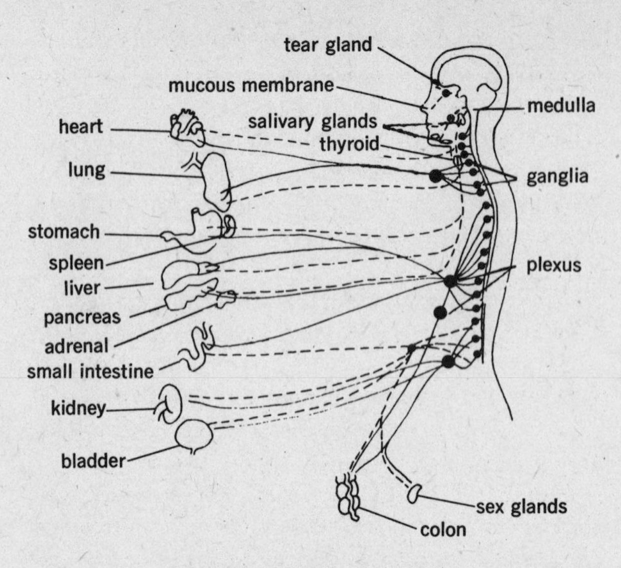

Autonomic Nervous System

Sense organs. The sense organs are specialized to receive impulses from the outside. For this reason, they are known as *receptors.* They are sensitive to a specific type of stimulus. The sense organs are: (1) the eyes, which are sensitive to light; (2) the ears, which are sensitive to sound; (3) the olfactory nerves in the nose, for smelling; (4) the taste buds in the tongue, for tasting; (5) the receptors for pressure, pain, heat and cold, in the skin. There are also internal receptors such as those that give us the sensation of hunger when the stomach is empty, and a sense of balance in the muscles.

The eye. The eye lies enclosed in a *socket* of the skull, and is protected by the movable *eyelids.* The eyelids close at intervals to keep the eye moistened. Tears are produced by tear glands, and serve not only to moisten the eye but also to keep bacteria out, with the aid of the substance *lysozyme.* The eyeballs are moved by three pairs of muscles. The outside of the eyeball is covered with a tough white covering, the *sclerotic coat.* In the front of the eye, this coat is transparent and is called the *cornea.* Beneath the sclerotic coat is the *choroid* coat, made of black tissue which absorbs all light rays except those that fall on the retina; this layer also contains a rich supply of blood.

The path of light rays into the eye is similar to the entrance of light in a camera. Light passes through the cornea and enters the small opening called the *pupil.* The size of the pupil is controlled by the *iris,* a diaphragm that automatically opens wide in a dim light, and contracts to a small opening in bright light. The iris contains pigments that give a person brown or blue eyes. From the pupil, the light enters the crystalline *lens.* There is a liquid, *aqueous humor,* between the lens and the cornea. The lens focuses the light through the large part of the eyeball which contains the clear liquid, *vitreous humor,* and which helps give shape to the eyeball. The light then falls on the sensitive layer, the *retina.* Here, there are two types of receptors, the *rods* and the *cones.* The rods provide vision in dim light; they contain the light-sensitive substance, *visual purple,* which requires vitamin A for its production. The cones help provide color vision. The *optic nerve* contains fibers from the rods and cones which carry the light impulses to the *cerebrum.* Here, the *visual area* interprets the sensations received from both eyes at the same time; it also corrects the position of the images received, because when they fall on the retina they are upside down.

Eye defects. The lens of the eye automatically changes its curvature when a person looks away from a printed page to a distant object. This is accomplished with the aid of small muscles at both ends that lengthen or shorten the lens, and so provide a sharp focus on the retina. When a person has *myopia,* or is nearsighted, the lens focuses the image in front of the retina; distant objects are seen vaguely. This condition can be corrected by wearing glasses that are *concave.* In the farsighted condition (*hyperopia*) the image is focused behind the retina; the details of nearby objects are not sharp. This condition becomes especially prevalent among older people as the lenses of their eyes lose their elasticity. *Convex* glasses correct the condition. In *astigmatism,* the curvature of the lens is not even, resulting in a distorted image on the retina. The eyeglasses in this case have an uneven thickness, to provide the appropriate correction.

Color blindness is an inherited condition in which there is an inability to distinguish certain colors, such as red or green; instead, they appear gray. The condition is probably due to a defect in the cones of the retina. The details of this type of inheritance will be discussed in chapter 24.

The ear. The ear has three parts: (1) the external ear, (2) the middle ear, and (3) the inner ear.

Sound waves are received in this way: (1) They enter the external ear and pass through the *ear canal,* at the end of which the *eardrum* is located. (2) The vibrations are transmitted by it to the parts of the middle ear. Here, there are three small bones, the *hammer, anvil,* and *stirrup.* (3) They pass the vibrations along to the membrane of the inner ear. This chamber contains the *cochlea,* a snail-shaped coiled structure, in which there is a liquid that receives the vibrations. (4) The movement of this liquid stimulates the nerve endings of the *auditory nerve,* which carries the impulse to the *auditory area* of the cerebrum.

Other structures of the ear. The *Eustachian* tube connects the middle ear with the throat. It helps equalize the air pressure on the inside of the eardrum with that of the outside. When we go up rapidly in the elevator of a tall building, we can feel the pressure on the eardrum. By yawning, we are able to open the end of the Eustachian tube in the throat, and to equalize this pressure.

The inner ear contains, in addition to the cochlea, the three *semicircular canals.* These canals are at right angles to each other. They are filled with liquid, and have nerve endings that are connected with the cerebellum. Along with receptors in the skeletal muscles, they give us a sense of balance. Motion sickness in air, on land, or on sea results from a disturbance of equilibrium in this part of the ear. The drug, Dramamine, helps to overcome the condition.

The nose. The receptors of the *olfactory nerves* on the inside lining of the nose receive the stimuli of smell. The impulses are then transmitted to the cerebrum. Some animals, such as the dog, rely heavily on the sense of smell for information about their surroundings.

The tongue. The taste buds which receive stimuli for the four taste sensations of sweet, sour, bitter and salty, are located in different parts of the tongue. The senses of taste are related to that of smell. When we have a bad head cold, or hold our nose, taste is affected.

The skin. Different parts of the skin have different receptors for pressure, pain, heat and cold. The finger tips are more sensitive to touch than the back of the hand, because they contain more pressure receptors. All of the receptors send impulses over sensory nerves to specific areas of the cerebrum where they are interpreted.

Nervous system in lower animals. Protozoa, of course, do not have a nervous system. Their protoplasm possesses the property of irritability, and enables them to respond to stimuli in their environment. Some protozoa, such as *paramecium,* have been found to have a *neuromotor* mechanism, consisting of fine threads of protoplasm which are connected to the base of the cilia. They are believed to regulate ciliary movement; when they are cut experimentally, certain groups of cilia stop beating.

Hydra. This animal has a nerve net, a network of nerve cells extending throughout its body. There is no definite pathway for impulses. A strong stimulus may be spread in all directions, to make the animal contract. Stimuli are received by *sensory* cells, which are sensitive to touch or to dissolved substances in the water. They are located in both of the layers of cells, and send impulses to the nerve cells of the network, which are located at the base of the ectoderm layer. The impulses are transmitted either to muscle cells which contract, or to gland cells which secrete. The nerve net also coordinates the various activities of the tentacles in grasping prey, and in swallowing it.

Earthworm. In contrast to hydra, which has no central nervous system, the earthworm has a large *ganglion* or *brain* above the pharynx. It is connected by nerves to another large ganglion below the pharynx, from which a double ventral *nerve cord* extends back to the rear of the animal. In each segment, this nerve cord enlarges to form a double ganglion which is connected with all parts of the segment by nerves. There are no eyes, ears, or nose. However, there are *sensory* cells in the skin that receive impulses of light and taste, and transmit them to the ganglia, which in turn stimulate muscles or glands. The rate of transmission of an impulse is slow, only about five feet per second, compared to over 300 feet per second in man.

Insects. In insects, there is a *brain* located between the eyes, above the esophagus. A pair of nerves connects it with a ventral double *nerve cord* that extends through the body. Some of the *ganglia* are fused, so that there are two ganglia in the head, three in the thorax, and five in the abdomen. The sense organs include a pair of *compound eyes,* three *simple eyes,* ears, and a pair of *antennae.* In addition, the mouth parts have sensory hairs for taste. There are also taste receptors on the feet. The compound eyes have numerous lenses, and are sensitive to color. The simple eyes are sensitive to light, but do not form images. The antennae are organs of touch, smell, and taste. The ears are simple sound-receiving organs, whose location varies in different insects; in the grasshopper, they are on the sides of the abdomen next to the third legs; in the katydid, they are on the upper part of the first legs. The brain serves chiefly to receive impulses from the sense organs in the head. It does not seem to control or coordinate the other organs of the body. Highly complex patterns of behavior are shown by the social insects, such as bees and ants. However, they are inborn and instinctive, rather than based on intelligent self-direction or thought.

Fish. Like all vertebrates, fish have a *brain* located in the skull, and a spinal cord enclosed in the backbone. Beginning with the front end, the parts of the brain are: a pair of *olfactory lobes;* a small double *cerebrum;* two *optic lobes,* which are the largest part of the brain; *cerebellum; medulla.* The *eyes* have no eyelids, being kept moist by the water; there is a large pupil for allowing a large amount of light to enter the eyes. The eyes are connected with the optic lobes. The nostrils are not used for breathing; they open into small pouches which have nerve endings leading to the olfactory lobes, and provide the sense of smell. Fish cannot hear, although they have an inner ear for balance. The *lateral line,* which runs along the outside of the body from the head to the tail, is believed to be sensitive to stimuli of water currents pressure, and transmits impulses to the central nervous system through nerves.

Frog. The brain of a frog shows a larger *cerebrum* than that of a fish. It also has *olfactory lobes, optic lobes, cerebellum,* and *medulla.* The spinal cord extends through the backbone. Cranial and spinal nerves extend to the various parts of the body. The *autonomic nervous system* controls the internal organs. The *eyes* are large and bulge out of the head; they have a transparent eyelid in addition to a regular eyelid. The *ears* consist of a pair of eardrums in back of the eyes.

Other vertebrate brains. The brain of birds has a relatively large pair of optic lobes, for its well-developed sense of sight; and a large cerebellum, which helps coordinate its movement and balance in flight. The cerebrum becomes progressively larger in fish, amphibia, reptiles, birds and mammals. In mammals, it shows convolutions as well. Finally, in the primates it is larger than in any other order. Man has the largest cerebrum of all, accounting for his superior intelligence.

Comparison of the Nervous and Endocrine Systems. The nervous and endocrine systems are similar in that they both (a) secrete chemical messengers, and (b) play a major role in maintaining homeostasis. The nervous system is different in these ways: (a) The responses of nerves are much more rapid than those caused by endocrines; (b) The responses of nerves do not last as long as those caused by endocrines. (c) Nerve impulses are transmitted along neurons; hormones are carried in the plasma of the transport system.

SUMMARY OF THE NERVOUS SYSTEM

STRUCTURE		FUNCTION
Brain	1. Cerebrum	Thought; reasoning; memory; voluntary activities. Sensory areas for sensation. Motor areas for voluntary movement.
	2. Cerebellum	Coordination of muscles; balance
	3. Medulla	Controls breathing, heartbeat and certain reflexes (blinking, sneezing, swallowing, salivary secretion)
Spinal cord		Pathway for impulses to and from brain; center of reflex actions
Autonomic nervous system		Regulates internal activities—heartbeat, peristalsis, kidney action, glandular secretion, size of bronchioles and blood vessels.

CHAPTER REVIEW

Select the correct choice in each of the following statements.

1. A nerve cell is known as a (n) (A) axon (B) axon sheath (C) neuron (D) trichocyst (E) nerve

2. A part of a nerve cell that is not found in other tissue cells is the (A) nuclear membrane (B) matrix (C) chromatin (D) dendrite (E) xylem

3. A sense organ is connected with the spinal cord by a (n) (A) sensory neuron (B) motor neuron (C) efferent neuron (D) associative neuron (E) connective neuron

4. A chemical produced in an axon is called (A) acetylcholine (B) aster (C) auxin (D) EEG (E) ECG

5. A bridge between neurons is called a (n) (A) dendrite (B) synapse (C) terminal branch (D) end brush (E) axon sheath

6. A mass of cytons is called a (A) cranial nerve (B) synapse (C) ganglion (D) motor area (E) nerve fiber

7. The pathway of a reflex arc involves the following structures in this correct order: (1-spinal cord; 2-sense organ; 3-motor neuron; 4-sensory neuron; 5-muscle) (A) 4–2–1–3–5 (B) 3–2–1–4–5 (C) 3–1–2–4–5 (D) 2–4–1–5–3 (E) 2–4–1 3–5

8. The spinal cord serves as the center of (A) subconscious thought (B) reflex actions (C) habits (D) tropisms (E) synapses

9. All of the following functions are controlled by the cerebrum except (A) intelligence (B)

memory (C) sensations (D) digestion (E) voluntary movement

10. Muscular coordination is controlled by the (A) spinal cord (B) convolutions (C) cortex (D) cerebellum (E) cerebrum

11. The breathing rate is controlled by the (A) lungs (B) medulla (C) cerebrum (D) cerebellum (E) sensory area

12. The autonomic nervous system controls all of the following activities except (A) secretion of digestive juice (B) peristalsis (C) sweating (D) thought (E) diameter of blood vessels

13. All of the following are examples of receptors except (A) eyes (B) ears (C) cyton (D) nose (E) tongue

14. Visual purple is found in the (A) retina (B) cerebellum (C) cornea (D) sclerotic coat (E) lens

15. The sense of balance is maintained with the help of the (A) cochlea (B) hammer (C) anvil (D) stirrup (E) semicircular canals

ANSWER KEY

1-C	4-A	7-E	10-D	13-C
2-D	5-B	8-B	11-B	14-A
3-A	6-C	9-D	12-D	15-E

14 Behavior of higher animals

Lower animals such as protozoa, hydra and the earthworm, respond to stimuli *automatically*. That is, they do not think about them, nor are they taught about them. This type of behavior is said to be *inborn,* and *unlearned,* and is called a *reflex*. It is like a tropism in these respects. Higher animals also show reflex behavior and in some cases *instinctive* behavior, which like reflexes is inborn, automatic, and unlearned.

Animals with a well-developed nervous system show additional types of behavior which are acquired after birth, and are therefore said to be *learned*. Examples are: *conditioned responses,* and *trial-and-error learning*. Humans are capable of additional higher forms of acquired behavior, such as *habit formation* and *intelligent behavior*.

Reflex. If you accidentally touch a flame, your finger is automatically pulled back. The mechanism for this act was explained in the preceding chapter as follows: A sensory neuron carries the impulse from the receptor, at the end of your finger, to the spinal cord; here an associative neuron receives it; and sends it over a motor neuron to a muscle in your arm, which contracts, and pulls your finger away. There is no thought involved in the reaction. It is centered in the spinal cord. A moment later, you are aware of it, because impulses are sent up the spinal cord to the brain. Fortunately you do not have to think about such stimuli. The response is inborn and is a factor in your survival. Such a reflex protects you.

We show other reflexes, some of which may be centered in the brain: The pupil of the eye expands in dim light, and contracts in bright light; we start at a sudden noise; we blink when a cinder flies into the eye—and our eye tears; we sneeze when irritating dust enters the nose; the knee jerks when the knee cap receives a blow; we cough if some water accidentally "goes down the wrong pipe." A demonstration to show the automatic nature of the blinking reflex, for example, may be performed as follows:

1. Have someone hold a pane of glass in front of his face. Instruct him to look at the wad of paper you are about to throw at him.

2. Throw the paper at his eyes. Repeat several times. He will blink each time, even though the glass protects his face.

A newborn baby is a mass of reflexes. It shows many reflexes: It sucks a nipple, grasps a finger placed in its palm, cries, coughs, and starts at the sound of a sudden noise. As it grows older, the baby will change some of these inborn reflexes and will replace them with various types of learned behavior which it will acquire.

Instinct. A spider spins a web perfectly the first time. This is another example of an inborn, automatic unlearned act. This type of behavior is more complicated than a simple reflex. It consists of a series of reflexes, each one acting as a trigger for the next, and is known as an instinct. In this complicated type of unlearned behavior, the first reflex is probably the secretion of a silky thread from its special glands. This leads to the attachment of the thread between two twigs. Another thread is then spun and attached—and so on, until the web is finished, each step leading to the next reflex in the series.

Other living things may also act by instinctive behavior, including those with a well-developed nervous system. Some examples: The caterpillar of a moth spins a cocoon during a certain stage in its life history. Honey bees store honey in the cells of a hive which are constructed with great precision. The mud-dauber wasp collects mud and builds a nest containing chambers for its developing young. Birds migrate many miles each fall to spend the winter in a warm climate, and then return in the spring. Each species of bird builds a particular type of nest, selecting from a variety of materials, such as twigs, thread, feathers, moss, mud and hair. Beavers cut down trees by gnawing through their trunks, and then use the branches to build a complicated dam.

In all these cases, there are inborn pathways over which impulses travel in a series of reflex acts. Sometimes the instincts appears so complicated and well-directed that we are tempted to consider them as examples of intelligent, purposive behavior. But this is not so. The organization of a bee hive with its workers, soldiers, drones and queen, all specialized to perform specific jobs, is merely another example of the complex, inborn activity known as instinctive behavior.

Conditioned reflex. Inborn behavior, such as a reflex, can be changed or conditioned so that a new or acquired type of behavior results. The Russian physiologist, Ivan Pavlov (1849–1936) did this experimentally with dogs, for the first time. He noticed that saliva flowed when the dogs were being fed. This was a simple reflex; the stimulus was food, the response was the flow of saliva. For a number of days, he rang a bell at each feeding time. After a while, he observed that merely ringing the bell without food being present was sufficient to cause the flow of saliva. In other words, he had substituted a new stimulus to produce the original response. This was now a conditioned reflex; it was acquired or learned. Pavlov found that his dogs would also produce saliva in response to other stimuli, such as the flashing of a light. In fact, the dogs even associated with food the white-coated attendants who fed them, and would produce saliva when they saw them. The conditioned reflex was found to be temporary if it was not accompanied by the feeling of satisfaction that came with eating. When the bell was rung a number of times in the absence of food, less and less saliva was produced, until the secretion stopped altogether.

A **conditioned response** results when a more complex form of behavior is changed. You teach a dog its name by saying "Spotty" each time you feed it. The dog will then come in response to the sound of "Spotty," even if you do not feed it, because you supply the element of satisfaction in another way; you play with it, or take it outside, or pet it. Much of so-called animal training is really a form of conditioning. The animal trainer invariably praises or feeds the animal a snack after a performance, to continue to give it a feeling of satisfaction.

Human beings also show conditioned responses by associating a new stimulus with an original one. In school, if your lunch period follows the French period, you will find yourself getting hungry during the French lesson. If on a certain day the French period occurs earlier, because of an assembly, you may be surprised to find yourself salivating long before lunchtime. In other words, you have associated food with French. Other types of conditioned responses are common. A child will probably develop a fear of mice if its mother screams at the sight of one, and will grow up associating mice with fear; its friend may play with mice without fear, if raised in a different setting. A girl may tremble at the sight of a harmless snake for the same reason. A demonstration to show conditioning in a goldfish may be performed as follows:

1. Feed the fish sparingly, and tap the side of the aquarium with your finger at the same time. Continue to do this over a period of two weeks.

2. At the end of that time tap the aquarium without adding any food. The fish will be seen to rise to the surface of the water. It has associated the tapping with food, and now responds to it.

Trial-and-error learning. A form of learning that depends on trying out several ways of doing things, and then selecting the correct way, is learning by "trial and error." If you have a dozen similar keys and wish to open a closet door, you may try several wrong keys before choosing the correct one. The next day you may have to try over again until you find the right key. If you do this daily, after a while you will be able to select the right key at once. You have learned this by trial and error.

Other examples of trial and error and learning: It takes some time for a hamster or a white rat to learn the correct route through a maze, and to avoid the blind alleys; after a number of days of making mistakes it will go through in the least possible time. A baby learns to eat with a spoon by holding it in various ways, until it learns how to grasp it properly with its fingers. A boy learning to ski must first avoid crossing one ski over another as he moves along; after a number of falls, he begins to keep the skis parallel. Even though you may not know how to play the piano you will probably be able to sound out the tune of "America" by trying out several combinations of notes; if you continue to do this, after a number of trials you will be able to play it without an error. After playing with a simple jigsaw puzzle for several days, a child soon learns to put it together without an error.

A demonstration to illustrate progress in learning by trial and error may be performed in this way:

1. Cut a letter "L" that is one inch wide, out of a card, with the long arm 4 inches long and the short arm 3 inches long. Cut it diagonally into four uneven pieces as shown.

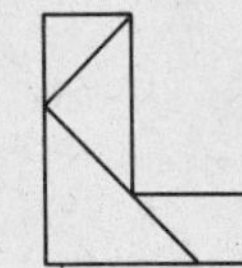

2. Time someone and see how long it takes him to put the four pieces together to form the letter.

3. Shuffle the letters and time him again. Do this a dozen times. Record the results in a simple table and plot them on a graph.

4. It will be seen that this type of learning takes place unevenly. The second trial may take less time than the first; the third, however, may take a little more time than the second; the fourth less time than the third, and so on, until learning is complete and no errors are made.

Habit. A learned pattern of behavior that is so automatic that it requires no thought, and is performed quickly and efficiently, is known as a habit. It is as automatic as a reflex, but it is not inborn. A secretary can typewrite so well that she need not look at the keys of a typewriter. She can work rapidly and correctly because she has established the right nerve pathways for the impulses to travel automatically. How has she done this?

Steps in forming a habit. (1) *Desire.* First, she had a desire to learn the habit. If a person is not interested in learning a skill, it will be difficult to acquire it; witness the slow progress of a reluctant piano pupil. (2) *Repetition.* Practice makes perfect. As she was learning to typewrite, she practiced and repeated the steps many times. During this repetition, the correct nerve pathways were being established for the impulses involved. (3) *Satisfaction.* As in other forms of acquired behavior, there must be a feeling of satisfaction resulting from the habit. She enjoyed this feeling of being able to typewrite successfully, and of doing her job properly.

Examples of other habits: Tying your shoelaces; riding a bicycle; writing; driving an automobile; playing an instrument; brushing your teeth; getting dressed; playing tennis. In these habits, there is no thought involved about the steps involved. They have become automatic.

Advantages of habit formation. A habit is performed quickly and accurately. It is done the same way each time. It also frees your mind to do other things, so that you are more efficient. These advantages may be demonstrated in the habit of writing as follows:

1. Divide a paper into two columns. In the left-hand one, write your name as many times as you can in thirty seconds.

2. Now do the same in the other column, but use your other hand.

3. By comparing your observations, it is apparent that when you were writing in the second column, you found yourself thinking about how each letter is formed, while in the first column, you were thinking about the writing itself. Also note the awkward appearance of the letters, the unevenness of the writing, the slow pace, etc.

Breaking a habit. There are certain habits that we form that we may wish to break, such as nail-biting or smoking. Since the nerve pathways are well-established for these habits, it becomes necessary to break up the pattern that led to them. (1) First of all, you must now have the desire to break the habit. (2) You should concentrate on it so that you do not find yourself unconsciously doing it. (3) You can enjoy the satisfaction of making progress in overcoming it. (4) You can lose the feeling of satisfaction in nail-biting by painting iodine over your nails so that you experience a bitter taste each time you attempt to bite them. (5) Substituting another stimulus is also effective: In smoking, instead of reaching for a cigarette, suck on a piece of candy; for nail-biting in girls, buying a simple manicure set and polishing the nails is effective. (6) In all these steps, it is important to allow no exceptions.

Intelligent behavior. Much of our behavior is based on voluntary activity, and is directed by our thinking. We have the ability to store memories, and to recall them. How this is done is not understood, at present. Apparently, neuron pathways in the cerebral cortex are established, which can be recalled after days, months, and even years. By putting together our experiences, memories, and reasoning ability, we are able to do abstract thinking and to create new ideas. In the trial-and-error learning demonstration of the letter "L," taken up earlier in this chapter, if you try to put the pieces together yourself, you make an interesting observation. You may not do it correctly and quickly the first time, but you will master the puzzle in much less time than the other person. Apparently, while he was putting the pieces together, you were noticing and remembering the correct arrangement. You were indirectly benefiting from his direct experience. When it was your turn, you were able to avoid some of his errors. Compare this with a hamster that watched a fellow hamster learn a maze. No matter how long he watched, it took him just as long to go through it correctly.

Studies in *psychology,* the scientific investigation of behavior, have shown that some of the other higher animals, including mammals and especially primates, are endowed with a certain amount of memory and even intelligent behavior. A dog will remember his young master upon his return from a summer at camp, and will greet him with joy. The behavior of the chimpanzee has been studied extensively by Dr. Yerkes and Dr. Kohler. In one experiment, a chimpanzee in a cage was allowed to become hungry. Some food was placed outside, just beyond his reach. A stick was placed nearby. He used the stick to reach the food. To test his reasoning ability still further, the chimpanzee was confronted with another problem—the food was now moved beyond the length of the stick. A second short stick was made available. After a number of vain attempts to reach the food with each stick, the chimpanzee happened to fit them together to make a long pole. He promptly proceeded to fetch the food without too much difficulty. He thus showed both memory and reasoning ability.

Human intelligence. Psychologists have devised tests that attempt to measure human intelligence. By means of a variety of questions they test such aspects of mental ability as verbal meaning, space perception, reasoning ability, use of numbers, memory, and speed of recognition. The *mental age,* M.A., is obtained from such tests. This is compared with the *chronological age,* C.A., to arrive at the *intelligence quotient,* or I.Q. The following formula is used to find the I.Q.:

$$\text{I.Q.} = \frac{\text{M.A.} \times 100}{\text{C.A.}}$$

Thus, a seventeen-year-old boy (C.A. = 17) who solves the problems of seventeen-year-olds (M.A. = 17), has an I.Q. of 100. If his M.A. is that of nineteen-year-olds, his I.Q. is 112. The range of I.Q. in a large number of unselected individuals will include some above 130, some below 70 and most between 90 and 110. Results of intelligence tests must be interpreted carefully, because the score may be influenced by such factors as cultural background, amount of education, language handicap, and emotional state.

Because he has the most highly developed cerebrum of all the animals, man is superior to all of them. He also has the power of speech for communication; the opposable thumb enables him to write and use tools. He has developed devices by which he can exceed animals, even in physical abilities in which they excel: He can see better than an eagle, using a telescope; he can fly faster and higher than a bird, in an airplane; he can hear a longer distance than a dog, using the telephone; he can exceed the strength of an elephant, with his machines; he can travel faster than a horse, in an automobile; he can travel under water more quickly than a fish, in a submarine; he can overcome a lion, using a gun. In addition, man has unlocked many of the secrets of nature, including the structure and the forces of the atom. The question now confronting mankind is whether man will use this knowledge intelligently for his benefit, or behave like a lower animal that has been given power it cannot control, and let it lead to his downfall.

Biological Clock. Some plants and animals, including man, are controlled by a type of inner biological clock which causes them to adjust regularly over the period of a day. For example, man shows a rhythm in his body temperature. It is highest in the afternoon, and lowest in the early hours of the morning; 98.6°F is only a daily average.

Rodents such as mice and rats, as well as certain insects such as cockroaches, are active at night and sleep during the day. Birds show daytime activity. Certain beach crabs on the ocean shore are active at low tide and stay in their burrows at high tide. Even when these crabs are moved inland to the midwest, they still show the same rhythm.

Among plants, legumes show a type of sleep-movement rhythm. They fold their leaves and lower them at night; during the day, they open and raise them. When such plants are kept indoors under continuous darkness and temperature, they continue with the same daily rhythm of sleep-movement.

SUMMARY OF TYPES OF BEHAVIOR

TYPE	CHARACTERISTICS	EXAMPLES
Tropism	Turning toward or away from a stimulus; inborn; automatic	Positive phototropism—leaves, stem, moth Positive geotropism—roots Negative geotropism—stem, leaves Positive hydrotropism—roots Positive thigmotropism—stems

SUMMARY OF TYPES OF BEHAVIOR (Continued)

TYPE	CHARACTERISTICS	EXAMPLES
Reflex	Inborn, unlearned, automatic; no thought involved	Knee jerk; eye pupil changes in dim and bright light; blinking; sneezing; pulling finger away from flame
Conditioned response	Learned; substitution of one stimulus for another	Dog salivating at bell; dog learning its name; child learning fear of mice; fish responding to tapping on aquarium
Instinct	A series of reflexes; inborn, unlearned, automatic	Spider spinning a web; honey bee specialization; bird migration; bird nest-building; beaver building a dam
Trial-and-error learning	Learned response acquired after many trials in which errors were gradually eliminated	Animal learning a maze; human learning a puzzle
Habit	Learned, automatic responses, developed through (1) desire (2) repetition (3) satisfaction	Typewriting, getting dressed, writing, tying a shoelace, roller-skating, driving a car
Intelligent behavior	Involves memory, reasoning and deliberation	Benefiting from someone else's experience; learning a lesson; building a bird house; making a painting

CHAPTER REVIEW

Select the correct choice in each of the following statements.

1. All of the following are examples of a reflex except (A) sneezing (B) blinking (C) thinking (D) coughing (E) contraction of the pupil of the eye

2. All of the following are characteristics of a reflex except (A) inborn (B) unlearned (C) automatic (D) acquired (E) inherited

3. The spinning of a web by a spider is an example of (A) an instinct (B) a conditioned reflex (C) an automatic, acquired act (D) a habit (E) trial and error learning

4. The scientist who pioneered in research on conditioned reflexes was (A) Schick (B) Banting (C) Pavlov (D) Pasteur (E) Reed

5. A conditioned reflex (A) is inborn (B) is unlearned (C) is acquired (D) occurs whenever saliva flows (E) occurs whenever a bell rings

6. A hamster goes through a maze the first time by (A) instinct (B) trial and error (C) habit (D) reasoning (E) conditioned reflex

7. Playing the piano is an example of a (n) (A) habit (B) inborn activity (C) instinct (D) conditioned reflex (E) reflex

8. All of the following steps are necessary for habit formation except (A) frequent exceptions (B) desire (C) repetition (D) satisfaction (E) practice

9. Working out a mathematics problem is an example of (A) an involuntary act (B) a voluntary act (C) an instinct (D) a tropism (E) reflex

10. The I.Q. is a measure of a person's (A) mental ability (B) mental age (C) chronological age (D) verbal quotient (E) space perception

ANSWER KEY

1-C	3-A	5-C	7-A	9-B
2-D	4-C	6-B	8-A	10-A

15 The habits of tobacco, alcohol and narcotics

The habits of smoking, drinking alcohol, and using narcotics are so different from the usual run of habits that they require special attention because of their pronounced effects on the health and welfare of the body.

Smoking. Boys and girls in their teens are often tempted to smoke. In their desire to appear grown up, they believe that smoking will make them seem more mature, and popular. Television and printed advertisements often help to heighten this impression. Opposition to smoking by parents and other adults has sometimes been based on emotional appeals. There are many facts about the effects of smoking, however, that should be known to anyone interested in the habit.

Tobacco. Tobacco smoke contains nicotine, tars, carbon monoxide, and other substances irritating to the lungs. Nicotine is a poison that is so powerful that about 60 milligrams would cause death if injected directly into the blood. In smoking an average cigarette, an individual absorbs about one or two milligrams of nicotine into his bloodstream, enough to produce certain effects on the body. It causes the blood vessels to constrict, resulting in a rise in blood pressure. The narrowing of the blood vessels leads to a lowering of the temperature in the fingers and toes by as much as 5°F. There is an irregular increase in the heartbeat. The breathing rate is also increased. The appetite is dulled and the process of digestion is delayed. The smoker becomes breathless more easily.

Some of these effects are recognized by athletic coaches, who prohibit smoking among the members of their teams. Doctors advise their patients who have heart trouble or other circulatory ailments to stop smoking. There is a high correlation between heavy cigarette smoking and coronary heart disease. In medical studies on this subject, death rates from coronary heart disease in middle-aged men were found to be from 50 to 150 per cent higher among heavy smokers than among those who do not smoke. When people with certain types of stomach ulcers give up smoking, they become cured; if they then start to smoke again, the ulcers return.

The tars in smoke are strongly suspected as being a cause of lung cancer. Many studies have linked the startling increase in the number of cases of lung cancer in recent years with the tremendous increase in the amount of smoking. Scientists connected with the National Cancer Institute believe that excessive cigarette smoking is one of the causes of lung cancer. A recent study by Dr. Wynder and Dr. Hoffman of Sloan-Kettering Institute for Cancer Research, concluded that the chances of a man over age fifty getting lung cancer were as follows:

1. Does not smoke—one chance in 29,411.

2. Smokes ½ to 1 pack a day—one chance in 1,686.

3. Smokes 1 to 2 packs a day—one chance in 695.

4. Smokes more than 2 packs a day—one chance in 460.

One medical group bulletin states: "Responsible public health officials now recognize almost universally that it's good medicine to tell people to quit smoking if they can, or to cut down on the number of cigarettes smoked, and especially not to start smoking in the first place."

Besides nicotine and tars, tobacco smoke also contains other irritants to the throat, tongue, nose and lungs. Steady smokers often have a persistent hacking cough that may eventually result in a weakening of the lungs and the development of asthma. They may also complain of a chronic husky voice and sore throat.

Dr. Raymond Pearl of Johns Hopkins University made a study of the life span of three groups of men—nonsmokers, moderate smokers, and heavy smokers—and for every 100,000 men in each group at the age of thirty, he concluded:

1. Non-smokers: 66,584 might be expected to live to be sixty years old.

2. Moderate smokers: 61,911 might be expected to live to be sixty years old.

3. Heavy smokers: Only 46,226 might be expected to live to be sixty years old.

With all this knowledge about the undesirable effects of smoking, people who have the habit often try to give it up. They usually have great difficulty doing so. However, those who do succeed invariably express satisfaction that they are through with it. They mention their improved appetite, keener sense of taste, better wind, and relief at being free from, rather than a slave to, a bad habit. Teenagers who want to start smoking should consider all the facts first, and then ask themselves: Is it worth it?

Alcohol. When a person drinks liquor of any type, alcohol is absorbed rapidly in his stomach and small intestine, and is circulated through the body. It has the following effects:

1. It is absorbed in relatively large amounts by the brain and nerves; it is not a stimulant, as is often claimed, but acts as a *depressant* on the nervous system, interfering with reflexes, judgment, and vision.

2. It is readily oxidized by the cells to release energy; however, unlike the nutrients, it does not furnish needed vitamins, proteins and minerals, and may lead to deficiency symptoms resulting from a poor diet.

3. It relaxes the smooth muscles in the blood vessels, enlarging their diameter; this results in a greater flow of blood through them, causing a loss of body heat from the increased flow of blood through the skin, and giving an individual a flushed appearance.

4. In large quantities, alcohol causes loss of muscular coordination, mumbling, loss of memory, and unconsciousness.

Drinking is a serious cause of automobile accidents. A driver who has had even one or two drinks has enough alcohol in his blood to affect the centers in the brain that control judgment, alertness, and coordination. Those who have had more alcohol than this cannot think clearly and act quickly enough, especially at high speeds, to drive safely. Studies have shown that more than one third of automobile accidents occurred among people who had alcohol in their bodies. The National Safety Council warns drivers: "It is unsafe to mix alcohol with gasoline."

Regular heavy drinkers of liquor have less resistance to disease. They are especially vulnerable to pneumonia. Because they neglect proper eating habits, they tend to become undernourished, and show a deficiency of the vitamin B complex. Another serious effect is cirrhosis of the liver. Life insurance company statistics indicate that they have a shorter life span.

Alcoholism is the name of the habit resulting from continued use of alcohol. It is considered to be a difficult disease to cure, since the individual shows an uncontrollable craving for liquor. Most alcoholics are emotionally maladjusted. It is thought that some type of diet deficiency may be responsible for their irrepressible urge to drink. The nationwide organization known as Alcoholics Anonymous (AA) consists of people who have successfully freed themselves of the habit, and who attempt to help others do likewise. The undesirable social effects of alcohol have led a number of states to supervise the sale of liquor carefully, especially to minors. Young people who drink liquor as a stimulus to a good time, face the possibility of serious accidents while driving, or the consequences of forming a degrading lifelong habit.

Drugs. Through the ages, doctors have used narcotics to deaden pain, lull the senses and bring about sleep. Used under their care, these drugs have been very useful. In 1680, the physician Suydenham wrote: "Among the remedies which it has pleased Almighty God to give to man to relieve his sufferings, none is so universal and so efficacious as opium." However, narcotics are sometimes used improperly, inducing habit-formation, and causing the user to become a drug addict. The chief drugs are: opium, heroin, morphine, codeine; marijuana; the barbiturates; amphetamines; and the hallucinogenic drugs.

Opium is derived from the dried juice of the seed capsule of the opium poppy plant, which is largely raised in parts of Asia. In Oriental countries, opium-smoking is the chief way of taking the drug. In this country, it is purified to form heroin, morphine and codeine.

Heroin has the appearance of a white powder. It is sold illegally by dope peddlers ("pushers") who generally dilute it with harmless powder and sell it at enormous profits. Drug addicts usually start by sniffing it, and then find it loses its "kick" unless they "mainline" it, or inject it directly into a vein. In an effort to help "junkies" (heroin addicts) overcome their habit, the narcotic methadone is being substituted under close medical supervision. As the treatment progresses, if the patient on methadone should take heroin, he notices no effect from it.

Morphine is a standard medical drug used by doctors to relieve extreme pain usually caused by accidents. Doctors are aware of its habit-forming properties and use it sparingly.

Codeine is a milder narcotic. It is used in some medicines to relieve severe coughing.

Marijuana is a drug present in the hemp plant. Cigarettes ("reefers") are made from its leaves, and may be the first step in drug addiction, as the user finds that he wants a stronger drug such as heroin to improve on the effects. Although it does not cause addiction, as heroin does, it is nevertheless considered to be psychologically addictive.

The **barbiturates** are synthetic chemicals that are not as strong as heroin but are just as vicious in their drug-addiction powers. The individual who starts out by taking them as sleeping pills may find it necessary to take stronger and stronger doses. Eventually, he may become an addict and suffer from mental depression, emotional instability and types of delirium. Addiction to the barbiturates is even more serious than that of morphine.

The **amphetamines** are known as "pep" pills. They induce a state of alertness, at first, so that the person does not feel the effects of fatigue. Truck drivers who drive at night have been known to take them, to keep awake. However, they require stronger and stronger doses, since the body develops a tolerance to them. Various side effects begin to appear, such as headaches, dizziness, irritability, and decreased ability to concentrate. The person may eventually have hallucinations.

The hallucinogenic drugs include LSD, mescaline and psilocybin. These drugs heighten sensory perception to the point where images and thoughts are weirdly distorted. It is impossible to predict their effects. They are supposed to be mind-expanding, but they can result in temporary or permanent insanity. LSD (lysergic acid diethylamide) has been shown to cause changes in the chromosomes, thus affecting heredity.

A *drug addict* is a person who has developed the habit of taking drugs. He may start by smoking a marijuana cigarette for a thrill. After a few such experiences, as his body gets used to the drug, he feels the need for something stronger, and starts sniffing heroin. This, too, is soon too weak, and he goes on to inject the drug directly in a vein. While under the influence of the drug, he feels a sense of well-being. His problems seem to diminish in importance, and the world seems very pleasant. When the drug has worn off, however, he faces reality again and finds the problems are still there. Worse than that, he soon begins to crave the drug. He must have it to feel at ease again. If he runs out of money, he will steal and rob, if necessary, to buy the drug. He is at the mercy of the dope peddlers. If he tries to drop the habit, he finds he cannot. He experiences "withdrawal" symptoms; he has sharp abdominal and leg pains, he cannot eat or sleep, he feels very chilly, he is very restless, he has nausea and diarrhea, he loses weight. The major purpose of his existence seems to be to obtain narcotics for his daily needs. Ill health, crime, degeneracy, and a low standard of living result. He needs medical help.

The hospital for drug addicts at Lexington, Kentucky, which is operated by the U.S. Public Health Service, tries to bring about a cure by very gradually reducing the amount of the drug taken, until finally no further doses are needed. However, addicts who have been cured, very often return, because the same conditions that led them into the habit are still there when they go home. For this reason, attempts are made to teach hospital addict patients a new trade, as well as a better understanding of themselves and their families. Teen-agers who fail to resist narcotics face the prospect of not only ruining their health but also their future chances of happiness and success.

CHAPTER REVIEW

Select the correct choice in each of the following statements.

1. Tobacco smoke contains (A) caffeine (B) nicotine (C) niacine (D) quinine (E) morphine
2. Smoking causes all of the following effects except (A) constriction of the blood vessels (B) increase in blood pressure (C) increase in breathing rate (D) regular decrease in heart beat (E) lowering of temperature in fingertips
3. There is a high correlation between cigarette smoking and (A) increase in appetite (B) increase in process of digestion (C) coronary heart disease (D) rickets (E) resistance to scurvy
4. Many scientists believe that cigarette smoking leads to (A) increase in pigmentation (B) resistance to colds (C) longer life span (D) baldness (E) lung cancer
5. Alcohol serves as a stimulant to (A) the brain (B) the nerves (C) vision (D) judgment (E) none of these
6. Alcohol is absorbed in relatively large amounts by the (A) brain (B) skin (C) lungs (D) legs (E) eyes
7. Alcoholism is (A) a custom (B) a disease (C) a preventive for pneumonia (D) a food substitute (E) socially desirable for shy teen-agers
8. All of the following are drugs except (A) heroin (B) morphine (C) codeine (D) chloroform (E) marijuana
9. Opium is obtained from the (A) cinchona tree (B) flax plant (C) poppy (D) bloodroot (E) Jack-in-the-pulpit
10. The drug addict who tries to drop the habit experiences (A) withdrawal symptoms (B) craving for liquor (C) sleepiness (D) increase in weight (E) a sense of well-being

ANSWER KEY

1-B	3-C	5-E	7-B	9-C
2-D	4-E	6-A	8-D	10-A

HOW WE FIGHT DISEASE

5

16 Study of bacteria

Bacteria were first discovered in 1683 by the Dutch microscopist, Anton van Leeuwenhoek. By means of his simple, homemade lenses, he was able to study various forms of bacteria and describe their activities.

CHARACTERISTICS OF BACTERIA

Bacteria are single-celled protists having a cell wall. Chromatin material is scattered throughout the cytoplasm, rather than being contained in a nucleus.

The average bacterial cell may be from 1 to 1.3 microns in length. (A micron, μ, is $^1/_{1000}$ of a millimeter, or $^1/_{25,400}$ of an inch.) A particle the size of a pinhead may contain as many as 8,000,000 bacterial cells. There may be hundreds of assorted types of bacteria on a fingertip.

Classifying bacteria. Bacteria are classified according to their shape:

Bacillus (i) —rod-shaped. Example, *Bacillus anthracis,* the cause of anthrax.

Coccus (i) —spherical-shaped. Example, *Streptococcus scarlatinae,* the cause of scarlet fever.

Spirillum (a) —spirally curved. Example, *Vibrio comma,* the cause of cholera.

Cocci may be arranged singly or in groups. A single coccus is a micrococcus; cocci in pairs are diplococci; in groups of four, they are tetrads; in cubes of eight, sarcinae; in chains, streptococci; and in grapelike clusters, staphylococci.

Spore formation. Certain bacteria have the ability to form a tough wall around their protoplasm. In this stage, they are known as *spores,* and can resist heat, low temperatures, dryness, and chemicals. When a spore germinates, the protoplasm emerges, grows, and multiplies. Spores may serve to carry bacteria through unfavorable periods; they may also constitute a stage in the life history of the bacteria. Few of the pathogenic (disease-producing) bacteria form spores.

Reproduction. Bacteria multiply by the process of

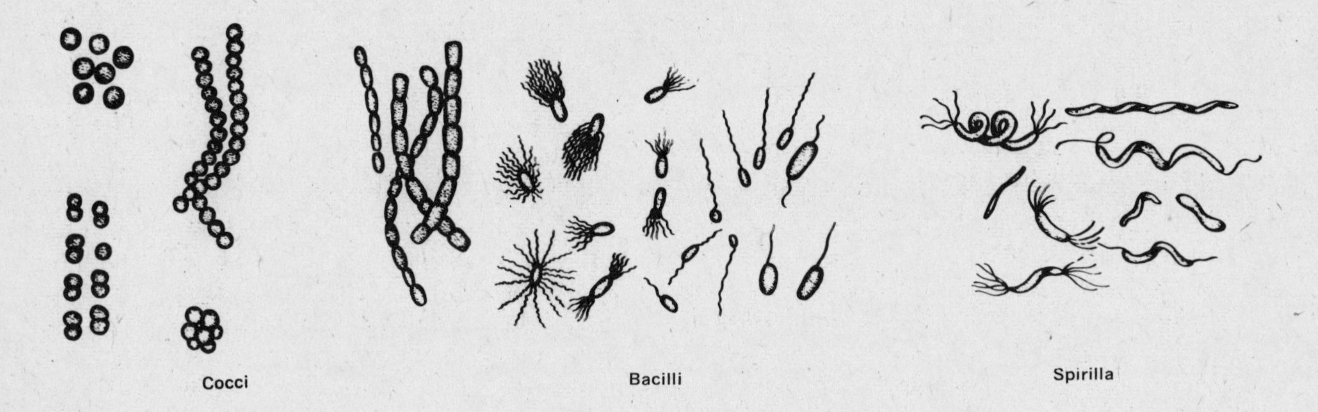

Types of Bacteria

simple fission, during which they split in half. If conditions of food and temperature are good, they may multiply every 30 minutes. At this rate, one bacterium can produce millions within a twenty-four hour period, as can be seen from the following table:

TIME	NUMBER OF BACTERIA
30 minutes	2
1 hour	4
1½ hours	8
2 hours	16
4 hours	256
8 hours	65,536
10 hours	1,048,576
12 hours	16,777,216, etc.

Some other characteristics. Some bacteria move by means of hairlike projections, *flagella,* which may be present either at one end, both ends, or all around the cell. Most bacteria need oxygen to live and are called *aerobic. Anaerobic* bacteria, such as the tetanus bacilli, can grow only in the absence of oxygen.

Certain bacteria are sometimes attacked by a virus parasite called *bacteriophage,* originally discovered by Twort and d'Herelle. The bacteriophage attaches itself to the bacterial cells and causes them to dissolve. It is specific in its action, and will distinguish between different strains of the same bacteria, attacking only certain types. For this reason, bacteriophage is used to type bacteria into various strains.

Growing bacteria. In the laboratory, bacteria may be grown and fed on various types of artificial culture media. The most common liquid culture medium is *nutrient broth,* containing beef extract, peptone, and water. This liquid is usually placed in test tubes covered with cotton plugs. When sterilized in an *autoclave,* or pressure cooker, for twenty minutes at 15 lbs. pressure, the liquid is clear. After bacteria are introduced, the liquid becomes cloudy. Other liquid nutrient media may contain milk, bile, stains, or glucose, depending on their specific use.

Another, more common culture medium is solid *nutrient agar.* This is a mixture of beef extract, peptone, water and agar-agar, a gelatinous seaweed extract. After it is sterilized, the melted nutrient agar is poured into either: (a) sterile test tubes which are kept at an angle so that the agar may cool and harden at a slant, after they are covered with a cotton plug, or (b) sterile petri dishes in which they form a solid layer about one-eighth inch thick. Other solid nutrient media may be made with gelatin, potato, blood, milk, potassium hydroxide, etc.

Studying bacteria around us. Sterile agar plates, that is, petri dishes containing nutrient agar, may be exposed to various conditions to show that bacteria are present everywhere. Thus, if there are seven plates to start with, six of them may be exposed as follows: A coin may be placed in one; a hair may be placed in another; a third may be exposed to the air for five minutes; freshly washed fingers may be touched to the surface; lips may be pressed against the surface; water from an aquarium may be added. The seventh agar plate should remain unopened, as the *control;* it is intended to show that all the dishes were sterile before they were exposed.

Colonies of bacteria. After twenty-four hours in an incubator at 37°C, colonies of bacteria will appear on the surface of the exposed plates. Each colony contains so many millions of cells, that the entire mass can readily be seen with the naked eye. It is sometimes hard to imagine that a colony a quarter of an inch in diameter originated from a single bacterial cell. A colony therefore consists of only one type of bacteria. Colonies may vary in size, shape, texture, luster, and color. Although most colonies are white, some appear red, as in the case of *Serratia marcescens,* or gray, yellow, green or purple.

A pure culture. A pure culture consists entirely of one type of bacteria. It may be obtained by transferring bacteria from one of the colonies to the surface of an agar slant in a test tube. The transfer is accomplished with the aid of a platinum or nichrome wire needle supported in a glass rod, in the following way:

1. Sterilize the needle by heating to red hot, both before and after using.
2. When cool, touch the end to the desired colony.
3. Remove the cotton plug from the test tube and pass the mouth of the test tube through the flame to sterilize it.
4. Insert the needle into the bottom part of the test tube, without touching the sides, place it in contact with the surface of the agar slant, and draw it upwards to the top.
5. Replace the cotton plug.

These steps should be performed carefully, in order to avoid contamination by bacteria from the air, or from other sources. After incubation for a day, a streak will be observed on the slant, with the naked eye. This streak consists of the one type of bacteria that was deposited by the needle.

MICROSCOPIC STUDY OF BACTERIA

Hanging drop method. It may be desired to observe

the growth of bacteria directly. This may be done by using a hanging drop slide, which can be prepared as follows:

1. Place a thin coat of vaseline around the depression of a concave slide.
2. Heat a platinum or nichrome wire loop until red hot.
3. When cool, insert the loop into a liquid culture of bacteria, withdrawing a small quantity of it in the loop.
4. Touch the loop to the center of a clean cover slip, thus depositing a small drop of the culture.
5. Invert the cover slip over the center of the slide so that the drop of fluid hangs in the center of the depressed part of the slide.
6. Press the edges of the cover slip into the vaseline to make an airtight seal which will prevent evaporation.
7. With the low-power objective of the microscope, focus on the edge of the drop. Then switch to high power, using the edge of the drop as a guide.

Living bacteria can be observed within the drop. A hanging drop slide is also useful for observing motility, size and shape of bacteria.

Staining method. Bacteria can be observed more readily when they are stained with a weak solution of various aniline dyes, such as methylene blue, crystal violet, etc. The following method may be used to stain bacteria:

1. Place a drop of water in the center of a slide.
2. Heat a straight wire needle until red hot.
3. When cool, touch lightly to the desired colony.
4. Spread the bacteria evenly through the drop of water on the slide to form a thin film.
5. Allow the slide to dry in the air, and then pass it quickly through a flame three times. This fixes the bacteria to the slide.
6. Cover the bacterial film with the stain.
7. After one minute, wash the excess stain off with water.
8. Blot the slide dry with filter paper.

Under the microscope, the bacteria will be seen bearing the color of the dye. Special stains are employed to study specific features of bacteria, such as flagella, spores, and granules.

Gram's stain. Probably the most important stain of all is Gram's stain. This is a differential stain which helps identify and classify bacteria. Bacteria that take a blue or violet color are called Gram-positive. Most cocci are Gram-positive. On the other hand, Gram-negative bacteria are stained pink. Most bacilli are Gram-negative, except for the diphtheria, tuberculosis, and spore-forming bacteria.

Helpful bacteria. Most bacteria are harmless. Some are quite useful to man in a number of different ways.

1. **Soil bacteria.** The soil is teeming with numerous types of bacteria. Bacteria of decay attack and digest dead animal and plant remains in the soil, releasing chemical elements and compounds which enrich the soil. Soil composted largely of decayed plant material is known as *humus.*

2. **Bacteria in the nitrogen cycle.** Among the compounds formed during the bacterial decay of plant and animal proteins is ammonia. *Nitrifying bacteria* (*Nitrobacter group*) in the soil change ammonia to nitrates which are valuable for protein synthesis by green plants. *Nitrogen-fixing bacteria* (*Rhizobum leguminosarum*) live within the swellings, or *nodules,* on the roots of plants of the legume family, such as beans, peas, alfalfa, and clover. These bacteria fix the free nitrogen of the air present in the soil into nitrates. In the free state, nitrogen cannot be used by plants, but as nitrates, it can be taken up and used. The farmer recognizes the value of these bacteria by practicing crop rotation as a method of increasing the supply of nitrogen compounds in his soil.

3. **Bacteria in the dairy industry.** The flavor of butter depends to a great extent on bacterial activity. The cream used in making butter is inoculated with a culture of desirable lactic acid bacteria which produce the desired flavor.

Swiss cheese, Cheddar cheese, and several other types of cheese receive their flavor, texture and color from the action of certain bacteria which are added to milk under carefully controlled conditions. The holes in Swiss cheese are produced as the result of gas formation while the cheese is still in the liquid condition.

Buttermilk, yogurt, and acidophilus milk are examples of fermented milk that have been treated with certain strains of bacteria, such as *Lactobacillus acidophilus.*

4. **Making of vinegar.** The word vinegar comes from the French and means sour wine. The alcohol of wine or hard cider is changed to the acetic acid of vinegar by fermentation due to the action of acetic acid bacteria (*Acetobacter* group). The bacteria grow on the top of the liquid in a thick, jelly-like layer called "mother of vinegar."

5. **Silage fermentation.** Chopped hay, cornstalks, beet leaves, etc., are packed by the farmer into the tall, cylindrical silo connected with his barn. The bacteria contained in the plant material carry on fermentation. This action results in an improvement in the flavor of the fodder, and at the same time preserves it against putrefaction.

6. **Flax and hemp.** Flax is the source of linen. The useful fibrovascular bundles or fibers in the flax stem are held together by certain "pectins."

During the retting process, these binding materials are removed. This is accomplished when the flax is placed in pools of water, where bacterial action takes place. After about ten days, the fibers are removed, dried, cleaned and are then ready for spinning. Hemp is prepared in the same way.

7. Other uses of bacteria. *Tobacco* leaves must be "cured" before they can be used; bacterial action takes place which produces the desired changes in the leaves. *Meat* that is cured has a tenderer quality and flavor due to the action of bacteria. The production of *sauerkraut* from shredded cabbage, and *pickles* from cucumbers depends on the fermentation action of bacteria. Natural *sponges* are prepared for commercial use by the action of bacteria in decaying and removing the protoplasmic material from the fibrous structure of the sponge.

Harmful bacteria. Certain bacteria are harmful to man and his welfare.

1. Food spoilage. Many foods, such as meat, fish, poultry, milk, and cooked vegetables and fruits spoil readily if allowed to remain at room temperature. Bacteria, yeasts and molds find in these foods excellent condition for growth. In order to protect our food as well as our health against these microorganisms, we use various methods of food preservation, such as the following:

a. Refrigeration. Low temperatures inhibit the growth of bacteria. However, they do not kill many of them. When food is returned to normal temperatures, the bacteria come out of the dormant state and begin to grow actively again.

b. Freezing. The water in frozen foods is crystallized as tiny particles of ice. It is therefore not available for the bacteria. Meat, fish, vegetables, and other foods may be kept preserved for a long time in the frozen condition.

c. Pasteurization. This process of heating foods to temperatures below the boiling point in order to destroy harmful microorganisms was named after its originator, Louis Pasteur. He discovered that the spoiling of wine was due to the growth of undesirable bacteria. By heating the wine, he killed the bacteria without altering its taste. Milk is now pasteurized by heating to 145°F for a half hour, and then cooling rapidly; or, by the short-time method, it is heated to 162°F for fifteen seconds, and cooled rapidly. Either process destroys disease-producing bacteria of tuberculosis and typhoid fever, but not the harmless bacteria in milk. Pasteurization is also used to preserve grape juice and apple juice. Pasteurization is not the same as sterilization, because many bacteria may still be present. During *sterilization,* they are all killed.

d. Canning. When food is canned, it is sealed and heated so that all the bacteria are killed. Unopened cans may therefore remain in good condition for many years. A sound can has concave ends. If the ends bulge, it is usually a sign that the contents have spoiled as the result of action by gas-forming bacteria. It is unsafe to open such cans because of the danger of botulism, caused by the deadly anaerobic *Clostridium botulinus.*

e. Dehydration. Water is necessary for bacterial growth. When water is removed from food, it usually becomes powdery or flaky. In this condition, food occupies very little space, and may be shipped and stored readily with little danger of spoilage.

f. Salting. Adding large quantities of salt to meat or fish helps preserve them by inhibiting the growth of bacteria. *Plasmolysis* of the protoplasm of the bacterial cells takes place when they are in contact with a concentrated solution of salt. The water of the protoplasm passes out of the bacterial cells, leaving the contents shriveled.

g. Other methods of food preservation. *Smoking* of meat helps prevent the growth of bacteria because of the chemicals present in the smoke; it also alters the flavor. Some *spices,* such as cinnamon and cloves, destroy bacteria, thus giving foods a keeping quality. Certain *chemicals,* such as sodium benzoate, are added directly to foods to preserve them. However, their use is strictly supervised under the terms of the Pure Food and Drug Act.

2. Dentrifying bacteria. (*Bacillus denitrificans*) These soil bacteria play an undesirable part in the nitrogen cycle. They produce enzymes that convert ammonia, resulting from the breakdown of proteins by other soil bacteria, into free nitrogen. In this free state, nitrogen cannot be used by plants, and can therefore be considered as lost to them. This waste of nitrogen may be corrected by proper cultivation and aeration of the soil.

3. Bacteria and health. The greatest harm from bacteria comes from their effects on our health. The classic work of Pasteur and Koch helped establish the fact that certain diseases are caused by specific microorganisms. This is known as the *Germ Theory of Disease.*

Louis Pasteur (1822–1895). Pasteur made many basic contributions to the field of bacteriology. He investigated the "disease" that caused wine to sour, and proved that it was due to bacteria. The silkworm industry in France was threatened by a disease of silkworms called *pebrine;* he showed that a microorganism was causing this disease. His solution of the silkworm disease stimulated him to turn his attention to other diseases such as anthrax, chicken cholera, and rabies, in which he established successful principles of immunity.

Robert Koch (1843–1910). Koch was a physician who became interested in the importance of bacteria in disease. He devised many of our present techniques for culturing, staining, sterilizing, and handling bacteria. He was the first to use gelatin for a solid culture medium, thus making it possible to isolate bacteria in pure culture. In 1876 he discovered the bacteria that cause anthrax in cattle. Then in 1884, he announced his discovery of the bacteria that cause tuberculosis in humans.

Koch's postulates. Koch used certain steps in identifying the tubercle bacillus. This procedure is the basis for proving that a certain disease is caused by a specific germ, and is known as Koch's Postulates:

1. The same microorganism is found in all cases of the disease.

2. It must be isolated and grown in pure culture.

3. The original disease is produced when the microorganism is injected into healthy animals.

4. The same microorganism is obtained from the diseased animals.

Joseph Lister (1827–1912). Lister, an English surgeon, used Pasteur's discoveries about microorganisms as the basis for eliminating the terrible infections that accompanied surgery. He applied carbolic acid (phenol) directly to wounds, bandages, instruments, and the hands of the surgeon. This *antiseptic surgery* killed germs, but it was also harmful to the tissues. He then introduced *aseptic surgery,* which is now used in all hospitals; the instruments, dressings, gloves, etc., are sterilized, and the operating rooms are kept free of germs—they are aseptic.

CHAPTER REVIEW

Select the correct choice in each of the following statements.

1. Rod-shaped bacteria are called (A) bacilli (B) spirilla (C) cocci (D) plasmodia (E) streptococci

2. Bacteria may survive unfavorable conditions by (A) splitting in half (B) forming flagella (C) forming spores (D) forming a bacteriophage (E) forming tetrads

3. When bacteria are to be grown on an agar plate, they are (A) sterilized (B) pasteurized (C) stained (D) incubated (E) flamed

4. A type of glassware most commonly used for growing bacteria is (A) flask (B) beaker (C) petri dish (D) delivery tube (E) thistle tube

5. Gram-positive bacteria are usually (A) disease-resistant (B) antibiotic resistant (C) bacilli (D) stained pink (E) cocci

6. Nitrogen-fixing bacteria live on the roots of (A) peas (B) clover (C) alfalfa (D) legumes (E) all of these

7. Eight colonies of bacteria appeared on an agar plate after twenty-four hours; the number of individual bacteria present at the beginning were (A) 2 (B) 4 (C) 6 (D) 8 (E) 8 million

8. The flavor of all of the following is due to bacterial activity except (A) ice cream (B) buttermilk (C) Swiss cheese (D) butter (E) yogurt

9. Bacteria are useful in each of the following cases except (A) making of linen (B) making of rayon (C) making of vinegar (D) production of sauerkraut (E) preparation of commercial sponges

10. All bacteria are killed by (A) refrigeration (B) food preservation (C) freezing (D) pasteurization (E) canning

11. A scientist who established principles of immunity in anthrax and rabies was (A) Lister (B) Pasteur (C) Koch (D) Leeuwenhoek (E) Harvey

12. The correct sequence of steps in Koch's Postulates (1-injecting the germ to produce the original disease; 2-isolating the same germ in many cases; 3-obtaining the original germ from infected animals; 4-growing the germ in pure culture) is (A) 1–4–2–3 (B) 1–2–3–4 (C) 2–4–1–3 (D) 2–1–4–3 (E) 2–4–3–1

13. The basis of eliminating infections during operations is (A) pasteurization (B) dehydration (C) aseptic surgery (D) salting (E) cauterization

14. In practicing crop rotation, farmers rely on the desirable activity of (A) denitrifying bacteria (B) nitrogen-fixing bacteria (C) nitrifying bacteria (D) *Lactobacillus acidophilus* (E) acetic acid bacteria

15. All bacteria are classified as (A) algae (B) fungi (C) parasites (D) saprophytes (E) symbionts

ANSWER KEY

1-A	4-C	7-D	10-E	13-C
2-C	5-E	8-A	11-B	14-B
3-D	6-E	9-B	12-C	15-B

17 Germs vs. man

Germs are microscopic organisms that cause disease among animals and plants.

HOW GERMS AFFECT THE BODY. Human beings may be affected in a variety of ways by different germs.

1. **Production of toxins.** Toxins are poisons produced by germs. As the germs live on the tissues of the body, they give off these poisons. Toxins are powerful in minute quantities. It has been estimated that only 0.00023 gram (about one-fourth of a milligram) of the toxin produced by the tetanus bacilli would be fatal to a full-grown man. The toxin of the botulism bacillus is about a hundred times more powerful than this.
2. **Destruction of tissue.** Some germs settle in certain parts of the body and affect specific cells and tissues, as shown in the table below.

HOW GERMS ENTER THE BODY. Different germs may enter the body in various ways.
1. **Air.** Many germs are airborne and enter the body through the nostrils or the mouth. Examples: Common cold, influenza, measles, mumps, pneumonia, tuberculosis.
2. **Water.** Water that appears clear may actually be contaminated with germs of typhoid fever, dysentery or cholera.
3. **Food.** Germs may live on the nutrients in food and may be carried into the body when these foods are eaten. Examples: Typhoid (milk, shellfish); tuberculosis (milk); trichinosis (undercooked pork); botulism (improperly canned food).
4. **Break in the skin.** Germs may be carried into the body when the skin is cut or broken. Examples: Tetanus (punctured wound); skin infections; boils; rabies (bite of infected dog).
5. **Insect bite.** Some germs are injected into the body by the bite of an insect. Examples: Malaria (Anopheles mosquito); yellow fever (Aedes mosquito); black plague (flea); African sleeping sickness (tsetse fly); typhus (body louse).

HOW THE BODY PROTECTS ITSELF AGAINST GERMS. The body is protected against germs in a number of ways:

1. **The skin** is a tough, outer envelope which serves as an effective barrier against the entrance of germs.
2. **The tears** of the eyes contain *lysozyme,* which was discovered by Fleming (the discoverer of penicillin), and which helps keep the eyes free of infection.
3. The **mucous membranes** lining the inside of the mouth and the nasal passages have a sticky surface due to the secretion of *mucus.* Germs and dust are swept outward by *ciliated* cells.
4. **Gastric juice** in the stomach contains hydrochloric acid which kills many bacteria that are swallowed.
5. **Phagocytes** are the white blood corpuscles that engulf bacteria which have entered the tissues and the bloodstream. They are attracted to a break in the skin which has become infected. *Pus* is a thick liquid which may collect at such a place, consisting of lymph, white corpuscles, and bacteria.
6. **Lymph nodes** are located at various parts of the lymphatic system. They serve to screen germs from the lymph, and may become swollen during an infection.
7. **Antibodies** are chemicals produced by the body, which act against germs or other antigens (foreign materials). They are each specific for a certain type of disease germ and are not effective against

DISEASE	PART OF BODY DIRECTLY AFFECTED
Amebic dysentery	Intestines
Malaria	Red blood corpuscles
Poliomyelitis	Nerves of spinal cord and brain
Rabies	Brain and spinal cord
Ringworm	Skin
Trichinosis	Skeletal muscles
Tuberculosis	Lungs, bone, or skin
Typhoid fever	Intestines, spleen

others. There are several kinds of antibodies:

a. **Antitoxins** are antibodies that neutralize the toxins produced by certain germs. A person who recovers from an attack of diphtheria, for example, has produced a supply of antitoxins that will usually protect him against the disease for many years.

b. **Agglutinins** are antibodies which cause bacteria to stop moving and to clump together.

c. **Opsonins** are antibodies which act on bacteria, making them more easily ingested by white blood corpuscles.

d. **Lysins** are antibodies which dissolve bacteria.

8. **Interferon** is a protein produced by cells that are attacked by viruses, which protects neighboring cells against the germs. It was discovered in England, in 1956 by Dr. Alick Isaacs and Dr. Jean Lindemann. It offers great promise in fighting viruses, because it is effective not only against one specific virus, but virtually against any of them. At present, methods of triggering the production of interferon artificially are being investigated.

9. **Reticuloendothelial system (RES)** is a recently discovered internal system of the body which clears foreign particles such as germs, from the blood stream. It consists of phagocytes, and other cells which have the same function, but are fixed in the lining of the blood vessels in such organs as the liver, thymus, lungs, spleen, bone marrow and lymph nodes. It may also play an important role in antibody formation.

PROVIDING IMMUNITY AGAINST GERMS. *Immunity* is the ability of a body to resist disease. On the other hand, a person who becomes ill with a disease is *susceptible* to it. There are various types of immunity.

1. **Natural immunity** occurs when a person is born with the ability to resist a disease.

2. **Acquired immunity** results either after a person has recovered from a disease, or has been protected against it. Acquired immunity may be *active* or *passive*.

a. **Active acquired immunity.** Most often, a child who has had a case of the measles will not catch this disease again. While he was ill, he produced antibodies which helped him recover, and which are now present in his body to protect him against further attacks of the germ. Since he acquired this type of germ immunity by actively producing the antibodies, this is known as active acquired immunity.

There are other ways of helping the body acquire active immunity and to build a supply of antibodies. When you were vaccinated against smallpox, the mild virus of *cowpox* was scratched into the skin of your arm. This caused your body to produce a large enough supply of antibodies to protect you against smallpox for many years. In the Sabin vaccine against polio, the live virus, which was weakened with chemicals, is swallowed to produce immunity.

Active immunity may also be acquired by inoculating a person with dead germs, as is done to protect a person against typhoid fever. This antigen stimulates the production of antibodies which kill any typhoid germs that may enter the body. The same principle is employed in the use of the Salk vaccine against polio. Dr. Jonas Salk made the virus inactive by treatment with formaldehyde. Inoculation of the dead virus causes the formation of antibodies against the polio virus, thereby protecting the person against the disease.

Deadly toxins that are weakened with chemicals are called *toxoids*. If toxoid is injected, it stimulates the body to produce antitoxins against the specific disease. Active immunity may thus be acquired against diphtheria or tetanus. *Toxoid* has now largely replaced *toxin-antitoxin* which was formerly used to stimulate the production of antitoxins by the tissues of the body. In toxin-antitoxin, the toxin stimulates the reaction, while, at the same time, the antitoxin protected the body against the toxin. Toxoids are now preferred because fewer injections are needed, and the danger of a sensitive serum reaction is avoided.

In all cases of active immunity, the body is protected for a long time, because its tissues have actively produced an abundant supply of antibodies.

b. **Passive acquired immunity.** In the case of passive immunity, antibodies are injected into the person from some outside source, such as another person or a horse. During a case of diphtheria, for example, a doctor will inoculate a child with antitoxins, in order to neutralize the deadly toxins that are causing the symptoms. The child recovers without having produced his own supply of antibodies. In the case of a deep puncture wound, tetanus may be prevented by injecting antitoxins into the injured person. The antitoxins neutralize any toxins that may be produced by tetanus bacilli.

Passive immunity is temporary, and lasts only a few weeks. The tissues of the body are not stimulated to produce antibodies.

USING CHEMICALS TO PROTECT THE BODY

A. External use

Germs on the outside of the body may be killed by various chemicals.

1. **Disinfectants** are powerful chemicals that can destroy not only germs, but also spores. Examples: chlorine compounds, lysol.

2. **Antiseptics** are chemicals that inhibit or prevent the growth of bacteria. They are less injurious to tissues than disinfectants. For this reason, we use them for a cut or break in the skin. Example: tincture of iodine.

B. Internal use

Some chemicals or drugs are used internally to fight germs that have entered the body. The chief precaution in using such chemicals within the body, of course, is that while they are expected to kill the germs, they should not harm the living cells of the organism.

1. **Quinine** was originally used by the Indians of South America to treat malaria. They chewed the bark of the cinchona tree which contains the drug. Nowadays, *atabrine* and *chloroquine* are largely used to treat the disease.

2. **Arsenic compounds** were used by Paul Ehrlich (1854–1915) to treat syphilis. His 606th experiment to discover a chemical treatment for this disease finally yielded a compound containing arsenic and bismuth that would not harm the body tissues. It was called "606," or *salvarsan.*

3. **Sulfa drugs.** The first of these so-called wonder drugs, *sulfanilimide* was developed in the 1930's. Since then, hundreds of varieties have been made, including, sulfathiozole, sulfadiazine, etc. They are effective against bacteria that cause streptococcus infections, meningitis, pneumonia, boils, and blood poisoning. They interfere with the metabolism of these bacteria, causing them to die. Since sulfa drugs may have harmful effects on the body, they should be taken only under a doctor's care.

4. **Antibiotics** are chemical substances produced by living things, especially fungi, that stop the growth of germs.

a. **Penicillin** was the first of the antibiotics to be discovered. In 1929, Alexander Fleming, a British doctor, observed that a green mold (*Penicillium notatum*) was accidently growing in a petri dish containing staphylococcus colonies. Instead of discarding the contaminated dish, he investigated the clear bacteria-free zone immediately around the mold. He theorized that the mold was giving off a substance that destroyed the bacteria, and after additional investigation, named it penicillin. This drug was first used on a large scale to save the lives of wounded soldiers during World War II. It is valuable in treating pneumonia, venereal diseases, and infections caused by cocci. The Nobel Prize was awarded to Fleming, Florey and Chain for their work in developing penicillin.

b. **Streptomycin** was obtained from soil fungi by Selman Waksman of Rutgers University, who also was awarded a Nobel Prize. This antibiotic is useful in combating tuberculosis, urinary infections, and other diseases.

c. Other antibiotics are being developed all the time. *Chloromycetin* is useful against typhus and Rocky Mountain spotted fever, both caused by Rickettsia germs. *Aureomycin, terramycin,* and *tetracycline* are just a few of the other useful antibiotics.

d. Certain antibiotics (example, terramycin) have been found to stimulate the growth of farm animals, and are now added to the food of chickens, pigs, and cows.

e. Resistance to antibiotics and the sulfa drugs is sometimes developed by certain strains of germs. These immune germs survive and then increase in number. Doctors may then use combinations of the drugs to overcome the germs.

f. Allergic reactions are sometimes shown by people who have become sensitive to certain antibiotics. Because of the body's reaction to the antibiotics, they may develop hives or become quite ill after they have become protected against a specific disease germ. For this reason doctors now advise using antibiotics only when absolutely necessary.

5. **Isoniazid** and similar chemicals are very effective against the tuberculosis germ, and are often used in combination with streptomycin in the successful treatment of tuberculosis.

CHAPTER REVIEW

Select the correct choice in each of the following statements.

1. Poisons produced by bacteria are known as (A) toxoids (B) toxins (C) antitoxins (D) antibodies (E) toxin-antitoxins

2. A disease in which the red blood corpuscles are attacked is (A) rabies (B) amebic dysentery (C) polio (D) trichinosis (E) malaria

3. Germs that enter the body in milk may cause (A) pneumonia (B) tetanus (C) mumps (D) tuberculosis (E) influenza

4. A disease caused by the bite of an insect is (A) typhus (B) typhoid (C) rabies (D) measles (E) botulism

5. All of the following structures of the body protect it against the entrance of germs except (A) tears (B) mucous membrane (C) ciliated cells (D) white blood cells (E) red blood cells

6. All of the following are examples of antibodies except (A) agglutinins (B) opsonins (C) lysins (D) antibiotics (E) antitoxins

7. An example of passive acquired immunity is (A) vaccination against smallpox (B) inoculating dead germs against typhoid (C) use of Salk vaccine (D) inoculation of toxoid against diphtheria (E) inoculation of antitoxin in case of a puncture wound

8. A drug obtained from the bark of a tree is (A) salvarsan (B) tincture of iodine (C) quinine (D) sulfanilimide (E) tetracycline

9. All of the following are antibiotics except (A) penicillin (B) streptomycin (C) aureomycin (D) terramycin (E) riboflavin

10. Penicillin was discovered by (A) Ehrlich (B) Fleming (C) Waksman (D) Koch (E) Metchnikoff

ANSWER KEY

1-B	3-D	5-D	7-E	9-E
2-E	4-A	6-D	8-C	10-B

18 Conquest of disease

A disease is a condition in which the health of the body is impaired. Germs are responsible for many types of disease. These microorganisms may be bacteria, other fungi, viruses, protozoa, or worms. The successful conquest of disease germs helped raise the life expectancy of the average American to 69.7 years by 1959.

DISEASES CAUSED BY BACTERIA

TUBERCULOSIS. This disease was formerly known as the white plague and was a chief cause of death in this country at the beginning of the century. Robert Koch (1882) was the first scientist to prove that it was caused by a specific germ, the tubercle bacillus. The bacilli may be spread through droplets in the air from coughing or sneezing; by using contaminated articles; in impure milk; and by direct contact with a person having the disease. The bacilli may infect the lungs where they form tubercles, little masses of germs surrounded by protective layers of calcium and fibers. If the body's resistance is low, the germs spread from these tubercles to attack the rest of the lungs and other parts of the body.

Detection of the disease may be accomplished by:

1. Chest X-ray which helps reveal the presence of tubercles in the lungs.

2. Tuberculin test (Patch Test) in which tuberculin material obtained from the germs is placed in the skin. A red area a few days later indicates that the person is sensitized to the tubercle bacillus, and may now have, or once had the active germs in his body. It should be followed up with a chest X-ray.

3. Sputum test, in which a sample of coughed-up sputum is examined for the presence of tubercle bacilli.

Protection against the disease may be acquired by:

1. Providing rest and good nutrition to a person having the disease. This is the classic treatment first developed by Dr. Edward Trudeau at the Lake Saranac sanatorium.

2. Drug treatment, using streptomycin, isoniazid and similar chemicals. This treatment has been so successful that many tuberculosis sanatariums have now been closed, including the Trudeau sanatorium.

3. Pneumothorax, in which a badly infected lung is collapsed, to permit the damaged tissue to repair itself.

4. BCG (Bacillus Calmette-Guérin) vaccination with weakened germs. This is a method of providing active immunity. Its value is controversial among doctors.

5. Proper health measures can prevent the spread of the germ. Examples: Avoiding the use of common drinking cups or towels; covering a cough; no spitting; not kissing or handshaking with an infected person.

6. Pasteurization of milk. This is necessary because cows are susceptible to tuberculosis, and raw milk may contain the tubercle bacilli.

7. Proper rest and good nutrition. Overcrowding and slum conditions are favorite breeding grounds of the tuberculosis germ.

TYPHOID FEVER. All new members of the Armed Forces receive immunization against this disease, upon their induction. Impure water, milk and food may bring the germ into the alimentary canal where it multiplies and damages the intestines. It is carried out of the body with excretions, and may infect other people if sanitary procedures are poorly observed.

Human carriers such as the famous Typhoid

Mary, are immune to the disease, but carry the germs. If they work with food, they may cause an epidemic. The common housefly may also distribute the germs with its feet, if it crawls over food after having been in contact with contaminated material.

Detection. The *Widal test* is a useful method of diagnosis. A culture of typhoid germs is mixed with serum from a person's blood. If the person has typhoid, the agglutinins in his blood will cause the typhoid germs to clump together.

Protection

1. Inoculation of a sick person with the antibiotic chloromycetin.
2. Vaccination with dead bacilli produces active immunity against the disease.
3. Inspection of food handlers.
4. Control of the housefly.
5. Providing a good sewage system.
6. Proper purification of water.
7. Pasteurization of milk.
8. Not eating shellfish found in polluted water.

DIPHTHERIA. Until it was controlled, around the beginning of this century, diphtheria was a dread killer of children. The germs enter through the mouth and nose, and settle in the throat. They produce a thick membrane which may close over the opening of the trachea and produce death by choking. Roux discovered in 1888 that the bacilli also produce a toxin that spreads throughout the body, causing high temperature, damaging tissues, and resulting in death. Von Behring in 1892 showed that antitoxins manufactured in the body of an animal such as a horse would protect children ill with diphtheria. This antitoxin in the serum neutralized the toxins produced by the germs.

Detection. The *Schick test* indicates whether a person is immune or susceptible to diphtheria. A small amount of the toxin is injected under the skin. If the person is immune, the area remains unaffected, because the toxin is neutralized by the antitoxins in the body. A red spot indicates susceptibility.

Protection. 1. A child already ill with diphtheria is inoculated with antitoxin. This confers passive immunity. 2. All infants are now inoculated with toxoid, which consists of toxin that has been weakened with chemicals. This stimulates the body to produce its own antitoxins, resulting in active immunity. As a result of this immunization, diphtheria has become a rare disease among treated children.

TETANUS. A person who sustains a puncture wound runs the risk of tetanus, or "lockjaw." The tetanus bacilli are anaerobic, and multiply deep within the injured area, away from oxygen of the air. Toxins are given off, that cause the muscle to contract, and eventually death.

Immediate protection. 1. Puncture wounds resulting from barbed wire, gunshot, shrapnel, etc., are exposed to the air and washed with antiseptic. 2. Tetanus antitoxin is injected to neutralize toxins that might be produced by the tetanus bacilli.

Long-range protection. Toxoid is inoculated into healthy persons to confer active immunity. All members of the armed forces receive this immunization.

DISEASES CAUSED BY VIRUSES

Viruses are ultramicroscopic germs that were not visible until the perfection of the electron microscope. They are so small that they pass through the finest filters that keep bacteria out. They were first shown to be complex protein molecules by Wendell M. Stanley, who purified them in crystalline form. They do not grow on agar, as bacteria do, and must be cultured within living cells such as those of fertile chicken eggs. While they may be purified as protein crystals, they can also multiply within living cells. They are thus considered to be on the borderline between the living and the nonliving. Viruses cause a number of serious diseases, such as smallpox, polio, rabies, yellow fever, the common cold, mumps, and measles.

SMALLPOX. Until the early part of the nineteenth century, smallpox was the cause of many deaths. Those who survived bore pockmarks on their faces. Dr. Jenner (1749–1823), an English physician, observed that milkmaids were immune to smallpox after an attack by cowpox, a similar but mild disease. He rubbed some of the material from the sores of a cow ill with cowpox into a slight scratch on a boy's arm. After developing a small sore the boy was immune from smallpox. This was the first case of vaccination and the beginning of our conquest of the dread disease. In this process, the body develops active immunity against the mild virus of cowpox and the related powerful smallpox virus. Although smallpox is rare in the United

States, it is still a menace in many parts of Asia, where vaccination is not practiced.

POLIO. Infantile paralysis, or poliomyelitis, mainly affects children, although adults may also contract this disease, as was the case with Franklin D. Roosevelt before he became President. The three types of polio virus, Types 1, 2, and 3, are probably spread through the air, from person to person, although there is some evidence that they may also be spread through polluted water. Once in the body, the virus may damage the nerve cells of the brain and spinal cord that control the muscles. If the diaphragm and chest muscles are paralyzed, the patient must be kept breathing by means of an "iron lung." Muscles of the arms and lungs stop functioning if the neurons that control them are damaged. A serious attack results in permanent crippling and possibly death.

It is believed that most adults have had a mild case of polio at one time or another, which produced symptoms like those of a cold. This probably resulted in the building up of antibodies, and may explain why most grownups are immune to the disease. Children have not had time to build up resistance to the polio virus, and as a consequence have been the chief victims of the disease.

After many years of research, however, the tide against polio was finally turned, with the development of the Salk vaccine in 1953. The next year a mass inoculation experiment involving almost two million school children proved the effectiveness of the vaccine. It is now used as the chief means of protection against the disease. The vaccine perfected by Dr. Jonas E. Salk is composed of dead polio virus, which stimulates the production of antibodies when injected. A number of scientists, however, are convinced that the Sabin vaccine, consisting of weakened polio virus and taken by mouth, may be superior to the Salk vaccine in conferring active immunity.

RABIES. The bite of a mad dog may result in hydrophobia, commonly known as rabies. The germs present in the saliva of such an animal enter the body through the broken skin, and find their way along the nerves to the brain and spinal cord. Death results unless the person is given the Pasteur treatment. This consists of a series of inoculations of weakened virus, each becoming progressively stronger, and stimulating the body to produce more and more antibodies against the virus. In recent years, a duck-embryo vaccine has been used instead of the Pasteur treatment, because it is safer and less painful. To provide more immediate protection, a special horse serum may also be injected. A dog suspected of having rabies is usually isolated and observed for about ten days. If it should die, its brain is examined for Negri bodies, the presence of which confirms the activity of the rabies virus. The incubation period of the virus varies from two weeks to six or more weeks, depending on how far the bite is from the brain.

Louis Pasteur developed the rabies vaccine by drying the spinal cord of infected animals. This treatment weakened the deadly virus so that it was no longer capable of causing the disease. However, it was able to stimulate the body to produce the valuable antibodies which gave protection against the disease. The rabies treatment was successfully used on a young boy for the first time in 1885.

INSECT-BORNE DISEASES

Several serious diseases are transmitted by insects, which inject germs into human beings as they suck blood from them. Thus they are quite different from the housefly, which may carry the germs of typhoid fever and tuberculosis on its feet and deposit them as it crawls over food.

MALARIA. The female Anopheles mosquito receives the *Plasmodium,* the protozoan cause of malaria, when it bites a malaria victim and sucks up some of his blood. In her stomach, these plasmodia multiply by sexual reproduction, during which male- and female-type cells are produced. After these unite, the germs travel to the salivary glands of the mosquito.

The mosquito injects these germs the next time she bites a person, thus spreading the disease. Another phase of the plasmodium's life cycle now takes place. Each protozoan invades a red blood

INSECT	DISEASE TRANSMITTED	TYPE OF GERM
Anopheles mosquito	Malaria	Protozoa
Aedes mosquito	Yellow fever	Virus
Flea	Bubonic plague	Bacteria
Body louse	Tyhus fever	Rickettsia
Tsetse fly	African sleeping sickness	Protozoa

corpuscle and feeds on it. It multiplies rapidly until there are so many germs that the red corpuscle bursts. The protozoa then invade new corpuscles and the process is repeated. Each time the red blood cells burst, toxins are released into the bloodstream, causing the person to have the characteristic chills and high fever of malaria. The destruction of so many blood cells also causes anemia. The life cycle of the plasmodium in both the mosquito and humans was revealed through the persistent research of many scientists, especially by the French doctor, Laveran, in 1880, and the British scientist, Ronald Ross, in 1890.

Infected persons find relief by taking either quinine or some of the newer drugs such as *atabrine, chloroquine* or *primaquine.*

The most effective methods of preventing malaria have been to control the mosquito. Some ways of doing this are:

1. Spray DDT to kill the mosquito.
2. Drain swamps where mosquitoes breed.
3. Spread a thin layer of oil to suffocate the larvae "wrigglers," and pupae stages of the mosquito.
4. Stock ponds with small fish which feed on the mosquito larvae and pupae.
5. Remove rain barrels and other containers of standing water where mosquitoes might breed.
6. Screen windows and doors.

YELLOW FEVER. Another type of mosquito, the *Aedes,* was identified by Dr. Walter Reed at the end of the Spanish American War, as the carrier of the yellow fever germ. Using human volunteers who allowed themselves to be bitten by the female mosquitoes, he conducted controlled experiments that definitely established this mosquito as the carrier of the germ. Dr. Gorgas then waged a campaign of mosquito extermination that eliminated "yellow jack" from Panama, thereby permitting the successful construction of the Panama Canal. Previously, the French had been halted in their efforts to build a canal because of the enormous casualties inflicted by the Aedes mosquito.

BUBONIC PLAGUE. Fleas may carry the deadly *Bacillus pestis,* cause of the "black plague," among rats. When a person is bitten by one of these fleas, the germs are injected into his blood, thereby causing the disease. During the Middle Ages, approximately one-third of the population of Europe was wiped out by bubonic plague. Ships docking in this country from the Orient are required to install large circular shields on their hawsers to prevent any rats from coming ashore with germ-carrying fleas.

TYPHUS FEVER. Typhus may be common in overcrowded, unsanitary living conditions where the people do not change their clothing, nor wash too often. During World War II, it became a threat in bombed-out areas of Poland and Italy. The disease is spread by the infected body louse that lives within the clothing and sucks the blood of its host. The germ is a *rickettsia,* an organism that is intermediate in size between a virus and bacterium.

DDT powder dusted into the clothing of the people halted a typhus epidemic when the American Army entered Naples during the war. A vaccine composed of dead rickettsia has been used to confer active immunity. The antibiotic *chloromycetin* is useful against the rickettsia of both typhus and Rocky Mountain spotted fever. This latter disease is spread by a tick, a member of the spider group.

AFRICAN SLEEPING SICKNESS. Large regions of Africa have remained underdeveloped because of the prevalence of the tsetse fly, the carrier of African sleeping sickness. The germ is a protozoan, the *trypanosome,* which is transmitted by the fly when it bites a person.

The wild game of Africa, such as the various types of antelopes, are also attacked by the tsetse fly, and serve as a reservoir of the trypanosome.

SOCIAL DISEASES TRANSMITTED BY HUMANS

Venereal diseases (VD) are largely spread through sexual intercourse between people who are infected with either syphilis or gonorrhea. The infection generally takes place through the mucous membranes of the reproductive organs or the mouth. A pregnant woman who has one of these diseases may give birth to a deformed, blind, or dead baby. As a precaution, it is common practice in hospitals to treat the eyes of newborn babies with silver nitrate.

Syphilis. The cause of syphilis is a spirochete, an organism thought to be either a protozoan or a bacterium. Sores on the skin (chancres) are the first sign of the disease. Eventually, brain damage and insanity result. Most states now require a blood test (Wassermann test) as a requisite for the issuance of a marriage certificate. Paul Ehrlich's successful search for a chemical compound—salvarsan—to fight syphilis was a first victory against the disease. Penicillin is the latest weapon against it.

Gonorrhea. This social disease is caused by the gonococcus bacteria. It is characterized by running sores on the membranes of the infected organs.

Eventually arthritis and heart disease result. The sulfa drugs and penicillin are used to treat the disease. The best protection against both syphilis and gonorrhea is to follow the moral code of society regarding sexual behavior.

DISEASES CAUSED BY FUNGI

Athlete's foot. A parasitic fungus is responsible for this common foot condition. The spores may be picked up in swimming and gymnasium areas used by people with infected feet. The spores germinate in the moist areas between the toes, causing irritation of the skin. The best protection is to dry the toes thoroughly with the aid of a foot powder. Chlorinated foot baths leading to swimming pools help to retard the spread of the spores. X-rays are used for advanced cases of the infection.

Ringworm. Circular infected patches on the skin, or lesions of the scalp, may be caused by one or more species of fungus. This type of infection may be transmitted by wearing apparel or by contact with scales or hairs from lesions. General sanitary measures can prevent the infection from spreading.

DISEASES CAUSED BY WORMS

Trichina worm. Poorly cooked pork may contain tiny living roundworms known as trichina worms. When such food is eaten, the worms pass through the intestinal wall into the bloodstream. They travel to the voluntary muscles of the legs, arms, diaphragm and face, where they form hard-walled *cysts* around themselves. This infection produces pain, fever, nausea, and often more serious effects. Prevention of *trichinosis* depends on proper cooking of pork products to at least 150°F in order to kill any worms that may be present. Some foreign countries, and some of our states, require that all garbage fed to pigs be precooked, in order to prevent them from getting the infection. Besides man and pigs, other hosts of the trichina worm are rats and bears.

Hookworm. The tiny roundworms of hookworm disease may occur in the moist soil of rural regions of the South where unsanitary conditions prevail. They enter the body by boring through the feet of people walking barefoot. Entering the bloodstream, they travel to the lungs where they damage the tissue, and then pass up the bronchial tubes and the trachea to the throat. Here they are swallowed and finally reach the small intestine where they hook onto the wall. They live on the blood of their victim and cause bleeding, producing anemia and listlessness. They reproduce rapidly and their eggs pass out of the intestines with the wastes to infect the soil.

Treatment consists of feeding such drugs as tetrachlorethylene, to remove the worms from the intestinal tract. Preventive measures include the wearing of shoes and the installation of proper sanitary disposal systems for human wastes.

Tapeworm. Undercooked beef or fish may contain tapeworm larvae. In the intestine, they attach themselves to the wall by means of hooks and suckers. As they feed on the food, they grow rapidly, sometimes reaching a length of twenty feet. They produce millions of eggs which pass out of the body with the wastes, and which may be swallowed by grazing cattle.

Treatment with drugs makes the tapeworm loosen its hold on the intestinal wall so that it can be excreted from the body. Careful meat inspection and thorough cooking of meat are important preventive measures.

DISEASES NOT CAUSED BY GERMS

In 1900, the chief causes of death in this country were the infectious diseases. Since that time, these diseases have been so successfully studied and overcome that they are no longer a dread threat. Today, the chief causes of death are diseases not caused by germs—heart disease and cancer.

HEART DISEASES. As the leading cause of death, the heart diseases pose a challenge to both the research scientist trying to discover the predisposing causes, and to the average member of our fast-moving civilization to learn to live sensibly in view of presently known facts.

The following advice is being offered to adults by more and more doctors as a precaution against heart troubles:

1. Limit the ingestion of fats to a maximum of 30 per cent of the calorie intake. Recent evidence indicates that the deposit of *cholesterol,* a fatty substance on the linings of the arteries, (atherosclerosis), may lead to high blood pressure, and a greater chance of heart attack from the blocking of the

coronary arteries which feed the heart. Saturated fats in butter, cream and animal fats that remain solid at room temperature tend to raise the level of cholesterol in the blood. However, unsaturated fats such as liquid vegetable cooking oils tend to lower this level. Although a simple clear-cut relationship between cholesterol and heart disease has not been conclusively established, it seems advisable that one should cut down on foods containing saturated fats.

2. Quit smoking; or cut down on the number of cigarettes smoked; or especially do not start smoking in the first place. There is extensive evidence that smoking contributes to or accelerates the development of coronary heart disease. The president of the American Heart Association recently stated that death rates from coronary heart diseases in middle-aged men were 50 to 150 per cent higher among heavy cigarette smokers than among those who do not smoke.

3. Keep your weight down to normal. Extra weight produces a strain on the heart since blood must be pumped to and from the excess tissue. For every excess pound, the body builds nearly a mile of capillaries. This unnecessary load makes the heart work harder, and over a period of time weakens it. A review of mortality records of one life insurance company showed that for every 100 men of normal weight who died at the age of forty-five, 139 men who were definitely overweight died at the same age.

4. Exercise mildly every day. This helps to maintain muscle tone, lymphatic circulation, and good heart condition. Walking is considered good exercise. By contrast, strenuous "week-end" athletics may be harmful if a person is not accustomed to regular exercise. The heart specialist who treated former President Eisenhower after his heart attack while still in office, stated that daily bicycle riding, swimming, or other mild form of exercise was essential for a healthy heart condition.

5. Relax and get plenty of rest. Do not develop the habit of worrying or becoming anxiety-ridden.

Types of heart diseases. There are various conditions that produce heart disease.

1. **Coronary heart disease.** As a person grows older, the walls of the arteries lose their elasticity and become thicker. This "hardening of the arteries" (arteriosclerosis) causes an increase in the blood pressure. The coronary arteries which supply blood to the tissues of the heart itself are also affected. When this occurs, the heart muscle may not receive enough of a blood supply and severe pains of the chest may result during a heart attack.

2. **Coronary thrombosis.** A small blood clot (*thrombus*) may block one of the coronary arteries, thereby shutting off the supply of blood to a part of the heart. Extreme pains of "acute indigestion" result, which may be followed by death.

3. **Hypertension or high blood pressure.** Prolonged strenuous living without sufficient periods of relaxation, may be one of the causes of hypertension, or high blood pressure. The continuous strain on the heart may cause it to become enlarged, and finally to stop beating.

4. **Infections of the heart.** An attack of scarlet fever or a streptococcus infection in young children may result in *rheumatic fever*. There is a weakening of the valves of the heart and pains in the joints. Such children may need special care for many years before they can outgrow the condition. Infections of the heart lining may also occur among people who are suffering from venereal disease.

CANCER. Cancer is now the second leading cause of death. Its cause is unknown and has been one of the baffling mysteries confronting medical research. It is a wild growth of cells that starts in any part of the body. A *benign tumor* is an abnormal growth that stops growing and is harmless. A *malignant tumor* is a harmful growth that continues to grow and invades surrounding tissues and organs. *Metastases* are cancer cells that spread to other parts of the body by means of the lymph and circulatory systems, and continue to multiply there. At first, a cancerous growth may be undetected or may appear as a painless lump. It is for this reason that an educational campaign has been sponsored by the American Cancer Society to alert everyone to certain symptoms that should be reported to a doctor at once.

Seven danger signals of cancer

1. A persistent sore that does not heal.
2. A lump or thickening, especially in the lip or breast.
3. Irregular bleeding from any of the body openings.
4. A change in the size or color of a wart or mole.
5. Persistent indigestion.
6. Continuous hoarseness or difficulty in swallowing.
7. Marked changes in normal bowel habits.

Treatment. If detected early enough, cancer can be cured. Much progress has been made in its early detection and treatment. About 40,000 Americans are now saved every year, who would have died of cancer fifteen years ago. One out of every three cancer patients is now being cured. The effective methods of treatment are: surgery, x-ray, radium and radioactive isotopes.

Surgery involves the removal of the cancerous tissues. If a surgeon is undecided about the malignancy tissue, a *biopsy* is performed. While the patient is on the operating table, a small piece of tissue is removed and quickly examined microscopically by a pathologist. If typical cancer cells are present, the surgeon is notified within a matter of minutes and the operation proceeds.

Cancer cells are more sensitive than normal cells to radiations of X-ray, radium or radioactive isotopes, and can be destroyed by them without harm to the rest of the body. The amount of the dosage and the time of exposure are carefully controlled. In some cases, radioactive cobalt is injected directly into cancerous tissue. Cancer of the thyroid gland is treated by having the patient drink radioactive iodine. The concentration of this iodine in the thyroid gland serves as a source of destructive rays to kill the cancerous cells.

Recent research has yielded a number of interesting clues to the cancer riddle. The *Papanicolau* smear technique has helped diminish the danger of cancer of the uterus, by making possible simple microscopic examination of surface cells from the uterus. Female *hormones* are being used to treat cancer of the prostate gland in men. Some *chemicals,* including deadly mustard gas, have shown promise in inhibiting, if not eliminating malignancies. A search is being conducted to develop a simple *urine or blood test* that will reveal the presence of cancer cell by-products. There is increasing evidence that there may be a *virus* mechanism involved in the cancer growth pattern. Over twenty independent studies have shown an association between cigarette smoking and lung cancer. Recent growing evidence indicates that there may be a link between cancer and virus infection, in which the DNA makeup of the cell is changed. These and other fields of study being conducted at scientific centers like the Sloan-Kettering Institute for Cancer Research undoubtedly will yield important answers to the cancer problem.

OTHER DISEASES NOT CAUSED BY GERMS

NUTRITIONAL DISEASES. Some diseases are caused by a lack of proper amounts of nutrients in the body.

1. **Anemia.** Failure of the body to build hemoglobin found in red corpuscles, because of a lack of iron.
2. **Beriberi.** A nervous disorder accompanied by lack of energy, due to a lack of thiamin (vitamin B_1).
3. **Bleeding.** Failure of the blood to clot. May be caused by a lack of vitamin K.
4. **Goiter** (endemic). A swelling of the thyroid gland brought on by a lack of iodine in the diet.
5. **Night blindness.** Inability to see well in dim light, caused by a lack of vitamin A which is needed in the production of visual purple in the retina of the eye.
6. **Pellagra.** Inflammation of the skin and tongue, accompanied by mental breakdown caused by a lack of niacin.
7. **Rickets.** A deformity of the bones of children in which they remain soft, because of the lack of either calcium or vitamin D, or both.
8. **Scurvy.** A condition of hemorrhages, bleeding gums, and swollen joints brought on by lack of ascorbic acid (vitamin C).
9. **Sterility.** In rats, failure to produce young, caused by lack of vitamin E.
10. **Tooth decay.** Recent evidence points to the lack of fluorine in the diet as a major cause; calcium, phosphorus and vitamin D are also necessary for tooth formation.
11. **Xerophthalmia.** Drying condition of the eye and epithelial tissue, caused by a lack of vitamin A.

ENDOCRINE DISEASES. Some diseases are caused by the improper functioning of the endocrine glands.

1. **Cretinism.** A condition of mental and physical retardation in children caused by lack of thyroxin.
2. **Myxedema.** Extreme mental and physical sluggishness and overweight in adults brought on by a lack of thyroxin.
3. **Exophthalmic goiter.** A condition of nervous activity, protruding eyeballs and underweight, brought on by an excessive production of thyroxin.
4. **Rheumatoid arthritis.** One of the causes of this disease seems to be a lack of cortisone and/or ACTH.
5. **Dwarfism.** Caused by an undersecretion of the hormone of the pituitary gland that controls the growth of the long bones.
6. **Gigantism.** Caused by an oversecretion of the hormone of the pituitary gland that controls the growth of the long bones.
7. **Acromegaly.** Abnoral enlargement of the hands, feet and face in adults, caused by excessive secretions of the growth hormones of the pituitary gland.
8. **Addison's disease.** A condition in which the skin becomes bronze-colored, accompanied by loss of weight and weakness, caused by lack of cortin.
9. **Diabetes.** Inability of the body to oxidize and store glucose, caused by lack of insulin.

MISCELLANEOUS DISEASES. Another group of diseases are caused in various ways.

1. **Allergies.** A variety of conditions, including asthma, hay fever, skin rashes, sneezing and coughing, are caused by sensitivity to various substances called *antigens*. An antigen is a foreign substance which enters the body and causes it to produce antibodies against it. When the person is subsequently exposed to the antigen, an antigen-antibody reaction occurs, in which the white blood cells release histamines. These chemicals bring on the uncomfortable symptoms of the allergy. Various anti-histamine drugs are now available to bring relief to the allergy sufferer.
2. **Hemophilia.** Bleeder's disease which is hereditary and due to a recessive, sex-linked, gene.
3. **Silicosis.** Occupational disease of miners caused by the collection of fine sand particles in the lungs.
4. **Radiation sickness.** Burns and weakness caused by overexposure to nuclear radiation. Death may result.

SUMMARY OF DISEASES

DISEASE	CAUSE	HOW SPREAD	PREVENTION	TREATMENT
Bubonic Plague (Black Plague)	Bacillus	Bite of flea from infected rodent	Control of rats; vaccine	Anti-plague serum
Diphtheria	Bacillus	Airborne; contact	Active immunization with toxoid; Schick Test	Inoculation of antitoxin
Malaria	Plasmodium (protozoan)	Bite of infected female Anopheles mosquito	DDT and other methods of mosquito control	Quinine, Chloroquine, Atabrine Primaquine
Measles	Virus	Airborne; contact	Gamma globulin fraction of blood; isolation	Bed rest
Rabies	Virus	Bite of rabid animal	Pasteur treatment or duck-embryo vaccine before symptoms appear; immunization of dogs with vaccine	No specific treatment; horse serum for immediate protection
Poliomyelitis (Infantile Paralysis)	Virus	Suspected: contact, airborne, polluted water	Either Sabin vaccine-weakened virus, or Salk vaccine-dead virus; isolation	Rest; muscle therapy
Smallpox	Virus	Contact; airborne	Vaccination with weak virus; isolation	None
Pneumonia	Coccus (some types)	Droplet infection	Maintenance of high resistance	Penicillin, sulfa, immune serum
Tetanus	Bacillus	Puncture wound	Toxoid immunization	Antitoxin injection
Hookworm	Roundworm	Walking barefoot on infected soil	Wearing shoes; proper sanitary facilities	Drugs eliminate the worm
Tuberculosis	Bacillus	Droplet infection; contact; contaminated food and articles; feet of housefly	Chest X-ray; tuberculin test; sputum analysis; BCG vaccine; tuberculin testing of cows; milk pasteurization; good bodily health	Rest; good food; combination of streptomycin and isoniazid; pneumothorax

SUMMARY OF DISEASES (Continued)

DISEASE	CAUSE	HOW SPREAD	PREVENTION	TREATMENT
Trichinosis	Trichinella worm (round-worm)	Eating poorly cooked pork	Thorough cooking of pork; feeding steamed garbage scraps to pigs	No specific treatment
Typhoid fever	Bacillus	Polluted water and contaminated food; feet of housefly; carrier (human)	Immunization with dead bacteria; water and food purification; inspection of food handlers; fly control	Chloromycetin
Typhus fever	Rickettsia	Bite of infected body louse	Body cleanliness; DDT spray of clothing; vaccine of dead virus	Chloromycetin
Whooping cough	Bacillus	Droplet infection; contact	Isolation of patients; active immunization with vaccine of dead bacilli	Convalescent serum; terramycin or aureomycin in early stages
Yellow fever	Virus	Bite of infected female Aedes mosquito	DDT and other methods of mosquito control; vaccine of weak virus	No specific treatment

CHAPTER REVIEW

Select the correct choice in each of the following statements.

1. All of the following include types of germs except (A) bacteria (B) viruses (C) protozoa (D) toxins (E) worms

2. A chest X-ray is useful in detecting a case of (A) diphtheria (B) typhoid (C) tetanus (D) tuberculosis (E) typhus

3. All of the following are useful methods of providing protection against tuberculosis except (A) pneumothorax (B) use of Schick text (C) BCG (D) rest (E) pasteurization of milk

4. Oysters grown in polluted water may contain the germs of (A) typhoid (B) typhus (C) smallpox (D) yellow fever (E) scurvy

5. A test for syphilis is the (A) Wasserman test (B) Widal test (C) Dick test (D) Schick test (E) Heaf test

6. Antitoxins against diphtheria are obtained from (A) weakened virus (B) horse (C) toxoid (D) *Penicillium notatum* (E) people who are naturally immune

7. A disease caused by bacteria is (A) smallpox (B) rabies (C) polio (D) common cold (E) tetanus

8. A person being vaccinated against smallpox receives (A) live cowpox virus (B) dead cowpox virus (C) live smallpox virus (D) dead smallpox virus (E) weakened smallpox virus

9. A disease in which the nerves are damaged is (A) silicosis (B) tetanus (E) rheumatic fever (D) polio (E) yellow fever

10. Viruses can be seen with the aid of the (A) oil immersion microscope (B) synchotron (C) electron microscope (D) spectroscope (E) ultracentrifuge

11. A disease in which the brain is affected is (A) bubonic plague (B) pellagra (C) gonorrhea (D) rabies (E) xerophthalmia

12. The Anopheles mosquito spreads (A) typhoid (B) malaria (C) typhus (D) African sleeping sickness (E) amnesia

13. All of the following are effective ways to control mosquitoes except (A) spray DDT (B) spray antitoxin (C) drain swamps (D) remove rain barrels (E) screen windows and doors

14. A scientist who proved that the Aedes mosquito spreads yellow fever was (A) Laveran (B) Ehrlich (C) Roux (D) Ross (E) Reed

15. The black plague is transmitted by a (A) cockroach (B) mosquito (C) fly (D) flea (E) mad dog

16. A disease caused by a rickettsia is (A) typhoid (B) typhus (C) rickets (D) beriberi (E) hemophilia

17. Two diseases caused by protozoa are (A) malaria and yellow fever (B) malaria and African sleeping sickness (C) yellow fever and African sleeping sickness (D) malaria and pneumonia (E) yellow fever and pneumonia

18. All of the following diseases are caused by worms except (A) trichinosis (B) hookworm (C) tapeworm (D) ringworm (E) liver fluke

19. Eating undercooked pork may lead to (A) hookwarm (B) trichinosis (C) pellagra (D) scurvy (E) athlete's foot

20. The leading cause of death at the present time is (A) tuberculosis (B) cancer (C) heart disease (D) myxedema (E) pneumonia

21. All of the following steps are advisable as a precaution against heart disease except (A) exercise mildly every day (B) reduce smoking (C) increase ingestion of fats (D) relax (E) keep weight down to normal

22. Rheumatic fever may result (A) from an attack of chicken pox (B) from an attack of measles (C) from a dislocated joint (D) in weakening of the valves of the heart (E) in weakening of the long bones

23. Effective methods of cancer treatment include the use of all of the following except (A) cosmic rays (B) X-rays (C) surgery (D) radium (E) radioactive isotopes

24. A contagious disease is (A) anemia (B) cretinism (C) acromegaly (D) cancer (E) tuberculosis

25. Jenner introduced the method of making people immune to (A) cholera (B) rabies (C) smallpox (D) polio (E) Rocky Mountain spotted fever

ANSWER KEY

1-D	6-B	11-D	16-B	21-C
2-D	7-E	12-B	17-B	22-D
3-B	8-A	13-B	18-D	23-A
4-A	9-D	14-E	19-B	24-E
5-A	10-C	15-D	20-C	25-C

HOW LIVING THINGS REPRODUCE

6

19 Reproduction among lower organisms

One of the properties of protoplasm is its ability to reproduce more of its kind. In one-celled organisms, the entire cell reproduces. In higher organisms, there are specific structures that specialize in reproducing the individual.

SPONTANEOUS GENERATION

At one time it was believed that living things could arise from nonliving matter. Some examples of this idea: Frogs and mice originate from mud; threads of cotton turn into worms; flies develop from decaying meat. This belief of spontaneous generation was disproved for the first time by *Francesco Redi* during the seventeenth century. He conducted the following experiment to determine whether flies could develop from decaying meat:

1. He allowed meat to decay in each of three jars. He left the first jar open, covered the second jar with gauze, and sealed the third with parchment.

2. For several weeks, he observed each jar: In the first jar, flies flew in and out of the jar, and maggots appeared on the meat. In the second jar, flies flew around the gauze and formed their maggots on top of it, but there were no flies inside. No flies were present around the third jar.

3. He decided that, in the first jar, flies were attracted by the odor of decaying meat and reproduced on it. In the second jar, they could not reach the meat and reproduce on it. The sealed third jar did not attract them at all. Therefore, he concluded, flies originate from other flies, and not from decaying meat.

Later, the discovery of bacteria by Leeuwenhoek led to the belief that they were simple enough to originate spontaneously from favorable food conditions. Spallanzani in the eighteenth century, and Pasteur and Tyndall in the nineteenth century, conducted conclusive experiments to disprove this idea. Spallanzani and Pasteur showed that there were no bacteria in sealed flasks of broth that had been heated, but that when they were opened, bacteria of the air would enter and multiply. Pasteur drew out the necks of flasks in a pronounced curve, and left the ends open; bacteria had no way of entering, and the broth remained sterile. It was definitely decided that living things can only come from previously existing living things.

Although there is no spontaneous generation of present forms of living things, scientists postulate that the simplest form of life on earth probably arose billions of years ago when conditions were different, from combinations of complex molecules, and gradually evolved into life as we now know it.

Asexual reproduction. Reproduction in which there is only one parent is known as asexual reproduction. Examples are: (1) fission, (2) budding, (3) spore formation.

Fission. After a one-celled organism, such as an ameba, grows to its maximum size, it reproduces as follows: The nucleus divides in two; equal amounts of cytoplasm collect around each nucleus; the cell splits into two daughter cells that are equal in size. Each cell is independent; it grows to its maximum size, then splits again by binary fission. Other organisms that reproduce in this way are: paramecium, pleurococcus, spirogyra, and bacteria. In the paramecium, both the macronucleus and the micronucleus divide in half and are distributed to each of the two new cells.

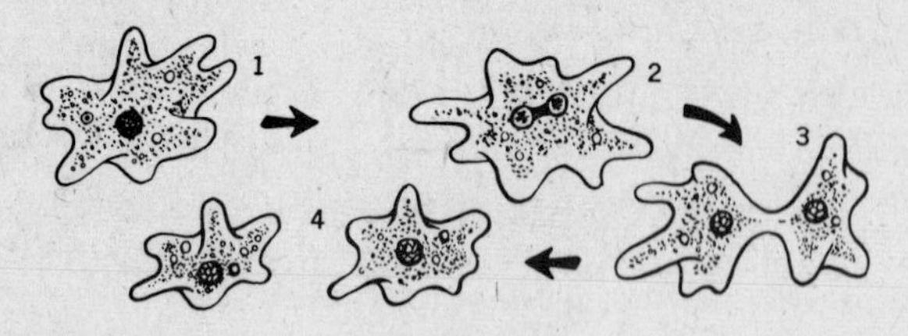

Fission in Ameba

Budding. Yeast cells are microscopic fungi that are useful in the baking and brewing industries, because they carry on *fermentation* of sugar. During this process, they produce carbon dioxide and alcohol. When a yeast cell reproduces, it forms a *bud;* the nucleus splits; a small amount of cytoplasm gathers around one of the nuclei; and a cell wall separates it to form a little cell attached to the large one. The two cells are *unequal* in size, and the bud remains attached to the larger cell. In time, it will grow and produce other buds.

The small animal, hydra, also reproduces by budding. When it is growing vigorously, a projection forms on its outer layer of cells, elongates, forms tentacles and a mouth, and remains attached to the parent animal. After a while, cells grow across its base and separate it from the parent.

Spore formation. If bread is kept in a moist, dark place, mold usually develops on it, unless the bread was made with calcium propionate, a chemical that retards the growth of molds. The bread mold (Rhizopus) consists of a mass of threads called the *mycelium.* Under the low power of a microscope, it can be seen to be made of stem-like structures called *hyphae.* At the end of a hypha, a ball-like swelling, the *sporangium,* or spore case, forms. As it develops, the nucleus divides many times. Each one of the nuclei becomes surrounded by a small amount of cytoplasm, and a tough cell wall. Hundreds of these tiny cells, or *spores,* are produced, and when the wall of the sporangium bursts, they are scattered about. Under favorable conditions of moisture, food and subdued light, each spore can germinate to form a new mold. Until then, the spores are resistant to extremes of temperature and to drying. They have been recovered from the stratosphere, more than seven miles up, and have germinated successfully when brought down and incubated.

Other fungi, including yeast, mushrooms and wheat rust, also reproduce by *sporulation,* or spore formation. In the blue-green mold from which penicillin is made (Penicillium), the spores are produced in little chains at the ends of finger-like projections from the hypha. Mosses and ferns also produce spores in the asexual phase of their life history. In mosses, the spores are found in capsules at the end of separate stalks; in ferns, they are usually located in the small round spots on the underside of the leaves.

Regeneration. If a hydra loses some of its tentacles, it grows new ones. If it is cut into several pieces, most of the pieces will grow into complete hydras. This ability to grow back a missing part, or to develop an entire organism from a part, is known as *regeneration.* A crab or a lobster can grow back a new claw. If a starfish is cut into several pieces, they can each regenerate into new organisms, especially if they contain parts of the central disc. Many experiments in regeneration have been performed with the small flatworm, Planaria, which is about a quarter of an inch long. It will regenerate into two new worms, if it is cut in half. Interesting results may be obtained by cutting it in different ways. In vertebrates, the power of regeneration is reduced. A salamander or a lizard can grow back a new tail. A human being can regenerate new skin in a cut, but cannot replace a limb, or part of one.

Sexual reproduction. Two parents, or two parent cells, are involved in sexual reproduction. Two cells or *gametes* unite to form a single cell, the *zygote,* which develops into a new individual. The two gametes are *alike* in the type of sexual reproduction called *conjugation;* they are *unlike* in *fertilization.*

Conjugation may be studied in spirogyra, bread mold, and paramecium.

1. **Spirogyra.** The cells of spirogyra are arranged in a long thread or filament. Each cell reproduces asexually by binary fission, resulting in the lengthening of the filaments. In the fall of the year, as the filaments lie parallel to each other in the pond, they undergo conjugation. Projections grow out toward each other from the cells of adjoining filaments. When they touch, their ends are dissolved by enzymes, forming a connecting tube between the cells. Now all the protoplasm of the cells in one filament moves through the tubes into the adjoining cells, and unites with them. The cells that move across are called *active gametes;* those that remain stationary, *passive gametes;* they are alike in size and shape. A single cell, the *zygote,* results from the union of the two gametes. A tough, resistant wall forms around it, and it is now a *zygospore.* Under the microscope, the two adjoining filaments

can be seen with connecting tubes between the adjoining cells; one filament has only empty cells, while the other contains dense zygospores. During the winter, the protoplasm remains protected within the heavy covering of the zygospores. In the spring, the wall of each breaks open, and a new cell emerges; it grows and reproduces asexually by binary fission to become a long filament. Scientists have succeeded in making conjugation proceed artificially by varying the concentration of chemicals in the water.

2. Bread mold. Besides forming spores asexually, bread mold may also reproduce by conjugation if two strains, called *plus* and *minus,* are growing near each other. A thread from one strain will grow toward a thread from the other. When they touch, a cell is formed at each end; these are the *gametes;* they are both alike in size and shape. The separating wall dissolves, and the protoplasm of both gametes unites to form a *zygote.* A zygospore results when a heavy wall grows around it. After a period of time, the wall of the zygospore breaks and a new hypha grows from it; it forms a sporangium and produces spores asexually. Conjugation occurs only when there are two different strains, a plus and a minus; it will not take place between two like strains.

3. Paramecium. During conjugation, two paramecia lie next to each other, and unite at their oral groove regions. Their macronucleus disintegrates, and their micronucleus undergoes a series of complicated changes. A portion of each micronucleus goes into the adjoining paramecium and fuses with the opposite micronucleus. The paramecia separate and divide actively by binary fission. The exchanged portion of the nucleus serves as a gamete. The resulting organisms contain nuclear material from the two original paramecia.

Fertilization. In fertilization, the uniting gametes are different from each other in size, shape and activity. It is the chief method of sexual reproduction among animals and plants.

1. Hydra. In addition to reproducing asexually by budding, hydra reproduces sexually, usually in the fall. A large bulge forms on the body wall; it is called the *ovary,* which will contain one large cell, the egg, which is well supplied with food. Another bulge on the body wall of either the same animal or another one, called the *testis,* forms, and contains many small *sperm* cells. The egg is the female gamete or sex cell; the sperm is the male gamete or sex cell. The testis opens and releases the sperm cells into the water. One of them enters the egg, and unites with it to form the *zygote,* or *fertilized egg.* This divides a number of times to form an embryo, or developing hydra. It drops away from the parent hydra, and forms a hard membrane. In the spring, the embryo completes its development into an independent young hydra. A hydra which contains both ovaries and testes is said to be a *hermaphrodite.*

2. Earthworm. The earthworm is also a hermaphrodite, since it has both ovaries and testes. However, sexual reproduction takes place only between two different individuals. During mating, they lie next to each other with the head ends pointing in opposite directions. Each worm receives sperm cells from the other in special sacs called sperm receptacles. Following this exchange, the worms separate. A swelling near the middle of the worm secretes a ring of mucus which glides toward the front of the worm. As it moves over the opening containing the eggs a number of eggs pass into it; farther along, sperm cells come out. The eggs are fertilized by the sperm in this mucous ring, or cocoon, which finally comes off the worm at the head end and is left in the soil. The fertilized eggs or zygotes develop into young earthworms, which move about and then leave the cocoon.

3. Insects. Insects are either male and have testes, or female with ovaries. During mating, the male introduces sperm, into a sac called the sperm receptacle, in the female's body, where they are stored until egg-laying time. As the eggs pass by this sac on the way out of the female insect's body, they are fertilized by the sperms. Each egg has a large amount of stored food, on which the developing embryo feeds. Some insects such as the butterfly go through four stages in their life history; egg—larva (caterpillar)—pupa (chrysalis)—adult. This is known as complete metamorphosis. Others, such as the grasshopper, have only three stages: egg—nymph—adult. This is known as incomplete metamorphosis.

5. Mosses. Lower plants also reproduce by fertilization. Mosses develop from asexual spores that are produced in spore cases. Each spore germinates on the moist ground and produces a small green filament (protonema). This becomes the moss plant. Egg cases are formed on the upper part of the plant, each containing an egg. Sperms are produced nearby, on the same plant or on other plants, and move through the moisture to the egg cases, where they fertilize the eggs. From the fertilized egg, a new "plant" develops, with a stalk and a spore case, deriving its nourishment from the original moss plant beneath it. There are thus two stages in the life cycle of the moss: the spore-making stage, or *sporophyte,* which is asexual, and which consists of the stalk and spore case; and the gamete-producing stage or *gametophyte,* which is the moss plant, and in which fertilization occurs. This alternation of a sporophyte phase and a gametophyte phase is known as *alternation of generations.*

6. **Ferns** also go through an alternation of generations. Spores are produced asexually on the under surface of the leaves, in rows of little spots. Each of these spots contains a number of *sporangia,* or spore cases. When these are ripe, they burst, scattering the spores. A spore will germinate in a moist place and instead of developing into a new fern, it forms a small heart-shaped plant, the *prothallus.* This is the gametophyte; it is less than a half inch in size, and can carry on photosynthesis. It forms egg cases with egg cells in them, and sperm cells. The sperms swim in the moisture to the egg cases, and one sperm will unite with an egg to form a fertilized egg. The new fern plant, or sporophyte, develops from this fertilized egg. It forms leaves and then produces more spores asexually. Thus, the spore-forming stage, or fern plant, is asexual; it alternates with the gamete-forming stage, or prothallus, which is the sexual phase of its life history.

CHAPTER REVIEW

Select the correct choice for each of the following statements.

1. All of the following scientists helped disprove the idea of spontaneous generation except (A) Redi (B) Spallanzani (C) Trudeau (D) Pasteur (E) Tyndall

2. Most protozoa reproduce by (A) budding (B) fission (C) sporulation (D) parthenogenesis (E) regeneration

3. Two organisms that reproduce by budding are (A) yeast and hydra (B) yeast and spirogyra (C) hydra and spirogyra (D) yeast and pleurococcus (E) hydra and pleurococcus

4. All of the following organisms produce spores except (A) bread mold (B) Penicillium (C) paramecium (D) moss (E) fern

5. A lobster can grow back a new claw by the process of (A) sporulation (B) metamorphosis (C) parthenogenesis (D) budding (E) regeneration

6. The uniting cells in sexual reproduction are called (A) zygotes (B) zygospores (C) spores (D) gametes (E) buds

7. Two organisms that reproduce by the process of conjugation are (A) spirogyra and ameba (B) paramecium and ameba (C) bread mold and spirogyra (D) bread mold and ameba (E) ameba and yeast

8. A zygote is a cell that (A) results from the union of two gametes (B) unites with another zygote to form a gamete (C) unites with another zygote to form a zygospore (D) results from the union of two zygospores (E) results from the union of a gamete and a zygospore

9. The simplest method of animal reproduction is (A) conjugation (B) fertilization (C) fission (D) sporulation (E) parthenogenesis

10. During the life history of a fern, the fern plant is (A) the sporophyte stage (B) the gametophyte stage (C) the prothallus stage (D) the structure that produces egg cells and sperms (E) developed from a spore

ANSWER KEY

1-C	3-A	5-E	7-C	9-C
2-B	4-C	6-D	8-A	10-A

20 Reproduction in higher plants

Flowers are the specialized organs of reproduction among the higher plants, or angiosperms. These plants are classified into various families largely on the basis of the structure and arrangement of the flowers.

Structure of a flower. Although different flowers vary in their appearance and structure, they tend to have the same basic parts:

1. **Petals.** These are the brightly colored parts that make up the most attractive feature of flowers. The group of petals in their circular arrangement or whorl is called the *corolla.*

2. **Sepals.** They are the small, green, leaflike structures which form an outer layer at the base of the petals. At one time, they enclosed the flower

bud. Together, they make up the *calyx*.

3. Stamens. Within the petals are many stamens, each of which consists of: a slender stalk, the filament; and the enlarged part on top, the *anther*, or pollen case, which contains the powdery *pollen*.

4. Pistil. This is located in the center of the flower, within the stamens. It contains an enlarged portion at the bottom, the *ovary*. Extending above is the slender *style*, at the top of which is the sticky *stigma*. The ovary contains one or more small structures, the *ovules*. When the flower matures, these ovules will become the seeds; the surrounding ovary will become the fruit.

The stamens and pistils are referred to as the *essential organs* of the flower, since they are directly concerned with reproduction. In some plants they are located in separate flowers. Thus, in corn, the stamens are borne in *staminate* flowers at the top of the plant, and make up the *tassel*. The pistils are found farther down on the plant, in *pistillate* flowers, making up the ear; the mass of their stigmas and styles makes up the *silk*. In the poplar family, which includes poplar and willow trees, the staminate and pistillate flowers are formed on separate trees, in structures called *catkins*. However, most plants have flowers containing both stamens and pistils, which are said to be *perfect* flowers.

Pollination. The transfer of pollen from the stamen to the stigma is known as pollination. If this takes place within the same flower, as in the pea or bean, it is called *self-pollination;* if pollen is transferred from the anther of one flower to the stigma of another flower, it is called *cross-pollination*. Cross-pollination may be accomplished in various ways, i.e., *insects* (bees, especially important in cross-pollinating apple trees); *wind* (corn and other grasses; gymnosperms); *man* (to improve varieties of flowers—this is also known as artificial pollination). Insects, as well as the hummingbird, are attracted to flowers to obtain the sugary liquid called *nectar* formed at the base of the flower. In this process of food getting, the pollen which becomes deposited on their bodies is distributed to other flowers.

Fertilization. Pollination is the first step in the fertilization of a flower. When the pollen grains are deposited on the stigma, they each begin to germinate, and form a *pollen tube*. A pollen grain is a single cell; after it has matured in the anther, it usually contains two nuclei. Now, as the pollen tube grows down through the style toward the ovule, one of these nuclei divides to form two nuclei. These two are the *sperm nuclei*, or male gametes; the other is the *tube nucleus*.

Within each ovule, one of the cells enlarges and is called the *embryo sac*. Its nucleus divides several times to form eight nuclei; one is the *egg nucleus;* two others unite to become the *endosperm nucleus;* the other nuclei are not significant.

As the pollen tube reaches an ovule, it enters through a tiny opening called the *micropyle*, and the sperm nuclei pass into the embryo sac. Double fertilization now takes place. One of the sperm nuclei unites with the egg nucleus to form a *fertilized* egg, or *zygote*. The other sperm nucleus unites with the *endosperm nucleus*. The tube nucleus disintegrates.

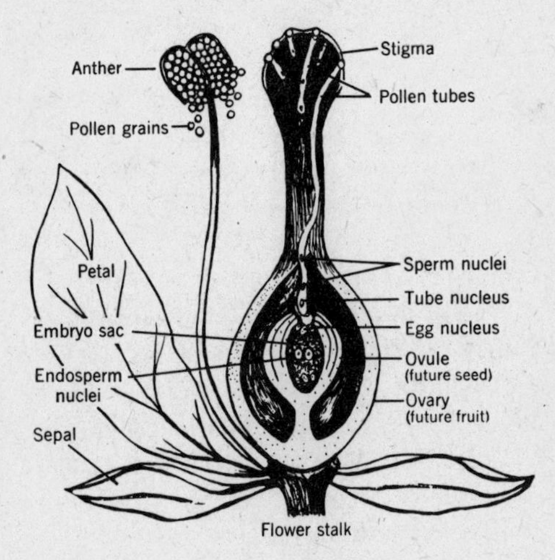

Flower Fertilization

Formation of the seed. Following double fertilization, the ovule becomes a seed. The fertilized egg undergoes numerous divisions to form an embryo, or baby plant. At the same time, the fertilized endosperm nucleus divides to form many large cells in which food is stored for use by the embryo later on, when the seed germinates. The outer cells of the ovule surrounding the embryo sac become the covering layer of the seed, or *testa*.

Parts of a seed. These parts can be seen by studying a large seed, such as the lima bean. It should first be soaked in water for several hours. On the outside of the testa, along the narrow end of the seed, a scar, or *hilum*, is easily seen; it shows the point of attachment of the seed to the ovary wall, or the pod. Near it, the tiny opening is the *micropyle*, through which the pollen tube entered the ovule. If the bean is split open, it can be seen to have two large *cotyledons*, each of which contains stored food. The remaining part of the embryo contains the *epicotyl* and its tiny plumules, which make up the upper part of stem and leaves of the plant, respectively; and the *hypocotyl*, which will develop into the lower part of the stem and the roots.

Formation of the fruit. After fertilization, while the ovule is maturing into a seed, several other changes are taking place in the flower. The petals, sepals, stamens, style and stigma shrivel and usually drop off. The ovary enlarges and becomes the fruit. In the apple, additional fleshy tissue is added to the ovary, providing the edible part. The dried-up remains of the sepals, styles and stamens can be seen at the bottom of an apple. A pea pod is also a fruit, since it is a matured ovary and contains seeds; the remains of the sepals can usually be seen at one end of the pod, and the dried-up style and stigma at the other. The tomato, cucumber and squash are also fruits, having developed from the ovary of a flower. A kernel of corn is considered to be a fruit containing one seed, since it developed from an ovary, in which there was one ovule; an ear of corn is thus a collection of many fruits.

In order for an ovule to become a seed and for an ovary to mature into a fruit, pollination and fertilization must take place. Unless the male and female gametes unite, a fruit does not "set," and the seeds do not develop. Recent research with chemical growth-promoting substances, however, has shown that ovaries can be made to develop into fruits artificially. *Gibberellin* and *auxins* are two such examples. When they are sprayed on a young tomato flower, the ovary grows and develops into a seedless fruit.

Seed dispersal. Plants have various adaptations by which seeds can be scattered or dispersed. (1) Fleshy fruits such as apples, cherries and watermelons are eaten by man, birds and other animals, and the seeds then discarded far from the original plant. (2) The wind scatters the seeds of dandelion and milkweed, each being equipped with a silken parachute; the seeds of the maple, ash and elm trees have broad blades by which they are blown about in the wind. (3) Some seeds such as those of the cocklebur have small hooks by which they become attached to passing animals. (4) The seeds of the witch hazel tree and the touch-me-not plant are popped away from the plant by the explosion of the fruit as it dries up. (5) The coconut can be carried to distant shores by floating in the water.

If there were no seed dispersal, the seeds would all germinate in one restricted area around the mother plant. They would then be so crowded that there would not be enough light, space, water and minerals for them all to survive. By being scattered about, there is a better opportunity for many seeds to develop into new plants.

Seed germination. After a seed has been formed, its embryo usually goes through a resting, or dormant, period, during which it can endure conditions of drying up and freezing weather. In the spring, with warmth and moisture available, the seed begins to sprout, or *germinate.* The embryo grows, obtaining its nourishment from the stored food in the seed. Roots develop first, to anchor the seedling and to obtain moisture. The stem and leaves grow up above the ground, and the young plant begins to make its own food. By differentiation, the various tissues of the plant develop.

Vegetative propagation. In some cases, flowering plants can reproduce more of their kind without the use of flowers, but from stems, leaves or roots. Since these are the vegetative parts of the plant normally used for nutrition, this method of reproduction is known as *vegetative propagation.* It is a form of asexual reproduction, since it involves only one parent.

1. **Cutting, or slip.** A stem may be cut and removed from a geranium or coleus plant, and placed in moist sand or in water. After a period of time, this *cutting, or slip,* will develop roots at the cut end of the stem. *Plant growth hormones* are now used to speed up the rate of root formation or cuttings. The cut end of the stem is dipped into a solution, or the powder itself, of the growth hormone. In this way, woody cuttings which normally do not produce roots have been induced to form them. If the leaves of some plants such as bryophyllum and begonia are placed on moist soil, new plants begin to develop from various points on the leaves; these are *leaf cuttings.*

2. **Runner.** A horizontal stem of the strawberry plant, called a *runner,* will grow along the ground. A short distance from the main plant, a new plant will develop, with roots that anchor it to the ground, and with a stem with leaves. In time, this new plant will also form runners.

3. **Layering.** In the blackberry and forsythia plants, some of their branches bend over and come in contact with the ground. They take root at these places, and new stems and leaves develop into independent plants.

4. **Rhizome.** In some plants such as ferns, snake plant, and Canada thistle, a thick underground stem, called a *rhizome,* at intervals sends up stems that produce leaves. Each of these becomes an independent plant.

5. **Tuber.** The white potato is really a fleshy, modified underground stem, known as a *tuber.* There are numerous buds, or *eyes,* on it, that can develop into new plants with roots, stems and leaves. A farmer plants a potato crop by cutting up potatoes in such a way that each piece has at least one eye. The young developing plant obtains its nourishment from the stored food in the potato.

6. **Bulb.** The tulip and onion *bulbs,* which are

formed underground, consist of short stems surrounded by fleshy leaves containing stored food. When planted, a bulb sends out roots from its base, while leaves and a flower grow upward, deriving their nourishment from the stored food in the bulb. Later on, as they manufacture their own food, the plants will store it in new bulbs that are formed next to the original one. Other plants that reproduce by bulbs are: hyacinth, gladiolus and lily.

7. **Fleshy root.** The carrot and sweet potato are modified roots containing stored food. If they are placed in contact with water, or are planted in the soil, stems and leaves will sprout upward, and roots will grow down.

8. **Grafting.** Man propagates certain desirable types of woody plants by *grafting*. This is done by selecting and cutting a stem or bud of the desired type of plant. This is the *scion;* it is attached to a closely related tree, the *stock,* on which an equal-sized branch has also been cut and notched. The place of attachment is securely bound and covered with melted wax, to prevent drying out or infection. The graft is performed in such a way that the *cambium,* or growing layer, of both the scion and the stock are in contact. In some cases, a bud is used as a scion, and is inserted under a slit in the bark of the stock. Soon, the graft takes, and the scion produces the desired fruit or flower, while obtaining its water and minerals from the stock. It is possible to buy an apple tree that will bear several different varieties of apple, such as Delicious, Macintosh, Winesap, Northern Spy, and Baldwin. Scions of each of these varieties were grafted onto the stock. The seedless orange is propagated by grafting, as are many desirable varieties of grapes, oranges, lemons, peaches and plums. In all these cases, the scion must be closely related to the stock on which it is grafted.

VALUES OF VEGETATIVE PROPAGATION

1. The plants are of the same type as the parents. They do not vary, as might usually be the case in sexual reproduction, where the characteristics of two parents are inherited.

2. Plants are reproduced much more quickly and in larger numbers, than if they were grown from seeds.

3. Seedless fruits, such as oranges and grapes can be maintained and propagated.

CHAPTER REVIEW

Select the correct choice for each of the following statements.

1. Pollen is produced in the part of the flower called the (A) corolla (B) calyx (C) stigma (D) anther (E) style

2. A fruit develops from the part of a plant called the (A) embryo sac (B) ovary (C) pollen tube (D) egg nucleus (E) endosperm

3. The ripened ovule of a pea plant is called the (A) seed (B) pistil (C) stamen (D) micropyle (E) essential organ

4. The structure which includes all the others is (A) ovary (B) ovule (C) style (D) pistil (E) stigma

5. The transfer of pollen from the stamen to the stigma of the same flower is known as (A) cross-pollination (B) self-pollination (C) fertilization (D) germination (E) seed dispersal

6. The female gamete of a flower is formed in the (A) pollen grain (B) pollen tube (C) stigma (D) style (E) embryo sac

7. All of the following are fruits except (A) pea pod (B) cucumber (C) radish (D) tomato (E) kernel of corn

8. A plant that reproduces by runners is (A) geranium (B) onion (C) lily (D) bryophyllum (E) strawberry

9. A tulip reproduces vegetatively by forming a (A) tuber (B) bulb (C) cutting (D) slip (E) rhizome

10. A tuber is an underground (A) stem (B) root (C) leaf (D) nodule (E) flower

11. Vegetative propagation is a form of (A) sexual reproduction (B) asexual reproduction (C) conjugation (D) fertilization (E) vegetative pollination

12. If a branch of a seedless orange is grafted to a tree that produces oranges with seeds, the (A) stock will bear both seedless and seed oranges (B) stock will bear only seedless oranges (C) scion will bear only seedless oranges (D) scion will bear only seed oranges (E) scion will bear both seedless and seed oranges

13. A bean seed contains all of the following except (A) seed coat (B) epicotyl (C) hypocotyl (D) hypha (E) cotyledons

14. All of the following vegetables are considered to be fruits except (A) tomato (B) cucumber (C) squash (D) carrot (E) corn

15. All of the following are advantages of asexual

reproduction except (A) offspring are the same type as the parents (B) cross-pollination is simpler with large flowers (C) there can be many offspring (D) reproduction is faster (E) seedless fruits can be propagated

ANSWER KEY

1-D	4-D	7-C	10-A	13-D
2-B	5-B	8-E	11-B	14-D
3-A	6-E	9-B	12-C	15-B

21 Reproduction in higher animals

The higher animals, or vertebrates, are alike in having special reproductive organs (gonads) that are devoted to perpetuating the species. These gonads produce two types of substances, (1) sex hormones and (2) gametes.

Secondary sexual differences. The male sex hormone, *testosterone,* and the female sex hormone, *estrogen,* are produced when the male and female animals mature. They are responsible for the appearance of secondary sexual differences between them. Thus, male birds have bright plumage and a singing voice, while the females have dull feathers and rarely sing. Among deer, the male has large antlers and is aggressive; the female lacks antlers and is gentle. The lion has a mane; the lioness does not. In humans, the male develops hair on his face and becomes deep-voiced; the female does not grow hair on her face, continues to have a high-pitched voice, and develops breasts.

Production of gametes. The male gonads are called *testes,* or *spermaries.* They have special cells that form the male gametes or sperm. In humans, a pair of testes is located in an outpocketing of the body wall known as the scrotum. The temperature here is 2-4 degrees lower than the body temperature, providing the best conditions for sperm manufacture and storage. Several glands, including the prostate, secrete a liquid which serves as a transport medium for the sperm.

Each sperm is a microscopic structure in which there is a tail, or flagellum, made of cytoplasm, and a head containing the nucleus. They move by the lashing of the flagella. Sperm cells are produced in extremely large numbers. They leave the body of the male through fine tubes called *sperm ducts.*

The female gonads, or *ovaries,* form the female gametes, which are also known as *eggs,* or *ova.* Each ovum is a single cell containing a nucleus, cytoplasm, and stored food. In practically all of the vertebrates except the mammals, there is a considerable amount of stored food in the eggs. As a result, they can readily be seen with the naked eye. They reach their largest size among the birds. After the eggs are formed in the ovaries, they enter the oviducts.

Fertilization and cleavage. When mating takes place, many sperms swim to the egg cells and surround them. One sperm cell penetrates the membrane surrounding an egg, and unites with its nucleus. A fertilization membrane forms around the egg and keeps the other sperms out. The union of sperm and ovum to form a *fertilized egg,* or *zygote,* is known as *fertilization.* Soon afterwards, the fertilized egg undergoes a series of cell divisions, developing into an embryo that has two, then four, then eight, and then more cells, as each cell divides.

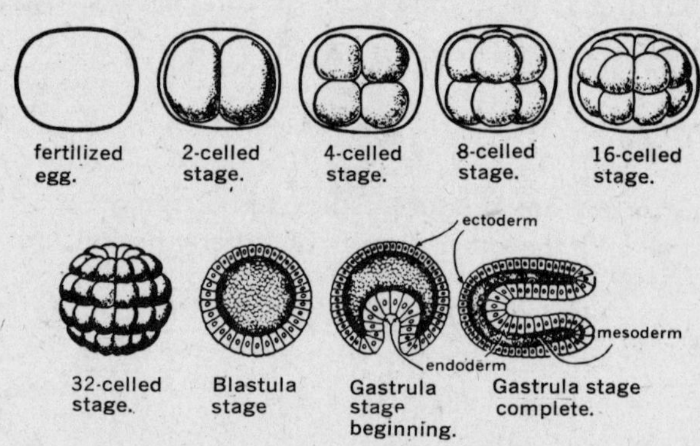

Early development of a fertilized egg.

This series of cell divisions is known as *cleavage.* In many animals, the embryo next takes the appearance of a hollow ball of cells one cell thick, known as the *blastula.* One part of the blastula then grows inward, forming a second layer. This is the cuplike, or *gastrula* stage. The outer layer is called the *ectoderm;* the inner, the *endoderm.* A third or middle layer then develops between these two, known as the *mesoderm.*

Differentiation. Up to this point, by cleavage, the gastrula has been produced consisting of many cells that are quite similar to each other. It is approximately equal in size to the original zygote. From this point on, the embryo elongates as the three layers of cells begin to *differentiate,* or form the different tissues and organs, as follows:

1. The **ectoderm,** or outer layer—epidermis of skin; and nervous system.
2. The **endoderm,** or inner layer—lining of alimentary canal, trachea, and lungs; pancreas, liver, and bladder.

3. The **mesoderm**, or middle layer—muscles, connective tissue including bone and cartilage, blood and circulatory system, kidneys, reproductive organs.

The organs of the body are developed from tissues derived from one or more of these layers. No organ originates exclusively in any one of the layers.

Experimental research by Hans Spemann (1869–1941) and others, into the basis of this differentiation, has disclosed that there are chemicals called *organizers* in the embryo, that appear to direct the development of the tissues and organs. In one of their experiments, some of the ectoderm cells that would form an eye were carefully removed from one frog embryo, and transplanted to another; it was found that the original tissue continued to form an extra eye on the top of the second embryo's head. Apparently, the ectoderm cells were being influenced by the presence of the chemical organizers. At an earlier stage of the gastrula, this does not happen; instead the transplanted tissue forms more of the same kind of tissue as the other cells of the second embryo. It has also been found that a certain part of the gastrula produces more organizer material than the rest, and appears to influence development to a greater extent. How and when the organizers are produced, the method of their action, and the other details of differentiation, are fascinating lines of current research in embryology.

Reproduction in fish. In most fish, fertilization takes place in the water, outside of the body. The female sheds many eggs, called *roe,* at spawning time. In some species such as perch, a hundred thousand eggs are laid by one fish. The male deposits *milt,* a liquid containing countless sperms over the eggs. The sperms move about and fertilize the eggs. Those eggs and sperm that do not unite fail to develop, and disintegrate. After cleavage, an embryo called a *fry* develops. It is nourished by a large *yolk sac* attached to the lower part of its body, containing the stored food of the egg. As the fry grows, the yolk sac disappears, and the young fish begins to hunt for its own food. Despite the large number of eggs produced by one female, comparatively few of her offspring reach adulthood; there is a constant struggle for existence against various predatory animals, as well as adverse environmental conditions.

In some fish, including certain tropical fish, fertilization takes place in the oviduct, within the female's body. The developing fertilized eggs may be retained for varying lengths of time before being laid. In "live-bearers" such as the guppy, the young fish develop entirely within the female's body, obtaining their nourishment from the stored food in the eggs. Most fish pay no attention to their offspring. A few, such as the stickleback and the sea horse, take special care of the eggs. The salmon migrates long distances from the ocean to the quiet fresh water upper regions of rivers to reproduce. The young salmon then migrate back to the ocean where they spend most of their lives. The reverse type of migration occurs in eels; they live in fresh water, and migrate to the ocean to breed.

Reproduction in frogs. In the spring, the female frog lays a mass of eggs which have been formed in her ovaries, and which pass to the outside through the oviduct. As the eggs leave, the male frog deposits sperm over them. In a very short time, a clear jelly-like material around each egg swells and holds the mass of eggs together. Fertilization occurs when a sperm unites with each egg to form a zygote. Cleavage and differentiation take place, but instead of a young frog being formed, a *tadpole* is produced. It has external gills for breathing, and a tail for movement; there are no legs. At first, it is nourished by the stored food of the egg. Soon internal gills replace the external gills. The tadpole feeds on algae and other plants. It has a two-chambered heart.

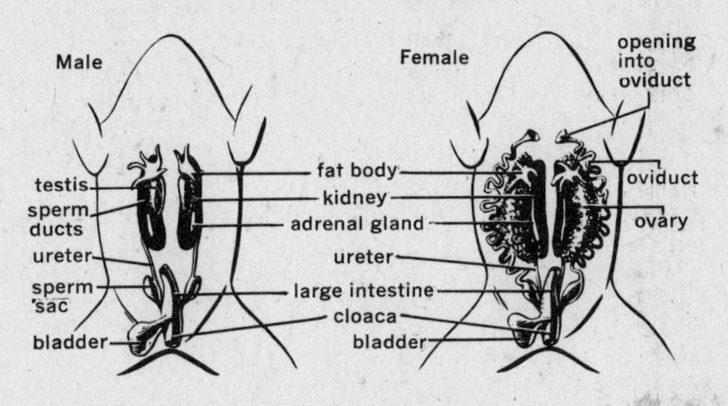

Frog Reproductive Organs

After several weeks or months, the tadpole goes through a series of changes called *metamorphosis,* and becomes a frog. In the bullfrog, this takes place over a two-year period. First, hind legs and then forelegs appear. The tail gradually shrinks. Lungs form and the gills disappear. The heart becomes three-chambered. The digestive system becomes modified for an animal diet. The frog leaves the water and lives on land. As in the case of most fish, frogs and other amphibia give no care to their offspring.

Parthenogenesis. Although an egg will not develop unless it has been fertilized by sperm, some lower animals reproduce by *parthenogenesis*—in which eggs develop without sperm. This is true of aphids, or plant lice; water fleas, and others. In bees the drone, or male, develops from an unfertilized egg.

Artificial parthenogenesis has been accomplished

by scientists experimenting with the eggs of sea urchins and frogs. Jacques Loeb stimulated frog eggs to go through cleavage and eventually to form frogs without fertilization by sperms; he used various stimuli, such as pricking the membrane with a needle, and treatment with salt solutions or acids. Dr. Gregory Pincus was successful in producing "fatherless rabbits" by removing the ova from female rabbits, treating the ova with salt solutions and implanting them in other female rabbits. The baby rabbits that developed were females, and were subsequently mated to produce normal offspring.

Reproduction in birds. A hen's egg consists of the following parts: (1) an outer shell; (2) a shell membrane within the shell; (3) albumen, or white of the egg; (4) the central yolk, which is the ovum, or "true egg," and which contains, in addition to stored food, the tiny living material, including a nucleus and cytoplasm. A bird's egg is therefore a tremendous single cell. The ovary contains many ova in various stages of development. One ovum at a time, with its yolk mass, leaves the ovary and enters the single oviduct. It takes a little more than half a day for the finished egg to be laid; during this time, various glands in the wall of the oviduct secrete, in turn, the albumen, shell membrane, and shell.

Fertilization in birds and other land mammals is internal. During mating, sperms are introduced into the oviduct, which provides a moist environment for the sperm to move. They move to the upper end, where one of them fertilizes the ovum as it enters the oviduct from the ovary. The embryo begins to develop at once, and is in the late gastrula stage when the egg is laid. If fertilization does not occur, the egg that is laid will not develop into a bird. The mother bird *incubates* the eggs for a period of time (twenty-one days in chickens) by sitting on them and providing the warmth needed for the embryo to develop.

During this time, the embryo is being nourished by the yolk. It is surrounded by a set of membranes which provide a favorable environment for development. These membranes include: (a) *the chorion,* which lines the shell, and surrounds the other membranes; it serves as a moist membrane for the exchange of gasses. (b) the *allantois,* which collects metabolic wastes from the embryo, and with the chorion, serves to exchange oxygen and carbon dioxide; (c) the *amnion* which contains the amniotic fluid surrounding the embryo, and protecting it from shock; (d) the *yolk sac,* around the yolk, containing blood vessels that bring food to the embryo. Reptile eggs have a similar set of membranes.

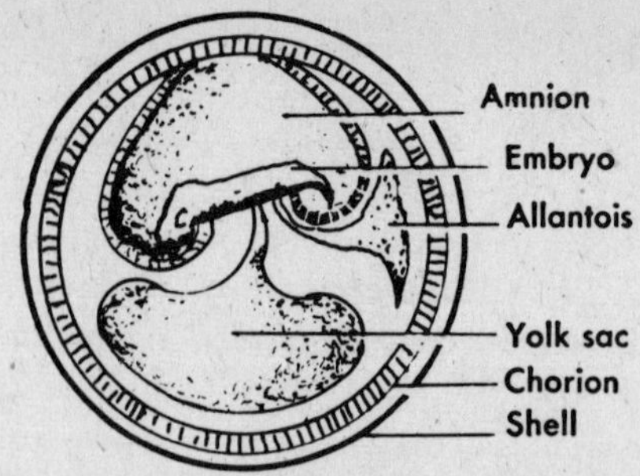

Embryonic Membranes in a Bird's Egg

When mature, the baby bird pecks its way out of the shell. In some birds, such as the chicken, the young chick is able to run about and obtain its own food a few hours after hatching. In the songbirds, the young are helpless, with little body covering to protect them. The mother bird makes countless trips to feed them with insects and worms; she continues to sit on the nest to keep them warm. After about two weeks, the young birds are sufficiently developed to learn how to fly and obtain their own food.

Reproduction in mammals. Mammals give birth to their young alive. Their embryos develop and are nourished internally; they are said to be *viviparous,* as contrasted with *oviparous* animals that lay eggs that develop outside the female's body (birds, amphibia, fish). Some fish, such as guppies, and some snakes retain the eggs in their oviducts until they hatch; the embryos are nourished by the food contained in the eggs. These animals are called *ovoviviparous.*

The mammal embryo develops internally, nourished by nutrients obtained from the mother's bloodstream. The ovum, or egg, has practically no stored food. Consequently, it is microscopic. The ova of a mouse and an elephant are about the same size. As in other animals, the eggs are produced in the ovaries. In some mammals, such as rabbits, pigs, mice, and dogs, a number of eggs are formed at one time. In other mammals, including humans, elephants and cows, usually only one egg is produced at a time. In humans, there are two ovaries located in the lower portion of the abdomen. They produce eggs in cavities called follicles. The release of an egg from a follicle is called *ovulation.* A woman is born with all the eggs she will ovulate.

When the egg leaves the follicle, it enters a modified part of the oviduct called the *Fallopian tubes.* If mating has occurred, sperm cells will be present here and will fertilize the eggs. If there are no sperm cells, the tiny eggs will continue out of the body of the female. If an egg is fertilized, it comes to rest in an enlarged part of the oviduct called the *uterus.* The lower part of the uterus leads to a muscular tube, the vagina.

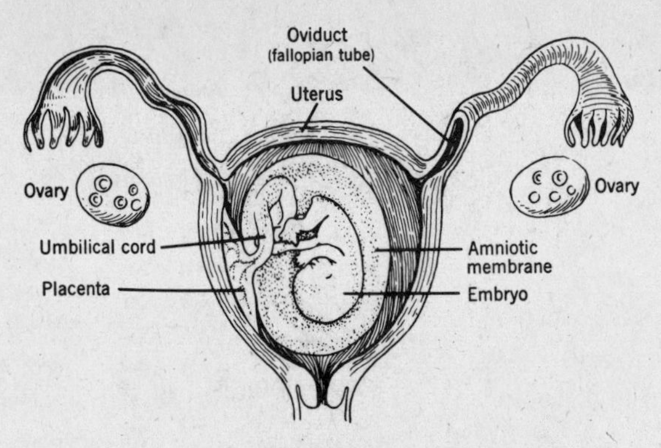

Mammal Development

In the uterus, the embryo (known as a *fetus* after it begins to take form) receives its nourishment from the mother through the *placenta*. This consists of a set of thick membranes composed of tissue of both the parent and the embryo, lying in intimate contact with each other. Both membranes are well supplied with many capillaries connected to the separate blood systems of both individuals. As blood flows through these two different sets of capillaries, food and oxygen diffuse from the mother's bloodstream into the embryo's bloodstream. The blood circulates out of the placenta through the *umbilical cord* to the embryo. The embryo's heart keeps the blood circulating to its cells. Wastes such as carbon dioxide and urea are sent from the embryo through the umbilical cord to the placenta, where they diffuse into the mother's bloodstream. There is no direct connection between the bloodstream of the mother and the embryo. Transport is accomplished by diffusion and active transport. During its development, the fetus lies suspended in the amniotic fluid, surrounded by the *amnion*. The fluid provides a watery environment and protection against shock.

The Menstrual Cycle. The human female has a reproductive cycle in which hormones from the pituitary gland, the ovary and the uterus interact. The pituitary hormones stimulate the development of the follicle in the ovary, and stimulate ovulation. The ovaries produce hormones that start the thickening of the uterus wall, and also control the production of hormones by the pituitary gland. The uterus secretes hormones that affect the production of ovarian hormones.

These hormones influence the progression of the various stages in the *menstrual* cycle. This cycle deals with the series of changes that take place in the preparation of the uterus wall for the receipt of the embryo. If the egg has not been fertilized, menstruation occurs. The stages in the menstrual cycle usually take about 28 days in most cases. They may be described as follows:

(a) Follicle stage (10-14 days). The follicle in the ovary is ripening, and the wall of the uterus is beginning to develop a rich supply of blood vessels.

(b) Ovulation. In the middle of the cycle, the egg bursts out of the follicle.

(c) Corpus luteum stage (4-5 days). The corpus luteum is the yellowish material that forms in the follicile after ovulation. During this stage, the lining of the uterus becomes spongy and thick with glands and blood vessels. If the egg has been fertilized, implantation of the embryo in the uterus wall takes place, about 6-10 days after fertilization, and the next stage does not take place.

(d) Menstruation. (2-6 days). If the egg was not fertilized, the thickened lining of the uterus breaks down, and is shed, along with the microscopic egg through the vagina.

The menstrual cycle starts at puberty, or sexual maturity, anywhere between the age of 9 and 18. It is temporarily suspended during pregnancy. It ends after about 35 years, at menopause, around the age of fifty. The duration of the monthly cycle may vary considerably from one female to another, and may be interrupted by illness.

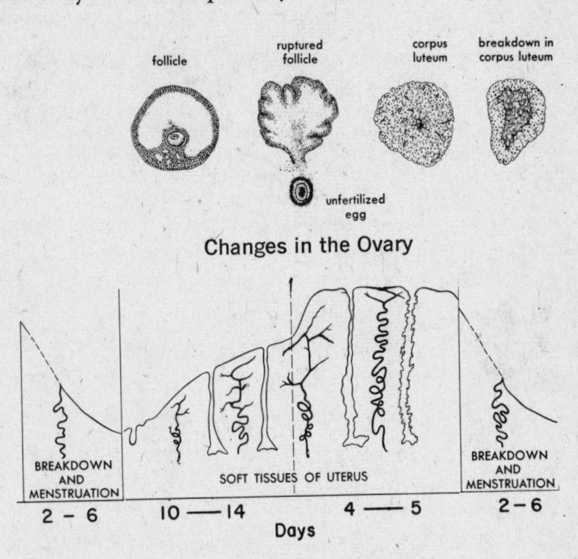

Changes in the Uterus Wall

The Menstrual Cycle

Development. The period of the embryo's development, or *gestation,* varies in different mammals; mouse — twenty days; cat — nine weeks; human — nine months; elephant — twenty months. At the end of that time, the young individual is born. The *navel* marks the place where the umbilical cord was attached. The young mammal is nourished on milk produced by the *mammary glands*. Some mammal young (the horse) are able to move about freely shortly after birth. Others (human, cat) are helpless for a period of time before they can walk and feed themselves. Mammals provide extended care for their young.

The duckbill platypus and the spiny anteater are unusual Australian mammals that lay eggs. When the young are born, they are fed on milk. Another group of Australian mammals, the marsupials, have a pouch in which the young are placed after they are born in an undeveloped condition; the kangaroo and the koala bear are examples. The only marsupial found outside of Australia is the opossum. Although they have internal development, pouched animals do not develop a placenta and the embryo does not receive direct nourishment from the parent in this stage.

Twins. Sometimes in human beings, two children, or twins may be born at the same time. They may be either fraternal or identical twins:

Fraternal twins develop from two eggs that were produced and fertilized at the same time. They are nourished from two separate placentas. They may be of the same sex or of different sexes. They are as alike as other members of the same family.

Identical twins develop from a single fertilized egg that started to undergo cleavage, and then split into two masses of cells, each of which developed into a separate organism. They are nourished from one placenta. They are exactly alike, and have the same sex.

It has been shown that twins appear in one out of 85 births; triplets have a frequency of $1:(85)^2$; and quadruplets appear in the ratio of $1:(85)^3$. In triplets, three separate eggs may be fertilized at the same time; or two eggs may be fertilized, with one forming identical twins and the other producing a fraternal triplet. It is also possible for identical triplets to develop from one fertilized egg; the developing egg mass splits in two, and then one of these splits again. Identical quadruplets probably originate from a fertilized egg that splits into two cell masses, which then split again, to form four similar offspring. The Dionne sister quintuplets that were born in Canada in 1934 are thought to be identical quintuplets, originating from one fertilized egg that split to form two cell masses; each of these then split again to form four cell masses; and one of these split again to form five all together.

SUMMARY OF VERTEBRATE REPRODUCTION

CLASS	NUMBER OF EGGS	WHERE FERTILIZATION OCCURS	WHERE DEVELOPMENT OCCURS	CARE OF YOUNG
Fish	Thousands or millions	Usually external	Usually external	Practically none
Amphibia	Hundreds or less	External	Water	None
Birds	One to a few	Internal	Internally for about half a day; then external incubation	Several weeks
Mammals	One to a few	Internal	Internal entirely, except for egg-laying mammals and marsupials	From a few weeks in mice up to 20 years in humans

CHAPTER REVIEW

Select the correct choice for each of the following statements.

1. The head of a sperm consists largely of (A) flagellum (B) cytoplasm (C) yolk (D) nucleus (E) stored food

2. One effect of the female sex hormone is (A) bright plumage of birds (B) dull plumage of birds (C) singing voice in birds (D) antlers of the deer (E) the lion's mane

3. The largest egg is present in a (A) mouse (B) horse (C) whale (D) sparrow (E) cow

4. The proper sequence of steps (1-gastrula; 2-zygote; 3-egg; 4-blastula; 5-two-celled stage) is (A) 5–2–3–4–1 (B) 3–5–2–4–1 (C) 3–2–4–1–5 (D) 3–2–5–4–1 (E) 3–2–5–1–4

5. The ectoderm develops into (A) circulatory system (B) nervous system (C) liver (D) muscles (E) bone

6. A young fry develops during the life cycle of a (A) fish (B) frog (C) bird (D) hydra (E) lobster

7. The egg-producing structure of an animal is called the (A) oviduct (B) ovum (C) ovary (D) ovule (E) ovipositor

8. A young tadpole is different from a frog in having (A) four legs (B) three-chambered heart (C) a tail (D) lungs (E) large, sticky tongue

9. The development of an egg without being fertilized by a sperm is known as (A) metamorphosis (B) regeneration (C) differentiation (D) parthenogenesis (E) incubation

10. The only part of a chicken's egg not formed in the oviduct is the (A) albumen (B) yolk (C) shell membrane (D) shell (E) shell pigment

11. The embryo of a mammal develops in the structure called the (A) ureter (B) uterus (C) urethra (D) pistil (E) ovary

12. The embryo of a cat receives its food (A) from the egg (B) from the yolk supply (C) through the micropyle (D) through the placenta (E) from the mother's small intestine

13. Fraternal twins (A) may be of the opposite sex (B) are always of the same sex (C) develop from one egg fertilized by two sperms (D) develop from two eggs fertilized by one sperm (E) are nourished from one placenta

14. A mammal embryo obtains oxygen from (A) its mother's bloodstream (B) its lungs (C) the amniotic fluid (D) oxidation of its food (E) its diaphragm

15. Testes are structures that produce (A) spores (B) sperms (C) seeds (D) ova (E) testa

16. The release of an egg from a follicle is called (A) fertilization (B) cleavage (C) metamorphosis (D) osmosis (E) ovulation

17. All of the following are membranes in a bird egg except (A) chorion (B) allantois (C) amnion (D) yolk sac (E) albumen

18. The human fetus is surrounded by (A) liquid (B) air (C) solid cells (D) fertilization membrane (E) nitrogenous wastes

19. All of the following organs produce hormones involved in the reproductive cycle except (A) testes (B) pituitary (C) pancreas (D) ovary (E) uterus

20. The correct sequence of stages in the menstrual cycle is (A) ovulation – follicle stage – corpus luteum – menstruation (B) corpus luteum stage – ovulation – follicle stage – menstruation (C) follicile stage – ovulation – corpus luteum stage – menstruation (D) menstruation – ovulation – corpus luteum stage – follicle stage (E) fertilization – cleavage – differentiation – ovulation – menstruation

ANSWER KEY

1-D	5-B	9-D	13-A	17-E
2-B	6-A	10-B	14-A	18-A
3-D	7-C	11-B	15-B	19-C
4-D	8-C	12-D	16-E	20-C

HOW LIVING THINGS INHERIT TRAITS

7

22 The basis of inheritance

A newly born individual contains features of both of its parents. The fertilized egg of a rabbit develops into a rabbit and not a cat; the fertilized egg of an apple flower develops into another apple tree, and not an oak. The reason for this is that the hereditary material present in the fertilized egg is distributed equally to all the cells of the new individual. The nuclear division by which this takes place is called *mitosis.*

Phases of mitosis. Mitosis can be observed when a fertilized egg divides during cleavage, and during the growth of any part of an organism, such as the tip of a root. The steps of mitosis can be observed in the following stages:

1. Interphase. Before undergoing mitosis, the nucleus contains a mass of *chromatin* material and a nucleolus. In animal cells, there is a small body, the *centrosome,* in the cytoplasm outside the nucleus. The centrosome contains a rod-like particle, called the centriole, which occurs in pairs.

2. Prophase. The chromatin material becomes visible as short stubby rods called *chromosomes.* These have already split lengthwise during the interphase, but remain attached together. This self duplication is known as *replication.* Each member of a double chromosome is called a chromatid. The chromatids are held together by a small structure called a *centromere.* The nuclear membrane disappears and the chromosomes are distributed through the cytoplasm. The nucleolus also disappears. In animal cells, the centrosome divides in two; each separates to opposite sides of the cell. Fine threads of cytoplasm called the *spindle* become attached to the centromere of the chromosomes from the centrioles. The radiating mass of spindle threads together with the centrosome from which they spread out is called the *aster.* In plant cells, these spindle fibers radiate from concentrated parts of the cytoplasm equivalent to the centrosomes.

3. Metaphase. The chromosomes become lined up in the center of the cell. The split between the chromatids appears more pronounced. The centromeres replicate.

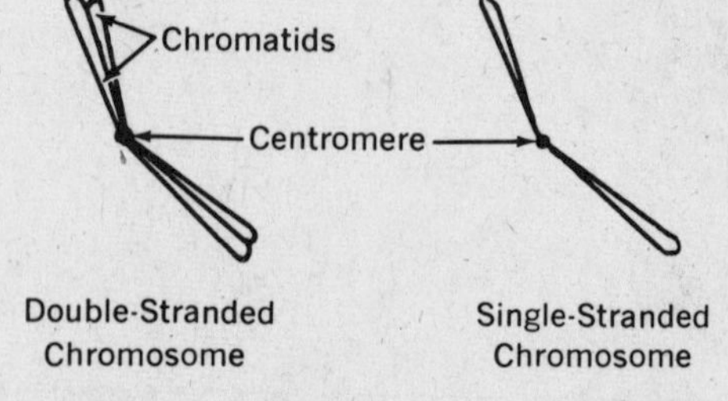

4. Anaphase. The split chromosomes begin to separate and to move toward the opposite sides of the cell. The number of chromosomes at each end remains the same as the number contained in the original cell.

5. Telophase. The chromosomes at each end of the cell collect to form a nucleus containing chromatin. A nuclear membrane forms. The spindle fibers disappear. The cell membrane *indents* in the middle of the cell to divide it into two. In plant

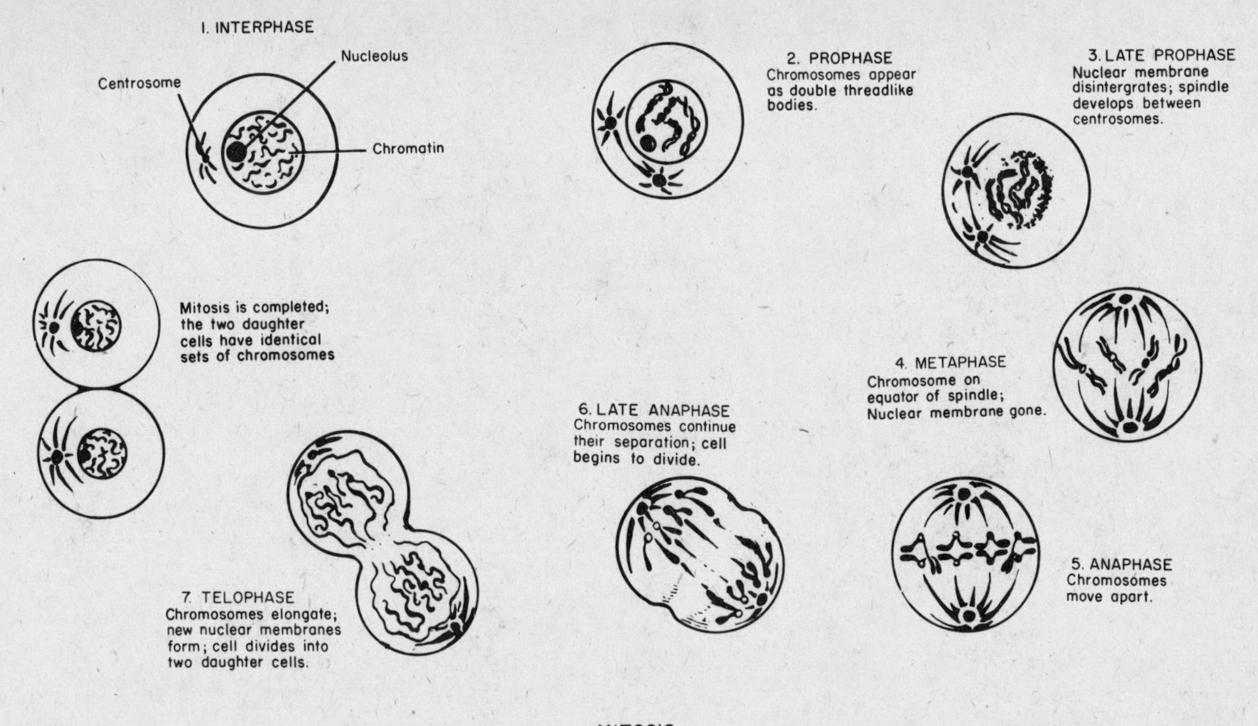

MITOSIS

cells, however, a *cell plate* forms in the middle of the cell, on which particles of cellulose are deposited, to form a cell wall.

After a period of time, each cell may divide again. Regardless of how many diversions take place, the resulting cells have the same number of chromosomes as the original. This is possible because *each chromosome always splits lengthwise into two identical chromosomes during mitosis.*

Chromosomes. Each species of plant or animal has a characteristic number of chromosomes in every one of its cells. Thus: corn—20; pea—14; cotton—52; frog—26; fruit fly—8; gypsy moth—62; human—46. In all cases, the chromosomes occur in pairs. The chromosomes in a pair are said to be *homologous;* one of the pair originated in the male gamate; the other in the female gamate. Each pair may differ somewhat from the others in size and shape. When the human fertilized egg containing 46 chromosomes splits, each of the two resulting cells receives 46 (or 23 pairs) of chromosomes. The adult human continues to have 46 chromosomes in every one of his cells. Chromosomes have been found to consist of nucleoproteins, which are compounds of nucleic acid.

Genes. Chromosomes carry the *genes,* which determine the hereditary characteristics of an individual. The genes are arranged lengthwise along a chromosome. They split each time the chromosome splits, so that the daughter cells always receive the same number as the parent cell. Dr. Thomas Hunt Morgan first suggested the existence of genes when he stated the *gene theory* in 1910. Since then, it has been found that the principal substance of which genes are made is a complex chemical called *DNA,* or deoxyribonucleic acid. The DNA molecule is large enough to be studied with the electron microscope. It appears to be coiled like a spring, or helix, and consists of two spiral chains of atoms.

Meiosis. The gametes, or eggs and sperm, are the only cells that do not have the same number of chromosomes as all the other cells of the organism. When they are formed, they receive half the number of chromosomes, the *haploid,* or *monoploid* number. Then, when fertilization occurs, the full number of chromosomes, the *diploid* number, is restored. In humans, the sperm and egg each have 23 chromosomes. The resulting fertilized egg then receives 46 chromosomes. Unless this reduction in the chromosome number of the gametes took place, the number of chromosomes in the fertilized egg would double, increasing the species number, each time. The process by which gametes are produced containing the monoploid number of chromosomes is known as meiosis. The production of sperm cells is known as *spermatogenesis;* the production of egg cells is called *oogenesis.*

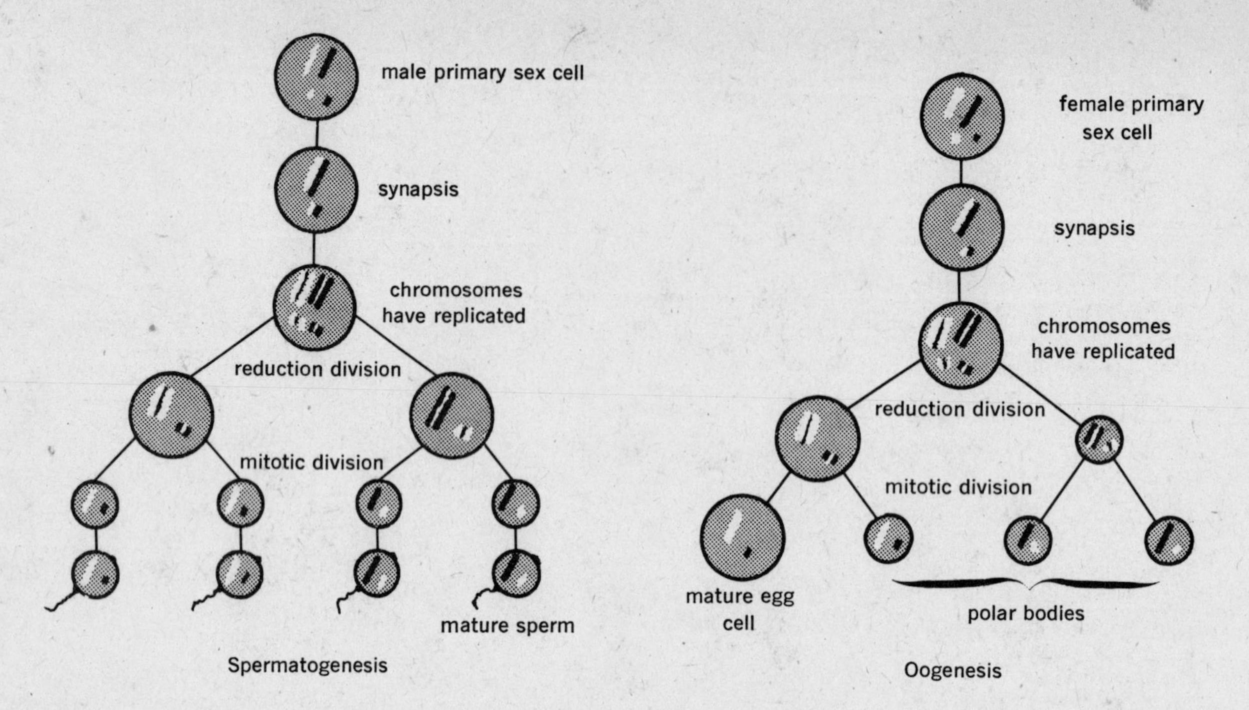

Spermatogenesis. The testes contain *primary sex cells;* these are the cells that will eventually become the sperm cells. They have the diploid number of chromosomes. When these cells are mature, the homologous chromosomes come together in pairs. This process is called *synapsis;* sometimes, during the pairing, the chromosomes twist about each other. Each of the chromosomes also replicates. The chromosomes line up in the center of the cell, the homologous members separate and the cell divides. The daughter cells now have the monoploid number. This division is known as reduction-division. Following this, the two cells split again, but this time by mitosis, in which the replicated chromosomes separate. There are now four cells with the monoploid number. These cells reorganize so that most of the cytoplasm becomes fashioned into a tail, and they become sperm cells.

Oogenesis. The ovary contains *primary sex cells* with the diploid number. When each cell matures, the homologous chromosomes come together in pairs, as in spermatogenesis. During this pairing, or synapsis, the chromosomes may twist about each other. Each of the chromosomes also replicates. The chromosomes line up in the center of the cell, the homologous members separate, and the cell divides by *reduction-division*. The two daughter cells now have the monoploid number. They are also unequal in size, one being practically the size of the original cell and the other being a tiny *polar body*. Both have the monoploid number. They divide again, this time by mitosis, in which the replicated chromosomes separate. The larger cell again divides unequally to form a large cell and a tiny polar body, while the other polar body forms two more. The large cell is the egg; the three polar bodies disintegrate.

Fertilization. As the result of meiosis, sperm cells and ova are produced, containing the monoploid (n) number of chromosomes. When a sperm fertilizes an egg, their nuclei unite, to form a zygote having the diploid (2n) number. The chromosomes now are in homologous pairs, again, one of each pair originating in the sperm, the other in the egg. The species number of chromosomes is thus restored. By cleavage and differentiation, a new individual is produced, having the diploid number in each of its cells. Each offspring has one set of chromosomes from the male parent and one set of chromosomes from the female parent. The hereditary characteristics of the individual are thus produced.

In flowers, the stamens have diploid cells which produce pollen containing monoploid nuclei. A monoploid egg nucleus is contained in the ovule. Following pollination, the pollen tube grows into the ovule. The sperm nucleus fertilizes the egg nucleus, resulting in a zygote with the diploid number of chromosomes. As the ovule develops into the seed, the zygote divides mitotically to produce an embryo with the diploid number. By differentiation and growth, a new plant is produced in which each cell has the diploid number of chromosomes.

CHAPTER REVIEW

Select the correct choice for each of the following statements.

1. Nuclear division is known as (A) binary fission (B) mitosis (C) differentiation (D) cleavage (E) fertilization
2. The rod like structures that appear during nuclear division are known as (A) spindle fibers (B) asters (C) nucleoli (D) chromosomes (E) centrosomes
3. Chromatin is to chromosomes as DNA is to (A) daughter cells (B) mitosis (C) genes (D) maturation (E) the electron microscope
4. The chromosome number of human gametes is (A) 12 (B) 23 (C) 46 (D) higher for eggs than sperms (E) higher for sperms than eggs
5. If the sperm cell of a fruit fly has 4 chromosomes, then the number of chromosomes in the body cells is (A) 2 (B) 4 (C) 6 (D) 8 (E) 16
6. Of the following cells, the only ones to have the haploid number of chromosomes is (A) skin (B) muscle (C) nerve (D) connective (E) ovum
7. The diploid number is restored as the result of (A) differentiation (B) fertilization (C) cleavage (D) reduction-division (E) maturation
8. Polar bodies are formed during (A) phototropism (B) oogenesis (C) spermatogenesis (D) the prophase stage (E) fertilization
9. During maturation, the chromosome number (A) is doubled (B) remains the same (C) is reduced (D) becomes diploid (E) becomes tetraploid
10. Reduction-division occurs during the process of (A) cleavage (B) differentiation (C) fertilization (D) meiosis (E) parthenogenesis
11. When a chromosome duplicates, it forms a pair of (A) centrosomes (B) spindles (C) chromatids (D) centromeres (E) nucleoli
12. The pairing of homologous chromosomes is called (A) meiosis (B) synapsis (C) reduction-division (D) cell plate (E) centriole
13. Spindle threads are attached to the chromosomes at the (A) asters (B) mitochondria (C) prophase (D) interphase (E) centromeres
14. In replication, a chromosome (A) splits (B) separates (C) twists about another (D) lines up in the center of the cell (E) reduces
15. In flowers, the monoploid number is found in the (A) seed (B) fruit (C) stigma (D) pollen (E) epicotyl

ANSWER KEY

1-B	4-B	7-B	10-D	13-E
2-D	5-D	8-B	11-C	14-A
3-C	6-E	9-C	12-B	15-D

23 Inheritance in plants and animals

The science of heredity is called *genetics*. It is one of the newer fields of science, having made its start in 1900. However, thirty-five years earlier, an Austrian monk named Gregor Mendel (1822–1884) described a series of experiments which laid the foundations of our knowledge of heredity. Unfortunately, his paper, "Experiments in Plant Hybridization," did not receive recognition until after his death.

MENDEL'S EXPERIMENTS

Mendel studied inheritance in garden pea plants. He selected seven contrasting characteristics and traced them from one generation to the next. He was careful to study only one of these traits at a time, thus avoiding the complex situation that frustrated previous scientists who had tried to trace the inheritance of many traits at a time, and had failed.

He started out by establishing *pure* lines. He obtained *pure tall* plants by allowing the flowers of tall plants to self-pollinate for several generations. These plants were about six feet tall. In like manner, he obtained *pure short* plants; they were only about a foot tall. He then cross-pollinated the two types of plants. Pollen from tall plants was placed on the stigma of flowers on short plants; of course, he first removed the stamens of the latter flowers, in order to prevent self-pollination. The flowers were

Table I

PURE PARENTS (P_1)			FIRST GENERATION (F_1)
1. Tall	x	short	All tall
2. Smooth seeds	x	wrinkled seeds	All smooth seeds
3. Yellow cotyledons	x	green cotyledons	All yellow cotyledons
4. Colored seed coat (gray or brown)	x	white seed coat	All colored seed coat
5. Inflated pods	x	constricted pods	All inflated pods
6. Green pods	x	yellow pods	All green pods
7. Axial flowers	x	terminal flowers	All axial flowers

then covered with paper bags, to keep stray pollen from being introduced. The reverse type of cross-pollination was also performed, in which pollen from short plants was placed on the stigmas of flowers on tall plants. Mendel collected the seeds and planted them. Would the plants be intermediate in height, short, or tall?

Law of dominance. Much to Mendel's intense interest, all the seeds developed into tall plants that were as tall as the original plants. None of these first generation (F_1) offspring appeared short, despite the fact that one of the parents had been short. These plants were called *hybrid* tall, since they had come from parents that were different in height, as compared with pure tall plants whose parents had both been only tall. When Mendel conducted similar experiments for the other characteristics, he obtained similar results; only one of the characteristics appeared, as is shown in Table I.

He called the characteristic that appeared, *dominant;* the one that did not appear was called *recessive*. He stated his results in the *Law of Dominance*: When organisms having pure contrasting traits are crossed, their offspring will show only one of the traits; the trait that appears is called the dominant trait, and the trait that does not appear is called the recessive trait.

Law of Segregation. Mendel next turned his attention to the hybrid tall plants. They were indistinguishable in appearance from the pure tall plants. He crossed hybrid tall plants, collected their seeds and in the next or second generation (F_2), he observed both tall and short plants. Although the recessive trait seemed to have disappeared in the first generation, it now reappeared in the second generation. He obtained similar results when he crossed plants that were hybrid for the other characteristics. When he counted all the plants, he obtained the results shown in Table II.

Mendel observed that there were three times as many dominant offspring as recessive. Then, when he crossed the short plants in the F_2 with each other, he obtained only short plants. When he crossed the tall plants, he noticed that some gave rise only to tall plants; others gave rise to a 3:1 ratio of tall and short plants. The same results were obtained for the other characteristics. On the basis of all this data Mendel summarized his findings in the *Law of Segregation:* When large numbers of hybrids are crossed, the factors segregate and then recombine to produce dominant and recessive offspring in a ratio of 3:1.

Genes and inheritance. Modern day knowledge of genes permits us to explain Mendel's results, and to predict the possible offspring of various crosses. Homologous pairs of chromosomes have pairs of genes which are called alleles. Pure tall plants have two genes for tallness (TT) in every cell; one gene was contributed by the sperm, the other by the egg, and the fertilized egg therefore received two genes for tallness. Pure short plants likewise have two genes for shortness (tt). The capital letter, T, is used for the dominant allele; the small letter, t, is used for the recessive allele. The method of

Table II

F_1 HYBRID PARENTS	F_2 RESULTS		RATIO IN F_2
1. Tall	787 Tall	— 277 short	2.84 : 1
2. Smooth seeds	5,474 Smooth	—1,850 wrinkled	2.96 : 1
3. Yellow cotyledon	6,022 Yellow	—2,001 green	3.01 : 1
4. Colored seed coat	705 Colored	— 224 white	3.15 : 1
5. Inflated pod	882 Inflated	— 299 constricted	2.95 : 1
6. Green pod	428 Green pod	— 152 white	2.82 : 1
7. Axial flower	651 Axial	— 207 terminal	3.14 : 1
Totals	14,949 Dominant	—5,010 recessive	2.98 : 1

transmission of the genes in the first generation may be depicted as follows:

Genes in the First Generation

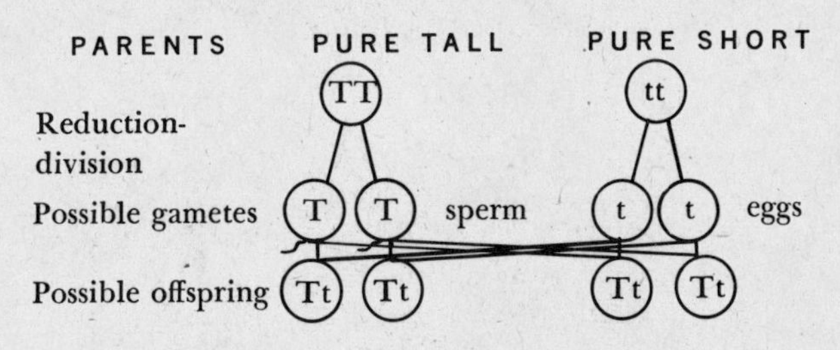

KEY

T = gene for tallness

t = gene for shortness

Results: Appearance (Phenotype) – All tall hybrid
Gene make-up (Genotype) – 100% Tt

The hybrid tall plants have a gene for tallness and a gene for shortness in every cell (Tt). Although the recessive gene for shortness is present, its effect is masked by the dominant gene for tallness. Such an individual possessing two unlike genes of an allelic pair is said to be *heterozygous* for the trait. The parents of the heterozygous tall plants are said to be *homozygous* for tallness and shortness since they possess alike allelic genes, TT and tt. The distribution of genes in Mendel's second generation results may be depicted as follows:

Genes in the Second Generation

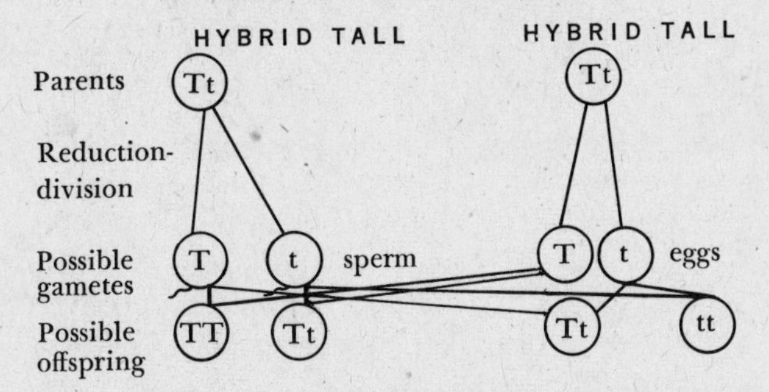

Results: Appearance (Phenotype) – 75% tall; 25% short
Gene make-up (Genotype) – 25% pure tall; 50% hybrid tall; 25% pure short (1 : 2 : 1)

In their appearance or *phenotype,* there are three times as many tall as short plants, or 75% tall and 25% short. The genetic make-up, or *genotype,* reveals a ratio of 1:2:1, or 25% pure tall, 50% hybrid tall, 25% pure short. In this example, a sperm with a gene for tallness may combine with either an egg containing a gene for tallness or with an egg containing a gene for shortness. Similarly, a sperm with a gene for shortness may combine with either type of egg. When many hybrid plants are crossed, both types of sperms will fertilize both types of eggs. The larger the number used, the closer will the results be to the ideal ratios of 1:2:1. If only a few individuals are used, however, the actual results will rarely give this ratio. *Chance* determines which sperm and which eggs will unite. A demonstration to illustrate how chance operates may be performed as follows:

1. Flip two pennies at the same time, and observe whether the results are two heads, a head and a tail, or two tails.
2. Tally the results under three columns labeled: TT, TH, HH.
3. Flip both coins together 100 times and tally the results.
4. Summarize the results. It will be seen that the ratio is approximately 1:2:1. Compare with the results of flipping the coins only four times. There may or may not be a ratio of this type, because of the small number of cases involved.

Using the Punnett square. A convenient method of working out different genetic crosses and the possible offspring, involves the use of the Punnett square. Example 1:

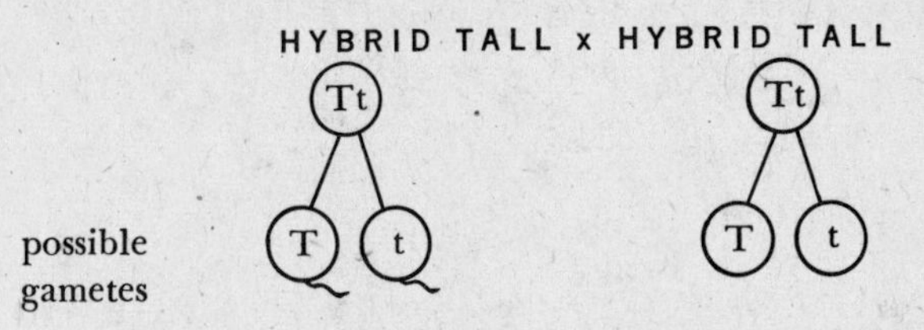

Key: T = tall
t = short

The possible male gametes are placed at the top of the square, one over each box; the possible female gametes are placed at the side of the square, one alongside each of the boxes. Each box now represents the type of zygote resulting from the union of the male gamete, indicated at the top, and the female gamete indicated at the side:

	T	t
T	TT	Tt
t	Tt	tt

Possible results:
25% TT, pure tall
50% Tt, hybrid tall
25% tt, short

Example 2:

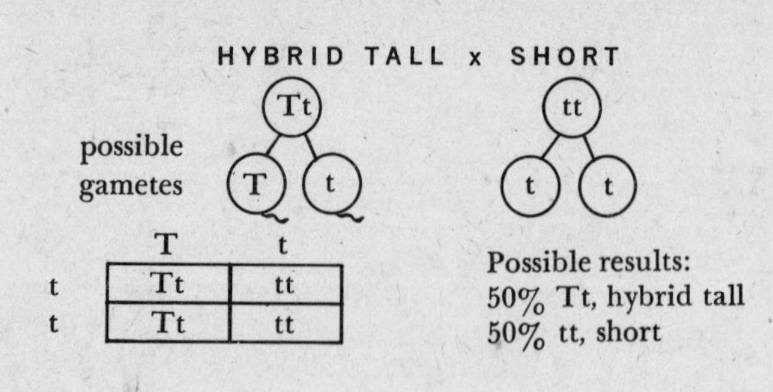

Example 3:

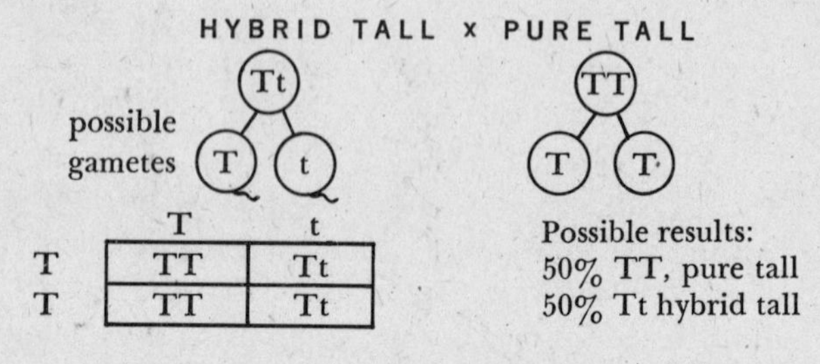

Example 4:

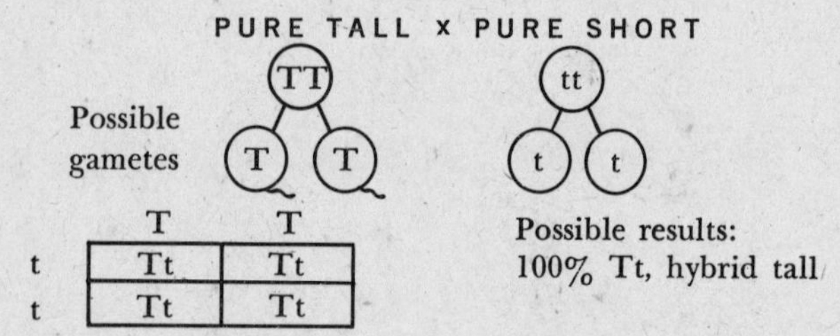

Law of independent assortment (Unit characters). Mendel was interested in determining whether characteristics are inherited independently from each other, or together. He started with plants that were pure for two contrasting traits; i.e., Tall plants with Yellow seeds x short plants with green seeds. The first generation plants were hybrid for both characteristics, or dihybrids; they appeared Tall and had Yellow seeds. When these dihybrids were crossed, he found that the characteristics were assorted independently, giving four different types of individuals: Tall Yellow, Tall green, short Yellow, short green. They were in a ratio of 9:3:3:1. Mendel stated this idea in his *Law* of *Independent Assortment,* or *Unit Characters:* "Characteristics are inherited independently from each other, and are not affected by each other." We now know that the genes for each of the characteristics studied by Mendel were, by coincidence, located on separate chromosomes, thus making this law possible.

We may study dihybrid inheritance on the Punnett square as follows:

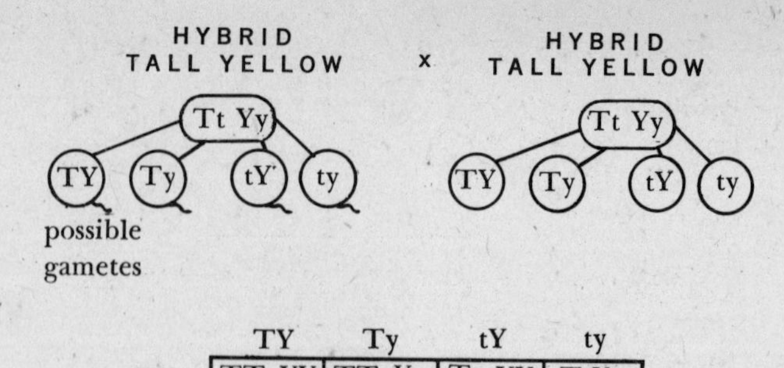

	TY	Ty	tY	ty
TY	TT YY	TT Yy	Tt YY	TtYy
Ty	TT Yy	TT yy	Tt Yy	Ttyy
tY	Tt YY	Tt Yy	tt YY	ttYy
ty	Tt Yy	Tt yy	tt Yy	ttyy

Key
T = tall
t = short
Y = yellow seed
y = green seed

Possible results:
9 Tall Yellow
3 Tall green
3 short Yellow
1 short green

Inheritance in other organisms. Since Mendel's day, the heredity of many plants and animals has been carefully studied. In general, Mendel's findings were also found to apply to them, as is shown in Table III.

Table III

ORGANISM	DOMINANT TRAIT	RECESSIVE TRAIT
Cattle	Hornlessness	horned
Guinea pig	Black fur	white fur
Mice	Pigmented coat	white coat (albino)
Chickens	White feathers	pigmented feathers
Wheat	Late ripening	early ripening
Corn	Yellow grain	white grain
Pea	Colored flower	white flower
Barley	Beardless	bearded

Test cross ("back cross"). An organism that is heterozygous has the same appearance, or phentotype, as the pure dominant. A test cross is performed to determine whether a dominant individual is homozygous or heterozygous. This is done by crossing the unknown with the recessive. There are two possible results: (1) If any of the offspring are recessive, it indicates that the unknown was heterozygous. (2) If none of the offspring are recessive, it indicates that the unknown probably was homozygous. This can be illustrated with a black guinea pig which may be either heterozygous black or homozygous black. It is mated with a white one. (1) If a white offspring is born, it shows that the black guinea pig was heterozygous. Thus:

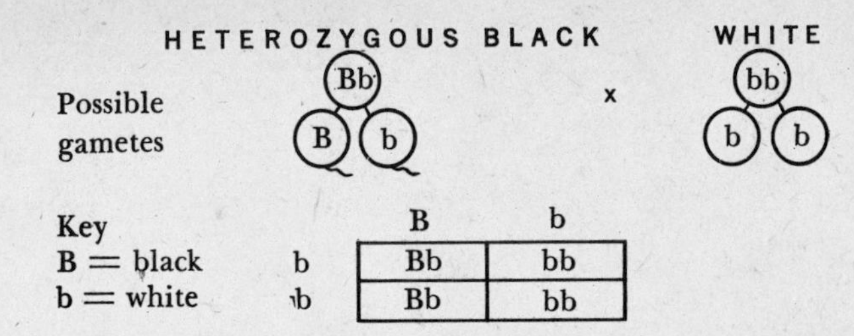

Key
B = black
b = white

	B	b
b	Bb	bb
b	Bb	bb

Possible results:
50% Bb – heterozygous black
50% bb – white

(2) If all the offspring are black, there is a good possibility that the black guinea pig was homozygous. Thus:

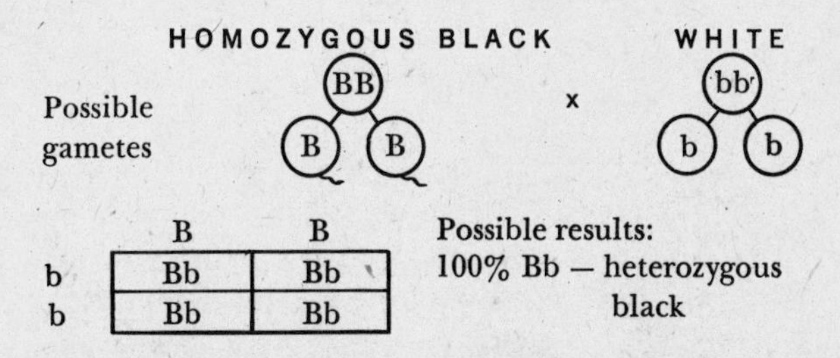

	B	B
b	Bb	Bb
b	Bb	Bb

Possible results:
100% Bb – heterozygous black

Incomplete dominance or blending. In some cases, neither characteristic is dominant, but both appear in the offspring, giving a blended effect. This is illustrated in the flowers of the Japanese four o'clock plant, and the snapdragon. A red flower crossed with a white will result in flowers that are pink.

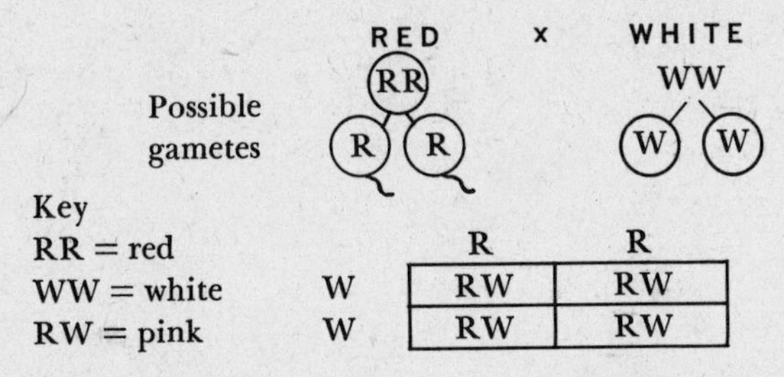

Key
RR = red
WW = white
RW = pink

	R	R
W	RW	RW
W	RW	RW

Possible results:
100% RW – heterozygous pink

If the heterozygous pink flowers are crossed, the offspring possess all three colors, in a ratio of 1:2:1, or 25% red, 50% pink, and 25% white.

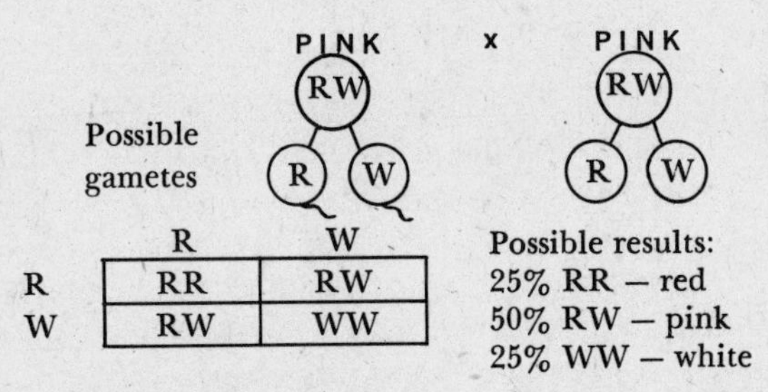

	R	W
R	RR	RW
W	RW	WW

Possible results:
25% RR – red
50% RW – pink
25% WW – white

Similar blending inheritance also occurs among Andalusian fowl. The pure types are black and white. When they are mated, the offspring appear blue. The blue individuals, in turn will produce black, blue and white offspring in a ratio of 1:2:1. In Shorthorn cattle, the homozygous types are red and white. The offspring have a patchy *roan* color that is a blend of red and white. When roan animals are mated, they will produce red, roan and white offspring in a ratio of 1:2:1.

Drosophila, or fruit fly. Probably the most useful of all organisms in genetic research has been the diminutive *Drosophila melanogaster,* the fruit fly. Among its advantages: it breeds in large numbers, every ten days; it is easily stored in half-pint bottles; it has only four pairs of clearly distinguishable chromosomes in each cell. Thomas Hunt Morgan received the Nobel Prize for his development ot the gene theory, based on research with Drosophila. He and his co-workers drew up chromosome maps showing the location of various genes on the chromosomes. Certain cells of the salivary glands have subsequently been shown to contain *giant chromosomes* about one hundred times larger than normal chromosomes. It is believed that the dark bands on these chromosomes coincide with the location of the genes.

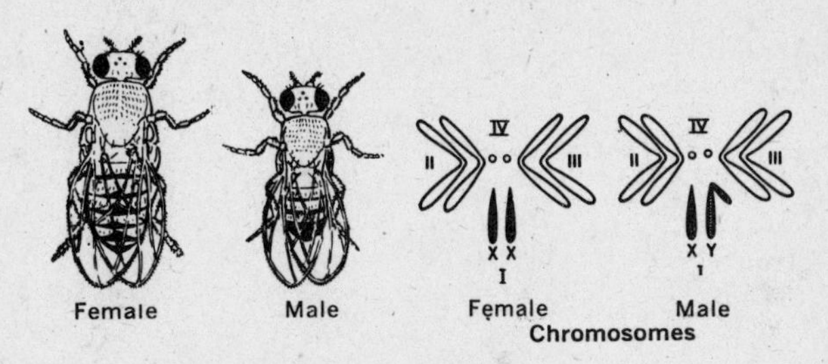

The Drosophila Fly

Linkage. Organisms have many genes distributed along the length of the chromosomes, like beads on a string. When a gene for one characteristic is inherited, many other genes on the same chromosome are also inherited with it. This is known as *linkage.* In Drosophila, for example, the genes for black body and short wings are linked, or inherited, together. This type of inheritance, therefore, is an exception to Mendel's Law of Independent Assortment, which was possible, because Mendel happened to study characteristics whose genes are located on different chromosomes.

Crossing over. Morgan also observed that during synapsis in the first part of meiosis, the chromatids of homologous chromosomes sometimes twisted about each other, and then separated, exchanging adjacent parts. This resulted in sets of genes *crossing over* from one chromosome to another, and producing new combinations of characteristics that were inherited together. The frequency with which this occurred was also the basis for drawing up the chromosome maps of the genes.

Gene Mutations. Genes sometimes change, altering the inheritance of the characteristics they control. Such a change is known as a *mutation.* The term was first used by Hugo De Vries (1848–1935) during his study of sudden changes in the evening primrose. Mutations have been found in many organisms, including white eyes, vestigal (very short) wings, and other traits in Drosophila; albinism (lack of all pigment) in white mice, frogs, and corn; hornlessness in cattle; seedlessness in oranges and grapes. Most mutations are probably harmful, since they upset the balance among the genes, and change the ability of the organism to survive in the original environment. This causes the organism to die out (*lethal* effect). However, some mutations are useful, and may improve the organism's chances of survival or its value to man. The cause of these mutations is not known, although it is suspected that cosmic rays from outer space may be responsible for the occasional change in a gene.

Mutations were produced artificially in 1927 by Herman J. Muller, who exposed Drosophila flies to X-rays. This treatment affected the genes in the reproductive organs, and resulted in offspring that showed changed characteristics. He received the Nobel Prize for this work. Other animals and plants have also shown mutations as the result of X-ray exposure. Recent research has indicated that mutations may also be brought about by other *mutagenic agents* such as ultra-violet rays, mustard gas, colchicine and radioactive isotopes. Dr. Muller and other scientists have predicted that atomic warfare, and possibly also atomic tests may cause undesirable mutations to appear among human beings. For this reason, as well as others, the birth rate and infant statistics in Hiroshima and Nagasaki are under observation by scientists.

Chromosomal Mutations. A change in the number of chromosomes, or in their structures is known as a chromosomal mutation. One example is *non-disjunction.* Sometimes chromosomes fail to separate from one another during meiosis. This results in gametes with slightly more or less than the normal monoploid number. After fertilization, the new individuals may have more or less than the normal 2n chromosome number, often leading to various defects.

Another example is *polyploidy,* in which an organism may have more than one entire set of chromosomes, leading to the triploid (3n), tetraploid (4n), etc, condition. A chemical, *colchicine,* which is used to treat cases of gout, has been found to cause the chromosome number of plant cells to double. Its effect is due to its interference with mitotic division, preventing the chromosomes from spreading apart after they have split. This results in the cells having twice the normal number, or the *tetraploid* number of chromosomes. Colchicine treatment has been used to produce giant marigold flowers.

Other types of chromosome changes have been studied, which also change the characteristics of organisms. In wheat and in jimson weed, for example, when tetraploid and diploid plants were crossed, the resulting offspring had the triploid number of chromosomes. In other cases, pieces of chromosomes become lost, or become attached to other chromosomes, during maturation. This changes the number of genes and alters the heredity of the offspring accordingly.

How genes produce their effects. Although the mechanism of gene inheritance has been thoroughly studied, relatively little is known about how the genes produce their effects. The following facts are known: Frequently, several pairs of genes *(multiple alleles)* act together in affecting a particular characteristic; thus in mice, gray coat color depends on the presence of two pairs of genes, while eye color of Drosophila is determined by fifty genes. Genes probably produce their effect by causing the production of enzymes which are needed to bring about certain chemical changes; thus, it has been shown that rabbits with black fur possess an enzyme which stimulates the production of black pigment, while white rabbits do not form this pigment; in another case, that of the mold *neurospora,* a gene controls the production of an enzyme that permits it to synthesize a particular member of the vitamin B complex. A pair of genes may affect more than one trait; for example, Drosophila flies with vestigial wings have a shortened life span, indicating that other parts of the body are also affected by the particular gene called vestigial.

Interaction of heredity and environment. The proper environmental condition is needed for genes to exert their effect. This can be demonstrated in the following manner: Germinate some corn seedlings in the dark, and some in the light. It will be observed that although the seedlings have genes for the formation of chlorophyll, they will remain white as long as they are kept in the dark. The control seedlings that were germinated in the light will show the normal condition of green color. If the first group of seedlings are now transferred from the dark to the light, they will become green.

A certain type of Drosophila inherits a condition of curly wings at a temperature of 25°C. But when these flies are raised at a temperature of 16°C, their wings have the normal straight appearance. If they are mated, the offspring will continue to have straight wings at this temperature. However, if the temperature is raised to 25°C, the offspring will have curly wings.

The Himalayan rabbit inherits a color pattern in which the animal is white, with black ears, nose, paw and tail. However, if some of the white fur is shaved off and an ice pack applied to the spot, the new fur that grows in will be black. If the fur on a paw is shaved off, and the paw kept warm in a wrapping, the fur grows back white. Apparently, the genes produce white fur in the warm parts of the body, and black fur in the extremities which have a lower temperature.

CHAPTER REVIEW

Select the correct choice for each of the following statements.

1. The foundations for the study of heredity were first laid by (A) Morgan (B) Mendel (C) Mendeleeff (D) Muller (E) Metchnikoff

2. In breeding experiments, paper bags are placed over flowers to (A) keep them warm (B) keep them from blowing away (C) prevent self-pollination (D) keep stray pollen away (E) protect them from excessive sunlight

3. The Law of Dominance is illustrated in the garden pea by (A) homozygous tall x heterozygous tall (B) heterozygous tall x heterozygous tall (C) homozygous tall x homozygous tall (D) pure short x pure short (E) homozygous tall x pure short

4. In pea plants, all of the following are examples of dominant traits except (A) wrinkled seeds (B) tallness (C) yellow cotyledons (D) axial flowers (E) inflated pods

5. When hybrids are crossed, the genotype of the offspring is (A) 1:1 (B) 3:1 (C) 1:2:1 (D) 4:1 (E) 3:2

6. If a pair of hybrid black guinea pigs are mated, and there are four offspring, their appearance may be (A) all black (B) 3 black: 1 white (C) 2 black: 2 white (D) 1 black: 3 white (E) any of the above

7. A cross in which ¾ of the offspring appear dominant is (A) Tt x TT (B) TT x tt (C) Tt x tt (D) Tt x Tt (F) TT x TT

8. When Mendel crossed hybrids, he derived the (A) Law of Dominance (B) Law of Genes (C) Law of Incomplete Dominance (D) Law of Unit Characters (E) Law of Segregation

9. The possible gametes of TTYy will contain the genes (A) TY, Ty (B) TT, Yy (C) TT, TY (D) TT, Ty (E) TT, YY

10. A test cross is performed (A) only with hybrids (B) only with pure types (C) to determine whether an organism is heterozygous or homozygous dominant (B) only between recessives (E) only between homozygous dominants

11. In the four o'clock flower, cross-pollinating pink flowers gives (A) 100% pink (B) 100% red (C) 100% white (D) 50% pink, 25% red, 25% white (E) 50% red, 50% white

12. Drosophila is useful for all of the following advantages except (A) it breeds every ten days (B) it produces many offspring (C) it has no linked characteristics (D) it has four pairs of chromosomes (E) it has giant chromosomes in its salivary glands

13. A change in genes is called a (A) genotype (B) phenotype (C) cross-over (D) back-cross (E) mutation

14. In peas, if 50% of the offspring are short, and 50% are tall, the parents were probably (A) TT x tt (B) Tt x tt (C) Tt x TT (D) Tt x Tt (E) TT x TT

15. Genes are *not* found in pairs in a(n) (A) sperm cell (B) fertilized egg (C) muscle (D) epithelial cell (E) zygote

16. In non-disjunction, chromosomes fail to (A) replicate (B) reduce (C) separate (D) fertilize (E) recombine

17. The exchange of genes between homologous chromosomes is called (A) polyploidy (B) chromatids (C) crossing-over (D) test cross (E) roans

18. All the following are mutagenic agents except (A) X-rays (B) colchicine (C) carbon dioxide gas (D) mustard gas (E) ultraviolet rays

19. A gamete with the monoploid number of chromosomes that unites with a gamete having the diploid number, results in a zygote that is (A) monoploid (B) diploid (C) triploid (D) tetraploid (E) none of these

20. In the Himalayan rabbit, low temperatures cause the growth of (A) white fur (B) black fur (C) white paws (D) white nose (E) black genes

ANSWER KEY

1-B	5-C	9-A	13-E	17-C
2-D	6-E	10-C	14-B	18-C
3-E	7-D	11-D	15-A	19-C
4-A	8-D	12-C	16-C	20-B

24 Human inheritance

The laws of heredity apply to man as well as to other livings things. However, relatively little is known about human inheritance for the following reasons: (1) it takes about twenty-five years for one generation; (2) small numbers of offspring are produced; (3) planned experiments are not possible; (4) description of ancestors on the basis of records or memory is not too reliable.

Eye color. It is known that brown eyes are dominant to blue. Therefore, blue-eyed parents have only blue-eyed children. Brown-eyed parents may have blue-eyed children if they are both heterozygous for brown eyes. However, if they are both homozygous for brown eyes, or if one parent is heterozygous and the other is homozygous, there will be no blue-eyed children. This can be shown as follows:

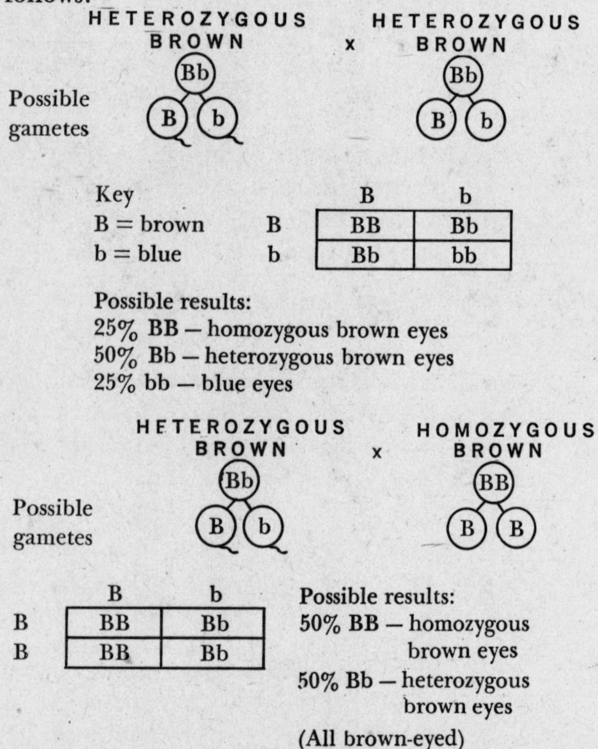

However, it is well to remember that we are considering only possibilities. If there are four children in a family, do the above ratios apply? The answer is — possibly, but not necessarily. When both parents are heterozygous for brown eyes, there may actually be two or more blue-eyed children — or none at all. Four is a small number statistically speaking, and the law of chance decides which combination of gametes will take place. The different shades of eye color, from dark brown to light gray, lead to the conclusion that several genes are involved in the determination of eye color.

Other human traits. Some other traits among human beings are summarized in the following table:

DOMINANT	RECESSIVE
Dark hair	light hair
Cury hair	straight hair
Normal pigmentation	albinism
Free ear lobe	attached ear lobe
White forelock	normal hair color
Rh positive blood factor	Rh negative blood factor
Normal color vision	Color blind vision
Bitter taste reaction to P.T.C. (phenyl-thiocarbamide)	No taste reaction to P.T.C.

The blood types A, B, AB, and O are inherited in predictable ways, based on the action of multiple alleles. There are three alleles involved. The allele for type A, I^A, and the allele for type B, I^B, are both dominant to the third allele, *i*. They are not dominant to each other, but may appear together in type AB. When they are both absent, type O results. The possible genotypes for the various blood groups may be summarized as follows:

BLOOD GROUP	GENOTYPE
A	I^AI^A; or I^Ai
B	I^BI^B; or I^Bi
AB	I^AI^B
O	ii

Rh positive blood factor is dominant to Rh negative. When a father is Rh positive and the mother is Rh negative, the developing child is usually Rh positive. As was mentioned in Chapter 12, this leads to complications later on, because there is a reaction in the mother's bloodstream against the Rh positive blood of the developing child. This is serious if the mother becomes pregnant a second time. Unless the doctor is prepared to act quickly the baby may die when it is born. For this reason, prospective mothers are now typed to see whether they are Rh negative.

A demonstration to show inheritance of taste reaction to the chemical phenylthiocarbamide (P.T.C.) may be performed as follows:

1. Distribute a piece of P.T.C.-impregnated paper to several people, and have them chew it. (P.T.C. taste papers may be obtained from the American Genetic Association, Washington, D.C.)

2. Some will experience an extremely bitter taste; others will report no taste sensation, whatever.

3. The tasters may be able to test the reactions of members of their family. Since the ability to taste P.T.C. is dominant, there will be more tasters than non-tasters.

Phenylketonuria (PKU) is a condition of feeble-mindedness caused by the lack of a dominant gene that normally leads to the conversion of the amino acid phenylaline into the amino acid tyrosine. The presence of the homozygous recessive genes leads to a failure to produce a liver enzyme that causes this conversion. Phenylaline accumulates, and causes brain damage. Fortunately, a simple PKU blood test has been developed for infants, which identifies the condition. The disease can be corrected by feeding the baby a special formula containing very little phenylaline.

Another type of hereditary defect, *sickle-cell anemia,* is also inherited as a recessive trait. When the oxygen supply is low, the red blood cells take on a sickle shape and break apart, leading to an early death. The hemoglobin molecules in this condition has been found to differ from normal hemoglobin by only a single amino acid out of a total of several hundred; otherwise the two hemoglobin molecules are identical. Apparently, a mutation occurred in the allele for the particular amino acid in the hemoglobin molecule. This condition is common in certain parts of Africa. In heterozygous individuals, the disease may be present in mild form. It has the advantage of providing resistance to malaria that is higher than usual. In this respect, the defective gene has survival value, and has continued to be inherited.

Tracing inheritance. The inheritance of a human trait for several generations is traced on a pedigree chart as shown in Table IV.

Although the mother and father have brown eyes, two of their three children have blue eyes. Their blue-eyed daughter, Mary, married blue-eyed Charles and had blue-eyed children. Their brown-eyed son, John, married brown-eyed Anne and had only brown-eyed children. It is obvious that the mother and father are heterozygous for brown eyes. Since Mary and her husband are recessive for blue eyes, all their children will have blue eyes. On the other hand, son John and his wife are brown-eyed and have five brown-eyed children; since five is a small number statistically speaking, it is not possible to state whether they are homozygous or heterozygous for brown eyes.

Sex inheritance. As was mentioned earlier, chromosomes occur in pairs which are alike in males and females. One pair of chromosomes, however, has been identified as the sex chromosomes. The other chromosomes are called *autosomes.* Thus, of the 46, or 23 pairs of chromosomes in the cell, 22 pairs are autosomes. One pair are the sex chromosomes. In females, they are alike and are known as X chromosomes, or as XX. In males, they are different, one being smaller than the other. The latter is known as the Y chromosome, and the other the X. Males thus have XY chromosomes. When gamates are formed, these chromosomes separate by reduction-division into separate cells. All eggs have one of the X chromosomes. Sperm cells, however, may have either an X or a Y chromosome. This makes two kinds of combinations possible:

1. If a sperm with an X chromosome fertilizes an egg, the offspring will have XX chromosomes and will be a female.
2. If a sperm with a Y chromosome fertilizes an egg, the offspring will have XY chromosomes and will be a male.

From this, it follows that the sex of a child is determined at the time of fertilization. Also that the sperm determines whether a child will be male or female. Sex chromosomes have also been identified in Drosophila and other organisms. In birds and moths, however, the male has two identical sex chromosomes (ZZ), while the female has two unlike chromosomes (ZW).

Sex-linked characteristics. The genes carried on the sex chromosomes are inherited in a certain pattern that is linked with the sex of the individual. It appears that the Y chromosome carries very few genes. Therefore, a female will have two genes for a characteristic in the XX chromosomes; the male will have only one gene in his single X chromosome. Some sex-linked characteristics in humans are *color-blindness, hemophilia* (bleeder's disease) and certain types of *baldness.* Sex-linkage has also been extensively studied in Drosophila for such traits as red and white eye color and about two hundred other characters.

Color-blindness refers to the inability to see cer-

Table IV

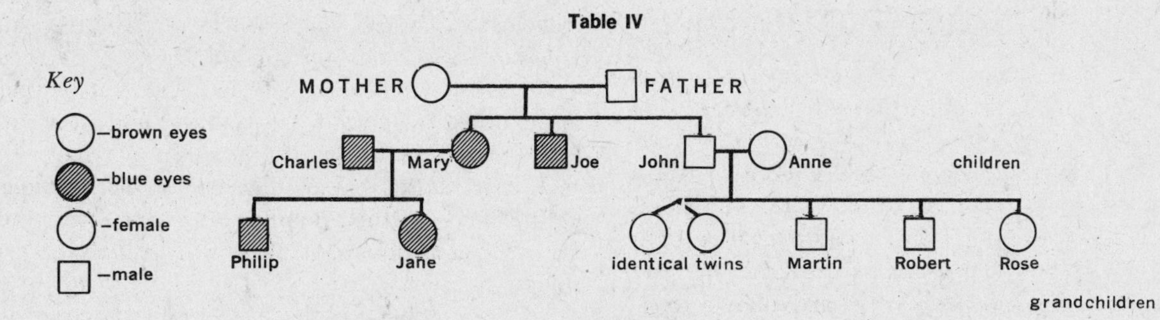

tain colors such as red and green, and in some cases, any colors at all. Normal color vision is dominant. The inheritance of color blindness may be traced as follows:

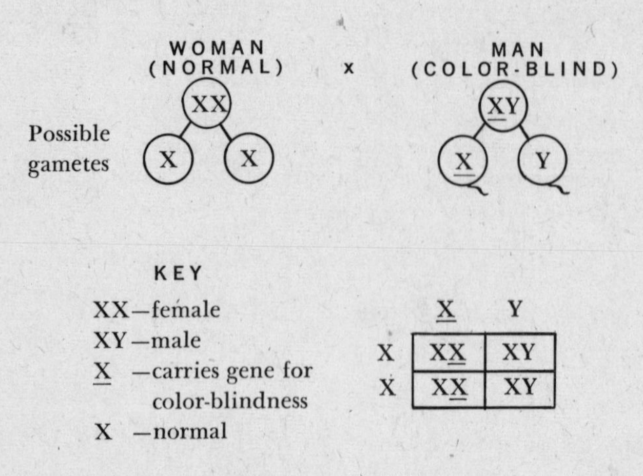

Possible results:
Female: All normal heterozygous
Male: All normal

The children of a color-blind man and a normal (homozygous) woman are thus normal. The daughters, however, are heterozygous, or *carriers.* To show the effect of one of these females bearing children to a normal male:

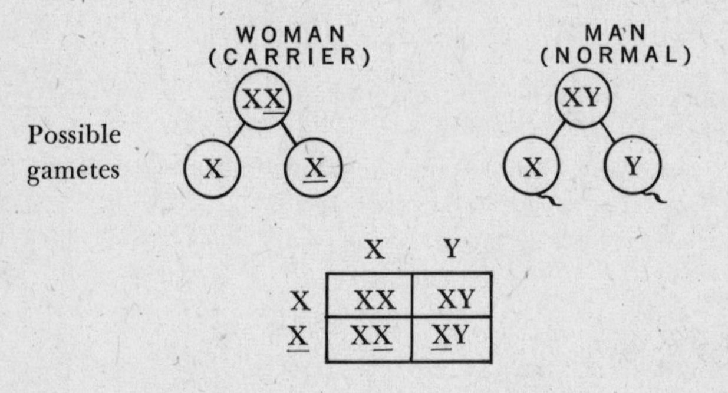

Possible results:
Female: 50% normal, homozygous
50% normal carrier (heterozygous)
Male: 50% normal
50% color-blind

In this case, there is a 50–50 chance that the sons will be color-blind. It can be seen that there is more likelihood that color-blind people will be males. There are relatively few color-blind females. The same pattern of inheritance applies to hemophilia.

Hemophilia is the inability of the blood to clot. As a result, a hemophiliac may bleed to death. This, too, is a recessive trait. Its inheritance follows the same pattern as that for color-blindness.

Chromosomal aberrations. Through the recent perfection of techniques for growing human cells in tissue culture, it has been possible to examine the chromosomes with increasing detail. Certain diseases, or conditions affecting humans have been identified as being caused by abnormalities, or *aberrations,* in some of the chromosomes. Mongolian idiocy (Down's syndrome), in which there is physical and mental retardation, has been traced to the presence of an extra chromosome in chromosome pair number 21, resulting from non-disjunction. Such an individual has a total of 47 chromosomes. Another condition resulting in an extra chromosome is Klinefelter's syndrome, where a male has XXY chromosomes, and is underdeveloped and mentally retarded. Other variations in chromosomes are being given increased attention as an important factor in many physical and mental diseases, and in behavior disorders.

Eugenics. Eugenics is the science of improving the human race by applying our knowledge of genetics. It was founded by Sir Francis Galton (1822-1911). Eugenics does not have much appeal among modern biologists because human heredity is too complex for specific guidelines to be followed. A review of the subject is included here for its historical value. Eugenists have based their recommendations on a study of pedigrees over a number of generations in such families as the Edwards, Darwins, Jukes and Kallikaks. However, their conclusions have not been definite because of the environmental effects involved.

The eugenic program. In an effort to eliminate undesirable traits, eugenists recommend the following program: (1) Feeble-minded people should be segregated in institutions and thus prevented from marrying. (2) Feeble-minded people should be prevented from having children by a simple operation that interferes with normal gamete activity (*sterilization*). (3) Cousin marriages should be discouraged, especially if there is a record of some defective trait in the family, because there is a greater chance of two recessive genes being inherited together in closely related people. (4) People of high ability should be encouraged to have large families (birth selection). (5) People should be taught the principles of eugenics.

Weaknesses of the eugenic program. Criticisms have been made of this program because: (1) Environmental conditions frequently determine the development of desirable or undesirable qualities. If the members of the Jukes family had had the social and economic advantages of the Edwards family, and vice versa, would they have turned out as they did? (2) Since feeble-mindedness is inherited as a recessive trait, most feeble-minded children are born to normal parents; the elimination of feeble-mindedness is thus a complicated matter. (3) It is dangerous to classify people as desirable or undesirable, because this decision may be based on political or social grounds rather than on scientific information. (4) Not enough is known about human genetics.

Euthenics. Euthenics is the study of improving the human race by improving the environment. When environmental conditions have been bettered, the standard of living has been raised and the quality of the people has been improved. There has been less disease, immorality, and poverty. The euthenists claim that less time is needed to improve the human race if a program is adopted including: (1) slum removal and better housing; (2) compulsory education; (3) better educational facilities; (4) old age pensions; (5) improved working conditions in factories and mines; (6) better leisure time, facilities; (7) health insurance; (8) compulsory medical examination as a condition for marriage; (9) unemployment insurance; (10) laws preventing pollution of water, food and air.

Since man is a product of both his heredity and environment, the key to an improvement in the human race is undoubtedly a combination of eugenic and euthenic measures.

Identical twins. Identical twins have the same sets of genes, since they originated from the one fertilized egg. If such twins were separated from birth and raised under different environments, this would amount to a controlled experiment as to the comparative effects of heredity and environment. Prof. H. H. Newman of the University of Chicago conducted studies on a number of identical twins that had been separated in this way. He found that: (1) They resembled each other greatly, showing that their genetic make-up controlled their physical appearance. (2) In some cases, the I.Q. was higher for an individual raised in a more favorable background, indicating that environmental factors help in the expression of native intelligence.

CHAPTER REVIEW

Select the correct choice for each of the following statements.

1. If two parents who are hybrid for brown eyes have four children, the eye colors of the children may be (A) all brown (B) 3 brown: 1 blue (C) 2 brown: 2 blue (D) 1 brown: 3 blue (E) any of the above combinations

2. All of the following human traits are recessive except (A) curly hair (B) albinism (C) Rh negative blood (D) color-blindness (E) light hair color

3. In the accompanying pedigree of a family, in which brown eye color is designated as ○ , and blue eye color as ● , (A) Alice and Joseph are pure for brown eyes (B) Mary and Charles are hybrid for blue eyes (C) Martin and Robert may be hybrid or pure for brown eyes (D) John and Anne are pure for brown eyes (E) Philip and Jane are hybrid for blue eyes

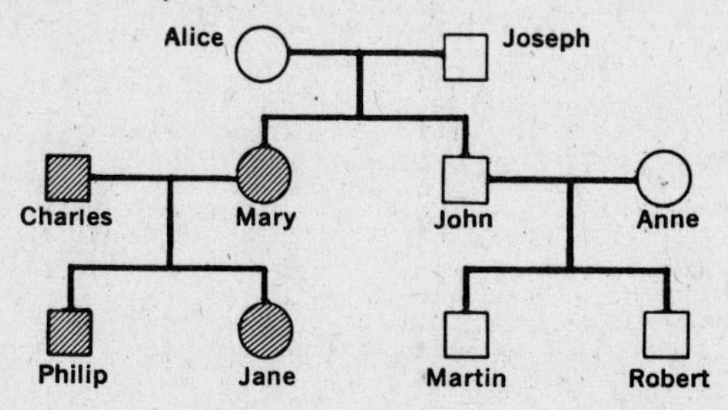

4. The human sperm cell contains (A) 46 chromosomes (B) 44 chromosomes and 2 X chromosomes (C) 44 chromosomes, and 1 X, and 1 Y chromosome (D) 23 chromosomes (E) 23 chromosomes, and 1 X, and 1 Y chromosome

5. If a man who is color-blind marries a woman who is pure normal for color vision, the chances of their sons being color-blind is (A) 0 (B) 100% (C) 50:50 (D) 75:25 (E) 25:50:25

6. A son with hemophilia will most likely result from parents represented as (Key: X-chromosome with gene for normal; $\underline{X}$-chromosome with gene for hemophilia) (A) XX x $\underline{X}$Y (B) XX x X$\underline{Y}$ (C) XX x $\underline{XY}$ (D) XX x XY (E) X$\underline{X}$ x XY

7. The blood group alleles, I^Ai, are for blood group (A) O (B) A (C) B (D) AB (E) M

8. PKU is caused by the lack of a gene for (A) a hormone (B) brain cells (C) an enzyme (D) liver cells (E) intelligence

9. The scientist who founded eugenics was (A) Eijkman (B) Gorgas (C) Jukes (D) Galton (E) Dalton

10. The hemoglobin of people with sickle-cell anemia differs from normal hemoglobin by one (A) lipid (B) monosaccharide (C) disaccharide (D) fatty acid (E) amino acid

ANSWER KEY

1-E	3-C	5-A	7-B	9-D
2-A	4-D	6-E	8-C	10-E

25 Modern genetics

In recent years, evidence has accumulated showing that the nucleic acid, DNA, (deoxyribonucleic acid) makes up the gene, and determines heredity. Some of this evidence is based on transformation, bacteriophage activity and transduction.

Transformation. The first indication of DNA action was found in the pneumococcus bacteria that cause pneumonia. These bacteria cells have a covering or capsule around them. They are similar to other pneumococci that do not have a capsule and that do not cause the disease. In the investigations of Frederick Griffith in 1928, the latter, harmless bacteria were injected into mice along with dead bacteria of the capsule type. The mice developed pneumonia! When the live bacteria were withdrawn from the sick mice and examined, they were found to have capsules. They had been transformed by something from the dead bacteria.

Later, it was found that merely using an extract from pneumococcus bacteria with a capsule, was enough to make the non-capsule bacteria in a petri dish start growing capsules. This new characteristic was inherited by the daughter cells of the transformed bacteria indicating that their heredity had been altered. DNA was shown to be the material in the extract, that changed the pneumococcus characteristic. This change in heredity brought about by the transfer of dissolved DNA is called *transformation*.

Bacteriophage. A bacteriophage is a virus that attacks bacteria. Like many other viruses, its structure consists of a central core containing DNA surrounded by a protein covering. To determine whether both of these parts, or only one of them attacks a bacterial cell, a classic experiment was performed in which the DNA core was tagged with radioactive phosphorus and the protein coat with radiosulfur. When the virus attacked a bacterial cell, the DNA entered through the tail end, which became attached to the cell; the empty shell of the virus was left behind, and was shown to contain the radiosulfur.

Once inside, the DNA replicated to form more DNA molecules. The radioactive phosphorus was shown to be in the bacterial cells. The DNA mole-

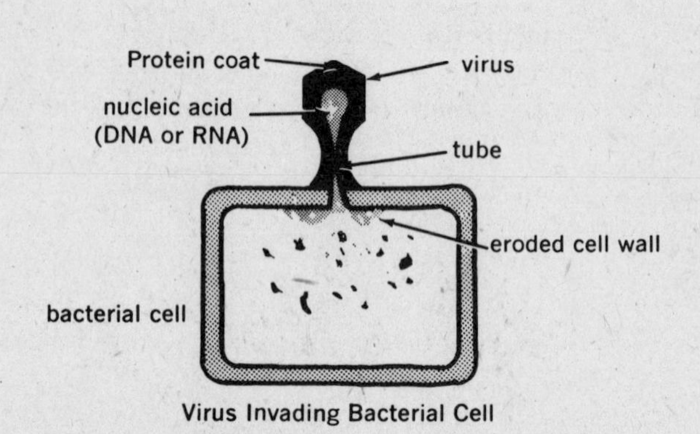

Virus Invading Bacterial Cell

cules formed coats of protein around themselves, to complete the structure of many aditional new viruses. These were released when the bacterial cell burst. Thus, it was shown that the hereditary material of a virus enters a cell as DNA and redirects its metabolism to form additional DNA molecules with protein coverings.

Some viruses, including the tobacco mosaic virus, may contain RNA, ribonucleic acid, in their core, instead of DNA.

Transduction. In transduction, it was found when a virus attacks a bacterial cell, and reproduces itself at the expense of the cell, some of the DNA of the bacterial cell may be incorporated into the viruses that are formed. Then, when the bacterial cell bursts, these viruses attack other bacteria, and bring with them the DNA from the original bacteria. This is another indication that the genetic material carried from bacteria to bacteria is DNA.

DNA structure. The DNA molecule is composed of thousands of smaller units called nucleotides. A nucleotide consists of three parts: (1) a 5-carbon sugar, deoxyribose; (2) phosphate, (3) and a nitrogen base which may be either adenine, thymine, guanine or cytosine. Thus, there are four types of DNA nucleotides, depending on which nitrogen base they contain.

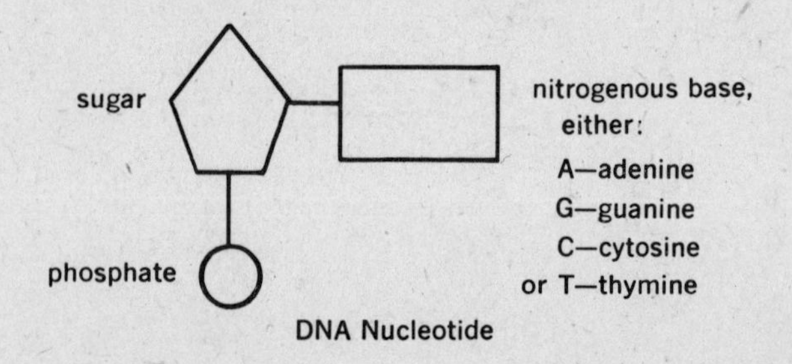

DNA Nucleotide

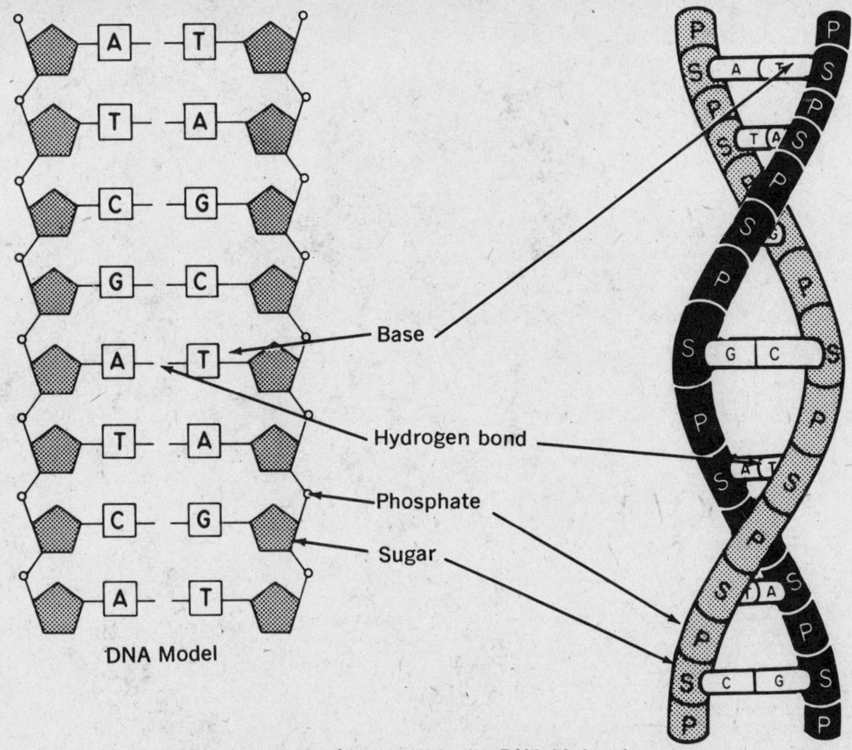

Arrangement of Nucleotides in DNA Molecule

Watson-Crick model. The arrangement of the nucleotides was described in 1953 by James D. Watson and Francis H. C. Crick in their now famous model of DNA. It was based on X-ray diffraction data supplied by Maurice H. F. Wilkins. All three scientists received the Nobel Prize in 1962. According to the model, the DNA molecule has the following characteristics. (1) ladder-type organization; (2) twisted in a double spiral, or helix; (3) the upright parts are made of deoxyribose sugar, S, and phosphate, P; (4) the rungs are made of paired nitrogen bases; (5) these bases are always paired as follows: adenine with thymine, A-T; and cytosine with guanine, C-G; (6) there is a relatively weak hydrogen bond between each of these bases; (7) a single molecule of DNA is made up of thousands of these nucleotides arranged in spiral fashion.

DNA replication. DNA molecules are present in the chromosomes. When the chromosomes replicate during mitosis and meiosis, each DNA molecule also replicates, to form two exact DNA molecules. *Replication,* or the production of an identical copy, probably proceeds as follows: (1) A stimulus, yet unknown, causes the rungs of the DNA ladder to break at the relatively weak hydrogen bonds holding the nitrogen bases together; (see diagram [a]) (2) the spiral ladder is now present as two half ladders; (3) there are numerous free nucleotides present in the cell, which move into position to form a bond with the matching nucleotide of the half ladder; (see diagram [b]) (4) the bases are matched, so that adenine (A) bonds with thymine (T), and cytosine (C) with guanine (G); (5) gradually a new upright portion is added to the half ladder to complete two sections of a new spiral ladder; (see diagram [c]) (6) each half ladder has now added an exact duplicate of the other, giving two identical DNA molecules.

(a),

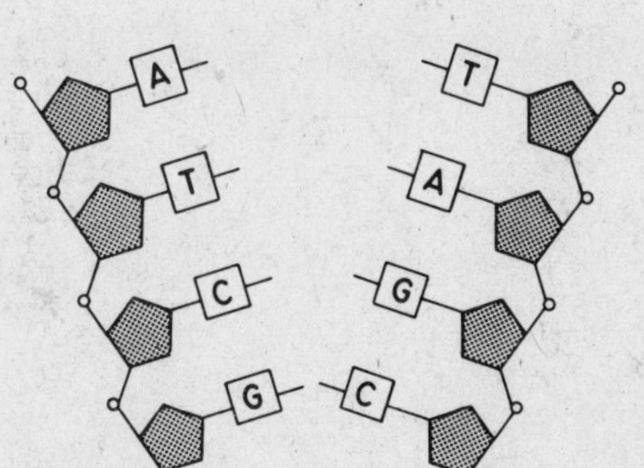

Beginning of DNA replication

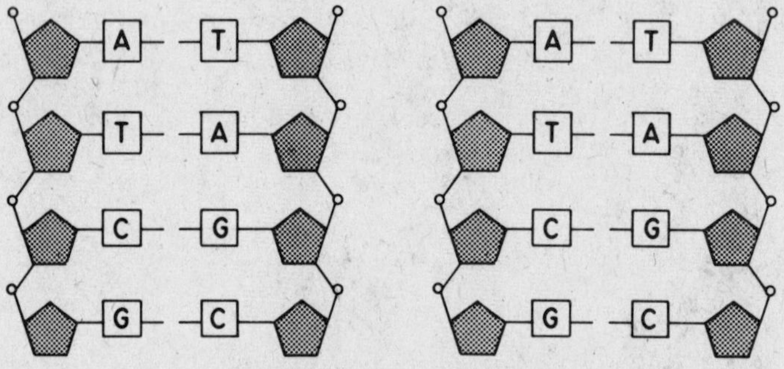

DNA Replication

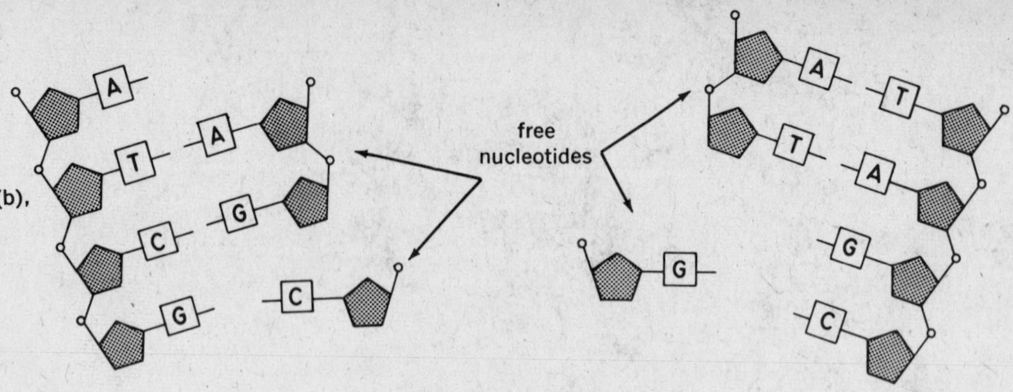

RNA, ribonucleic acid. DNA contains the hereditary code for the various characteristics of an organism. It also directs the formation of specific enzymes which are involved in the cell's activities. DNA is located in the nucleus of the cell, and sends its instructions for making enzymes into the cytoplasm through the nucleic acid, RNA, ribonucleic acid.

RNA, like DNA, is composed of nucleotides. However, it shows these differences from DNA: (1) its five-carbon sugar is ribose, not deoxyribose; (2) RNA has the same nitrogen bases, adenine (A), cytosine (C) and guanine (G), but its fourth base is uracil (U) instead of thymine (T); (3) the RNA molecule is a single strand, while the DNA molecule has a double strand, or in some cases, a single strand.

Two different types of RNA in the cell are: (1) *messenger RNA,* which carries the code from DNA in the nucleus to the ribosomes in the cytoplasm; (2) *transfer RNA,* which picks up amino acid molecules in the cytoplasm and transfers them to the ribosomes where they are formed into proteins.

Protein synthesis. Protein synthesis consists of the building up of complex proteins from combinations of amino acids. This takes place in the ribosomes, which contain most of the RNA of the cell. The formation of enzymes, which are proteins, probably takes place in the following manner: (1) a portion of a DNA molecule, which is a gene, serves as a template, or pattern, for the synthesis of messenger RNA from free RNA nucleotides in the nucleus; (2) some of the weak hydrogen bonds between the nucleotides break, and a portion of the DNA strands separate; (3) the free nucleotides line up next to the appropriate DNA nucleotides — cytosine (C) with guanine (G), and uracil (U) with adenine (A). (see diagram [a]). RNA contains a uracil instead of a thymine base; (4) the newly formed RNA molecule, which is a "reverse copy" of the DNA which produced it, separates from the DNA strand, moves through a pore in the nuclear membrane, and enters the cytoplasm.

(5) The new messenger RNA becomes located in a ribosome; the order of arrangement of its nucleotides was determined by the arrangement of nucleotides on the DNA molecule; (6) small transfer RNA molecules pick up specific amino acid molecules in the cytoplasm, and line up with messenger RNA; (7) each transfer RNA molecule has a group of three nucleotides, and its bases fit the appropriate bases of the messenger RNA; i.e., if transfer RNA has nucleotides containing the sequence UGC, it will fit in with a messenger RNA section of bases that are ACG; U (uracil) always bonds with A (adenine), and C (cytosine) always bonds with G (guanine). (See diagram [b]).

(8) The arrangement of the nucleotides on messenger RNA dictates the order in which the amino acids are lined up and bonded together into polypeptide chains; (9) a new protein is formed in this way, ready to be used by the cell (see diagram [c]) (10) the transfer RNA separates and messenger RNA is available to dictate the synthesis of more protein molecules.

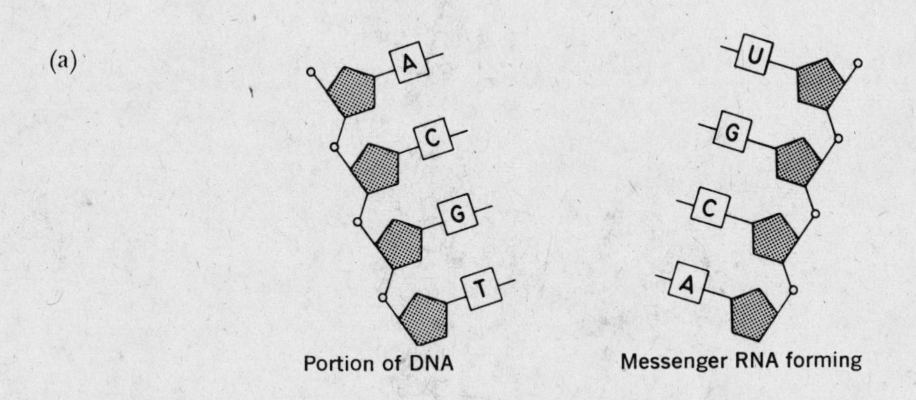

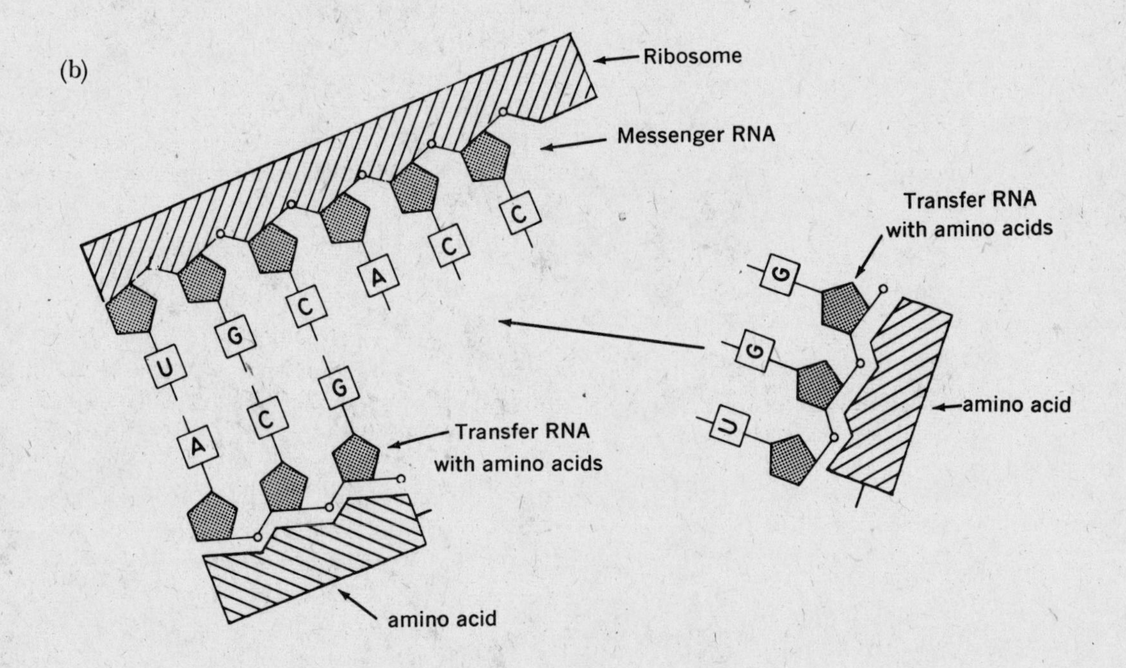

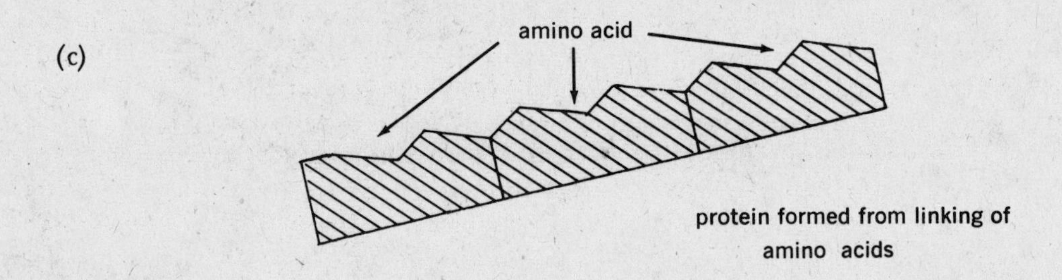

Protein Synthesis

The genetic code. The sequence of the nucleotides in a DNA molecule is important in determining the formation of proteins, through its directions to messenger RNA. Thus, if the order of nucleotides bases is adenine-guanine-thymine-cytosine, the complementary arrangement in messenger RNA is uracil-cytosine-adenine-guanine. This dictates the type of transfer RNA, and the particular amino acid that it carries, that will line up on a part of messenger RNA. All living things contain DNA, and the same four types of nucleotides. However, the DNA of a human is different from the DNA of a maple tree. The arrangement of the nucleotides on the DNA molecule is thought to be responsible for the difference.

The genetic code based on a four-letter alphabet is as follows:

DNA – A, T, C, G
RNA – A, U, C, G

These letters are arranged in multiples of three in the formation of the various amino acids. Some examples:

AMINO ACID	RNA CODE
phenylaline	UUU
alanine	CCG, UCG
histidine	ACC
tryptophan	UGG
lysine	AAA, AAG, UAA

Gene action. The genes, which are made up of DNA, have two principle types of action: (1) they pass on copies of themselves by replication, to all the cells of an individual, following the formation of the zygote; (2) they control the activities of the cell by producing specific enzymes. The current *one gene-one enzyme* hypothesis is based on the idea that a single gene governs the synthesis of a specific enzyme in the cell. An example already discussed, deals with the disease PKU, in which a simple recessive gene affecting the production of a certain enzyme in the liver, results in a form of feeblemindedness.

Experimental evidence of single action was provided by Dr. George W. Beadle and Dr. Edward L. Tatum. They exposed the mold, *Neurospora,* to X-rays. The spores of the mold were then grown in test tubes containing a simple culture medium. Some of the spores did not germinate, indicating that mutations had been induced by the X-rays. Apparently these new spores were now unable to produce a vital amino acid from the culture medium.

The scientists then added one of the different amino acids to each of 20 test tubes of the medium. The spores were found to grow in one of these, the test tube containing the amino acid arginine. This led to the conclusion that the X-ray treatment had affected a single gene that was normally responsible for the production of an enzyme involved in the synthesis of arginine. The mold was no longer able to form arginine, and could not grow, unless it was supplied artificially. Additional mutants were also discovered that required other amino acids, such as ornithine or citrulline. The Nobel Prize was awarded to Beadle and Tatum in 1958.

Gene Mutations. A mutation may occur as the result of a slight change in the DNA molecule. This change may affect just one nucleotide. Yet, as a result, there may be a change in the structure of a single enzyme. This in turn, may affect an important reaction dealing with the life of the organism.

It has been shown that the ability of *Neurospora* to grow depends on its ability to synthesize certain amino acids. The loss of ability to make arginine, in the experiments of Beadle and Tatum for example, prevented the mold from growing. Such a mutation would be harmful, and the organism would not be able to survive. However, other mutations could make an organism better able to compete in the struggle for existence.

Mutations can also result from a change in the sequence of the nitrogen bases in the DNA molecule; also from additions or subtractions in these bases.

The hemoglobin molecule in cases of sickle-cell anemia differs from normal hemoglobin by only a single amino acid out of a total of over 300. The sickle-cell type of hemoglobin has the amino acid valine instead of the normal amino acid known as glutamic acid. It is thought that this change occurred as a gene mutation, changing one base in the genetic code for glutamic acid (UAG) into the code (UUG) for valine.

Cytoplasmic inheritance. Although inheritance is directed by DNA in the nucleus, recent evidence indicates that there may be some forms of inheritance in the cytoplasm. Chloroplasts and mitochondria appear to divide during cell division. The chloroplasts have been found to contain small amounts of DNA. This leads to the possibility that they may contain separate genes which are regulated by the genes in the nucleus.

The chloroplasts in a plant appear to be inherited from the cytoplasm that accompanies the egg nucleus. If the female parent has leaves that are pale green, or variegated with blotchy areas of color, the descendents will have leaves of the same appearance.

Another example of cytoplasmic inheritance occurs in a certain paramecium, *Paramecium aurelia.* It produces a "killer" strain which secretes a poison into the water that kills strains of paramecia sensitive to it. When the two strains were crossed in conjugation, some unexpected results were obtained. The descendents of killers remained killers, and the descendents of sensitives remained sensitive. Under normal conditions, the descendents would have been heterozygotes, and either killers or sensitives. It was concluded that the killer trait was determined by the parent cytoplasm. Later, it was found that the secretion of the killer substance was due to particles in the cytoplasm that contained DNA, and were self-duplicating.

CHAPTER REVIEW

Select the correct choice for each of the following statements:

1. The first indication of DNA action was found in (A) protozoa (B) bacteria (C) algae (D) Drosophila (E) corn
2. The change in heredity brought about by the transfer of dissolved DNA is called (A) transduction (B) transformation (C) transpiration (D) trichinosis (E) trypanosome.
3. Bacteria may be attacked by viruses called (A) cocci (B) bacilli (C) spirilla (D) bacteriophage (E) staphylococcus.
4. The central core of a virus is to its covering as (A) protein is to DNA (B) DNA is to protein (C) lipid is to DNA (D) DNA is to lipid (E) lipid is to protein.
5. Of the following, the base which cannot be present in DNA, is (A) AUCG (B) ATCG (C) ATTA (D) CGTA (E) GCAT
6. During replication of DNA, a nucleotide base that would bond with cytosine is (A) adenine (B) thymine (C) uracil (D) cytosine (E) guanine
7. RNA is different from DNA in that it has (A) a ribose sugar (B) an extra adenine base (C) an extra thymine base (D) a double strand (E) no similar nitrogen bases.
8. All the following are involved in the steps of protein synthesis except (A) messenger RNA (B) transfer RNA (C) ribosome (D) centriole (E) DNA code
9. Neurospora has been useful in illustrating (A) DNA action (B) RNA action (C) single gene action (D) test cross (E) replication
10. Killer paramecia are a good example of (A) parasitism (B) cytoplasmic inheritance (C) meiosis (D) the DNA code (E) messenger RNA

ANSWER KEY

1-B	3-D	5-A	7-A	9-C
2-B	4-B	6-E	8-D	10-B

26 Improving the species

As early man first began to domesticate animals and raise plants, he gradually improved them to serve his purposes. Modern man now applies his knowledge of genetics toward improving the various species for the purpose of breeding better plants and animals. Probably the largest research center for this purpose is maintained by the U.S. Department of Agriculture at Beltsville, Md.

AIMS OF BREEDING

Plants and animals are bred for the following purposes:

1. **Improved yield.** For example, dairy cattle today produce twice as much milk as at the beginning of the century. Prize hens lay almost an egg a day, which is about three times the yield of fifty years ago. The productivity of other domesticated animals and plants has likewise been increased over the years.

2. **Improved quality.** Some examples: The butterfat content of milk has been improved. The strength of the fiber in cotton has been increased. Although the weight of hogs has been raised, there has been an increase in the leanness of pork.

3. **Increased resistance to disease.** Kanred wheat has been developed, which is resistant to the fungus parasite known as the wheat rust. An improved variety of cattle immune to Texas fever

has been produced, which can endure the intense summer heat and insect pests of the gulf coast area.

4. New varieties for special purposes. Special breeds of dogs, larger and more attractive daisies, faster racehorses, new fruits such as the plumcot and pink grapefruit, and showy varieties of tropical fish have been bred for purposes of attractiveness, uniqueness, speed, endurance, special utility, and so on.

Methods of the breeder. Several methods have been used in breeding these better animals and plants. In some cases, new types have actually been made to order.

1. Selection. The oldest method was simply a matter of selecting the best type for breeding in each generation. The various pure breeds of dogs have been produced in this way, each being selected for a particular purpose. The best desired types were then *inbred,* or mated to one another, in order to maintain the line. We now know that this method essentially resulted in selecting the best combinations of genes, and keeping them by inbreeding. Selection may sometimes give extreme results, as in the case of the Boston bulldog which has been selected for its flat face to such an extent, that it has difficulty in breathing through its nose. As an example among plants, cotton plants resistant to the fungus wilt disease have been improved by selection. In practicing selection among plants, self-pollination is carried on so that no other traits will be introduced. Eventually, the type is fixed by selection, and pure lines are established that will breed true.

2. Hybridization. By hybridization, or cross-breeding, the breeder combines desirable qualities of two different organisms into one. An example of this is the improvement of the shorthorn cattle in the Southwest. They were useful because of their good beef qualities, but they became sickly as a result of the Texas fever. They were crossbred with Brahman cattle that were imported from India; these cattle did not have the same good beef qualities, but were immune to Texas fever. The offspring showed various qualities; good beef but susceptible; poor beef but immune; poor beef and susceptible; good beef and immune. The latter type was selected and bred for several generations, until the new Santa Gertrudis breed was developed, containing both desired qualities.

Luther Burbank (1849–1926) is probably the best known plant breeder. As part of his work, he created new plants by combining qualities of different plants: the plumcot was the result of crossing the plum and the apricot; the white blackberry resulted from a cross of the Lawton blackberry with a wild pale type; the Shasta daisy came from a cross of three daisies, the American, which is sturdy, the English, which has large petals, and the Japanese, which is brilliantly white.

One result of hybridization is known as *heterosis,* or hybrid vigor. In this case, the hybrid surpasses the parents. The outstanding example is hybrid corn. This is produced from two separate inbred pure strains that are cross-pollinated. Hybrid corn combines the desirable genes from both parents. It is more vigorous and produces more high quality ears of corn. The yield from hybrid corn has been increased about 35 percent. Because of its high yield, the introduction of hybrid corn into the devastated countries of Europe at the close of World War II helped avert a famine.

The mule is an example of a hybrid of superior vigor and strength resulting from a cross of two species, the horse and the donkey. The cattalo is the hybrid of the bison and domestic cattle, also having unusual vigor. In both of the above cases, however, the hybrids are sterile. Additional curiosities of little economic values are the zebroid, the cross between the zebra and the horse, and the tiglon, resulting from a cross between a tiger and a lion.

3. Saving the best genes. When a breeder develops a useful variety, he values the outstanding individuals in it, because they contain the best genes. A prize-winning racehorse will continue to be valuable even after it has grown too old to race. It is used for breeding other racehorses. One of the most famous, and most valuable for this purpose, was Man o'War; his offspring continued to win many prizes. A purebred bull which carries genes for good milk production is worth thousands of dollars. Records are kept of purebred dogs, and are passed on to new owners so that they may know the pedigree of their pets. The American Hereford Cattle Breeders Association keeps a registry of purebred Hereford cattle. The American Guernsey Cattle Club, which was founded in 1877, maintains a record of purebred Guernsey dairy cows.

A practical way to make use of useful genes is in the technique of *artificial insemination.* The sperm of prize bulls, for example, is stored and frozen. It is used to inseminate or fertilize prize cows, and is available long after the death of the bull.

4. Useful mutations. Mutations, or gene changes, that occur spontaneously in domesticated plants or animals may introduce new favorable features. Breeders are always on the lookout for them. For example, the seedless or navel orange originated in Brazil. It was introduced to California in 1873, and has been propagated by grafting ever since. The pink grapefruit is another mutant that has been maintained by grafting. The nectarine is a peach with a smooth skin that arose as a mutation. A very useful mutation among cattle is the *polled,* or hornless condition, which is a dominant characteristic. The present polled Hereford strain of cattle was bred by mating horned Hereford cows to mutant polled bulls. The Ancon breed of sheep originated from a mutant that had very short legs.

Native american plants and animals. The explorers and early settlers in this country came across plants and animals they had never seen before. The turkey is a native American bird; Benjamin Franklin suggested using it as our national symbol instead of the eagle. Columbus found corn being grown by the Indians in Cuba and on the mainland of South America. Potatoes were brought back to Europe by the early Spanish explorers. Tomatoes, beans, squash and pumpkins also originated in the New World.

Advantages of plant breeding. The breeding of plants offers certain advantages over animal breeding: (1) Plants reproduce in large numbers. (2) There is thus a greater chance of favorable mutations appearing. (3) By vegetative propagation (grafting, cutting, bulbs, etc.) the same genetic make up can be obtained in the new individuals, thus making them true to type. (4) By vegetative propagation, more plants can be obtained in a shorter time. (3) Inbreeding is easily practiced by simply self-pollinating the flowers.

Importance of the environment. The best results of improved breeding are of little value unless the proper environment is provided. A prize cow will be prevented from giving a record-breaking supply of milk if she is poorly fed, or becomes sick. The yield from choice hybrid corn seeds will be low unless the proper fertilizer and soil conditions are provided by the farmer. In short, the best results are obtained by a combination of the right genes and a good environment.

CHAPTER REVIEW

Select the correct choice for each of the following statements.

1. All of the following are aims of the breeder except (A) improved yield (B) improved quality (C) increased susceptibility to disease (D) increase resistance to disease (E) new varieties

2. A new variety of fruit produced by hybridization is the (A) seedless grape (B) seedless orange (C) seedless grapefruit (D) plumcot (E) Delicious apple

3. In practicing selection among plants, the breeder carries on (A) self-pollination (B) cross-pollination (C) crop rotation (D) strip cropping (E) soil fertilization

4. Inbreeding is carried on by (A) crossing unrelated animals (B) crossing related animals (C) cross-pollinating flowers (D) hybridizing plants (E) using hydroponics

5. Combining the desirable qualities of two different organisms into one is referred to as (A) inbreeding (B) self-breeding (C) pure line breeding (D) hybridization (E) conjugation

6. All of the following organisms resulted from cross-breeding except (A) Ancon sheep (B) cattalo (C) Santa Gertrudis cattle (D) mule (E) Shasta daisy

7. An advantage of the plant breeder over the animal breeder is that he can use (A) vegetative propagation (B) mutations (C) binary fission (D) alternation of generations (E) inbreeding

8. Two plants that originated in the new world are (A) corn and wheat (B) wheat and tomato (C) tomato and rice (D) rice and corn (E) corn and tomato

9. A useful animal mutation is the (A) long-horned Texas steer (B) Boston bulldog (C) Guernsey cow (D) Hereford cattle (E) polled cattle

10. One result of hybridization is (A) a pure line (B) heterosis (C) gene change (D) inbreeding (E) improved environment

ANSWER KEY

1-C	3-A	5-D	7-A	9-E
2-D	4-B	6-A	8-E	10-B

HOW LIVING THINGS HAVE CHANGED

8

27 The record of prehistoric life

PALEONTOLOGY is the study of prehistoric life as revealed by fossils. The record in the rocks shows that these earlier organisms were different from those of today; they were also simpler.

Rock formation. *Geology* is the study of the earth and its rocks. It is estimated that the earth is about four billion years old. Originally, it was extremely hot, with a surface of molten material. After it cooled, three main types of rocks were formed:

1. **Igneous rock** was the earliest type of rock, being formed from the molten material. An example is granite. Igneous rock is still being formed today from the molten lava of volcanoes.

2. **Sedimentary rock** is formed from sediment that is carried into the sea by rivers. This sediment results from the effects of erosion and weathering in breaking up rocks into smaller particles, and eventually forming soil. Under the pressure of the water over the ages, the material in the sediment becomes cemented together, turning it into sedimentary rock. In this way, sandstone was formed from sand, shale from clay, and limestone from mud or the remains of shells or coral. Sedimentary rocks usually appear in layers or strata.

3. **Metamorphic rock** is formed from previously existing igneous or sedimentary rocks whose structure has been transformed by great pressure, heat and other factors. Example: marble.

Fossil formation. A fossil is the remains of a living thing that existed long ago. Fossils are practically always found only in sedimentary rocks, since they would be destroyed by the high temperature of igneous rocks and the crushing pressures of metamorphic rocks. Fossils may be formed in a number of ways:

1. **Petrifaction.** An animal or woody plant that is covered with undisturbed sediment under water for thousands of years, gradually becomes petrified or turned into stone. This happens as minerals replace the hard parts of the bones or wood, while the sediment itself is changing into sedimentary rock. The petrified bones of dinosaurs and the petrified wood of trees were formed in this way. Subsequent changes in the earth's surface and erosion have brought them within reach of paleontologists.

2. **Imprints.** Footprints or leafprints that were left in soft mud or sand have been preserved as imprints. The soft soil dried up and later changed into sedimentary rock. Dinosaur tracks are a good example. Leaf and stem imprints are sometimes found in coal. Coal itself was formed from the remains of trees under the surface of the water, following thousands of years of pressure.

3. **Casts or molds.** Animals such as snails were buried in mud under water for ages. As the organism decayed, its place was taken by minerals,

leaving a cast of the original animal in the surrounding sedimentary rock.

4. Freezing. The preserved bodies of prehistoric mammoths have been found in frozen Siberia, where the intense cold prevented bacterial decay from taking place.

5. Tar pits. The La Brea tar pits in Los Angeles have yielded the skeletons of countless animals that were trapped in the sticky material from which they could not escape. Skeletons of the saber-tooth tiger and other prehistoric animals have been found there.

6. Amber. Prehistoric ants, wasps and other insects were trapped in the sticky resin of evergreen trees which gradually hardened into amber. Their bodies were protected from bacteria and the air, and did not decompose.

Age of rock and of fossils. There are several ways of determining the age of rock and of fossils:

1. Radioactive elements. Uranium is a radioactive element that gradually changes into lead. By analyzing the proportions of uranium and lead in a rock, it is possible to calculate its age. On this basis, the age of the earth has been estimated to be in the vicinity of four billion years. Another radioactive element, an isotope of carbon, C-14, has a half-life of 5,760 years. That is the period of time in which it loses half of its radioactivity. Using C-14, scientists have been able to compute the age of wood, charcoal and bone remains of early man, and his culture.

2. Position of strata. Since sedimentary rock is laid down in layers, or strata, it is evident that the lowest strata are the oldest. Fossils found in them are older than fossils located in upper strata.

3. Rate of sedimentation. It has been estimated that some types of sediment turn into sedimentary rock at the rate of one foot in 900 years. By measuring the thickness of a stratum, its age can thus be arrived at.

4. Saltiness of the ocean. Another method of estimating the age of the earth is to analyze the salinity of the ocean. It continues to increase as rivers flow into it carrying dissolved salt.

Geologic eras. By studying the various rock strata and the fossils in them, geologists have drawn up a history of the earth. It is divided into six *eras,* each of which is subdivided into *periods.*

1. Azoic era. This is the earliest era, when the earth was cooling and when the mountains and oceans were being formed. It lasted about one to two billion years. There was no life on the earth at this time.

2. Archaeozoic era. During this era, the simplest forms of life probably appeared. Some of the rocks of this era contain carbon, an indication that living things most likely existed. This era lasted for over 600,000,000 years.

3. Proterozoic era. Traces of algae, bacteria, shelled protozoa and the burrows of worms have been found in this next era. Fossils are relatively rare, indicating that the organisms probably did not have hard parts. Life continued to exist only in the water. This era lasted about 600,000,000 years.

4. Paleozoic era. There are abundant fossils of animals and plants in this era. In the earliest period (Cambrian Period), there were sponges, jellyfish and corals. An interesting type of lobster-like animal, the trilobite, appeared, formed many species throughout the era, and then became extinct. In later periods, mollusks appeared, and then the vertebrates. Fish became well established. The amphibia made their appearance as air-breathing vertebrates. Toward the end of this era, primitive reptiles began to appear. In the meanwhile, insects, including giant dragonflies with a wingspread of over two feet, flourished. Mosses and ferns grew on land. This era included the age of the coal-forming plants, the Carboniferous Period. The era lasted about 400,000,000 years.

5. Mesozoic era. This is the "Age of Reptiles," during which the huge dinosaurs appeared. Among them were the herbivorous *Brontosaurus,* the carnivorous *Tyrannosaurus rex,* and the armored *Stegosaurus.* They eventually became extinct, possibly because the climate changed, or because their eggs were destroyed by the newly-risen mammals that began to appear toward the end of this era. The fossil bird, *Archaeopteryx,* also appeared toward the close of the era. It was about the size of a crow, and had birdlike features, such as feathers and wings; it also had reptilian features, such as numerous teeth in the beak, an elongated tail containing about twenty vertebrae, and claws at the ends of the wings. Evergreen trees also made their appearance at this time. The Mesozoic Era lasted about 125,000,000 years.

6. Cenozoic era. About 60,000,000 years ago, the present "Age of Mammals" began. Most of the modern mammals and flowering plants began to appear. Rather complete fossil records have been found showing the various changes that have taken place in such mammals as the horse, camel, elephant and pig since the beginning of the Cenozoic Era. The fossil history of the horse reveals the following record:

1. *Eohippus,* the dawn horse, lived during the earliest period of the era. It was about the size of a cat, had simple teeth, and had four toes on its front legs and three toes on its hind legs.

2. In higher rock layers, fossils are found of a

somewhat larger animal, with a foot consisting of a large middle toe and two smaller toes on each side of it that just reach the ground.

3. In still higher strata, the fossils show an animal about the size of a pony, in which the middle toe of each foot is larger, and the side toes do not reach the ground. The teeth are becoming more complex.

4. Modern horse has a foot which is supported by one large toe and which contains the remains of the two side toes as splints, visible only in the skeleton. The teeth have a complex grinding surface.

5. The Indians were unfamiliar with the horses that the Spaniards brought with them to explore the New World, and regarded them with wonder. However, the ground under them contained the fossil remains of the many ancestors of the horse. For some reason, the animal became extinct in this part of the world during the last stages of the Cenozoic Era. After it was reintroduced, the modern horse flourished throughout the continent.

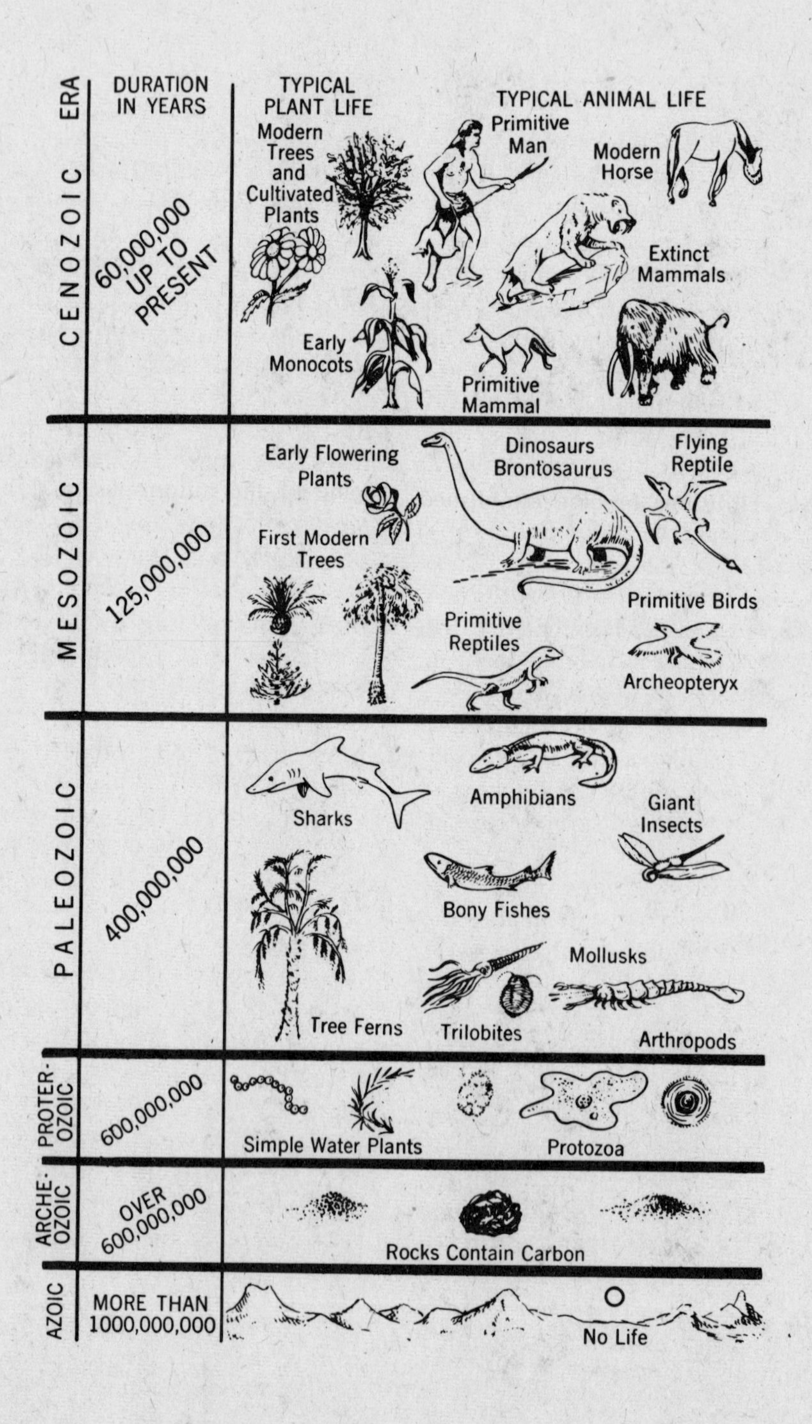

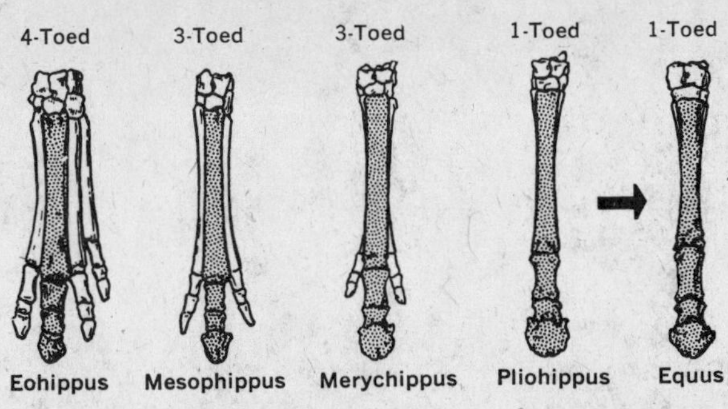

The evolution of the forelimb in the horse. The single toe of the modern horse *(Equus)* is the sole remainder of four toes present in Eohippus.

Within the last million years, several ice ages occurred, the last about 25,000 years ago. A primitive type of man appeared about a million years ago. Sometimes this last period of time is referred to as the Psychozoic Era, the era of man's supremacy.

Evolution. The changes in living things that have taken place during the various eras since the earliest beginnings of life, are known as *organic evolution.* Modern living things have developed from simpler organisms of the past. This process of change has taken place over countless millions of years. It is still going on.

Origin of Life. In the early history of the earth, conditions were very different from those of today. The earth itself may have been a molten mass of rock. As it cooled, conditions become established for the emergence of a primitive form of life. This was quite different from the process included in the theory of spontaneous generation, in which fully-formed living things were supposed to arise suddenly from non-living matter. The beginning of a very elementary form of life must have taken place gradually over a period of millions of years, under a far different set of conditions,

Heterotroph hypothesis. An explanation of how life originated is offered in the heterotroph hypothesis. This hypothesis was suggested by a group of scientists, including the Russian biochemist Alexander I. Oparin. In his book, *The Origin of Life* (1936), he stated these points:

1. The atmosphere of the primitive earth is supposed to have contained hydrogen, methane (CH_4), ammonia, water vapor and carbon dioxide. There was little or no oxygen. The atmosphere itself was formed from gases given off by frequent volcanic eruptions.

2. Heavy rains washed these gases into the early oceans, lakes and ponds, where they were mixed with dissolved minerals.

3. In the "hot thin soup" of these bodies of waters, the molecules of these substances were acted upon by various forms of energy, such as: solar radiation (including X-rays and ultraviolet radiation), cosmic rays, lightning discharges, the earth's heat, and radioactivity in the rocks.

4. In this energy-rich environment, chemical bonds were formed among the dissolved molecules, resulting in the production of larger organic molecules, such as amino acids and sugars. These organic molecules interacted to form more complex organic molecules, such as polypeptides and proteins.

5. Some of the large, complex molecules formed *aggregates,* or *clusters* of molecules. These aggregates (also known as coacervates) incorporated molecules from the ocean as food. In this form, they were *heterotrophs.*

6. As the aggregates became increasingly complex, and highly organized, nucleic acids were formed, with the ability to reproduce. Once they could reproduce, the aggregates are considered to have been alive.

7. The respiration of the heterotrophs must have been anaerobic, and they obtained their energy by fermentation. In this process, they added quantities of carbon dioxide to the atmosphere.

8. Some of the heterotrophic aggregates developed a method of using carbon dioxide to manufacture their own food, and so became pioneer *autotrophs.*

9. Through the food-making activities of the autotrophs, oxygen was added to the atmosphere. Some of the autotrophs, and heterotrophs developed methods of using oxygen to obtain energy from food, in the beginning of aerobic respiration.

10. The evolution of the first primitive forms of life to the complex organisms of today took place gradually over a period of two or more billion years.

Confirmation of the heterotroph hypothesis. Recent findings in ancient rocks provide chemical evidence that life existed on earth 2.7 billion years ago. Since the crust of the earth is believed to have formed about 4.7 billion years ago, it is possible that life evolved within two billion years after the earth took on its present form.

Laboratory experiments of Dr. Stanley Miller, working under the guidance of Nobel prize-winner, Harold Urey, have led to the artificial production of amino acids. He boiled water continuously in a flask containing methane, ammonia, hydrogen and water vapor. These materials were subjected to electric discharges. After a week of duplicating the earth's primitive conditions in this way, Miller found that a number of organic molecules, including amino acids, had been formed in the flask.

Other scientists have been successful in using such sources of energy as heat, and gamma radiation to form chains of amino acids, as well as sugars, ATP, purines and pyrimidines.

CHAPTER REVIEW

Select the correct choice for each of the following statements.

1. The study of prehistoric life is called (A) eugenics (B) euthenics (C) paleontology (D) abiosis (E) physiotherapy

2. It is currently estimated that the age of the earth is about (A) one million years (B) four million years (C) one billion years (D) four billion years (E) one light year

3. Fossils may be found in (A) limestone (B) sandstone (C) shale (D) none of these (E) all of these

4. The replacement of bone or wood by minerals is known as (A) amber formation (B) transmutation (C) petrifaction (D) vulcanization (E) metamorphosis

5. A prehistoric animal preserved in tar pits was the (A) brontosaurus (B) tyrannosaurus (C) trilobite (D) coelacanth (E) saber-tooth tiger

6. An animal whose body was preserved for thousands of years in ice was the (A) triceratops (B) mammoth (C) prehistoric ant (D) giant grizzly (E) eohippus

7. The most reliable method of estimating the age of the earth is through the study of (A) volcanic ash (B) igneous intrusions (C) living volcanoes (D) uranium-lead-deposits (E) craters on the moon

8. A form of animal life that probably arose after fishes was (A) sponges (B) molluscs (C) trilobites (D) amphibia (E) jellyfish

9. The Mesozoic Era is referred to as the age of reptiles because during it, (A) dinosaurs first appeared (B) dinosaurs became extinct (C) reptiles reached their greatest development (D) reptiles destroyed all other types of vertebrates (E) reptiles were the only living things on earth

10. Eohippus is considered to be an ancestor of the (A) camel (B) horse (C) elephant (D) pig (E) whale

11. The heterotroph hypothesis was proposed by (A) Darwin (B) Oparin (C) DeVries (D) Lamarck (E) Metchnikoff

12. According to the heterotroph hypothesis, the correct sequence in the origin of life was (A) heterotrophs — aggregates — autotrophs — organic molecules (B) organic molecules — aggregates — heterotrophs — autotrophs (C) organic molecules — autotrophs — heterotrophs — aggregates (D) aggregates — autotrophs — organic molecules — heterotrophs (E) heterotrophs — autotrophs — organic molecules — aggregates

13. The remains of *Archaeopteryx* indicated that birds are most closely related to (A) flying insects (B) flying mammals (C) flying fish (D) reptiles (E) amphibia

14. The number of toes on each leg modern horse walks on is (A) 1 (B) 2 (C) 3 (D) 4 (E) 5

15. All of the following statements about evolution are true except (A) living things have changed (B) modern living things developed from simpler organisms (C) evolution has taken place over many millions of years (D) evolution is still going on (E) evolution has ceased

ANSWER KEY

1-C	4-C	7-D	10-B	13-D
2-D	5-E	8-D	11-B	14-A
3-E	6-B	9-C	12-B	15-E

28 Evidences of evolution

There are many lines of evidence that establish the fact that animals and plants have changed through the ages, progressing from simple to more complex forms.

Fossil evidence of evolution. The record of prehistoric life discussed in the preceding chapter has revealed the following points: (1) The oldest layers of rock contain only the simplest fossils. (2) The adjacent higher strata reveal similar fossils, but also new and slightly more complex organisms. (3) Successively more recent fossils show a continuing complexity, from simple invertebrates to more advanced invertebrates, to vertebrates to mammals. (4) Fossils of the most complex and most up-to-date plants and animals are present only in the most recent strata of rock. (5) There are rather complete fossil series extending over millions of years, for animals like the horse and the elephant, showing step-by-step progressive changes from a primitive type to the modern animal of today. (6) The relationship between two separate classes of animals, reptiles and birds, is shown by *Archaeopteryx,* which has characteristics of both classes.

Evidence from the study of heredity. Mutations in plants and animals are going on all the time. Man has been able to make use of some of these gene changes in breeding better varieties. In fact, by means of radiations from isotopes and X-rays, some animals and plants have been made to develop mutations artificially. The use of chemicals such as colchicine has also changed the chromosome number, resulting in new types. In addition to these sudden changes in the characteristics of animals and plants, other changes have been produced as the result of breeding better and newer varieties. Improved fruits, vegetables, cattle, etc., that did not exist two generations ago, are now well-established.

Evidence from comparative anatomy. Comparative anatomy is the comparative study of the structure of different organisms. The arm of man and the ape, the flipper of a whale, the leg of a cat, the wing of a bird, and the leg of a frog are clearly used for different purposes. Yet the skeletal structure of these forelimbs reveals a similar bone arrangement, with the same name being applied to their parts. Such structures which are similar in structure, but not necessarily in function are called *homologous* structures. Similarities are also evident when a comparison is made of the brain and nervous system, the muscular system, digestive system, and circulatory systems of various vertebrates.

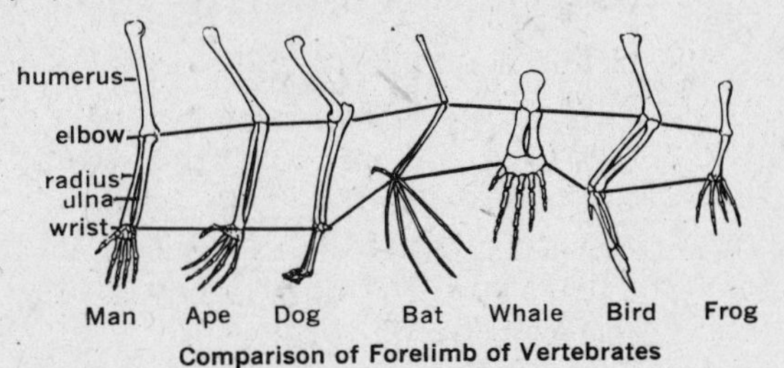

Comparison of Forelimb of Vertebrates

These similarities are explained by the presence of similar genes which all of these animals have inherited from some distant common vertebrate ancestor. Since the time of that common ancestor, changes have occurred, resulting in the evolution of new types of animals. However, these present-day animals still have enough genes in common to retain the similarities noted.

Evidence from vestigial structures. Vestigial structures are useless parts of animals that have no function. In man, some examples are: the appendix, scalp and ear muscles, third eyelid (nictitating membrane) in the inside corner of the eye, and the coccyx, or tailbones. In herbivorous animals, the appendix is used for digestion; many mammals use their scalp and ear muscles, and some rare people are even able to wiggle their ears; amphibians, reptiles and birds have a functional third eyelid which covers the entire eye. The presence of vestigial organs indicates descent from an ancestor that once used them. Although they are no longer used today, there are still genes present for them, and they appear in the animal's body.

Other examples of vestigial organs: Whale—small, vestigial hipbones, indicating descent from vertebrate ancestors that had hipbones and hind legs. Horse—splints of two toes on either side of the large toe; the series of fossils from Eohippus supply evidence that these splints are the remains of toes that were once used.

Evidence from embryology. Embryology is the study of the development of the embryo. The larvae of certain molluscs and annelids are remarkably similar in having a ciliated band around them. In both phyla the larvae, which are called *trochophores,* possess other similar structures. Beyond this stage, differences appear and development proceeds along different lines. Because of these early similarities, it is probable that the two phyla had a common ancestor. Although many changes have occurred since then, there are still genes present for the similarities during the larval stage.

A comparison of the embryos of a fish, salamander, turtle, chick, rabbit, and man shows: (1) They appear very similar in the early stages. (2) They all have a tail and gill slits during their development. In fish, gill slits contain gills; although land animals breathe by lungs, their embryos develop gill slits during their early development, which then disappear later on. The embryo of chick and man have a tail which disappears. The various embryos become less and less alike as they continue to develop.

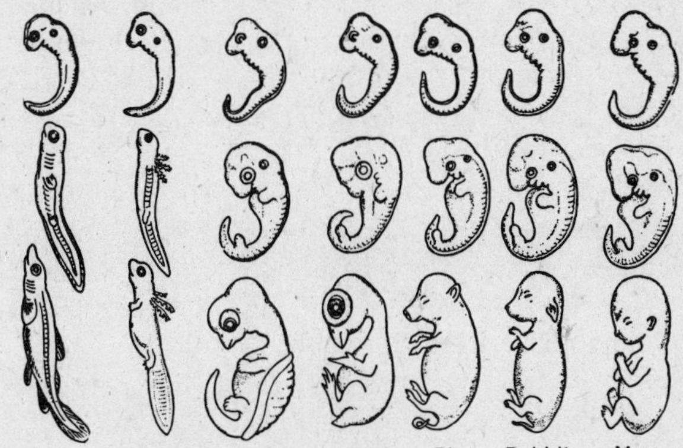

Embryological development of vertebrates

The similarity of these embryos in the early stage indicates that these animals are descended from a common ancestor, and that they still have some similar genes. The more closely related animals are, the greater the similarity; the less closely related they are, the earlier in their development will differences appear. The "theory of recapitulation" attempts to summarize these observations of embryonic development by stating that as the individual organism develops, it repeats the evolutionary development of the species, or that "Ontogeny recapitulates phylogeny."

Evidence from classification. A "tree of life" arrangement of animals and plants, based on their similarities, gives us the system of classification described at the beginning of this book. Like animals are grouped together in groups. Within a genus, the animals have many characteristics in common, indicating a common descent. Branching out, from genus to family, to order, to class, and to phylum, the animals have fewer and fewer similarities, but enough to group them together. The further apart the phyla are, the greater the differences. They all show a relationship, and a progression from simple, one-celled animals to more complex forms of life.

Evidence from geographic distribution. Geographic distribution refers to the fact that plants and animals tend to spread in all directions. They do this until they are stopped by barriers, such as climatic conditions, oceans, mountains and deserts. If they become isolated for a long period of time as the result of such barriers, and cannot interbreed with other animals, they will eventually become different from them. These differences come about as the result of mutations which occur from time to time, and the random recombination of genes. The longer the period of separation, the greater the differences.

These effects of isolation are illustrated in the native mammals of Australia, which are all pouched, such as the kangaroo, koala and wombat. At one time, Australia was connected with the mainland of Asia and the primitive pouched mammals of the time wandered freely between the two areas. After Australia was isolated by the ocean as the result of geologic changes, its pouched mammals continued to evolve to the present types. On the mainland, however, they evolved into mammals that gave birth to their young. The two lines of development within and outside of Australia occurred over a long period of time, demonstrating the ever-changing nature of life.

Similarly, Charles Darwin found distinct differences among the birds and other animals that inhabited the various Galapagos Islands, which are about 600 miles west of Ecuador. There were thirteen different species of finch, a small bird. Some lived in trees and ate insects. Others lived close to the ground and sought seeds. Their beaks were different in shape and size, varying from thin, pointed beaks to broad and heavy beaks. Each of these species resembled one another more than they resembled the related species on the mainland of South America. This was so because migration from the continent was rarer than from one island to another. The longer apart the animals had been, the more pronounced the differences between them.

Evidence from comparative biochemistry. The similarities among living things which are illustrated by the facts of comparative anatomy, are strikingly substantiated by biochemical and physiological resemblances. The *precipitin test* establishes a delicate scale of blood relationships. The test is performed as follows: (1) A small amount of human serum is injected into a rabbit, stimulating it to produce antibodies against human serum. (2) The sensitized rabbit serum is mixed with human serum. (3) A white precipitate forms. (4) If this sensitized rabbit serum is mixed with serum from a chicken, there is no reaction and the liquid remains clear.

It has been found that a precipitate also forms in decreasing degrees when the serum of a chimpanzee, an orang-outan, and a gorilla are added to rabbit serum which has been sensitized to human serum. In like manner, the serum of a dog and a wolf show precipitations with serum sensitized to dog serum. The precipitin test also links birds with reptiles, whales and porpoises with cows and pigs, and horseshoe crabs with scorpions. The geological records indicate similar relationships in these various cases.

Other biochemical similarities point to close relationships among organisms. Thus, antibodies for diphtheria and for tetanus, which are produced by a horse, may also be used in the human body. Various hormones produced by cattle, such as insulin and ACTH, are used by doctors to treat diseases in man. Among plants, the potato plant is attacked by the potato beetle, which will not eat the leaves of any other cultivated plant. However, it will eat the leaves of wild plants that are related to the potato.

Recent discovery. A few years ago, an unusual fish, the *coelacanth,* was caught in the depths of the Indian Ocean between the island of Madagascar and Africa. This fish was believed to have died out 70 million years ago. It is considered a link between fish and amphibia. Several living specimens of this "living fossil fish" have been found.

Conclusion. A study of all the evidences leads to the inescapable conclusion that evolution has taken place over the ages, and is continuing to take place at the present time. These evidences are listed below:

Fossils
Heredity
Comparative anatomy
Vestigial structures
Embryology
Classification
Geographic distribution
Comparative biochemistry
Recent discoveries of "living fossils"

CHAPTER REVIEW

Select the correct choice for each of the following statements.

1. All of the following areas provide evidence for evolution except (A) euthenics (B) heredity (C) comparative anatomy (D) fossils (E) vestigial structures

2. The flipper of a whale is most similar in structure to the (A) claw of a lobster (B) claw of a crab (C) tentacle of an octopus (D) elephant's trunk (E) arm of man

3. An example of a vestigial structure is the (A) horse's foot (B) appendix of man (C) brain of man (D) ear muscles of a dog (E) claws of a cat

4. The presence of gill slits in a rabbit embryo indicates (A) that rabbits breathe by gills in the gastrula stage (B) that rabbits descended from amphibia (C) common ancestry of vertebrates (D) inheritance of acquired characteristics (E) that the theory of regeneration may be true

5. Animals with the greatest number of similarities are grouped together in a (A) phylum (B) family (C) genus (D) class (E) order

6. Early in its embryonic development, the embryo of a chicken has (A) gills (B) fur (C) hair (D) tail (E) fins

7. A native animal of Australia is the (A) koala (B) rabbit (C) mongoose (D) dormouse (E) horse

8. Darwin studied similarities among species of birds on the (A) Island city of Venice (B) Thousand Islands (C) Islands of the West Indies (D) Galapagos Islands (E) Isles of Langerhans

9. The precipitin test helps establish evidence for evolution in the field of (A) anatomy (B) comparative biochemistry (C) embryology (D) plant and animal breeding (E) geographic distribution

10. Progression from simple to complex forms is summarized in the term (A) sedimentation (B) petrifaction (C) evolution (D) catastrophism (E) erosion

ANSWER KEY

1-A	3-B	5-C	7-A	9-B
2-E	4-C	6-D	8-D	10-C

29 Theories of evolution

That evolution has taken place appears to be an incontrovertible fact. However, *how* it took place is open to speculation. There are several theories that attempt to explain how changes have occurred among plants and animals.

Lamarck's theory of use and disuse. Jean Lamarck (1744–1829) was one of the earliest biologists to recognize that living things have changed. In 1809, he attempted to explain how evolution has taken place by his theory of *use* and *disuse,* which is based on the following points:

1. When an animal uses an organ to adapt itself to its environment, the organ becomes well-developed and enlarged. New organs arise according to the needs of the organism.

2. If an animal does not use an organ as it adjusts to its environment, the organ is undeveloped and remains small.

3. These *acquired characteristics are inherited,* so that the offspring will have either a well-developed organ that has been used, or a smaller organ if it has not been used. After many generations, the descendants will be changed as the result of inheriting well developed organs – or as the result of the disappearance of unused organs.

Lamarck explained the development of the long neck of the giraffe as being the result of constant stretching to eat the leaves of trees. The cumulative inheritance of slightly greater neck length in each generation resulted in the unusually long neck of today. According to this theory, fish found in the waters of caves are blind because their ancestors did not use their eyes; after many generations of inheriting weak eyes, they eventually became blind.

Evidence against Lamarck's theory. Lamarck's theory is not accepted because of the lack of experimental evidence to substantiate his claims for the inheritance of acquired characteristics. His basic error was that he gave purpose to evolutionary change. August Weismann (1834–1914) conducted an experiment to test this idea. He cut off the tails of twenty-two generations of mice but found that the size of the tail was not affected. He stated that changes in the body cells (*somatoplasm*) are not passed on to the next generation. Only changes in *germplasm,* consisting of the reproductive organs and their gametes, are inherited in the next generation. He formulated the theory of *"Continuity of Germplasm"* to summarize his ideas that germplasm is passed on from one generation to the next, but that the somatoplasm portion of an individual dies and is not passed on. According to him germplasm is immortal.

The chief present-day support for Lamarck's theory came from the Russian agriculturist, Lysenko, who claimed that environmental effects are inherited. Practically all other scientists remain unconvinced that acquired characteristics can be inherited.

Darwin's theory of natural selection. Charles Darwin (1809–1882) spent twenty-five years gathering facts about evolution before he finally published his epoch-making *The Origin of Species* in 1859. While a young man in 1831, he had traveled to many part of the world as the naturalist on the ship *Beagle*. The observations he made at that time led him to realize that evolution had taken place. When he first announced his theory, there were many people who were not ready to accept the idea of evolution. Since then, it has had profound effects on most fields of biological, philosophical, and social thought. By a coincidence, a similar theory was proposed to Darwin by Alfred Wallace, whereupon Darwin presented both theories to the world of science. Darwin's theory includes the following points:

1. **Overproduction.** Plants and animals reproduce in such large numbers that the earth would soon be covered with them if they all survived. The roe of a single codfish may consist of ten million eggs; the common oyster may shed as many as eighty million eggs in a season; protozoa could, in a few weeks, produce a mass of offspring many times the size of the earth. Darwin estimated that a pair of elephants, the slowest breeding animals known, would have nineteen million descendants in eight hundred years.

2. **The struggle for existence.** The tendency toward overproduction is checked by the struggle for existence carried on by living things. They compete for food, space, water and other requirements in the environment. This rivalry is not only among members of the same species, but also among different species. The organisms that cannot survive die out. Of all the eggs produced by a codfish, only two need develop to maturity if the number in the ocean is to remain constant.

3. **Variation.** All living things differ in size, color, strength, speed of reactions, and in countless other ways. Some individuals are therefore better equipped in the struggle for existence than others, because they possess more favorable variations.

4. **Survival of the fittest by natural selection.** Those organisms that are most fit for a particular environment will survive. The others will die out. Sometimes, the animal that is a little faster, or a little stronger, or a little larger, will survive; the slower, weaker, or smaller animal will be killed off by nature. However, the strongest or largest do not always survive, as is seen in the case of the dinosaurs. Fitness is comparative, depending on the requirements of the environment. On some islands which have strong winds, there are very few winged insects; those with wings are blown out to sea and perish, leaving the "weaker" ones without wings as those most fit to survive in that environment. When an animal is introduced into a new country, it may be extremely successful, because of the lack of natural enemies. Thus, the rabbit overproduced in immense numbers after it was brought to Australia. The Japanese beetle has likewise become a menace to cultivated plants in the United States.

5. **The origin of new species.** By the gradual accumulation of favorable variations, generation after generation, the various species eventually changed. When the variations were unfavorable, the species died out.

Darwin's theory explains the development of the long neck of the giraffe in this way: The short-necked ancestor of the giraffe produced many offspring whose necks varied slightly in length. Those with a slightly longer neck had the advantage of being able to feed on the leaves of trees. In the struggle for existence, they were the fittest, and survived. In each of the succeeding generations, the longer-necked animals continued to survive. Gradually, only giraffes with long necks came to be present on the earth.

Difficulties of the theory. Darwin was the first to recognize certain limitations of his theory, and included a full chapter, chapter VI, of his famous book on this subject. He could not distinguish between variations that are inherited and those that are not; the science of genetics was still forty years away. Also, the mere act of selecting the fittest does not in itself create new variations. Natural selection operates only after the variations have appeared. Another problem is that minor variations do not seem important enough in the first establishment of new organs; an organ would have to be of some size before it could be significant in the survival or the organism.

De Vries' theory of mutations. One answer to the question of how variations arise was supplied by Hugo De Vries (1848–1935), a Dutch botanist. In 1901, he stated that new species arise as a result of sudden changes, called mutations, in their hereditary make-up. He based his conclusions on a study of variations in the evening primrose. His theory added to Darwin's theory by clearing up the nature of variations and their role in the change of species. The main points of his theory are: (1) Mutations occur that may be beneficial or harmful to a species. (2) By natural selection, the beneficial mutants survive, while the harmful ones die out. (3) New species arise by the accumulation of useful mutations.

The mutation theory would explain the development of the present-day giraffe with the long neck as follows: At one time, the ancestor of the giraffe had a short neck. As the result of a mutation, one of its offspring was born with a slightly longer neck. It had the advantage of being able to reach the leaves of trees, and thus survived. This mutation was passed on to succeeding generations. Again, a mutation occurred producing offspring with a neck that was still a little longer than the others. Over millions of years, many such mutations resulted, giving rise to giraffes with longer and longer necks, which were better fitted to survive.

Modern theory of evolution. Since Darwin's time, additional research has given us greater insight into the process of evolution. Today, the basis for variations is explained by gene mutations, chromosome mutations, and the recombination of genes. New understanding of how species change has come through the study of population genetics.

Population genetics. Our early knowledge of genetics came by tracing the inheritance of genes in individuals. Now, studies are made of the distribution of genes (gene frequency) in a population.

In biological language, a *population* includes all the members of a species inhabiting a certain location. Examples: the paramecia in a pond; a grove of pine trees; dandelions in a lawn; people in a town. All of the individuals in the population share the same *gene pool.* The gene pool is considered to be the sum total of all the genes in the population. These genes are exchanged as the individuals interbreed, to give many recombinations.

The Hardy-Weinberg Principle. The two scientists named in this law studied populations and arrived at the conclusion that the gene pool of a population remains constant from generation to generation under the following conditions: (1) the population is large; (2) there are random matings; (3) there are no new factors such as mutations or migration. Under these theoretical conditions, a species would tend to remain practically the same. In other words, this principle applies to populations that are in equilibrium for their alleles.

The genotypes in a population in which there has been random matings distribute themselves as follows:

GENOTYPE	FREQUENCY
AA	p^2
Aa	2pq
aa	q^2

In this relationship, p = the frequency of the dominant allele (A); q = the frequency of the recessive allele (a); p^2 = the frequency of the homozygous dominant (AA); q^2 = the frequency of the homozygous recessives (aa); 2pq = the frequency of the heterozygous (Aa) individuals in the population.

The possibilities of such matings can be shown in the usual way:

	A (p)	a (q)
A (p)	AA (p^2)	Aa (pq)
a (q)	Aa (pq)	aa (q^2)

The total genotypes in the population can be indicated by the expression

$$p^2 + 2pq + q^2 = 1$$

Suppose we consider the frequency of genes for brown (*B*) and blue (*b*) eye color. Assume that 64% of a given population has brown eyes and 36% has blue eyes. Can we determine how many of the brown-eyed people are homozygous or heterozygous?

We know that the freqency for gene *b* is expressed as q. If $q^2 = 0.36$, the square root of q = 0.6. Since q is the frequency of the recessive allele, we can say that 60% of the alleles are recessive, *b*. The other 40% are dominant, B; p therefore = 40%.

The possibilities can be expressed as follows:

POSSIBLE SPERM

40% 60%

POSSIBLE EGGS

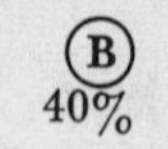

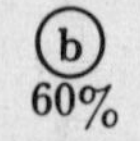

40% 60%

	.4B	.6b
.4B	.16BB	.24Bb
.6b	.24Bb	.36bb

Possible results:

16% BB – homozygous brown eyes
48% Bb – heterozygous brown eyes } 64% brown eyes
36% – blue eyes

In other words, of the 64% who are brown-eyed, 16% are homozygous for brown eyes, and 48% are heterozygous for brown eyes.

Population changes. Gene pools tend to be unstable, because the theoretical conditions needed for the Hardy-Weinberg principle are not always met. As a result, changes occur in the population. If environmental conditions change, some of the genes may provide an advantage, i.e., protective coloration. By natural selection, those individuals that have the favorable genes will survive and reproduce. Soon, there will be a higher proportion of individuals with these genes, and a lower proportion with the unfavorable alleles.

Isolation. When members of a population are isolated into smaller groups, they are prevented from breeding freely with each other. In such small populations, changes in the gene pool take place more readily. Random breeding establishes a new gene balance in each group that may be different from the others. Any mutations would most likely be different in each of the groups, giving rise to distinct characteristics.

Development of new species. As a result of the interaction of many factors new species arise. The members of the new species are no longer able to interbreed with the orginal species. Some of the factors may be listed as follows: (1) variations – the individuals in a population share the gene pool, and reflect many variations; (2) changes in existing environment – those traits with high survival value will increase in frequency; (3) migration to new environments – under new conditions, the gene pool changes as there is a smaller group of individuals for random mating; (4) natural selection – those individuals with favorable characteristics for the particular environment will pass their assortment of genes along to their offspring; (5) isolation – when groups are isolated, they are not able to breed at random, and differences become accentuated on both sides of the barrier. The Kaibab squirrels on the north rim of the Grand Canyon are similar to the Abert squirrel on the south rim; but have become different, probably because of the separation between them. Another type of isolation is biological, caused by the differences in breeding time, i.e., between two groups of toads; their populations are kept apart, leading to the development of two species.

As a group of organisms adapts to a new environment, they may change in different directions. The evolutionary pattern which takes place as they fit into the different environmental niches is known as *adaptive radiation*. The finches on the Galapagos Islands illustrate this. Some evolved into insect eaters and live in trees. Others became seed eaters and live close to the ground. Within each type there were additional variations.

Recently, environmental conditions for house flies changed and they adapted accordingly. The introduction of the insecticide, DDT, was effective in killing practically all that were exposed to it. However, a few flies had a genetic makeup that made them immune and they survived. DDT became less and less effective as more and more of these resistant flies survived and reproduced. Thus, a new strain of flies developed, with different genes for survival.

A similar situation resulted in the development of new strains of bacteria resistant to antibiotics. The occurrence of mutations led to the survival of strains of staphlococcus that were not killed by penicillin.

CHAPTER REVIEW

Select the correct choice for each of the following statements.

1. The first scientist to present a theory of evolution was (A) De Vries (B) Weismann (C) Lamarck (D) Darwin (E) Lysenko
2. The theory of use and disuse was based on (A) gene changes (B) chromosome changes (C) changes brought about by the environment (D) mitotic changes (E) mutations
3. According to Lamarck, evolution occurred as the result of (A) natural selection (B) the theory of recapitulation (C) overproduction (D) blending inheritance (E) inheritance of acquired characteristics
4. According to Weismann, changes in (A) somatoplasm are inherited (B) germplasm are inherited (C) somatoplasm and germplasm are inherited (D) body cells are inherited (E) body cells and germplasm are inherited
5. The name of Darwin's famous book was (A) *The Beagle* (B) *The Continuity of Germplasm* (C) *Brave New World* (D) *The Origin of the Earth* (E) *The Origin of Species*
6. All of the following are parts of Darwin's theory except (A) metamorphosis (B) survival of the fittest (C) natural selection (D) struggle for existence (E) variation
7. In order for the population of codfish to remain constant, the number of eggs produced by one codfish that must survive is (A) 1 (B) 2 (C) 4 (D) one million (E) ten million
8. In the struggle for existence, animals survive that are always the (A) largest (B) strongest (C) fastest (D) heaviest (E) fittest
9. Most scientists believe that species have changed as the result of (A) use and disuse (B) inheritance of environmental variations (C) mutations (D) changes in somatoplasm (E) heterosis
10. The scientist who proposed a revision of Darwin's theory was (A) Redi (B) Banting (C) Mendel (D) De Vries (E) Ehrlich
11. A population in an area includes (A) all the living things (B) all members of the same phylum (C) all members of the same species (D) all the living things in relation to their physical environment (E) all the autotrophs and heterotrophs
12. According to the Hardy-Weinberg principle, the gene pool may remain stable if there are (A) random matings (B) many mutations (C) frequent migrations (D) selected matings (E) random mutations
13. In the Hardy-Weinberg principle, $p^2 + 2pq + q^2 = 1$, q^2 represents the frequency of the (A) homozygous dominant (B) heterozygous dominant (C) heterozygous recessive (D) homozygous recessive (E) blended genes
14. If in a given population, 36% of the people have blue eyes, the percentage of the recessive allele is (A) 24 (B) 36 (C) 40 (D) 60 (E) 64
15. All of the following are factors in the development of new species except (A) variation (B) asexual reproduction (C) sexual reproduction (D) isolation (E) natural selection

ANSWER KEY

1-C	4-B	7-B	10-D	13-D
2-C	5-E	8-E	11-C	14-D
3-E	6-A	9-C	12-A	15-B

30 Man's development

Man, like other living things, has changed through the ages. The first types of man appeared about a million years ago. Humans have many similarities to the apes, indicating that both had a common ancestor. Relatively few fossils of man have been found; hence our knowledge of man's evolution is not complete. In addition to the few fossil fragments, early man is studied from his cultural remains, including tools and weapons (artifacts), refuse heaps, carvings and paintings in caves. Present-day *Homo sapiens* probably did not come upon the scene until about 25,000 years ago.

Java man. In 1891, at Trinl, Java, Dr. Dubois found the top of a skull, a left thigh bone, and some molar teeth that were recognized as belonging to a primitive type of man. The eyebrow ridges were prominent. The brain was smaller than modern man's, but larger than that of an ape. The brain

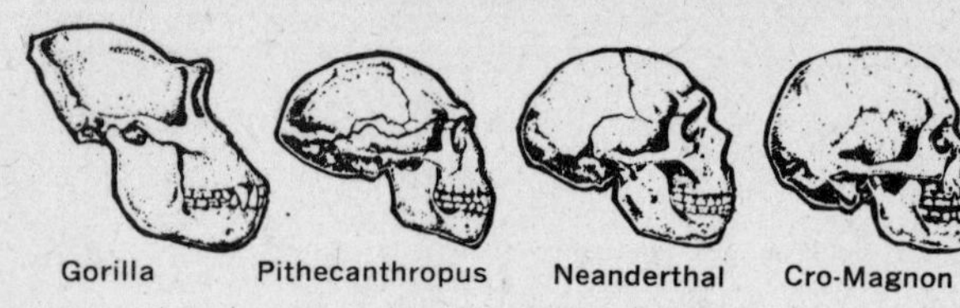

Skulls of Primates

capacity of an ape is less than 580 cubic centimeters; Java man's was 940 cc.; modern man's is about 1,500 cc. Java man had a heavy neck and a receding chin. He walked more or less erect. He is called *Pithecanthropus erectus,* the ape man that walked erect. He lived about 500,000 years ago.

Peking man. In 1926 and in 1929, a number of fossil remains of a primitive type of man that lived in the same period as Java man were found near Peking, China. He received the name *Sinanthropus pekinensis.* The cranial capacity was larger than Java man's, being about 1,000 cc. It is believed he had the power of speech, and used fire. He also used crude tools.

Heidelberg man. This type of man is known from a whole lower jaw which is massive and has very human teeth. He lived about 200,000 years ago.

Neanderthal man. *Homo neanderthalensis* probably appeared about 150,000 years ago, and became extinct about 25,000 years ago. He had projecting eyebrow ridges, a low, retreating forehead, a massive jaw, and thick, heavy bones. His brain capacity was about the same as modern man's. His average height was about five and a half feet. Many fossil remains have been found in various parts of Europe. His later tools were chipped and polished, indicating a rather advanced form of culture.

Cro-Magnon man. Many skeletons have been found of Cro-Magnon man, who lived about 25,000 years ago. He was of slightly taller stature than modern man, with a brain capacity that was 1,550 cc., a little larger than that of today. He had a high forehead and a well-developed chin. Cro-Magnon man's tools were made of polished stone, bone, and ivory. He sewed skins for clothing. He has left us paintings of deer, bisons and mammoths on the walls of caves in the southern part of France. He is believed to be the same species as modern *Homo sapiens.* He overlapped the race of the slower-witted Neanderthal man, and may have led to his extinction.

Recent discoveries. The current excavations by Dr. Louis Leakey in the Olduvai Gorge of northern Tanzania, Africa, have yielded the fossil remains of the other types of primitive man. *Zinjanthropus* man (1959) is supposed to have lived one million years ago and used stone tools. *Homo habilis* (1961) may have lived, 1,750,000 years ago. The fossils indicate that he was small-brained, walked erect, and used tools. In 1967, Dr. Leakey and his wife Mary reported the discovery of fossils of a still older type of man, in Kenya, *Kenyapitheus,* whose age was established by radioactive potassium argon dating as being 19-20 million years. Many years earlier, in 1948, the Leakeys had discovered the fossil remains of a primitive type of man they named *Proconsul* which may have lived 20-30 million years ago. It is likely that additional fossil remains will be found to fill in the gaps in the geologic history of man.

Modern man. There are now over three billion human beings on earth, all belonging to the same species, *Homo sapiens.* They are similar in (1) their ability to walk erect, (2) in having a well-developed cerebrum, (3) in having a hand with the thumb opposite the other fingers, (4) in having the power of speech for communication, (5) and in their other physical and physiological characteristics. Although they all show variation in physical make-up, they are more similar than different. On the basis of some of their differences, notably skin, eye, and hair pigmentation, hair texture, stature, skull shape, and facial features, they are classified into three main *stocks,* Caucasoid (white), Negroid (black) and Mongoloid (yellow). Each of these is further classified into races.

Racial characteristics. *Anthropology* is the study of man's origin and his classification. One of the characteristics studied is the *cephalic index,* representing the shape of the head. This is obtained by measuring the width of the head, and dividing the result by the length of the head. Heads with a cephalic index of 75 or less are considered long, or *dolichocephalic,* while those above 80 are broad, or *brachycephalic.*

Another characteristic which helps to distinguish races is blood group. The four blood groups, A, B, AB, and O, are found in practically all races. A person with type A blood, for example, can be Negro, Chinese, English, or Navajo Indian. The various races differ in the relative number of persons which possess each of the four blood groups. Therefore, a blood transfusion may be made as conveniently between two type A persons of different races as between members of the same race. In fact, it would be fatal for a person of one race to receive blood only from a member of the same race, if the blood type were different.

Confusions about race. Confusion is often created by considering a race as being pure. Because of migrations and intermarriages, the genes for the various characteristics are carried in varying quantities by many individuals of different races. The Nordic race, for example, is found in the northern part of Europe, especially in the Scandinavian countries. Yet one finds many individuals there who do not possess all of the characteristics of the Nordic race. A study of the Swedish army revealed that only about 11 per cent of the men were tall, blond, and blue-eyed. Many Nordics may be short, have dark hair, and have round heads. To avoid the confusion arising from the use of the word "race," some anthropologists suggest replacing it with the expression *racial type*. This would help convey the idea that races are not pure and that a person is classified on the basis of having an average of certain characteristics. There may be so much variation within racial types that there are Hindus who are darker than some Negroids, and there are Chinese who are taller than some Nordics.

Another source of confusion sometimes arises from the tendency to consider racial type and national boundary as being related. Thus, Italy is considered as consisting of Mediterranean racial types. However, there are parts of northern Italy in which many of the people have Nordic characteristics and others have Alpine traits. Also, there is no Italian race; the people who live in Italy have the characteristics of many racial types.

Similarly, other cultural traits, such as religion and language, have no connection with racial types. There is no Jewish race, since Judaism is a religious conviction. Also, there is no Latin race, based on languages derived from Latin, such as French, Spanish and Italian. There is no Aryan race, either, since the term Aryan also refers to language.

Racial differences have led some people to believe that some groups are superior to others. However, there is no scientific evidence to support this idea. All groups seem to have equal mental ability. Since the various characteristics are inherited independently of each other, there can be no linkage of intelligence with hair color, stature, eye color, etc. When human beings from various parts of the world are compared, it is evident that they are similar in practically all of their characteristics (i.e., they all have a four-chambered heart, an appendix, a large cerebrum, two kidneys, 206 bones in the adult skeletal system, etc.), and differ in only a few ways.

The future of mankind may depend on the ability of *Homo sapiens*, man the wise, to cooperate with his fellow members of the human race.

RACES OF MAN

STOCK	RACE	SKIN COLOR	HAIR COLOR	HAIR TEXTURE	HEAD FORM	STATURE
Caucasoid	Nordic	White	Light	Wavy	Long	Tall
	Alpine	White	Medium	Wavy	Broad	Medium
	Mediterranean	White-olive	Dark	Wavy	Long	Medium-short
	Hindu	White-brown	Dark	Wavy	Long	Medium
Negroid	Negro	Black	Black	Woolly	Long	Tall
	Melanesian	Black	Black	Woolly	Long	Medium
	Pygmy	Black	Black	Woolly	Medium	Very short
	Bushman	Black	Black	Woolly	Long	Very short
Mongoloid	Mongolian	Yellow	Black	Straight	Broad	Short
	Malaysian	Brown	Black	Straight	Broad	Short
	American Indian	Brown-red	Black	Straight	Broad	Tall-medium

CHAPTER REVIEW

Select the correct choice for each of the following statements.

1. The reason for man's similarities to the ape is that (A) man descended from the apes (B) apes descended from man (C) a mutation of the apes became man (D) a mutation of man became the apes (E) man and the apes had a common ancestor

2. Our knowledge about primitive man is based on all of the following except (A) widespread fossil remains (B) refuse heaps (C) carvings (D) paintings in caves (E) artifacts

3. The first types of man appeared about (A) 25,000 years ago (B) 100,000 years ago (C) 1,000,000 years ago (D) 10,000,000 years ago (E) 100,000,000 years ago

4. A type of man that appeared after Neanderthal man was (A) Java man (B) Peking man (C) Heidelberg man (D) Cro-Magnon man (E) *Pithecanthropus erectus*

5. *Homo sapiens* is thought to have been derived from (A) Heidelberg man (B) *Sinanthropus pekinensis* (C) *Pithecanthropus erectus* (D) Cro-Magnon man (D) Peking man

6. All of the following are true of Cro-Magnon man except (A) his tools were made of polished stone (B) his brain capacity was smaller than that of modern man (C) he sewed skins for clothing (D) he made paintings of animals on the walls of caves (E) he was slightly taller than modern man

7. All of the following races are classified in the Caucasoid stock except (A) Nordic (B) Alpine (C) Hindu (D) Mediterranean (E) American Indian

8. Human beings are similar in all of the following ways except (A) they walk erect (B) they have a well-developed cerebrum (C) they have an opposable thumb (D) they have the power of speech (E) they have the same amount of skin pigmentation

9. The study of man's origin and classification is known as (A) eugenics (B) euthenics (C) anthropology (D) paleontology (E) physiology

10. A Navajo Indian with type A blood can receive a transfusion of blood (A) only from another Navajo Indian (B) from a Mexican (C) from a Negro (D) from an American cowboy (E) from all of the above

ANSWER KEY

1-E	3-C	5-D	7-E	9-C
2-A	4-D	6-B	8-E	10-E

HOW TO MAINTAIN THE BALANCE OF LIFE

9

31 The web of life

Man and the other living things on earth live in an entangling relationship with each other. They do not exist in isolated fashion. They are interdependent and each forms a strand in the web of life. *Ecology* is the study of living things in relation to each other and to their environment.

Structure of life The pattern of life may be said to be organized according to the following arrangement:

Atoms
↓
Inorganic molecules
↓
Nucleic acids
↓
Viruses
↓
Cells
↓
Tissues
↓
Organs
↓
Organ systems
↓
Organism

Organisms do not live alone. There is a dynamic equilibrium which exists in the interactions of organisms with other organisms and with their environment. The structure of life for an individual may therefore be shown in its wider relationships as follows:

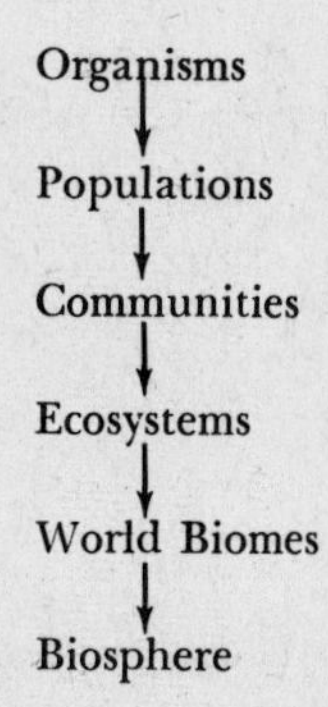

Population. All the members of a species inhabiting a given location make up a population. The members of a species are more or less alike. They are capable of breeding with one another. Examples: sunfish in a lake; trees in a pine grove; people in a city.

Community. All the plant and animal populations interacting in a given environment make up a community. There are aquatic communities which inhabit a pond, a river, a lake, or an ocean. There are also terrestrial communities in a field, a desert, a cave or a forest.

Ecosystem. An ecosystem is a self-sustaining, dynamic community of plants and animals in relation to their physical environment. In such a system, matter is transferred, and used over and over again in a series of cycles, i.e., water, carbon, oxygen, hydrogen, and nitrogen. Light, temperature and energy are also involved in the interrelationships of organisms with their physical surroundings.

The living things in an ecosystem affect each other and depend on each other in a variety of ways, in food chains and food webs.

World biome. A biome is a major ecological grouping of organisms on a broad geographical basis. It is determined largely by the major climate zones of the world. The land biomes are classified according to the climax vegetation they contain. Aquatic biomes are not classified on this basis because the distribution of plants is not fixed. Since more than 70 percent of the earth's surface is covered by water, aquatic biomes are a widespread influence on the content of life.

Biosphere. The biosphere is the thin layer of life at the surface of the earth, including the biologically inhabited soil, water and air. It is also thought of as including a system of relationships between the living things and the materials and energy surrounding them.

Physical factors in the environment. The non-living (abiotic) factors that affect living things include: light, temperature, water supply, oxygen supply, minerals, pH and soil or rock (substratum). Light is needed by green plants for food-making. The depth in the ocean to which plants can grow is limited to the area of penetration of sunlight below the surface.

Temperature affects the metabolism of living things, and consequently determines their geographical distribultion. The various climatic zones have different varieties of living things that are best adapted to live at their temperatures.

Plants may be classified according to their ecological need of water as follows:

1. Hydrophytes — plants that live in water or in marshes. They usually have large leaves that float on the water; they have reduced root systems. Examples: water lilies, water weeds, cattails.

2. Xerophytes — plants that live in dry conditions. They have deep root systems, much water-storage tissue, a thick epidermis, and modified leaves or spines. Examples: cactus, sagebrush, yucca tree.

3. Mesophytes — plants that live on a medium amount of water. They have a well-developed system of roots, stems and leaves, Examples: plants of field and forest, including grasses, maple trees, and clover.

Practically all living things need oxygen. The depth to which organisms can live in the soil is limited by the penetration of air. Likewise, the supply of oxygen in the ocean waters is a limiting factor that determines the depth at which marine organisms can survive. The mineral content of the soil and water has an important influence on the nature of the living things that can grow in a particular location. Likewise, the pH of soil affects the growth of plants; lime must be added by gardeners and farmers to counteract the effects of soil that is too acid.

The substratum is the soil or rock in which organisms live. Clay, sand and rocky soils have different communities which can live in them. Soil containing much organic matter is teeming with organisms, as compared to sand.

Cycle of materials. Important elements such as nitrogen, carbon, oxygen, hydrogen, and phosphorus are constantly being cycled between organisms and their environment. These elements pass from inorganic sources to organic forms and back again to inorganic forms. Much of this cycling is accomplished through the actions of decomposers, scavengers and saprophytes.

The nitrogen cycle. The chief features of the nitrogen cycle are as follows:

(a) During protein synthesis, green plants bind nitrogen into their protoplasm.

(b) Animals eat plants and synthesize animal protein from plant protein.

(c) Plants and animals produce nitrogenous wastes. When they die, their bodies are used as sources of energy by decomposers, bacteria of decay.

(d) The nitrogen of the nitrogenous wastes and the proteins is released and converted into nitrates in several ways; it is converted to nitrates by decay bacteria and nitrifying bacteria; nitrogen-fixing bacteria also produce nitrates from gaseous nitrogen; the nitrates are made available to plants for protein synthesis.

(e) Denitrifying bacteria return molecular nitrogen to the atmosphere.

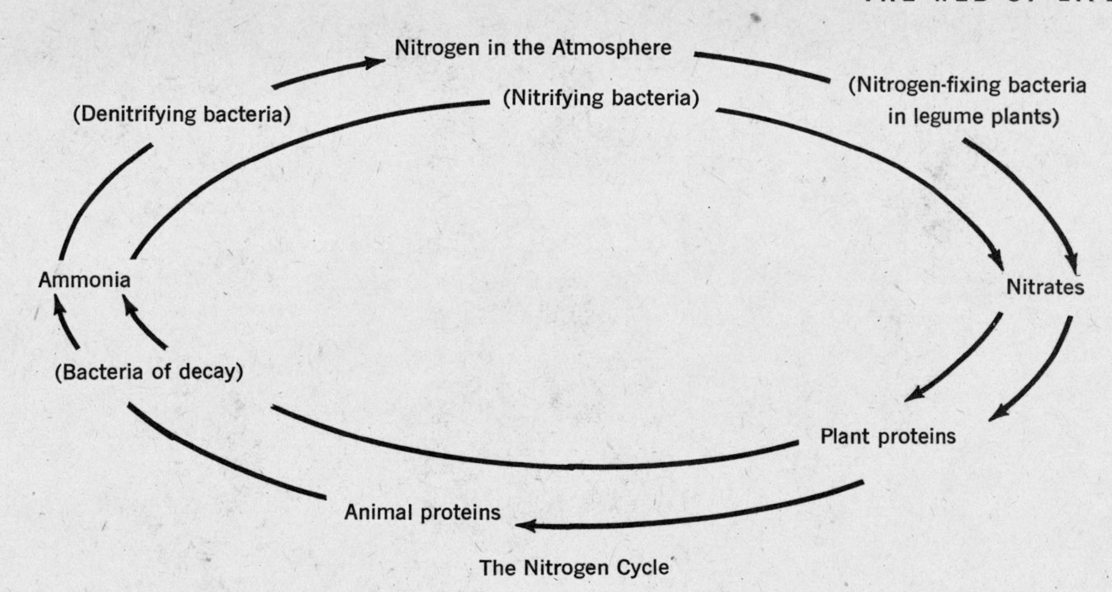

The Nitrogen Cycle

The carbon-hydrogen-oxygen cycle. The elements carbon, hydrogen and oxygen are present in all living things. By the process of respiration, carbon dioxide and water are released. When plants and animals die, carbon dioxide and water are given off through the action of decomposers such as bacteria of decay and fungi. When green plants carry on photosynthesis, they take in carbon dioxide and water, and give off oxygen. In the burning of coal, petroleum and wood, oxygen is used up and carbon dioxide and water are given off. Through these and other activities, the elements carbon, hydrogen and oxygen are kept in circulation between the biotic (living) and the abiotic (non-living) environment.

The phosphorus cycle. Elements such as phosphorus and calcium are cycled from soil, rock or water to organisms and back again. The cycle for phosphorus is typical. This element is present in compounds which become dissolved in water. Plants take in the phosphates. Animals eat plants. Their wastes include some of these phosphates, which are returned to the soil, or to the ocean. When the plants and animals die, decomposers break them down through decay, and release the phosphorus compounds, which are returned to the soil or ocean for further circulation.

Food relationships. Practically all living things depend for their existence on green plants which can make their own food by photosynthesis. Green plants are autotrophs and are considered to be **independent organisms**. Organisms which cannot synthesize their own food are heterotrophs, and are dependent.

Saprophytes are plants that live on dead organic matter, i.e., bacteria of decay, mushrooms. **Herbivores** are plant-eating animals, i.e., cow, horse. **Carnivores**, or meat-eating animals, eat the flesh of other animals, which directly, or indirectly, obtained their food from plants. They are of two types, (a) *predators,* which will kill other animals, i.e., lion, hawk, and (b) *scavengers* which eat dead organisms that they did not kill, i.e., vulture, jackal, snail. The snapping turtle is an example of an animal that may be both a predator and a scavenger.

Omnivores are animals that eat both plants and animals, i.e., man, rat. When plants and animals die, their remains are used as food by smaller organisms, **decomposers**, such as bacteria of decay and fungi.

Symbiosis. There are many organisms that live intimately together in a close association that may or may not be beneficial to them. Although the relationships are not always clearcut, the types of symbiosis may be described as follows:

(a) **Commensalism** — one organism is benefited and the other is not affected.
Examples: Barnacles live on the hide of a whale, obtaining a habitat as well as a means of transportation. The remora fish attaches itself to the bottom of a shark by means of a suction pad on its head, and feeds on food scraps left over by the shark.

(b) **Mutualism** — both organisms mutually benefit from living together. Examples: A lichen is made up of both a fungus and an alga; the fungus provides moisture, while the alga makes food by photosynthesis. Nitrogen-fixing bacteria live in the nodules on the roots of legume plants such as clover; the bacteria convert nitrogen into nitrates, while the plants provide a habitat and nutrition for the bacteria.

(c) **Parasitism** — one organism, the parasite, attaches itself to another, called the host and benefits at its expense. Examples: The athlete's foot fungus grows on the skin of man. Tapeworms and other types of worms, attach themselves in the intestines of animals and absorb digested food. The lamprey eel, a primitive fish without jaws, has a circular, sucking mouth lined with hooklike teeth, by means of which it attaches itself to trout and other fish, and sucks their blood. Disease germs, such as viruses and bacteria, live within the body of an organism, and may cause serious illnesses.

Food chain. A food chain represents the different links along which food is passed from one organism to another. It starts with green plants which can make their own food in the presence of sunlight. Sunlight serves as the source of energy for photosynthesis to take place. The green plant may then serve as a source of food for herbivores, which, in turn are eaten by carnivores, and for decomposers, which bring about decay in animals, as well.

The links in a food chain may be illustrated by referring to the well-known jingle by the 18th century satirical writer, Jonathan Swift:

Big fleas have little fleas
Upon their back to bite 'em,
And little fleas have lesser fleas
And so, ad infinitum.

In the oceans, great numbers of algae synthesize carbohydrates, and then convert them into proteins. Protozoa and other minute invertebrates feed upon the algae. Together, they make up the floating mass of microscopic plant and animal life called *plankton.* Small fish eat the plankton, and in turn are eaten by larger fish. Eventually, the food chain extends from the microscopic algae to whales, sharks, and man. Without the aquatic green plants there could be no aquatic animals. Aquatic animals also depend on green plants for their supply of oxygen. In turn, the carbon dioxide given off by animals, is used by the green plants in photosynthesis.

A terrestrial food chain may involve the following sequence: green plants, such as corn, make food – mice feed on the plants – snakes prey on the mice – hawks eat snakes as well as mice – bacteria of decay and fungi live on all of the organisms when they die.

Classification in a food chain. The various links in a food chain may be classified as follows:

(a) Producers – green plants are the producers, since they synthesize organic compounds which serve as food.

(b) Consumers – organisms that feed directly on green plants are called *primary,* or *first-order* consumers. Animals of this type are known as herbivores. *Secondary,* or *second-order,* consumers are carnivores with sharp tearing teeth, which prey on the primary consumers. There may be *third-order consumers* which are predators on the secondary consumers, and frequently on the primary consumers.

(c) Decomposers – break down the wastes and dead bodies of producers and consumers into simpler compounds. These simpler compounds are then cycled, or returned to the environment where they are used over again by other living things.

The sequence in the terrestrial food chain mentioned previously, can thus be rewritten in this way: green plants (producers) – mice, (first-order consumers) –snakes (second-order consumers) –hawks (third-order consumers) on snakes; (second-order consumers) on mice – decomposers.

Food web. Food chains overlap into food webs, because there are many kinds of producers which may be eaten by different kinds of primary consumers (i.e., insects, rabbits, etc.); these herbivore consumers may shift from one plant to another. Carnivores, which live on herbivores, may likewise shift from one source of prey to another, and may be of different types (i.e., frogs, foxes, owls). An ecological balance is achieved when the numbers of the various populations are kept at a relatively stable level in relation to each other.

Energy flow. Green plants have the ability to store the radiant energy of the sun as chemical energy in the bonds of the organic compounds they synthesize. When green plants are eaten as food, this energy is taken into the consumers and used by them for their life activities. The various pathways into which the food energy is transmitted make up the energy system of the food web.

There is a decrease in the total amount of energy as it is passed along from the producer to the consumers. Each member of the food web uses up some of the energy as it carries on its various metabolic activities. In the sequence of energy transfers, the amount of usable energy runs down. In addition, since the consumers must seek and find their prey, they may not always be successful in obtaining food. In other words, the various members of the food web pass on less energy than they received.

Pyramid of energy. Since there is a loss of energy at each feeding level, it can be seen that the producers contain the greatest amount of energy. This amount decreases at each level. Consumers are generally larger than the animals they live on. Consequently, many organisms at the lower feeding levels are required to furnish sufficient food energy for a single organism at a higher feeding level. This idea is referred to as the pyramid of numbers.

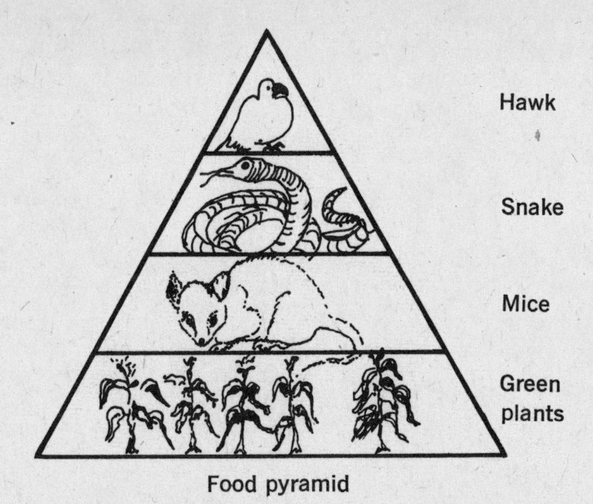

Food pyramid

Niche. The specific environment of a particular species is known as its *niche.* If two different species occupy the same ecological niche, they will compete with each other for food and reproductive sites. The species that reproduces faster will eliminate its competitor. As a result, one species is established per niche in a balanced community. All the members of that species use the same kinds of food, and occupy similar reproductive sites.

The giant saguaro cactus of the Southwest harbors two kinds of owls, the elf owl and the screech owl, in holes made in its stem by woodpeckers. Although both species of owl may occupy adjoining holes, they are not in the same niche. The elf owl eats mainly small insects. The screech owl eats larger insects, as well as mice and scorpions; it also breeds earlier in the year. Thus, these two species occupy niches that are different.

Ecological succession. Communities change as the conditions affecting them change. This may be illustrated in the history of a pond. A small pond may be covered with water lilies, which make up the dominant species. Other types of plants may grow in it, including elodea and bladderwort. Frogs, tadpoles, snakes, water beetles and small fish may also inhabit the pond.

As the plants in the pond die, their remains build up until they form a semisolid base at the edge of the pond. Cattails and other marsh plants begin to grow there. Mosses and ferns soon invade the area. As the amount of soil increases, a new community develops which will be replaced later on by shrubs and willow trees. Other low plants, shrubs, and trees, including goldenrod, wild carrot, and sumac, succeed this community. The amount of available light is changed as poplar and pine trees replace the earlier community. The soil is becoming increasingly enriched and changed.

This orderly sequence of communities that replace each other in a given area is known as *ecological succession.* It may take a few years or several hundred years, for the pond to disappear, depending on the circumstances of size, source of water, location, surrounding topography, etc. The replacement of communities continues until a climax community is reached.

Climax community. Continuing with the succession of communities in the pond area, a community of oak and hickory trees is next in the order of succession. Other species are also present, but these are dominant. As the amount of shade increases in the forest, their seedlings have difficulty surviving, and maples and beeches appear, to become the successful types. Their seeds are able to sprout, and survive, thus resulting in a permanent or final community. This now is a *climax* community. There is an equilibrium between the various living things in it, and it may continue indefinitely. Each species reproduces at a rate that maintains itself at about the same numbers from year to year.

A climax community receives its name from the dominant types of plants which characterize the situation. Besides the maple-beech climax, which is the typical climax in the northeastern part of the country, other climaxes are: Cypress and white cedar, in wet areas of Louisiana; beech-magnolia, in the South; tall grasses on the prairies; scrub oak on a mountaintop; pine forests in New Jersey.

If a climax community is destroyed by man through unlimited lumbering practices or by fire, or by a hurricane, a succession of communities starts all over again. The maple-beech forest does not come up at once. Low shrubs and trees will be the pioneer dominant forms for many years until they are eventually replaced by the succeeding types of tall trees leading once more to the maple-beech climax, perhaps a century later.

Often, the climax community is not replaced. Climax communities may vary in the same area. In New York State, for example, hemlock-beech-maple trees are often the climax community at high elevations. At lower elevations, the climax trees are oak and hickory.

The succession that ends in a climax community can be traced back to the *pioneer organisms* that first populated a given area. Pioneer organisms, such as lichens, first appear on bare rock. They help build up soil by breaking down the surface of the rock and releasing its minerals. Spores of mosses that are blown around may land in such a very thin layer of soil. As mosses grow, their remains help build up the soil. After a while, other plants will appear, and the succession of communities is on its way.

World biomes. Through the study of biogeography, which deals with the distribution of plants and animals in the various areas of the world, a greater understanding can be reached of the major biomes. A biome is a major ecological grouping of organisms.

Terrestrial biomes. The major plant and animal associations on land are determined by the chief climate zones of the earth. These climate zones are distinguished from each other on the basis of factors such as temperature, amount of rainfall and amount of solar radiation. There are also modifications caused by local land and water conditions. The land biomes take their characteristics and names from the climax vegetation in an area. Since green plants are the producers of food, the major plant associations on the earth determine the kind of animals that will inhabit a particular locality.

The major land biomes can be classified as follows:

(a) Tundra is the treeless region in the far north. It has an arctic climate of severe cold, with a long dark winter, and a mild summer having continuous daylight. The underlying part of the soil is permanently frozen. The upper part thaws out temporarily during the summer, to form numerous bogs and ponds. Mosses, lichens, including reindeer moss, and small plants grow actively during the summer. Birds, the snowshoe hare, caribou, flies and polar bears are representative of animal life.

(b) Taiga includes the northernmost forests that extend in a broad zone just below the tundra, across Europe, Asia and North America. These forests contain coniferous trees, such as pine, spruce, fir and hemlock. Some of the animals are moose, black bears, wolves, rodents and birds.

(c) Temperate deciduous forests contain trees that shed their leaves during the cold winters and grow them back during the warm summers. They occur in the eastern United States, England and central Europe. Characteristic trees; maple, oak, elm, beech and birch. Deer, pumas, squirrels and foxes are some of the animals.

(d) Tropical rain forests occur in regions of high temperature and ample rainfall. The warm, humid climate of the tropics encourages a rich abundance of plant life. The physical conditions are relatively constant throughout the year. There are many climbing vines and numerous insects.

(e) Grasslands have less rainfall than the deciduous forests. Because of the recurrence of droughts, trees are generally not able to develop in these areas. Most of the plants are of the grassy type, and serve as food for the grazing animals.

(f) Deserts have the driest conditions of all. The days are hot and the nights cold. The cactus and other plants are adapted to conserve water. Rodents and snakes may be common.

Effects of latitude and altitude. The land biomes take their characteristic largely from their latitude, or their distance from the equator. The further away from the equator—or, the greater the latitude—the colder the climate. Altitude also has a similar effect. The higher up one goes on a mountain, the greater the change in temperature. The vegetation also changes accordingly. The change is similar to that in traveling north. In the tropical zone, the tops of high mountains have conditions resembling those of the arctic zone. The similarity between latitudinal and altitudinal life zones can be shown as follows:

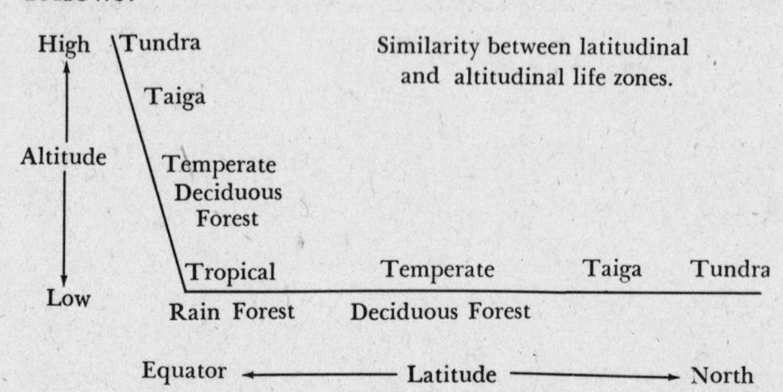

Similarity between latitudinal and altitudinal life zones.

Aquatic biomes. More than 70 percent of the earth's surface is covered by water. Variations in temperature are not extreme. Water absorbs large quantities of heat. It absorbs this heat slowly, and loses it slowly. Much of the solar heat that reaches the earth is absorbed by water, thus keeping the temperature of the earth relatively low. Aquatic areas have the largest and most stable ecosystems on earth. The living things are affected by such physical factors as the amount of dissolved oxygen, temperature and the amount of dissolved minerals.

(a) Marine biome. The oceans of the world make up a huge continuous body of water. They absorb and hold large quantities of solar heat, and regulate the earth's temperature. They contain a relatively constant supply of salts, which originated in the land masses. The greatest amount of food production on earth takes place in the oceans, along the edges of the land masses.

The coastal waters are fairly clear, and photosynthesis takes place where light can penetrate. Since water absorbs much light energy, photosynthesis takes place near the surface. The deeper regions are too dark for photosynthesis.

Near the surface, the basic food is the *plankton,* consisting of microscopic forms of life such as diatoms and other algae, and protozoa. Plankton is the basis of the food chains of the ocean. At the lower depths of the ocean, the food consists of a "rain" of dead organic matter that comes down from above. At the bottom, bacteria and scavengers break down the complex organic molecules into simpler molecules. The mineral content is thus released.

Through the circulation of the water, these dissolved nutrients are brought to the surface where they are used over again for food.

(b) **Fresh water biomes** include lakes, rivers, ponds and swamps. Life in fresh water is affected by similar factors as those in marine environments, namely, available oxygen, temperature, transparency of the water, depth, and salt concentration. The low concentration of salts in fresh water compared to the higher concentration in the living cells, causes a diffusion gradient or difference to exist. The semipermeable cell membrane controls the passage of water and salts by active transport. Excess water in protozoa is eliminated by specialized structures, i.e., contractile vacuoles. Homeostatic mechanisms are also present in other fresh water inhabitants.

Balance in Nature. Living things live in a delicately balanced equilibrium with each other. When this balance is upset, the animals and plants are seriously affected, and many of them die. Man has been the greatest cause in upsetting the balance of nature. One reason for this has been man's increasing population. At present, the earth's population totals 3½ billion. The United Nations Demographic Yearbook predicts that at the present rate of increase, this figure will double by the year 2005. Concern over this trend is based on the limitations of the food supply. The economist, Robert Malthus, pointed out at the end of the 18th century that the food supply increases at an arithmetic ratio, or gradually, while the population increases geometrically, that is, it doubles itself. He stated that populations are kept in check by disease, famine and war. There are many regions on earth today where people live close to a starvation level.

Ways of controlling the "population explosion" are being studied seriously to cope with the future problems. Birth control methods, including "the Pill," and the intrauterine device, are being publicized as desirable approaches. In Great Britain, a group of distinguished scientists issued a "Blueprint for Survival" early in 1972, recommending that the nation's population be reduced from 55 million to 30 million in the next 150 years. They also urged that other measures be adopted as well, such as the recycling of resources and the use of agricultural and industrial techniques that do not threaten the stability of the environment.

The widespread use of pesticides (chemicals used to destroy insects, weeds, fungi and other pests) and insecticides (insect-killing chemicals like DDT) has resulted in heavy losses of fish and wildlife. These chemicals have also been found to be harmful to man himself. Another threat to man, as well as to other living things is the growing pollution of the air and the water.

Man's increasing urbanization has also interfered with the ecological balance. He has not always realized how he affects other living things. Some examples:

1. A farmer may dislike hawks because they may occasionally make off with a chicken. However, hawks are useful to the farmer because they hunt and kill mice, rats and rabbits, which destroy his crops. If hawks were eliminated, mice, rats and rabbits would increase in such large numbers that they would ruin crops. Hawks are thus natural enemies of these small mammals; their numbers are kept fairly constant by the available food supply.

2. On the Kaibab Plateau of Arizona, pumas, wolves and coyotes prey on the deer. One year, a drive was made to kill off these enemies of the deer. The result was that the deer multiplied in such large numbers that they did not have enough food. Many of them died of starvation. Others chewed the bark of the trees, killing them. When the trees died, the shelter and food of the birds and small mammals were removed, and they were affected. It was predicted that if enough trees were killed, the soil would not be held together by their roots, and erosion would take place. The following year, the natural enemies of the deer were allowed to increase in numbers, in order to preserve the balance of nature.

3. A few pairs of European rabbits were released in Australia in 1862. This was the first time these mammals had been brought to Australia. Within a few decades there were millions of rabbits. Without natural enemies to keep them in check, they destroyed grasslands and crops. The rabbits became a national menace. In 1950, a deadly virus disease of the rabbits, *myxomatosis,* was introduced by scientists to control the rabbit scourge.

4. The island of Jamaica was troubled with rats in the sugar-cane fields. To control them, various animals were introduced, including the mongoose. Unfortunately, it was found that mongooses, which kill rats, also killed birds, snakes and lizards. Without these natural enemies, the cane beetles multiplied, and now do as much damage as the rats once did.

5. In 1916, the first Japanese beetles were found in New Jersey. They are not considered a menace in Japan, where they are held in check by natural enemies. But in this country they spread rapidly and soon were responsible for the loss of millions of dollars' worth of fruit crops, lawns and flowers. Efforts are now being made by the Department of Agriculture to discover insect, bacterial and fungus enemies of the beetle.

6. Man has brought other insect troubles to him-

self by upsetting the balance of nature. Elm trees are now rapidly being destroyed by the Dutch elm disease, which is caused by a fungus introduced from Europe in the 1920's and is spread by the elm bark beetle. The gypsy moth was brought to Massachusetts in 1869 by a man engaged in research on silkworms; through his carelessness, the insects escaped, and have become a serious pest of trees. The European corn borer entered this country before 1917, in broomcorn being imported from Italy for use in broom factories in New England. It now causes losses amounting to hundreds of millions of bushels of corn annually.

7. When DDT is spread widely in order to control insect pests, it may have harmful effects on birds, fish, and other wildlife. These animals may either be deprived of their insect source of food, or they may be poisoned by the DDT. In addition, useful honey bees may also be killed. A direct result of this is that the apple crop may be reduced, since bees are a major agent in cross-pollination of apple flowers.

8. In the Tonopas area of the Colorado Rockies, the sheep raisers were convinced that their sheep were being menaced by coyotes. Although this predator actually preys on sheep only occasionally, and lives mostly on smaller animals such as rabbits and gophers, its numbers were drastically reduced by the use of poison. When this happened, the rabbits and gophers multiplied in such numbers that they ruined thousands of acres of pasture land. With this lesson in ecology, the cattle growers have now formed the Tonopas Grassland Protective Association, with a campaign to protect the coyotes!

Extinction of animals. Because of the spread of civilization, the balance of nature has been so upset by man that over a hundred different kinds of animals have become extinct. Only a little more than a hundred years ago, the passenger pigeon was so numerous that the famous naturalist and painter, Audubon, described their flight as darkening the sky. Uncontrolled hunting gradually led to their destruction, and in 1914, the last passenger pigeon died. Other animals that became extinct during the past century: heath hen, Carolina paroquet, sea mink, Merriam elk. The ivory-billed woodpecker, North America's largest woodpecker, has not been seen since 1952, and may also be extinct.

A bird that became extinct as far back as 1693 was the *dodo*. This bird, which is now known only in museums, was about three feet tall, and could not fly. It made little effort to escape when sailors hunted it on its native island of Mauritius in the Indian Ocean. The introduction of pigs, which ate its eggs, led to its extinction on the island.

Animals threatened with extinction. In this country, tremendous herds of buffaloes, or American bisons, once dominated the Great Plains. As the result of uncontrolled hunting and the advancing front of civilization, countless numbers were killed. At the beginning of this century, only a few hundred remained alive. Conservation laws were passed protecting them. At the present time, the survivors are slowly increasing in number.

Some birds on the verge of extinction are: the whooping crane, a large white bird, which now numbers about thirty-eight individuals as compared with fourteen in 1938. The California condor, America's largest bird, having a wing span of 10 feet, which now totals about sixty; the trumpeter swan, our largest water bird, which was recently down to about seventy-five birds, and is now slowly beginning to increase in numbers. The best-known mammal facing extinction is the grizzly bear; it is estimated that fewer than 800 grizzlies remain. Other endangered mammals are the sea otter, sea cow, elk, and Key deer. Among rapidly disappearing fish are the lake sturgeon, Great Lakes whitefish, and lake trout.

CHAPTER REVIEW

Select the correct choice for each of the following statements.

1. Plankton is composed of (A) microscopic life (B) fish (C) whales (D) sharks (E) man

2. Green plants are the basis of life of (A) herbivorous animals (B) carnivorous animals (C) omnivorous animals (D) saprophytes (E) all of the above

3. The study of living things in relation to each other and their environment is known as (A) conservation (B) eugenics (C) endocrinology (D) ecology (E) evolution

4. A cactus is an example of a(n) (A) hydrophyte (B) xerophyte (C) mesophyte (D) saprophyte (E) pteridophyte

5. All the members of a species inhabiting a given location make up a (A) community (B) population (C) ecosystem (D) biome (E) biosphere

6. The next to last successful members of a maple-beech climax are (A) cattails (B) popular and pine trees (C) oak and hickory trees (D) sumac trees (E) willow trees

7. All of the following are examples of climax communities except (A) cypress and white cedar (B) beech-magnolia (C) tall prairie grasses (D) water lilies (E) scrub oak

8. Rabbits became a menace in Australia because (A) they transmitted rabbit fever to the kangaroos (B) their holes in the ground were a hazard to livestock (C) in the absence of natural enemies, they multiplied tremendously (D) they caused myxomatosis among newborn infants (E) they removed valuable nutrients from the soil

9. Japanese beetles are a worse pest in the United States than in Japan because here (A) the soil is better (B) there are fewer earthquakes (C) there are few natural enemies (D) there are more plants (E) there is a longer summer

10. All of the following animals have become extinct except the (A) passenger pigeon (B) whooping crane (C) dodo (D) ivory-billed woodpecker (E) heath hen

11. All of the following are physical factors of the environment except (A) light (B) temperature (C) water supply (D) predators (E) pH

12. The transfer of materials between organisms and the environment is known as a (A) niche (B) cycle (C) synthesis (D) free gas (E) solution

13. All of the following are examples of consumers except (A) carnivores (B) herbivores (C) saprophytes (D) green plants (E) parasites

14. An example of a parasite is (A) barnacle on a whale (B) nitrogen-fixing bacteria in clover roots (C) remora on a shark (D) vulture (E) disease germ

15. The correct sequence in a food chain is
(A) mice – green plants – snake – hawk
(B) mice – snake – green plants – hawk
(C) green plants – mice – snake – hawk
(D) green plants – hawk – snake – mice
(E) hawk – green plants – snake – mice

16. In a pyramid of energy, the greatest amount of energy is present in the level represented by (A) producers (B) first-order consumers (C) second-order consumers (D) third-order consumers (E) decomposers

17. An example of a pioneer organism is (A) pond lily (B) cattail (C) lichen (D) fern (E) wild carrot

18. The northernmost biome is (A) grassland (B) desert (C) taiga (D) tropical rain forest (E) tundra

19. At the equator, a taiga type of growth (A) is impossible (B) occurs only in winter (C) occurs in deep alleys (D) occurs in the mountain heights (E) occurs only in deserts

20. Compared to land masses, the temperature of the marine biome (A) is below freezing (B) is more stable (C) is less stable (D) absorbs less solar radiation (E) has greater extremes

ANSWER KEY

1-A	5-B	9-C	13-D	17-C
2-E	6-C	10-B	14-E	18-E
3-D	7-D	11-D	15-C	19-D
4-B	8-C	12-B	16-A	20-B

32 Conservation of our resources

America is called a land of plenty. Our country enjoys a high standard of living, there is an excess of food, and we enjoy the benefits of many luxuries. Yet we are in danger of losing our valuable natural resources because of waste and poor planning. We have found that our use of pesticides and insecticides has backfired in the form of contaminated crops and water supplies. The dangerous effects of these chemicals has been well depicted in the best-selling book, *Silent Spring,* by Rachel Carson.

In addition, the threat of pollution is affecting our use of the air and the natural waters. Combustion in industrial plants and automobile engines is adding poisonous wastes to the air that we breathe. Smog is a serious condition affecting the health of large city dwellers. The water supply is polluted by sewage, industrial wastes, silt, fertilizers, and salt water. The Great Lakes, which once had pure, fresh water, are in danger of becoming mere sewage basins in which few living things can survive.

We are warned that we must practice *conservation,* the planned use and preservation of our natural resources, especially (1) forests, (2) soil, and (3) wildlife.

FOREST DESTRUCTION

The destruction of our forests has been taking place ever since the first settlers arrived. The major causes of this loss have been:

1. **Excessive cutting.** The early settlers cut down trees in order to clear the land for crops and pastures. As our country grew, the original virgin forests were cut down to the point where many of them disappeared, especially in the East. Since the forests were so plentiful at first, and grew down to the water's edge, little thought was given to their value. Later on, when lumbering became well established, poor timber practices consisted of cutting down one forest and then moving on to another. There was no plan for replanting trees for the future.

2. **Forest fires.** Every year, there are more than 200,000 fires, which burn over 23 million acres of our forests. The damage to human lives, wildlife, soil, watersheds, recreation areas, and future growth of trees is incalculable. The cost of the bill amounts to about $60 million annually. The worst forest fire in American history was the Peshtigo Fire in Wisconsin in 1871; 1,280,000 acres were burned, 1,500 people were killed, and towns and settlements were destroyed. Forest fires are chiefly caused by human carelessness, involving smoking, campfires, and small fires to clear land which get out of control. Lightning and railroad train sparks also are causes.

3. **Insects and diseases.** Certain insects are serious enemies of trees. Some of them bore into the bark, destroy the living cambium cells, and stop the flow of sap. Examples: Spruce beetle, pine bark beetle, and elm bark beetle. Others eat the leaves, stopping the manufacture of food and causing the trees to starve. Examples: Tussock moth, spruce budworm, and gypsy moth. In a six-year period, beginning in 1942, four billion feet of magnificent spruce forests were destroyed by the spruce beetle in the Rocky Mountains of Colorado. These and other insects destroy far more trees than do forest fires.

Certain deadly fungus diseases also destroy trees. Among these: Dutch elm disease fungus, which is carried by the elm bark beetles as they bore into the bark. White pine blister rust destroys the valuable white pine tree; this fungus also lives on two alternate hosts, currants and gooseberries. The chestnut blight has wiped out the American chestnut tree from Canada to the Gulf states. Heart rot fungus attacks the heartwood of oak, ash, and most types of evergreen trees.

Value of Forests. Forests supply us with many useful products. The paper of this book, as well as that used in newspapers, magazines, cardboard boxes, and writing paper, originates in wood pulp. The structure of various parts of buildings requires immense amounts of lumber. Furniture, telephone poles, turpentine, rayon, cellophane, railroad ties, acetone, and methyl alcohol are but a few of the products derived from wood.

Forests serve as a home for wildlife, including such fur bearers as the raccoon, fox, opossum, and mink; game animals as rabbits, squirrels, deer, ruffed grouse; birds such as thrushes, warblers, woodpeckers, hawks. When the woodland is destroyed, these creatures disappear.

The woods are also used for recreation by many Americans, who make 21 million visits a year to the national parks alone. They come to picnic, fish, hunt, ski, swim, hike, ride, look and sit. The first national park to be established was Yellowstone Park in 1872.

An essential value of forests is their role in preventing erosion and floods. The leaves and branches intercept rain, breaking the effect of its fall on the soil below. The organic material of the soil (humus), consisting of decayed plant and animal matter, acts as a spongy layer, and soaks up the water; this prevents it from running downhill and from washing away the topsoil. So the forests also act as a watershed, retaining water and yielding it slowly for run-off streams.

Forest conservation measures. The national forests were established in 1891. There are now national forests in thirty states. Together with state forests, they serve as demonstration areas of good forestry practices; they also protect watersheds and wildlife, produce valuable forest products, and provide places for recreation. The 230 million acres of national forests are administered by the U.S. Forest Service, a branch of the Department of Agriculture.

Forests are now looked upon as a growing crop. Selective cutting is practiced, by which mature trees, as well as dying or diseased trees, are cut down. This process gives young growing trees an opportunity to mature and produce seeds. After older trees are cut down, they are replaced by young trees which are carefully spaced and cared for. As the trees are felled, they are prevented from damaging other trees. Like any other crop, trees must be harvested or they go to waste. Lumbermen are actively engaged in forest conservation to protect the trees we now have, and to ensure a continuous growth of more trees for the future. As a result, we are today finally growing more wood than we are using.

Reforestation is another practice that is restoring woodlands. To date, over eight billion trees and shrubs have been grown in nurseries and have been planted in needed areas. In addition, seeds have been planted directly in the woodlands.

Forest rangers are trained to manage and protect forests. They patrol the forests from fire towers that are distributed throughout the woods, and from airplanes and helicopters. At the first sign of a fire they hurry into action. They use various methods to put out the fire. One of the first steps is to make a fire line, or barrier, down to the soil, all around the fire. This is done to keep the fire from spreading. Then all the burning material within the fire area is extinguished. In remote forest areas, smoke-jumpers, or parachute fire fighters, are dropped by airplane to put out fires while they are still small. Fire breaks or fire lanes are built in the forests to divide them into small blocks. A backfire may be started within a fire line, to travel back toward an oncoming fire, and to burn out the potential fuel. One of the most important weapons of fire-fighting is an educational campaign to make the public aware of the need for exercising great care with matches, cigarettes and campfires.

A continuous program of research is carried on to control the insect and fungus menace to forests. Airplane spraying of DDT is a standard method of insect control. This method is most effective against insects that eat leaves. Genetics has supplied a new tool against forest diseases. Experiments are being conducted to breed trees that are resistant to fungus infections. Thus, the American chestnut has been hybridized with the Chinese chestnut to produce a promising resistant timber tree.

SOIL EROSION

In recent years such important books as *Our Plundered Planet,* by Fairfield Osborn, and *Deserts on the March,* by Paul B. Sears, have called attention to the threat that exists to our national survival in the critical problem of soil erosion. When the country was first settled, there was an average of about nine inches of topsoil; since then, about one third of our topsoil has been lost. It takes about five hundred years for an inch of topsoil to form under ideal conditions, by the slow decay of organic material and rock. This valuable resource can be lost in only a few years.

Erosion is the wearing away of the topsoil by the action of water and wind. *Gully erosion* is common on slopes, and is caused by the cutting action of running water in small channels that become ever wider. *Sheet erosion* results when a sloping area is so full of water that the topsoil is lifted up in a sheet and floated away. *Wind erosion* results when the wind carries off dry topsoil which is not held together by the roots of plants. Erosion has been brought about by several factors:

1. **Forest destruction.** The roots of trees hold soil together. The soil acts as a sponge and absorbs much water during a rainfall. When the forests were cut down, the soil was not held together, and was washed away by the action of falling raindrops and running water. With the loss of topsoil, floods resulted, washing away still more topsoil and doing considerable damage.

2. **Cultivation of soil.** As farmers plowed up the soil and planted such crops as wheat and corn, the binding action of roots on the soil was weakened, leading to water and wind erosion. In addition, the mineral content of the soil was depleted by the same crop being planted year after year. The Dust Bowl in the 30's had dust storms in which the valuable topsoil was blown away from countless farms in the Southwest.

3. **Overgrazing.** Poor grazing practices on fields and in forests led to the loss of grass and young trees, opening the way to eventual soil loss.

Soil conservation measures. A number of practices help to overcome the loss of valuable topsoil:

1. **Forest conservation.** The protection of forests, including reforestation, is an important step in conserving soil.

2. **Windbreaks and shelterbelts.** Trees have been planted around farms as a *windbreak* to break and reduce the eroding action of the wind. *Shelterbelt* rows of trees have also been planted in strips across the prairies, to reduce the blowing action of the wind and to reduce excessive evaporation.

3. **Cover crop.** During the winter, fields of corn, or other cultivated clean-till crops are planted with rye grass, so that the soil will be held together and improved. The perennial lespedeza and kudzu plants have converted severely eroded land in many parts of the South to excellent grazing areas in which the soil particles are bound together. The Department of Agriculture demonstrated that in one area of Texas on which corn was grown, more than 20 tons of soil an acre were lost, along with 13½ per cent of the rainfall; when Bermuda grass was grown, the soil loss was only 0.02 of a ton an acre, and the water loss only 0.05 per cent of the precipitation. The Soil Conservation Service recently planted a cover crop in the state of Washington, to convert a river that caused floods in the spring and ceased to flow during late summer, into a continuously flowing river that no longer floods and

no longer carries away vast quantities of productive soil.

4. Contour plowing. Fields are plowed along the contour of sloping land, instead of up and down the slopes. The furrows hold back the water after a rain, and allow it to soak into the ground instead of running downhill and carving away the soil.

5. Strip cropping. Cultivated crops such as cotton or corn are planted on slopes in alternate strips with grasses and legumes. Any soil that is washed away from the cultivated rows is held by the cover crop below, and is not lost.

6. Terracing. On steep slopes, terraces or steps of soil and rocks are built across the hill and planted with soil-binding plants. This method holds the water back and prevents the loss of soil.

7. Crop rotation. Land that is planted with the same crop year after year (corn, for example), loses its mineral value and its capacity to hold water and soil; the yield of the crop is also reduced. By means of a three-year rotation of corn, oats and clover, the yield of corn is increased, while the loss of soil is reduced by the cover of oats and clover. In addition, the nitrogen content of the soil is increased by the action of nitrogen-fixing bacteria in the nodules of the clover. Other legume crops that are useful besides clover are alfalfa, lespedeza, and vetch.

8. Soil improvement. As plants grow, they remove valuable minerals from the soil. The farmer restores them to his soil by adding chemical fertilizers containing important compounds of nitrogen, phosphorus, magnesium, potassium, etc. When soils become too acid, he adds lime. He also spreads animal manure to increase the organic and mineral content of the soil. He further improves the soil by plowing into it the remains of his crops, such as cotton, wheat, or corn, and by plowing up fields of legumes and grasses; they are known as *green manures*. The resulting increased humus not only provides a greater mineral content, but it also increases the water-holding capacity of the soil.

Strip Cropping

Contour Plowing

Terracing

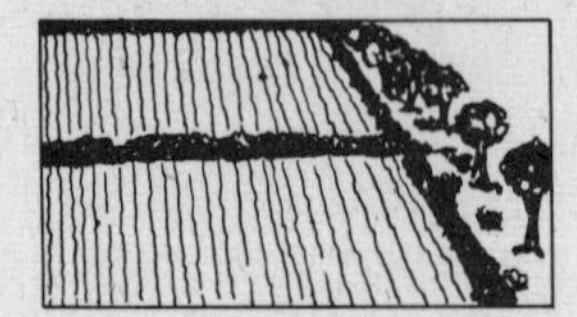
Shelterbelts: Windbreaks

Methods of Soil Conservation

Also, when the surface is covered with leaves and other plant material, the splashing action of falling raindrops in causing the breaking up of the soil is reduced.

OUR VANISHING WILDLIFE

With the spread of civilization, the once abundant wildlife of our country has been threatened, including mammals, birds, fish, and flowers and shrubs. Specific factors leading to their destruction have been: the eradication of breeding and feeding grounds such as forests, prairies and swamps; uncontrolled hunting and fishing; stream pollution by sewage and industrial wastes.

Wildlife conservation measures

1. Wildlife refuges. *Sanctuaries* have been set aside in which the wildlife is protected. From time to time, when there is a surplus of animals, open seasons are permitted for controlled hunting and fishing. Agencies such as the U.S. Fish and Wildlife Service set up the regulations which help maintain a balance in the animal populations. Windbreaks and stripcropped areas also provide refuge for small mammals and birds. In addition, small ponds are built by farmers in soil-conservation districts, which are stocked with fish and which serve to attract ducks; about 87,000 of these ponds are built annually.

2. Hunting and fishing limitations. Game laws regulate the season when hunting and fishing may be carried on. There are limits as to the number that may be killed, the size and number of the fish caught, the protection of females, and the types of animals that may be hunted. Recently, a limited number of 400 hunters were picked at a public drawing from 8,000 applicants for the first open season since 1936 on moose in New Brunswick, Canada; at that time, the number of moose had been so depleted that they had to be fully protected until the present time when their numbers were found to be on the increase again.

3. Bird laws. Birds are protected by a number of state and federal laws. Songbirds, which destroy countless insects, are fully protected. Ducks and geese may be hunted only during a restricted open season. Migrating birds are protected by treaties with Canada and Mexico.

4. Fish protection. Such fish as trout, bass, salmon, and perch are raised in fish hatcheries. When the young fish are large enough, they are released in streams and ponds. This helps to replenish the supply. Erosion control also protects fish by preventing the water from becoming polluted by silt, which covers their food and interferes with spawning. The salmon is assisted in its migration upstream by the construction of *fish ladders* along dams.

5. Wild flower protection. As forests, swamps and fields were destroyed by the advancing front of civilization, many attractive wild flowers were threatened. Some of them have been picked to the point of extinction. State laws have been passed to prohibit the picking of such attractive flowering plants as dogwood, lady's-slipper, trailing arbutus, Jack-in-the pulpit and Dutchman's breeches.

6. Fur farming. Mink, silver fox, and muskrat are now being raised on thousands of fur farms. Chinchillas are also being raised in this way. It is predicted that more and more of our furs will be produced on such farms, as the numbers of fur-bearing animals diminishes. As a result, there will probably be less fur-trapping of these animals, thus indirectly helping to conserve them.

CHAPTER REVIEW

Select the correct choice for each of the following statements.

1. All of the following are major causes of forest destruction except (A) excessive cutting (B) forest fires (C) conservation (D) insects (E) fungus diseases
2. Two insects that are enemies of trees are the (A) elm bark beetle and Drosophila fly (B) Drosophila fly and honey bee (C) honey bee and gypsy moth (D) elm bark beetle and monarch butterfly (E) elm bark beetle and gypsy moth
3. A tree that has been practically wiped out by disease is the (A) American chestnut (B) Norway maple (C) sycamore (D) weeping willow (E) staghorn sumac
4. Decayed organic matter in the soil is known as (A) aqueous humor (B) humus (C) hilum (D) hydra (E) hyphae
5. Trees prevent floods because (A) their roots soak up excess water (B) their leaves store excess water (C) their xylem ducts store excess water (D) their roots bind the soil (E) they carry on excess transpiration
6. Forests are now looked upon as a (A) source of fires (B) growing crop (C) home for harmful insects (D) waste of valuable space (E) menace to wildlife
7. All of the following are useful in controlling forest fires except (A) fire lines (B) smoke-jumpers (C) smokescreens (D) fire breaks (E) backfires
8. When the country was first settled, the average thickness of the topsoil layer was about (A) 6 inches (B) 9 inches (C) 12 inches (D) 6 feet (E) 9 feet
9. Soil erosion is caused by (A) shelterbelts (B) windbreaks (C) reforestation (D) over-grazing (E) spraying with DDT
10. A useful cover crop is (A) Bermuda grass (B) wheat (C) corn (D) cotton (E) rice
11. Terracing is practiced to (A) hold water back (B) enrich the soil (C) increase the nitrogen content of soil (D) add green manure to the soil (E) keep rodents away
12. All of the following are examples of legumes except (A) clover (B) alfalfa (C) vetch (D) oats (E) lespedeza
13. All of the following are examples of good soil conservation measures except (A) contour plowing (B) cultivation of soil (C) cover crops (D) strip cropping (E) crop rotation
14. Songbirds are useful because they destroy (A) field mice (B) moles (C) snakes (D) earthworms (E) insects
15. A protected wildflower is the (A) daisy (B) dandelion (C) clover (D) flowering dog-wood (E) aster

ANSWER KEY

1-C	4-B	7-C	10-A	13-B
2-E	5-D	8-B	11-A	14-E
3-A	6-B	9-D	12-D	15-D

NEW FIELDS IN BIOLOGY

10

33 Biology and atomic energy

The explosion of the first atomic bomb at Alamogordo, New Mexico, on July 16, 1945, ushered in a new era for mankind. Man had learned to harness the tremendous forces that hold the atom together.

Radioactivity. In 1896, Becquerel discovered that a piece of uranium ore affected an unexposed photographic plate next to it; he concluded that it gave off radiations. Pierre and Marie Curie discovered radium and polonium, two other radioactive elements. Three kinds of radiations are given off by radioactive elements: (1) *Gamma rays,* which are like X-rays, and are very penetrating. (2) *Alpha particles* which are the nuclei of helium atoms (containing 2 protons and 2 neutrons). (3) *Beta particles,* which are electrons. These rays are constantly being given off by the radioactive elements. As a consequence, these elements are unstable and are breaking down all the time. Uranium decays or breaks down to other radioactive elements, including radium, and finally ends up as lead, a stable element. This process takes over 4½ billion years. During that time, half of the uranium loses its radioactivity. Radium has a half-life of 1,590 years; carbon-14 has a half-life of 5,760 years. The *half-life* is the time required for one half of a radioactive element to disintegrate. Some radioactive elements have a half-life of a second.

Splitting the atom. The nucleus of an atom is held together by a powerful force called *binding energy,* which resists the separation of its protons and neutrons. If the nucleus is broken, this energy is released. Atomic particles such as protons, alpha particles, neutrons, and electrons, have been used as bullets to split the nucleus. These particles have been generated by such atom-smashing machines as the cyclotron, Van der Graaf generator, synchotron, and cosmotron. When common elements are bombarded by these accelerated particles, they become radioactive.

The impact of a neutron on the nucleus of the relatively unstable uranium-235 causes a reaction known as fission, resulting in: (1) production of immense quantities of energy, according to Einstein's equation, $E = mc^2$ (E, energy; m, mass; c, speed of light); (2) formation of nuclei of two simpler elements, barium and krypton; (3) emission of gamma radiation; and (4) release of one to three additional neutrons.

The release of these neutrons may lead to a *chain reaction,* since each splits other nuclei of U-235, which in turn yield more neutrons, and so on. In the *atomic bomb,* this occurs within a fraction of a second, when the right amount, or *critical mass* of U-235 is concentrated in one place. The uncontrolled chain reaction of the atomic bomb produces immense amounts of energy. In a *nuclear reactor* or *atomic pile,* the speed of this chain reaction is controlled by rods of the metal cadmium, which can absorb the neutrons. A man-made element, *plutonium,* may also be used in the process of atomic fission.

The *hydrogen bomb* acts by *fusion* rather than by *fission.* It is a *thermonuclear reaction,* requiring the enormous heat of an atomic fission bomb to set it off, involving temperatures of more than 2 million degrees centigrade. In this process, nuclei of the isotopes of hydrogen, deuterium and tritium are fused to form nuclei of helium, a reaction that is going on in the sun and the stars. It may also be brought about by the fusion of hydrogen with the element lithium, to form helium. The H-bomb is many more times powerful than the A-bomb.

Hazards arising from the use of nuclear energy. The tremendous force of the atomic bomb was demonstrated toward the end of World War II, when some 250,000 Japanese in Hiroshima and Nagasaki were killed or injured by just two such bombs. Since then, the continued testing of atomic and hydrogen bombs has caused serious concern, because of the hazards of fall-out radiation effects and genetic effects.

Fall-out. Within a fraction of a second after the explosion of a nuclear bomb, a huge mushroom-shaped cloud rises, sucking up thousands of tons of dirt and debris. This displaced material may rise as high as 25 miles, to be spread by winds over wide areas. *Fall-out* consists of radioactive dust and water particles that fall back to earth. Some of the larger particles fall back to earth within the first few days, and are capable of causing *radiation sickness,* due to excessive exposure to radiation. This condition is characterized by such effects as skin eruptions, loss of hair, vomiting, diarrhea, destruction of blood cells, bleeding, fever, interference with the endocrine system, and general weakness; in extreme cases, death results. Clouding of the lenses of the eyes (cataracts) is another result of high exposure to radiation.

The finer radioactive dust remains suspended in the air for a long time, and may fall back over a period of months. One of the fission products it may contain is strontium-90, which has a half-life of twenty-eight years. It is dangerous because, when it falls back to the ground, it is absorbed by plants through their roots. Cows that eat such plants then pass the radioactive element on through their milk. Humans, especially children, may absorb it in their bones, where the radiations can damage the adjacent cells and tissues; leukemia and bone cancer may result. Strontium-90 may also be found in vegetables and fruits. Ways are being sought to reduce the dangers of this radioactive isotope. Progress has been made in filtering strontium-90 from milk. By washing vegetables, it has been shown that the strontium-90 content can be reduced considerably. Other fission products contained in fall-out have a half-life ranging from a few days to several years, and are consequently not as much of a threat as the longer-lived Sr-90.

However, it was recently discovered that the threat of iodine-131, which has a half-life of only 8 days, is not to be ignored. In 1968, studies made on children on Rongelap Island, near Bikini Atoll, where a hydrogen bomb was exploded in 1954, showed that 17 of 19 of the children had developed thyroid gland damage. The iodine-131 had become concentrated in their thyroid glands and had remained long enough to be harmful. The children were stunted in growth, and their blood contained large amounts of iodine-rice protein, which was not thyroxin. The children were treated with thyroid extract to prevent further stunting and non-cancerous nodules were removed from their thyroid glands by surgery.

Radiation limits. Almost all of the fall-out radiation consists of gamma rays. These are like X-rays and have great penetrating ability. They are measured in units of *roentgens* (r). Persons exposed to 450r in the course of a few minutes have a 50-50 chance of surviving. An exposure of 600r within a one-day period would undoubtedly be fatal. If the whole body absorbs as much as 50r in a single day, it is not considered to be a fatal dose unless it is repeated on successive days. In cancer treatment, thousands of roentgens are safely given to a small confined area of the body for a limited period. The average human is normally exposed to daily background radiation from cosmic rays and radioactive rock. Taking this into account, the Committee on the Biological Effects of Atomic Radiation of the National Academy of Sciences recommends that the average dose of radiation accumulated during the first thirty years of life should not exceed 10r of man-made radiation.

Large-scale exposure of the body, however, requires immediate hospitalization. The present treatment includes complete rest and the avoidance of fatigue and chills. Whole-blood transfusions are given until the bone marrow has had time to regenerate and begin producing blood cells. There is also danger of infection resulting from the destruction of the germ-fighting white blood cells; antibiotics are therefore given. Experimental treatment now being tried out includes the injection of bone marrow. Another experimental field of research is the development of antiradiation pills that would protect a person against strong doses of radiation.

Disposal of radioactive wastes. Radioactive wastes must be disposed of without leading to undesirable radiation exposure. In some cases, they are enclosed in concrete containers and sunk in the ocean; however, there is always the danger that there may be leakage of these materials. They are also being buried deep in carefully selected burial sites in various parts of the country.

Genetic effects. The future of the human race may

be affected by radiation injury to the reproductive organs. Permanent sterility can be produced by high exposure. With lower dosages, temporary sterility may result, followed by repair of the damaged tissue in these organs. If the molecular structure of the genes is affected, mutations may result that can cause the appearance of defective human beings in the next generations. Dr. Herman J. Muller, who received the Nobel Prize for his pioneer research in producing mutations in Drosophila by means of X-rays, has predicted that this may be the most serious consequence of fall-out radiation. Another scientist, Dr. Roberts Rugh, has pointed out that embryos are especially sensitive to radiation, with consequent danger to the central nervous system. He predicts that the effects may not be evident for generations. Some experimental evidence also indicates that there may be a slight shortening in the life span as a result of irradiation. However, there are other scientists who claim that the risk from fall-out is slight. They point to the fact that most radioactivity from fall-out dies out after two or three weeks.

Detecting and measuring radiations. It is possible to detect the presence of radioactive material by the use of certain instruments, the best known of which is the *Geiger counter*. It contains a tube from which some of the air has been removed. There is a wire through the center connected to a battery, an amplifier, and a loudspeaker. When radiations enter the tube and come in contact with the air molecules, a tiny amount of current is caused to flow, which produces a clicking sound; a mechanical counter can record the number of clicks, thus measuring the degree of radioactivity. The presence of *cosmic rays,* which emanate from outer space, provides *background radiation,* causing an uncontrolled small number of clicks on the Geiger counter. A much more active clicking series of sounds is heard in the presence of radioactive material.

Another detecting device consists of a small *film badge,* which is worn by staff members of radiation laboratories. The film is inspected regularly to see if it has been darkened by radiation. The *dosimeter,* or pocket meter, resembles a fountain pen; it contains a charged fiber which is discharged by radiation. Other devices include the Wilson cloud chamber, electroscope, and scintillation counter.

Radiation research scientists use great care in shielding themselves against radiations. They wear special laboratory clothing, lead-impregnated aprons, and rubber gloves. From behind thick walls they operate remote-control arms to manipulate containers of radioactive materials; precautions are taken to dispose of radiological wastes safely; special signs indicate the location of radiation areas; health physicists inspect facilities and personnel for overexposure to radiation.

Peacetime uses of atomic energy. Many valuable uses have been found for radioactive isotopes in medicine, physiology, plant research, genetics and practically every other field of biology. When elements are made radioactive, either in an atomic pile, such as at Oak Ridge, Tenn., or in a cyclotron, such as the California Institute of Technology, they can be used as *tracer elements* (*tagged atoms*). That is, they can be traced through the body by their radiations with the use of a Geiger counter. They can be traced not only through the bloodstream and the various organs, but also through various chemical compounds in the metabolism of the body. Some biological uses of atomic energy are:

1. **Cancer treatment.** For many years, a standard method of using radiations in cancer was by means of X-ray or radium treatment. Now, the relatively inexpensive and readily available cobalt-60 is largely being used for this purpose. Radioactive gold is also being used to treat deep-seated cancer. Radioactive iodine, I-131, is used for cancer of the thyroid gland; when it is swallowed in an "atomic cocktail," the iodine is absorbed practically entirely by the thyroid gland, where its radiations reduce the activity of the cancer cells. If it is suspected that metastases of such a cancer have spread to the other parts of the body, a Geiger counter can help locate places where the iodine-131 has been absorbed by the malignant cells.

Phosphorus is an element that is concentrated in bone marrow; since this is the region where blood cells are formed, the overproduction of white blood cells (leukemia), or of red blood cells (polycythmia) is treated with radioactive phosphorus-32. Another use of phosphorus-32 is as a tracer to label the dye fluorescein; this dye is readily absorbed by brain tumors, thus permitting such a tumor to be located more accurately by means of a Geiger counter.

2. **Artificial mutations.** Ever since Muller first announced in 1927 that he had successfully produced mutations in Drosophila with X-rays, scientists have been trying to develop useful mutations artificially. Now, gamma rays and neutrons are used in addition. At Brookhaven National Laboratory on Long Island, many experiments are being conducted to produce mutations in apples, peaches, grapes, carnations, corn, and other plants. Most of the mutations obtained are undesirable and are discarded. A few are kept for further development. In the "gamma field" plants are arranged in concentric circles around a container of cobalt-60, which serves as a central source of gamma rays. The plants

closest to the center receive the highest doses of radiation. Some useful mutations have been obtained at various laboratories by this and other methods of radiation treatment, such as: an improved strain of *Penicillium,* yielding larger amounts of penicillin; a variety of oats that resists rust, a parasitic fungus; a new variety of peanut yielding a 30 per cent higher yield per acre; an earlier-ripening peach; a hardy variety of rice; a red carnation flower derived from a white flower; a giant tetraploid snapdragon flower with twice the number of chromosomes per cell; and a normally red dahlia flower in which half the petals are white. In Sweden, similar research has yielded an improved variety of mustard plant, the Primex variety, with a greater yield of oil for use in making margarine and mayonnaise; the Ray variety of pea with a greater yield, through having more branches; and a hardier variety of barley that can be grown in the north, and ripens earlier.

3. Understanding life processes. The use of tracer elements gives a greater understanding of how living things carry on their activities. For example, it is known that oxygen is given off during photosynthesis—but does it originate from the CO_2 or the H_2O taken during this process? By giving plants H_2O in which oxygen-18 was used, it was determined that oxygen is released from the water, and not the carbon dioxide. Also, the various intermediate steps in the making of carbohydrate are being traced by using carbon dioxide containing carbon-14. At least ten different compounds in this process have been shown to appear within the first few seconds of photosynthesis activity.

Through the use of tracers, it has been shown that the atoms of the human body go through a constant reshuffling. In the course of a year, there is a complete exchange of practically all of these atoms. Even the structure of such "solid" parts of the body as the teeth and bones are included in this dynamic replacement process. In other words, as far as atoms are concerned, you are not the same person you were a year ago.

Through the use of radioactive iron, it has been determined that the red blood cells live about 127 days, and then are removed and destroyed in the liver and spleen. The iron is used over again to form new red blood cells. Iron is normally stored in the body as the compound *ferritin.* The intestinal lining or mucosa controls the absorption of additional iron. When the amount of iron is normal, no more new iron is absorbed. When the level is reduced, additional iron is taken into the body. It has been shown that a person who is anemic absorbs more iron if it is taken before a meal than during or after a meal when there is a good deal of food in the digestive tract. It is now known that when fats are eaten, they are not immediately used for energy; instead they are deposited in the fatty tissues while the older body fats are used for energy. Stored fats are constantly being exchanged and moved around in the body. Proteins are also undergoing a dynamic, ever-changing composition.

It has also been found that radioactive salt which was injected into the bloodstream appears on the surface of the body in sweat in a little over one minute.

4. Surgical use. The circulation of the blood is revealed with the use of radioactive sodium-24 as a tracer. If table salt containing this isotope of sodium is injected, sensitive radiation counters can follow it in the circulating bloodstream. In case of a severe injury, it is possible to determine whether circulation has been stopped or is continuing, by locating the radioactive salt. This information can help a surgeon decide whether an amputation is required. It also helps locate any obstruction in the circulation. The pumping action of the heart can similarly be studied with sodium-24.

5. Uses in agriculture. Radioactive phosphorus-32 can be used in phosphate fertilizers to determine the best time and place for using it; also to determine how much is actually taken up by plants. Thus, it is now known that certain crops, such as tobacco, cotton, corn and sugar beets take up phosphate only during the early stages of their growth; phosphate that is applied later is wasted. By contrast, potatoes need phosphate during their entire growth. The use of radioisotopes has recently revealed that plants are capable of absorbing fertilizers through their leaves and branches. Even in freezing winter weather, radioactive phosphorus can be absorbed through the surface of branches of a tree, and transported to the buds. This knowledge has now led to the practice of spraying fruit trees with fertilizers. It was once thought that leaves were covered with an impervious cuticle; now it has been shown that leaves are well adapted to absorption. It is now believed that spraying the leaves with phosphorus is much more effective than applying fertilizer through the soil; as much as 95 per cent is used by the plant in the former case; only about 10 per cent in the latter. Also, the distribution through the plant is faster when absorbed through the leaves. Calcium, on the other hand, is better applied through the soil. Plants have been found to vary in this respect.

On the basis of this new knowledge about fertilizers, farmers are capable of saving much money and effort. Up to now, they have been spending about a billion dollars a year on fertilizers.

The use of tracers in chicken feed has shown that

eggs are made from food eaten over a month earlier. The proteins used as part of the egg come from the chicken's body, and are used for egg production after they have been replaced by new protein. On the other hand, the calcium of the eggshell comes from food eaten within the past twenty-four hours.

In Honolulu, research with carbon-14 on sugar cane plants revealed that the sugar made by a single leaf in an hour was distributed through the entire 11-foot sugar cane plant within three days.

Radioisotopes are proving of great value as markers in tracing the movements of insects, and in suggesting methods of controlling harmful ones. The feeding habits of bees is being studied with radiophosphorus. The migration habits of the cotton boll weevil, spruce beetle, and other insects harmful to crops is being followed. The screw worm is the parasitic maggot of a fly that can wipe out whole herds of cattle, hogs and sheep. This damaging insect in the Southeastern part of the United States now seems to be coming under control as the result of releasing male flies sterilized by radiating the developing pupae with cobalt-60; since they are sterile, they cannot produce another generation. As a result, many months have elapsed over a two-year period with only a single report of screw worm damage.

6. Food preservation. Foods have been irradiated in order to sterilize or pasteurize them. In sterilization, the food is packed in cans and thoroughly irradiated to kill all the bacteria. Such food remains fresh as long as the can is unopened. However, there is some undesirable change in color, texture, and flavor of food treated in this way. A low dose is applied to "pasteurize" food that is to be kept for only a short time. Thus, irradiated potatoes remain fresh after eighteen months at 47°F; irradiated oranges were still fresh after seventy days at about the same temperature.

People have become so accustomed to eating food in its preserved form that they prefer it to the natural product; an example is pasteurized milk. Irradiated ham and shrimp taste more like the fresh products than do canned ham and shrimp; so do chicken, pork and bacon. However, irradiated beef has a disagreeable taste and odor. Varying results have so far been obtained with vegetables; some approach their natural taste, smell and appearance. Another problem in this field is to eliminate all radioactivity resulting from the treatment of the food.

The sterilization of pork by irradiation can render it safe against trichinosis by destroying the trichinae larvae.

The value of irradiated foods is of interest to the Armed Forces, because the possibilities of shipping and stockpiling foods without refrigeration would greatly simplify military supply problems.

7. "Atomic calendar." The Nobel Prize in Chemistry for 1960 was awarded to Dr. Willard F. Libby for his method of using radioactive carbon-14 in establishing the date of various organic substances. He determined that it decays at a fixed rate; it has a half-life of 5,760 years. After 11,520 years its radioactivity has been reduced to one-fourth, after 17,280 years to one-eighth, and so on. The age of objects up to 50,000 years old has been determined in this way, including wooden remains from tombs, charcoal and bones from archaeological sites of ancient man, trees from the Ice Age, and the Dead Sea scrolls. In the process of dating the ratio of nonradioactive carbon to radioactive carbon in the organic material is measured.

Dr. Libby showed that the production of radiocarbon was a common occurrence in the earth's atmosphere. This happens when nitrogen atoms are changed to atoms of carbon-14 by collision with neutrons from cosmic rays. When it unites with oxygen, this radiocarbon forms small amounts of radioactive carbon dioxide. As a plant absorbs the radioactive carbon dioxide during photosynthesis, it became incorporated into plant and then animal matter. Even after the organism dies, the organic remains continue to contain the radiocarbon.

Dr. Libby now plans to use tritium to trace the age and source of water supplies from 50,000 to 300,000,000 years of age.

CHAPTER REVIEW

Select the correct choice for each of the following statements.

1. Early studies of radioactivity were made by (A) Becquerel and Libby, (B) Curie and Libby (C) Libby and Muller (D) Becquerel and Curie (E) Curie and Muller

2. All of the following atomic particles have been used as bullets to split the nucleus, except (A) protons (B) chromatin (C) alpha particles (D) neutrons (E) electrons

3. Strontium-90 is dangerous because it (A) is absorbed by the bones (B) is absorbed by the

thyroid (C) has a half-life of 5,760 years (D) cannot be detected (E) remains in food that has been preserved by irradiation

4. Carbon-14 has a half-life of 5,760 years; in 17,280 years, its radioactivity will have been reduced to (A) one-half (B) one-third (C) one-fourth (D) one-eighth (E) one-sixteenth

5. The impact of a neutron on the nucleus of uranium-235 causes all of the following except (A) production of immense quantities of energy (B) release of cosmic rays (C) formation of nuclei of barium and krypton (D) emission of gamma radiation (E) release of additional neutrons

6. It is now believed that fall-out may cause radiation sickness only within (A) the first few days (B) the first six weeks (C) the first six months (D) the first year (E) the half-life period of strontium-90

7. Another instrument besides a Geiger counter that is useful in detecting radiations is the (A) dosimeter (B) calorimeter (C) anemometer (D) electrophoresis apparatus (E) chromatograph

8. A scientist who believes that defective human beings may appear in future generations because of fall-out radiation is (A) Becquerel (B) Einstein (C) Muller (D) von Braun (E) Geiger

9. One of the effects of high radiation exposure is (A) fertility (B) sterility (C) night blindness (D) near-sightedness (E) heterosis

10. Background radiation results (A) from a defective Geiger counter (B) in damaged Geiger counters (C) in very active clicking sounds on a Geiger counter (D) from cosmic rays (E) in radiation sickness

11. A radioactive isotope that is now widely used in cancer treatment is (A) cobalt-60 (B) deuterium (C) tritium (D) oxygen-18 (E) sodium-24

12. Radioactive isotopes have been useful in all of the following ways except (A) producing artificial mutations (B) cross-pollination of ferns (C) "atomic clocks" (D) photosynthesis research (E) tracing metastases

13. All of the following combinations are correct except (A) radioactive iron—life of red blood cells (B) phosphorus-32—leukemia (C) radioactive gold—cancer (D) radioactive silver—thyroid gland (E) sodium-24—blood circulation

14. A radioisotope useful in establishing the age of ancient objects is (A) nitrogen-14 (B) potassium-39 (C) lead-208 (D) aluminum-27 (E) carbon-14

15. Radioactive phosphorus is useful in all of the following ways except (A) to treat the overproduction of white and red blood cells (B) location of brain tumors (C) determination of the best time for the use of fertilizers (D) application of fertilizers to leaves (E) determining the age of the Dead Sea Scrolls

ANSWER KEY

1-D	4-D	7-A	10-D	13-D
2-B	5-B	8-C	11-A	14-E
3-A	6-A	9-B	12-B	15-E

34 Man and space

The year 1961 will be remembered as the year in which human beings went into space and orbited the earth. The feat was first accomplished on April 12, when Yuri Gagarin of the Soviet Union traveled around the earth in eighty-nine minutes, at an altitude of between 112 and 203 miles. John Glenn was the first American astronaut to go into orbit, the following February. Since then, more than two dozen astronauts, most of them American, have traveled in space for varying lengths of time. The actual landing of a man on the moon took place on July 20, 1969. The Apollo 11 astronauts, Neil Armstrong and Edwin E. Aldrin, successfully accomplished this historic adventure and brought back some of the moon rocks for analysis.

As scientists solve the mechanical and biological problems involved in space flights, more extensive flights will continue to be carried out. Among the biological problems encountered are: (1) Acceleration, (2) Weightlessness, (3) Radiation, (4) Physiological effects, (5) Psychological effects.

Acceleration and deceleration. When a rocket takes off for outer space, it achieves speeds up to 17,500

miles an hour in a very short time. The astronaut experiences violent vibration and noises during this take-off. The increase in speed, or *acceleration,* exposes the astronaut to great gravitational pressure, which is measured in terms of G. At rest, a person is subject to a force of 1 g, which is equal to the force of gravity. An acceleration of 4 G produces a pressure, against a person who weighs 140 pounds, of a powerful force equivalent to four times his weight, or 560 pounds. Pilots of high-speed planes, making a quick turn or a sudden climb after a power dive, have experienced such great G forces that have caused them to "gray out" or "black out." This condition is due to the failure of the blood to reach the head quickly enough, under the pressure of the G forces; if severe enough, unconsciousness results.

At the Aviation Medical Acceleration Laboratory of the Naval Air Developmental Center in Johnsville, Pa., there is a 50-foot centrifuge containing a gondola which can spin at great speeds to produce accelerations up to 40 g. By means of motion cameras, TV equipment and other measuring devices, the reactions of human subjects and animals are studied. So far, it has been observed that a human in a sitting position can easily stand up to 3 G. At 4 G, he has difficulty keeping his head erect, and can hardly move his limbs; breathing is difficult, chest pains develop, his heart is pressed against his backbone, and there may be temporary loss of vision. At 6 G, he becomes unconscious.

It has been determined that in a lying position, a human is able to endure great G pressure. A space suit with inflated sections has been developed that presses against parts of the body in such a way that blood is prevented from being drained away from vital organs. Using these aids, an astronaut can endure a force of more than 9 G. After the first five minutes of acceleration, the astronaut finds relief from the G forces.

However, there is a similar problem upon return to the earth, this time resulting from the sudden decrease in speed, or deceleration. As the space vehicle comes within 60 miles of the earth, it encounters shock waves, heating the air around it to 11,000°F. At approximately the same time, the decelerating force may reach a force up to 11 G.

Weightlessness. Moments after enduring the phases of violent acceleration, noise, and vibration, the astronaut finds himself in the strange world of weightlessness. There is a sensation of floating or falling which is capable of producing a panic reaction unless the astronaut has had proper training. There is difficulty in achieving muscular coordination involving judgment of distance; in simply attempting to reach out for an object a few inches away, one might find himself knocked off balance by the effect of using too much exertion for the weightless condition. Eating may also present a problem; food might not enter the esophagus properly without the effect of gravity to bring it to the point of being acted on by peristalsis, and instead, it might be sucked into the trachea by the breathing action. Water has to be taken in by sucking, since it does not pour without gravity. Food is taken in as soft material resembling baby food, from plastic squeeze tubes with straws. There is also the possible hazard of exhaled carbon dioxide remaining in the mouth and nose and being inhaled again.

During extended flights, his shoes would have magnetic soles. These would be powerful enough to keep the astronaut attached to metal surfaces in a stationary position, while in the weightless state.

Exposure to radiation. Once he goes into orbit, the astronaut no longer benefits from the protecting blanket of the earth's atmosphere. He is exposed to the full influence of primary cosmic rays, which originate somewhere in space. These particles, which consist mostly of protons, travel with tremendous energy and can penetrate through tissue, destroying thousands of cells in their path. Some scientists believe that if one of these primary cosmic particles were to hit a vital part of the brain, the astronaut would suffer serious loss of ability. When black mice were successfully sent aloft in a rocket recently, and returned, they were found to have a number of white and gray hairs several weeks later; apparently the pigment-producing cells of the hair were damaged by the cosmic radiation. According to present experiments, exposure of twenty-four hours' duration does not produce severe health hazards; however, the effects of exposure of a week or more during longer space trips might be very serious.

Another hazard is in the two Van Allen radiation belts in space, which surround the earth, extending from a distance of 800 miles to 50,000 miles, over the equatorial and temperate regions. Not much is yet known about these bands of high radiation intensity, but space travelers would have to be shielded against them. They probably consist of streams of electrons and protons which can react with the walls of a space vehicle to produce concentrated showers of X-rays. As a result, as much as half of the weight of a space ship may have to be devoted to providing a radiation shield. Some scientists think that the presence of a thin skin of a light metal such as aluminum over the surface of the vehicle may help avoid the formation of these X-rays.

It has been suggested that one way of helping an astronaut overcome exposure to radiation would be to have some of his bone marrow removed ahead of time and stored until after an exposure. Then it could be reinjected into him. Previous results with victims of radiation have shown dramatic response to such injections of bone marrow.

Another serious problem is exposure to the ultraviolet rays from the sun. A fresh exposure at the beach has given painful memories to many people. In space, the intensity is much greater. The windows of the space vehicle and of the helmet would have to be darkened to protect the eyes and face.

Physiological effects. Within the limited atmosphere of the space capsule, the astronaut has to be provided with conditions resembling those on earth. The capsule is pressurized, so that the body can survive in the near vacuum of space. It is known that under reduced air pressure, the dissolved nitrogen in the bloodstream comes out of solution and forms bubbles, producing a serious condition like the bends, which also affects deep-sea divers. Since there is no oxygen in space, it must be carried along, probably as liquid oxygen. Carbon dioxide must be removed, as must also the liquid wastes from the lungs, skin, and kidneys. Solid wastes must also be provided for. An efficient air-conditioning system is needed to equalize the extremes of temperature resulting from the intense heating effect of the sun on one side, and the freezing condition on the shaded side.

The clothing consists of a special pressure or space suit developed at the request of NASA (National Aeronautics and Space Administration). It weighs over 31 pounds. It is made of double-walled rubberized airtight material; the inner wall is perforated to permit the skin to give off moisture and heat. Air circulates through the suit. For ventures outside of the space capsule, the suit carries 10 extra layers of temperature-resistant nylon covered with thick felt, as protection against the low temperatures, as well as against possible micrometeroroid puncture. Two extra-clear plastic visors cover the face. They are treated to prevent heat leakage from the suit into the vacuum of space, and to protect the astronaut against the glare of the sun.

Electrical wires plugged into the suit, and sensors attached to the astronaut's skin, measure heartbeat, breathing and temperature. Approximately 150 other measurements of pressure, noise, vibration, and acceleration are also radioed to the ground.

On prolonged trips, the astronaut might also have to take into account the "day-night" cycle that normally regulates basic metabolic rates of various functions of the body, i.e., the regular rise and fall of sugar content in the blood, body temperature, and endocrine gland activity. The response of these rhythms is normally related to the presence of daylight. In space, the equilibrium of the body's activity could be upset unless there were artificial control of the light within the space vehicle.

Temperature on the outside of the vehicle may reach 2,600°F, but the astronaut is protected by the double-walled insulated hull. The inside temperature may rise as high as 120° F, which is easily tolerated by a man in an air-conditioned suit, where the flow of cool oxygen keeps the temperature down to 85°.

During the space flight of astronaut Richard F. Gordon, who took a "walk" outside the space capsule in Sept. 1966, his heartbeat was measured as racing at 180 per minute, while his breathing rate rose 40-50 a minute. His spacecraft life support system was not able to absorb the moisture fast enough, and he could not wipe the blinding sweat out of his eyes, By comparison, astronaut Charles Conrad, who remained inside the capsule, had a heart beat of 120, and a respiration rate of 22. Since EVA (extra vehicular activity), or activity outside the spacecraft, may be part of a future astronaut's normal assignment, these observations indicate some of the problems that remain to be solved.

Other studies of astronauts after they returned from space showed that they experienced various physiological effects such as: low blood pressure, inner ear disturbance, loss of red blood cells and lowered calcium content in their bones. These effects seemed to level off after about a week.

During the Apollo 15 moon expedition, in the summer of 1971, it was found that astronauts James B. Irwin, David R. Scott and Alfred Worden experienced a considerable loss of potassium from their bodies. This element is important in the normal functioning of the muscles, including those that make up the heart. Its loss apparently caused the irregular heart beat that affected the astronauts. As a result of this experience, during the Apollo 16 moon trip in April, 1972, electrocardiograms were sent back to earth from space for each of the astronauts, John W. Young, Thomas K. Mattingly and Charles M. Duke. These telemetered records were carefully studied by heart specialists during the entire expedition. The astronauts were also provided with potassium-rich diets, and were equipped with a kit of lidocaine and other heart drugs to treat any cardiac emergencies that might develop.

The supply of food has been suggested as coming not only from tubes of food concentrates, but also from algae such as Chlorella. One pound of dried Chlorella, equal to about five pounds of live Chlorella, provides enough calories for one man per day. Also, these algae liberate enough oxygen for three men, if illuminated continuously. The space vehicle might be equipped with tanks of these algae.

Psychological effects. An astronaut projected into space is living in potentially hostile surroundings, only arm's distance from the vacuum of space. His space vehicle might be threatened by meteors; he might be subjected to primary cosmic radiation. These factors could easily produce a subconscious state of tension. The strange appearance of the sky, and the distance from the earth, could also induce a sense of isolation. As he becomes fatigued, the cramped area of the capsule might seem to become too restricting, and the conditions surrounding him might assume increasing importance. From studies of other explorers who navigated alone, it is expected that there will be periods of monotony, weariness, exhilaration, and possibly hallucinations. Studies have shown that complete isolation from incoming sound, light, or touch can give men hallucinations in only forty-eight hours.

The recent discovery that the earth travels through space in a heavy halo of dust may indicate another psychological hazard to the astronaut. The sound of these particles striking the outer wall of the vehicle will be a constant reminder that a larger particle capable of puncturing the wall may be coming. A puncture would be a serious problem that could affect the pressure, the oxygen supply, and the very life of the astronaut.

Astronaut. In view of all the problems and hazards that face a traveler in space, it would seem that the demands are almost beyond the limitations of human beings. Space travelers, or astronauts, are selected from among many candidates meeting rigid requirements. They are in excellent physical condition, with unusual psychological stability. They are engineers who have flown as jet test pilots; they are under age forty, and no taller than five feet eleven. They each possess an intellect far above the average.

Extraterrestrial life. Are there living things on other planets in our solar system, or outside of it? It has been suggested by some astronomers that life exists on Mars. They think that the seasonal color changes there are supposed to be due to the growth and decline of larger areas of vegetation. Analysis of the atmosphere also seems to point to the presence of hydrocarbon-like materials, another indication of living activity.

Meteorites also provide clues about the possibility of life beyond the earth. Recent studies of these rocks which have fallen to earth from space, reveal that they contain amino acids similar to those from which proteins are formed on earth. But some of the amino acids in the meteorites are different, indicating that they are probably of extraterrestrial origin. These findings indicate that conditions may exist for the chemical formation of life elsewhere in the universe.

There are two problems of space exploration of the planets or of the moon. One is that in landing there, man will contaminate them with microbes from the earth; this would interfere with the accurate study of any original forms of life there. Secondly, there is the possibility of the contamination of humans by foreign organisms, and the return to earth of new diseases.

Scientists have been investigating the belief that possibly life exists somewhere in space among the countless stars that must have planets, just as our sun is a star with its planets. Project Ozma consists of an 85-foot radio telescope at Green Bank, West Virginia, which started tuning in on the sounds from outer space in April, 1960. This telescope has been directed at seven stars, including Alpha Centauri, which are within fifteen light years away. Thus, ideas that were once the imaginative thinking of science fiction writers are now commanding the talents of the most alert scientists.

The puzzle of extraterrestrial life is the subject of the popular book, *We Are Not Alone,* by Walter Sullivan, science editor of *The New York Times.* He concludes the book with these visionary words: "Yet the conclusion that life exists across this vastness (of the universe) seems inescapable. We cannot yet be sure whether or not it lies within reach, but in any case, we are part of it all; we are not alone!"

CHAPTER REVIEW

Select the correct choice for each of the following statements.

1. The increase in speed of a rocket as it takes off is known as (A) zero (B) acceleration (C) gravity (D) deceleration (E) blast off

2. A person at rest is subject to a force of (A) zero G (B) 1 G (C) 2 G (D) 3 G (E) 4 G

3. An acceleration of 4 G produces a pressure against a person weighing 150 pounds equivalent to a force of (A) 146 pounds (B) 150 pounds (C) 154 pounds (D) 450 pounds (E) 600 pounds

4. A person subjected to excessive G force may experience all of the following effects except (A) weightlessness (B) graying out (C) blacking out (D) difficulty in breathing (E) chest pains

5. In the weightless condition, an astronaut (A) may permanently lose up to 11 pounds (B) has no difficulty achieving muscular coordination (C) has difficulty achieving muscular coordination (D) finds his muscular coordination increased (E) finds it easy to drink water directly from a glass

6. People on the earth are protected from primary cosmic rays by (A) their clothing (B) developing a coat of tan (C) sunburn lotion (D) the atmosphere (E) clouds

7. All of the following space hazards threaten an astronaut except (A) electrons (B) Van Allen belts (C) protons (D) X-rays (E) light rays

8. If the pressurized atmosphere of a space capsule were to be suddenly reduced, an astronaut's bloodstream would develop (A) clots (B) bubbles of nitrogen (C) an excess of sugar (D) a deficiency of sugar (E) liquid oxygen

9. An astronaut's clothing consists of (A) a double-walled rubberized suit (B) a single-walled rubber suit (C) a suit sprayed with flat black paint (D) a perforated single layer of nylon (E) a perforated double layer of nylon

10. A planet that may have plant life on it is (A) Mercury (B) Jupiter (C) Mars (D) Alpha Centauri (E) Saturn

ANSWER KEY

1-B	3-E	5-C	7-E	9-A
2-B	4-A	6-D	8-B	10-C

THE VARIETY OF LIVING THINGS ON EARTH

11

35 How living things are classified

Biology is the study of living things. All together, there are probably at least two million different kinds of living things on earth. They live in a great variety of places; in the steaming tropics, in the frozen north, in the temperate zone; high on mountains, deep in the ocean, in deserts, in marshes, at the seashore, in crowded cities. To understand and study these living things, scientists have arranged them in groups. Aristotle first tried to classify organisms over 2000 years ago. The present system of classification was devised by the Swedish biologist Carolus Linnaeus (1707-1778).

Binomial Classification. Linnaeus used two Latin or Greek names for each plant and animal in his system of binomial nomenclature, a *genus* and a *species* name. Thus man is known as *Homo sapiens,* while the dog is *Canis familiaris,* and the house cat is *Felis domestica.* A species is considered to be a closely related group of plants or animals that are similar in structure and can interbreed. A genus consists of one or more species that show many similarities. Thus, the house cat species, *Felis domestica* is grouped in the same genus, *Felis,* as other species of the cat family, *Felis leo,* the lion and *Felis tigris,* the tiger. The genus name is spelled with a capital letter, while the species name usually begins with a small letter.

Modern Views of Classification. Linnaeus assumed that species do not change, and that they represent an ideal type. Today, scientists agree that members of a species show considerable variation from each other. Over a long period of time, they may also become extinct, or change into new species.

Taxonomy is the division of biology that deals with the classification of living things, based on similarities in their structure, development and evolutionary history. There is not universal agreement among all present biologists on a single system of classification. However, it is generally agreed that similar living things that are grouped together have descended from a common ancestor.

One of the most commonly accepted systems classifies living things into three Kingdoms, the Protist Kingdom, the Animal Kingdom and the Plant Kingdom. Each of these is subdivided into large groups called phyla (singular-phylum). The classification scheme follows this arrangement: Kingdom — phylum — class — order — family — genus — species — variety. The smaller the grouping, the more similar the types of individuals will be. Those that are placed in the same genus, for example, will be comparable in structure, and development; they undoubtedly are descended from a common ancestor, and have evolved rather recently, along somewhat different lines from each other.

As an example of the classification scheme, the house cat would be classified as follows:

Kingdom: Animal
Phylum: Chordata (animals with a notochord)
Sub-Phylum: Vertebrates (animals with a backbone of vertebrae)
Class: Mammalia (warm-blooded animals having hair and producing milk)
Order: Carnivora (meat-eating animals with long canine teeth)
Family: Felidae (cat family, with sharp, curved claws)
Genus: Felis (closely related members such as cat, lion, tiger)
Species: domestica (the house cat)
Variety: specific types of cat, such as the Persian cat, Siamese cat, Manx cat, etc.

CLASSIFICATION OF PROTISTS

Protists are classified into phyla of bacteria, protozoa, algae, slime molds and fungi.

PHYLUM	CHARACTERISTICS	EXAMPLES
1. **Bacteria**	Smallest cells known; have a cell wall; may form spores; no nuclear membrane; some have flagella	Bacilli, cocci, spirilla
2. **Protozoa**	Animal-like; single-celled; some form colonies, mostly free-moving.	Ameba, paramecium, euglena, malaria plasmodium
3. **Algae**	Contain chlorophyll; may be unicellular or in colonies	Spirogyra, diatoms; blue-green, brown, red algae
4. **Slime Molds**	Naked mass of protoplasm; many nuclei; no internal cell membranes; spores form flagellate cells which fuse into ameboid form	Physarum
5. **Fungi**	Plant-like; lack chlorophyll; spores	Yeast, molds, mushrooms

CLASSIFICATION OF ANIMALS

Animals are classified according to the following table which includes most of the invertebrate phyla and the vertebrate subphylum.

PHYLUM	CHARACTERISTICS	EXAMPLES
1. **Porifera**	Two-layered organisms with pores; attached	Sponges
2. **Coelenterates**	Two layers of cells; hollow digestive cavity, with tentacles at opening	Hydra, jelly fish, coral, sea anemone
3. **Platyhelminthes** (Flatworms)	Three layers of cells; mostly parasitic	Tapeworm, planaria, liver fluke
4. **Nematoda** (Roundworms)	Thread-like; digestive system includes an anus; many parasitic	Hookworm; Ascaris; Trichina worm

CLASSIFICATION OF ANIMALS (Continued)

PHYLUM	CHARACTERISTICS	EXAMPLES
5. Annelida (Segmented worms)	Long, cylindrical body with ringed segments; nervous system; circulatory system	Earthworm; leech
6. Echinodermata (Spiny-skinned)	Radial symmetry; spiny exoskeleton; marine dwelling	Starfish, sea urchin, sea cucumber
7. Mollusca	Soft-bodies, mostly protected by a shell	Clam, snail, octopus
8. Arthropoda	Jointed legs; exoskeleton; segmented body	
Class 1. Crustacea	Gills for breathing; two pairs of antennae; jointed legs with claws	Crab, lobster
Class 2. Arachnida	Eight legs; two body parts; no antennae	Spider, scorpion, tick
Class 3. Insecta	Six legs; three body parts (head, thorax, abdomen); one pair of antennae	Butterfly, bee, grasshopper
Class 4. Chilopoda	Each segment with one pair of legs	Centipede
Class 5. Diplopoda	Two pairs of legs per segment	Millipede
9. Chordata	Possess a notochord, supporting rod structure during life history; dorsal nerve cord; gill slits	Sub-phyla: Vertebrates; Amphioxus; Tunicates; Acorn worms.
SUBPHYLUM Vertebrate	Backbone enclosing spinal cord	
Class 1. Pisces (Fish)	Most have bony skeleton; gills; scales; two-chambered heart	Salmon, trout, cod
Class 2. Amphibia	Live in water in early stages, with gills; develop lungs; thin moist skin; three-chambered heart	Frog, toad, salamander
Class 3. Reptilia	Lungs; dry scales; eggs with a horny covering; cold-blooded; mostly three-chambered heart	Snake, lizard, turtle
Class 4. Aves	Feathers; wings; warm-blooded; four-chambered heart; eggs with shell	Sparrow, chicken, ostrich

CLASSIFICATION OF ANIMALS (Continued)

PHYLUM	CHARACTERISTICS	EXAMPLES
Class 5. Mammalia	Warm-blooded; hair; diaphragm; young born alive, fed on milk	Human, dog, bat, whale

CLASSIFICATION OF PLANTS

Plants are green and multicellular, and generally live on land. There are two phyla.

PHYLUM	CHARACTERISTICS	EXAMPLES
1. Bryophyta	Simple stems; rootlike and leaf-like structures; alternation of generations; produce spores	
Class 1. Hepaticae (Liverworts)	Flat, leaf-like structure; simple rootlike structures	Marchantia
Class 2. Musci (Mosses)	Usually erect with stem; capsule contains spores	Sphagnum
2. Tracheophyta	Have a vascular system; true roots, stems, leaves.	
SUBPHYLUM 1. Lycopsida	Creeping stems	Club moss; ground pine
SUBPHYLUM 2. Sphenopsida	Scale-like leaves; spores in cone-like structures	Horsetail; scouring rush
SUBPHYLUM 3. Pteropsida	Broad-leaved	
Class 1. Filicineae (True ferns)	Spores on leaves or special structures	Royal fern; Boston fern 
Class 2. Gymnospermae	Naked seeds, often in a cone; needle-like leaves, mostly evergreen	Pine; hemlock; spruce 
Class 3. Angiospermae	Seeds enclosed in fruit or nut; broad leaves, not evergreen	
Sub-class 1. Dicotyledons	Net-veined leaves; seeds with two cotyledons	Maple; rose; bean
Sub-class 2. Monocotyledons	Parallel-veined leaves; seed with one cotyledon	Corn; lily; grass

CHAPTER REVIEW

Select the correct choice for each of the following statements.

1. Linnaeus originated the system of classification which employed (A) bilateral symmetry (B) binomial nomenclature (C) continuity of germplasm (D) conservation of energy (E) valence of elements

2. The division of biology that deals with classification is (A) cytology (B) histology (C) botany (D) morphology (E) taxonomy

3. Of the following, the most closely related group of organisms is known as a (n) (A) genus species (C) order (D) family (E) class

4. The number of protist phyla is (A) 3 (B) 5 (C) 6 (D) 7 (E) 8

5. Classification is based on all of the following except (A) structure (B) development (C) evolutionary history (D) common ancestry (E) size

6. The correct sequence of the following groups (1-class; 2-family; 3-phylum; 4-order; 5-genus; 6-species) is
(A) 3–4–1–6–5–2 (B) 3–2–4–1–5–6
(C) 3–1–2–4–5–6 (D) 3–1–4–2–5–6
(E) 3–4–1–5–6–2

7. The Latin words of the name given to man, *Homo sapiens,* include the (A) genus and species (B) genus and family (C) family and order (D) order and class (E) genus and class

8. Of the following, the animal not included in the same genus as the house cat is (A) lion (B) tiger (C) dog (D) leopard (E) Siamese cat

9. Two invertebrate phyla are (A) protozoa and thallophyta (B) protozoa and mammalia (C) protozoa and arthropoda (D) protozoa and pteridophyta (E) coelenterates and bryophytes

10. Protists include (A) algae and mosses (B) algae and protozoa (C) protozoa and mosses (D) algae and sponges (F) protozoa and sponges

ANSWER KEY

1-B	3-B	5-E	7-A	9-C
2-E	4-B	6-D	8-C	10-B

36 The Protist Kingdom

Originally, living things were classified into either the Animal Kingdom or the Plant Kingdom. However, recent investigations have shown that certain groups of living things do not fit easily into these two kingdoms. Some types of protozoa appear to have characteristics of both. Organisms like bacteria and slime molds, are radically different from either kingdom. Consequently, there has been general agreement among some, but not all biologists, that a third kingdom, the Protista, should be set up.

The Protists include smaller living things whose classification is not clear cut. They are believed to have appeared early in evolutionary history. They are composed essentially of single cells, or colonies of cells. In some cases, they show a *multinucleate* condition, in which there are many nuclei distributed throughout the cytoplasm, without intervening cell membranes, or cell walls.

In general, their cells differ from those of higher animals and plants. They are such an uncertain group, that some of them (protozoa) may be considered by some biologists to be animals, while others (algae and fungi) may be considered to be plants. There are some types (flagellates) that are incuded in either kingdom, because they have characteristics of both — they possess chlorophyll, and move about by flagella.

Phylum 1 **Bacteria.** Bacteria are the smallest cells known, varying from 1 to 3 microns (μ) in size, compared to about 10 microns for most other cells (There are 1,000 microns in a millimeter). Some bacteria are the cause of serious diseases in man. Others are useful allies in the decay of organic substances in the soil, and in the dairy industry. Bacteria are so important in the affairs of man, that they will be considered in greater detail, in a later section of the book (Unit 6).

Phylum 2 **Protozoa.** This phylum includes about 15,000 different species. Most of them live in fresh water, some in the ocean, and some are parasitic.

Class 1 Sarcodina. The best-known member of this class is the *ameba.* It consists of a little mass of protoplasm without any particular shape. It moves by forming projections or *pseudopodia*

(false feet) and flowing into them. As it moves along, it surrounds and engulfs its food. Some members of this class form a skeleton of lime or silica. The chalk cliffs of Dover were once under water, and are composed of the remains of countless numbers of such shell-forming protozoa.

Class 2 **Flagellata.** These protozoa have long hairlike projections, *flagella,* which beat in whiplike fashion, and by which they move.

Euglena is unusual in having chloroplasts, a characteristic of plants. It also has a red eyespot that is sensitive to sunlight. The *trypanosome* causes African sleeping sickness and is spread by the bloodsucking tsetse fly. Another type of flagellate lives in the intestines of termites and digests the wood that these insects eat. Without them, the termites would not be able to make use of the wood, and would starve.

Ameba Euglena Vorticella

Class 3 **Infusoria.** The paramecium is undoubtedly the best-known member of this group. They all move by *cilia* and have two kinds of nuclei. The cilia are used both to move, and to create currents in the region of the mouth for feeding. As food enters the body, it forms into a *food vacuole,* where digestion occurs. Two *contractile vacuoles* at either end eliminate excess water. Another ciliate, *Vorticella,* has a row of cilia at one end of its bell-shaped body and a sensitive stalk at the other end which contracts when it is disturbed.

Class 4 **Sporozoa.** These are parasitic protozoa, such as the malaria *plasmodium,* which form spores. In this process, the nucleus divides many times; a little cytoplasm collects around each nucleus, and a number of different spores are produced. These spread. The plasmodium is carried by the female Anopheles mosquito. When it enters the human body, it invades and destroys red blood corpuscles.

Phylum 3 **Algae.** Algae are simple plants that contain chlorophyll, and so are capable of making their own food. About 20,000 different algae have been classified. A well-known type of green algae is *Spirogyra,* which is found in fresh-water ponds. It consists of single cells arranged in a long silky thread or filament; the chloroplast is present as a spiral ribbon. Another example is *Pleurococcus,* commonly found growing on tree trunks, and mistakenly called moss. Other common types include single-celled *desmids,* and *diatoms.* The latter have a hard, sculptured wall and are responsible for thick deposits of diatomaceous earth which have accumulated over the ages; this earth now is used for toothpaste, polishes and filters.

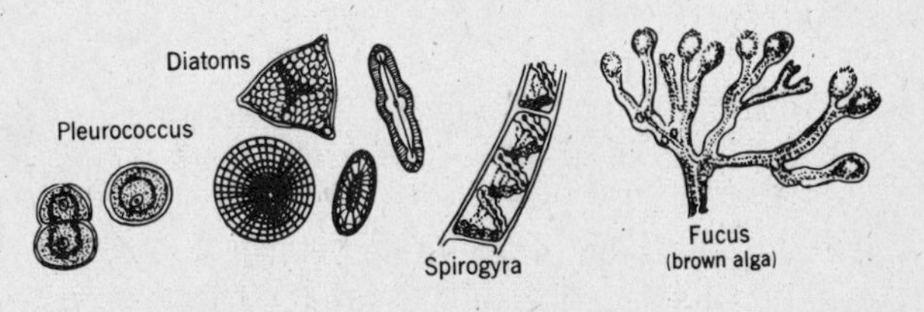

Types of Algae

In addition to the green chlorophyll, algae contain various other colored pigments. Blue-green algae are extremely simple in structure and include *Oscillatoria,* which swings to and fro slowly. Brown algae and red algae include most of the seaweeds. The rocks of the seashore are covered with deposits of brown algae. *Kelps* are brown algae that may reach a length of over a hundred feet. Agar-agar is a gelatinous substance used for growing bacteria that is obtained from red algae growing in the Pacific Ocean.

Algae make up a good part of plankton, the floating mass of microscopic food in the sea, upon which small animals and in turn, the larger fish of the oceans depend for their existence.

Phylum 4 **Slime Molds.** This phylum consists of organisms that have characteristics of plants and animals. They consist of a naked mass of protoplasm called a *plasmodium.* It spreads slowly over decaying logs, somewhat like a huge ameba. It may achieve a size of up to a foot in diameter. Its streaming cytoplasm contains many nuclei. Its mass is not subdivided into cells.

At times, the plasmodium begins to resemble a fungus. It forms spore cases, or *sporangia,* on short stalks. Spores are released, and in some cases, germinate into flagelate cells. These cells then fuse to form an ameboid, multinucleate mass, which becomes a new plasmodium.

The common slime mold *Physarum* may be found in the woods, growing in a moist area on a fallen log or decaying tree stump, as a glistening, bright-yellow plasmodium. It can be cultured in the laboratory on moist toweling, or mimeograph paper in a covered battery jar. It can be fed simply by sprinkling oatmeal powder over it.

Phylum 5 **Fungi.** Fungi are simple, plant-like organisms that do not contain chlorophyll, and therefore cannot make their own food. Many are *parasites,* attaching themselves to other living things as a source of food; they injure and sometimes kill their host. Others are *saprophytes,* obtaining their food from non-living organic matter. Examples — mushrooms, cheese molds and yeast.

About 65,000 species of fungi have been identified and arranged in several classes. Yeast plants, are single-celled. They are useful in the baking of bread and cakes, by making the dough rise. They also give beer its characteristic flavor. These activities result from its *fermentation* of sugar, giving rise to the products carbon dioxide and alcohol.

Another class includes larger fungi, the molds. Some of them are useful in the making of Roquefort cheese, and in the production of antibiotics (i.e., *Penicillium notatum*). Others attack foods such as bread and fruits, spoiling them. A few are parasitic, and cause athlete's foot, and ringworm. Molds reproduce by spores, and form a mass of thread-like structures (mycelium) as they grow.

Mushrooms, smuts and rusts constitute another group of fungi. Wheat and corn crops are seriously affected by rust and smut parasites. Mushrooms obtain their food from organic materials in the soil. The undersurface of a mushroom cap contains sections or gills where spores are produced. Poisonous mushrooms, or toadstools may be recognized by the ring or collar below the cap. Some mushrooms are edible, but usually, only an expert can tell whether a mushroom is poisonous or not.

Lichens really consist of two organisms, a fungus and one-celled algae, living together in a helpful partnership. The algae make food, while the fungus furnishes moisture and minerals. This mutually helpful relationship is known as *mutualism.* Lichens are found on the surface of rocks and logs. An important example of a lichen is reindeer moss, an important food for reindeers, which is not a moss at all.

CHAPTER REVIEW

Select the correct choice for each of the following statements:

1. Protista are (A) more complex than plants but simpler than animals (B) more complex than animals but simpler than plants (C) more complex than both plants and animals (D) simpler than plant and animals (E) too complex to classify.
2. Protists are classified separately from plants and animals because (A) they appeared later in evolutionary history (B) their classification is not clearcut (C) they are all multi-celled (D) Linnaeus did not know of their existence (E) they are harmful to man.
3. Some Protists (A) have characteristics of both plants and animals (B) are invisible (C) are plants at one time and then turn into animals (D) are animals at one time and then turn into plants (E) originated in other planets.
4. The organism which is *not* included in the same phylum as the others is (A) ameba (B) vorticella (C) paramecium (D) physarum (E) trypanosome
5. One-celled green protists are known as (A) bacteria (B) algae (C) fungi (D) rusts (E) slime molds.
6. The lichen is an example of (A) an alga and a fungus (B) a protozoan (C) bacteria (D) algae and a slime mold (E) algae and yeast.
7. All of the following are algae except (A) pleurococcus (B) diatoms (C) ameba (D) spirogyra (E) kelps.
8. Plankton is made up largely of (A) bacteria (B) bacilli (C) algae (D) cocci (E) slime molds.
9. All of the following are examples of fungi except (A) yeast (B) mushroom (C) smuts (D) rusts (E) vorticella.
10. The smallest known cells are in the phylum (A) bacteria (B) protozoa (C) algae (D) slime molds (E) fungi.

ANSWER KEY

1-D	3-A	5-B	7-C	9-E
2-B	4-D	6-A	8-C	10-A

37 The animal kingdom

The study of animals is known as *zoology.* About a million different species of animals have been named.

Phylum 1 **Porifera.** Sponges, or "pore animals" are simple, many-celled animals which do not move and which are attached to rocks or sticks. Water enters through many pores into the hollow central cavity, and eventually leaves through a large opening (excurrent pore) at the top. The current of water is created by the beating of flagella located in *collar cells* lining the cavities of the sponge.

As the water streams by them, these cells engulf particles of food and digest them. The continuous current of water passing through a sponge also supplies oxygen to the cells, and removes carbon dioxide and nitrogenous wastes.

Most sponges live in the ocean, although a few species are found in fresh water. Commercial sponges are collected in the warm shallow waters near Florida, the Bahamas, and in the Mediterranean Sea. When alive, they resemble slimy pieces of raw liver. They are prepared for commercial use through the decomposition of the soft protoplasmic material by bacteria. The final product, the bath sponge, consists of the fibrous skeleton. Other members of the phylum: The glass sponge, Venus flower basket; fresh-water sponge, Spongilia.

Phylum 2 **Coelenterata.** This group includes jellyfish, hydra, sea anemone and coral. These animals have a central cavity, and consist of two layers of cells, an outer *ectoderm,* and an inner *endoderm.* The only body opening, the mouth, is surrounded by a ring of tentacles containing stinging cells (*nematocysts*). When small animals are paralyzed by these cells, they are carried to the mouth by the tentacles. Digestion is carried on by the endodermal cells. Indigestible remains are eliminated through the mouth. There are also primitive muscle, nerve and gland cells in the body of these animals.

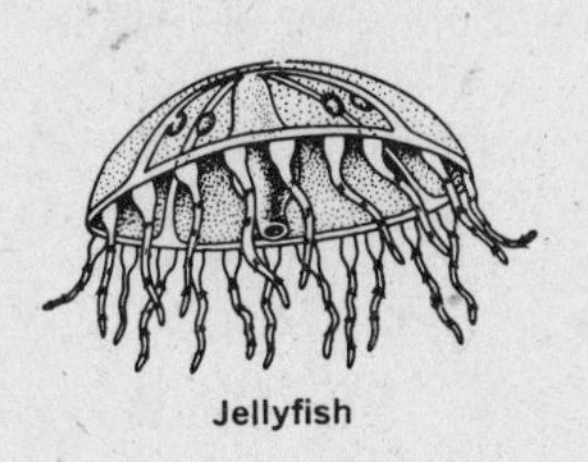

Jellyfish

Jellyfish vary in size, from some that are microscopic to others that grow as large as a man. They are made of 99 per cent water, and move by a pulsating movement. When they come in contact with a person swimming in the water, their stinging cells may cause a severe reaction. *Hydra* is a fresh-water coelenterate that looks like a piece of string, less than an inch long, with frayed ends where the tentacles are. When disturbed, it contracts to the size of a pinhead. It feeds on water fleas and small worms. The corals are colonial forms that take lime out of the water and secrete limestone on the outside of their ectoderm layer. Accumulations of this coral skeleton in tropical waters form coral reefs and even whole islands.

Phylum 3 **Platyhelminthes** (*Flatworms*). Beginning with the flatworms, all the higher animals have a third layer, the *mesoderm* between the ectoderm and the endoderm. This layer produces the muscles and the reproductive organs. Definite organs also begin to form, each composed of several types of tissues.

Planaria are small flatworms about a half an inch long that live in fresh water under stones. They have the beginning of a nervous system, with a concentration of nerve tissue in the head comprising a simple brain. They also have a digestive system, and both male, and female reproductive organs.

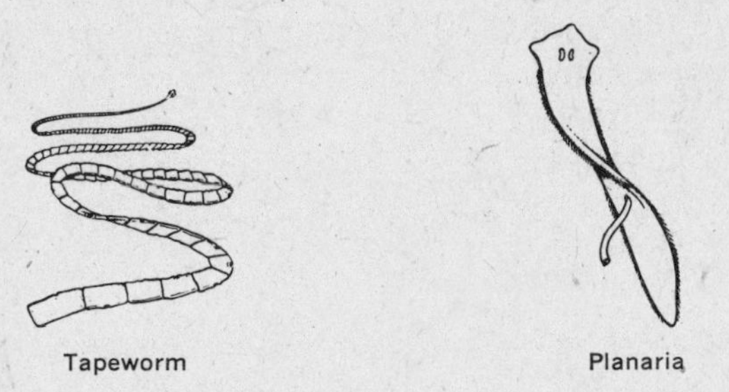

Tapeworm Planaria

The *tapeworm* is a long, flat, ribbon-like animal made up of sections, which lives as an adult in the intestines of man, and may reach a length of more than twenty feet. It attaches itself to the wall of the intestine by means of suckers and hooks on its head. It does not have a digestive system, but absorbs digested food from its host directly into its body. The mature sections drop off when full of eggs and are passed out of the body with solid wastes. Another host, a cow, may take these eggs into its body as it grazes, and may also become infected. Thus, the beef tapeworm lives in two hosts.

Another parasitic flatworm, the *blood fluke,* is a serious menace in the Orient, where it infects millions of people during rice-planting time. The worm bores through the skin of their legs as they stand barefooted in the shallow water of a rice field. It eventually settles in the blood vessels of the intestines. It causes body pains, severe dysentery, anemia, and general weakness. A snail is the intermediate host of this fluke. Another parasite, the Chinese liver fluke, has two other hosts besides man, a fish and a snail.

Phylum 4 **Nematoda** (*Roundworms*). The roundworms or threadworms, have a digestive canal with a mouth opening at one end, and an anus at the other. Many are parasitic on plants as well as on animals. The *hookworm* causes a serious health problem in the South, where it lives in the soil and bores its way through the soles of people walking barefoot. It travels through the bloodstream, heart and lungs before finally boring into the intestines where it attaches itself and feeds on blood. Victims show anemia, listlessness and even a retardation of mental and physical development.

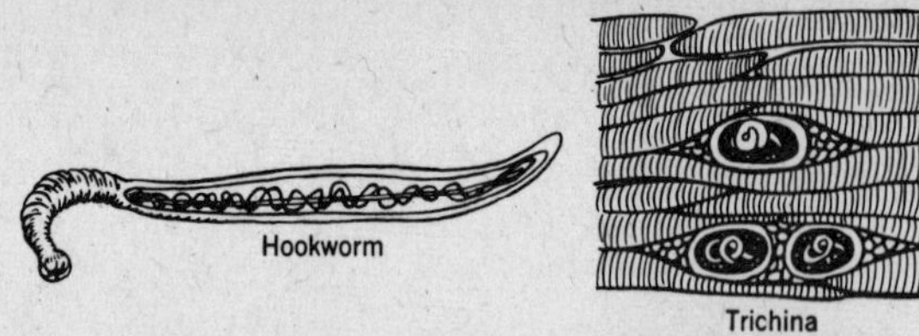

Trichina
(encysted in human muscle)

The *trichina* worm is a roundworm parasite that is the cause of *trichinosis*. It enters the body when a person eats undercooked infected pork. The worms bore into the skeletal muscles causing intense pain, fever and swelling. Another roundworm, the *filaria,* is carried by a mosquito, and blocks the lymph channels, causing tremendous swellings of the affected parts, and resulting in a disease known as *elephantiasis.*

Phylum 5 **Annelida** (*Segmented Worms*). The third group of worms are called segmented because their body structure is arranged in rings or segments. The digestive tube runs the length of the worm's body, and constitutes a tube within a tube. The muscles are well-developed for wriggling. The circulatory system has blood vessels with muscular walls for contracting and driving the blood along. The nervous system is well developed, including a brain, a long nerve cord, and nerve centers or *ganglia* in each segment.

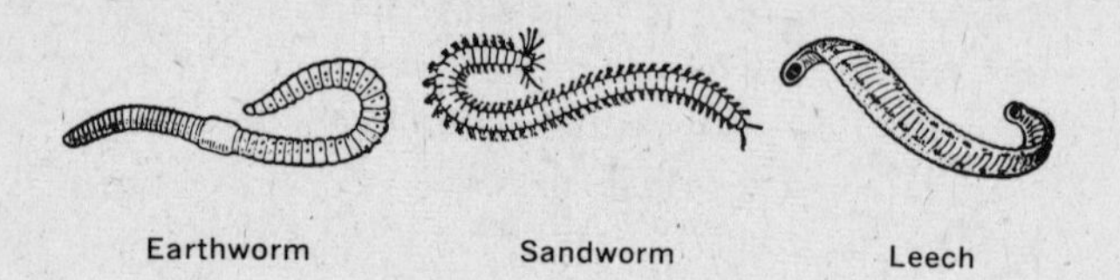

The *earthworm* is useful because of its habit of turning over the soil and making it porous. It does this by swallowing earth which passes through its digestive canal where the organic material is digested and absorbed. The *sandworm* (Nereis) is a marine relative of the earthworm. It has two pairs of eyes, and a pair of extensions or simple paddles on each segment for swimming. The *leech* has a sucker at each end of its body by which it attaches itself to an animal, and then sucks its blood. While it is feeding, it secretes the chemical *hirudin,* which prevents the blood from clotting.

Phylum 6 **Rotifers** are microscopic animals of little economic importance, that are somewhat more complicated than the flatworms in some ways, and less so in others. They are grouped in the phylum *Rotifera,* meaning wheel-bearers, and take their name from the beating crown of cilia at the head end. Although they are no larger than the one-celled paramecium, rotifers are many-celled. They possess a digestive tube with a mouth and an anus. The whole body can be folded together in telescopic fashion when they contract. They reproduce sexually. They are commonly found in pond water.

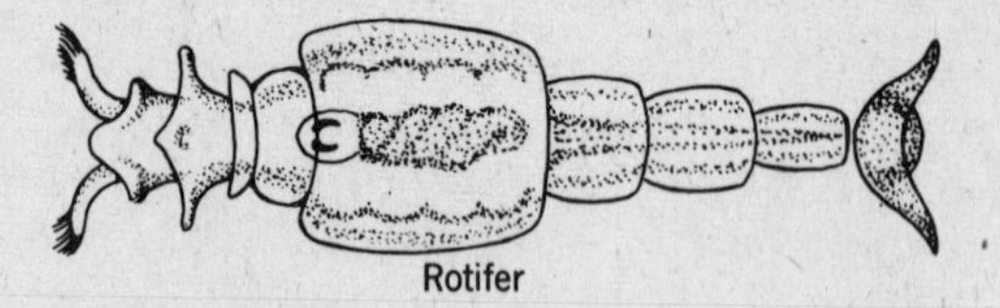
Rotifer

Phylum 7 **Echinodermata.** The body plan of these animals is different from that of other animals in that it shows radial symmetry. There is a central disc from which five or more arms radiate, like the spokes of a wheel. The *starfish* is a well-known example. Its mouth is located on the undersurface part of the disk. It has the characteristic spiny skin of the phylum. Locomotion is achieved by means of a water vascular system which has canals going into each of the arms.

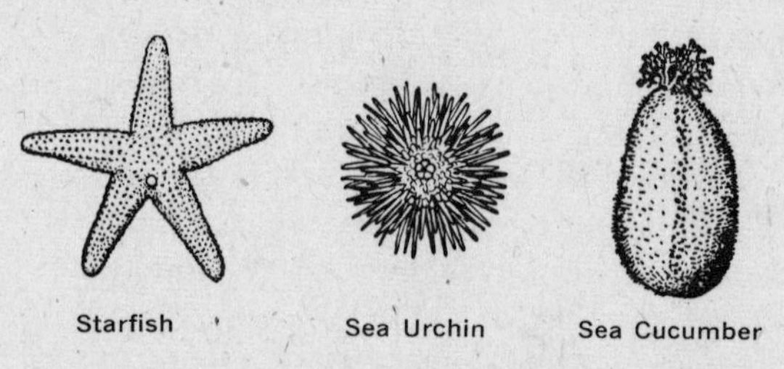

The starfish is quite harmful to the oyster industry. It feeds by slowly forcing the shell of an oyster open with its many small tube feet. Then it turns the lower part of its stomach inside out, extends it through its mouth, and envelops the soft body of the oyster, digesting it. Oyster fishermen used to hack starfish to pieces in order to get rid of them. However starfish have unusual powers of regeneration and can re-form a new organism from a piece containing only one arm and a small part of the disk.

Other members of this phylum: *Sea urchins,* have so many spines that they resemble a pin cushion; *sea cucumbers* are elongated, with finely branched arms around the mouth; *sea lilies* are deep-sea forms that have a stalk with branched arms at one end.

Phylum 8 **Mollusca.** The mollusks are soft-bodied animals, many of which have a shell. They have a muscular foot, and a heavy fold of tissue, the mantle, which secretes the shell. There are three major types; the snail, having a coiled shell; the oyster and clam, with two shells or valves—hence the name bivalve; and the octopus and squid, with a thin horny shell on the inside, and a foot containing tentacles. About 80,000 different species of mollusks have been described. Most of them live in salt water, while some are found in fresh water and on land. Clams, oysters and some of the other mollusks make up our large supply of invertebrate food.

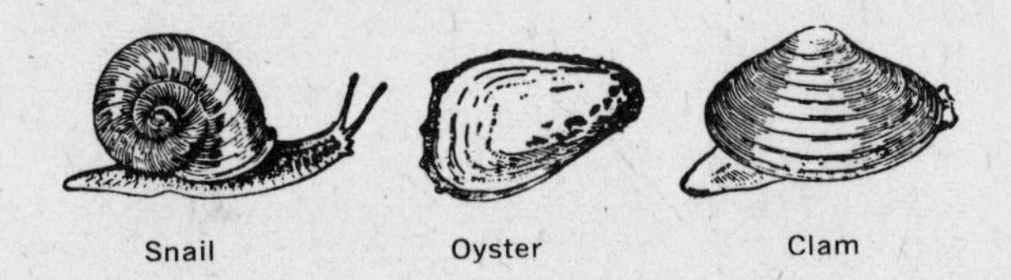

Snail Oyster Clam

Snails are generally scavengers scraping fragments of food as they glide along slowly on their foot. Some types are destructive to garden plants. Clams, oysters, and mussels feed by drawing in currents of water between the slightly open shells and straining the food particles. When disturbed, they shut their shells tightly closed by means of two large muscles at either end. People who eat "scallops," are really eating such muscles. If an irritating substance such as a grain of sand becomes located between the mantle and the shell of an oyster, a secretion is formed around it to produce a pearl. Buttons are made from the shells of clams.

The squid and octopus have suckers on their tentacles for seizing their prey. When attacked, they contract their ink sac, giving off a cloud of inky material. They can dart backward with considerable speed by expelling a jet of water. They have well-developed eyes, which, like man's, are built on the same principle as a camera.

Phylum 9 **Arthropoda.** This is the largest phylum of the invertebrates. Its members have three or more pairs of jointed legs, an outer skeleton (exoskeleton) made of *chitin,* and segmented bodies. They are arranged in five classes, including crabs and lobsters, insects, spiders, centipedes and millipedes.

Class 1 **Crustacea.** Examples: Crab, lobster, crayfish, barnacle, shrimp. The head and thorax are fused together and are distinct from the abdomen. These animals have two pairs of antennae or feelers. They breathe by means of gills. Their jointed legs may include large pincers for holding and crushing, as in the lobster. In addition to five pairs of jointed legs for walking, they also have additional appendages for swimming and eating. Because they have a rigid exoskeleton, they grow by molting, or shedding their skin.

Class 2 **Insecta.** There are more species of insects than all the other animals on earth put together. They are considered man's closest competitor because of: their success in living practically everywhere; the damage they do to his crops, clothing and buildings; the diseases they transmit; their enormous powers of reproduction.

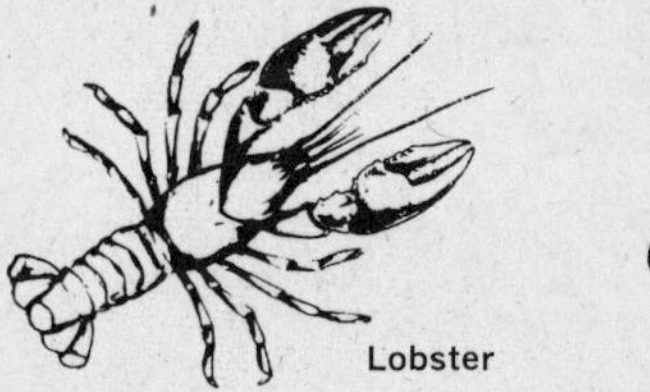

Lobster

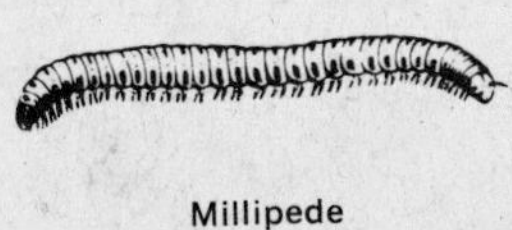

Millipede

Insects have three pairs of legs, one pair of antennae, and a body with three distinct parts—head, thorax and abdomen. They have compound eyes with many lenses, as well as simple eyes. Some of them, such as butterflies and bees, have four wings; flies and mosquitoes have two wings; fleas have no wings. Some (butterflies and mosquitoes) have four stages in their life history, or complete metamorphosis, including egg, larva, pupa, and adult. Others (grasshoppers and plant lice) have three stages or incomplete metamorphosis, including egg, nymph and adult. Some (bees, wasps, ants) live in social communities with a high degree of organization. Serious diseases are transmitted by: Anopheles mosquito (malaria); Aedes mosquito (yellow fever); body louse (typhus); flea (bubonic plague); tsetse fly (African sleeping sickness).

Class 3 **Arachnida.** Spiders have four pairs of legs, and two main body parts—fused head and thorax, and an abdomen. They have no antennae. Many of them have silk glands for spinning webs. Other members of this class: ticks, scorpions, horseshoe crabs.

Spider

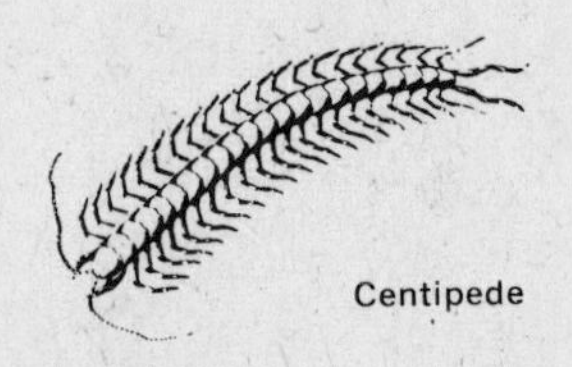

Centipede

Class 4 **Chilopoda.** The centipedes ("hundred-leggers") have one pair of legs on each segment. They can move rather rapidly. One pair of appendages contains poison claws. They are carnivorous, and live on earthworms and insects.

Class 5 **Diplopoda.** Millipedes ("thousand-leggers") have two pairs of legs for each segment. Despite the large number of legs, they move slowly. They are plant-eating scavengers.

Phylum 10 **Chordata.** Chordates are animals that have a notochord, which is a cartilage-like supporting structure along the back, in some stage of their life history. Above the notochord is a hollow nerve cord. They also develop gill slits. There are three small sub-phyla including animals known as: the lancelet (amphioxus); sea squirt (tunicate); and acorn worm. The fourth sub-phylum is the large group of vertebrates, animals that develop a backbone to replace the notochord.

Sub-phylum **Vertebrata.** Vertebrates are animals with a backbone or spinal column composed of sections called vertebrae, and enclosing a spinal cord. The brain is contained in a bony skull, or *cranium.* The vertebrates are classified into the following five important classes:

Class 1 **Pisces.** Fish are cold-blooded animals that live in either salt or fresh water. They breath by means of gills; as water passes over the gills, dissolved oxygen is taken in by the many tiny blood vessels located there and carbon dioxide is given off. They have a two-chambered heart, consisting of an auricle and a ventricle. Fish move by means of paired fins which are like the front and hind limbs of higher animals, and a tail. The streamlined shape of the body permits them to glide through the water with a minimum of resistance. Fish possess an air bladder which helps them rise, descend, or remain stationary. The surface of the body of most fish is covered with scales which are made slippery by mucous glands in the skin. The nostrils are used for smell.

Most fish lay eggs which are fertilized outside of the body. Codfish may produce several million eggs in one year. Migratory ocean fish such as the salmon travel long distances to breed in the fresh waters of rivers. By contrast, eels leave inland waters to breed in the ocean. Lungfish are unusual fish that bury themselves in the mud during the dry season and breathe with their air bladder. Bony fish are classified as the sub-class Osteichthyes. Another group of fish, which have a skeleton made of cartilage, include sharks and rays. They are classified in the sub-class Chondrichthyes.

Pisces

Amphibia

Class 2 **Amphibia.** The amphibians, such as frogs, toads and salamanders, have two stages in their life history. They develop from eggs laid in the water, and fertilized externally. During the *tadpole* stage, they breathe by means of gills. They later form legs and lungs, and live on land. This completes their *metamorphosis,* or life changes. The skin is thin and moist, permitting the absorption of some oxygen through it. The heart has three chambers, consisting of two auricles and one ventricle. The red corpuscles are oval cells with nuclei. Food, consisting of insects and worms, is caught by means of the sticky tongue which is flipped out from its place of attachment at the front part of the mouth. Toads, with a rough skin, do not cause warts, as some people believe, and can remain in dry places for a longer period of time than frogs or salamanders.

Class 3 **Reptilia.** Examples—snake, lizard, turtle, alligator. These animals are cold-blooded, and breathe by lungs. Their bodies are generally covered with dry scales. Fertilization is internal, but development of the young from eggs generally takes place externally. The heart is three-chambered, and becomes four-chambered among the alligators. Dinosaurs are extinct reptiles that dominated the earth millions of years ago. Although snakes have no legs, they can move fairly rapidly by wavelike contractions of muscles attached to the numerous ribs and scales. Most snakes are useful, judging by the many insects and mice they catch. The few harmful ones, such as the rattlesnake and the copperhead, have a pair of hollow fangs in their upper jaw through which poison is injected when they bite. Prompt first-aid treatment and antivenin are needed for snakebite.

Reptilia

Aves

Mammalia

Class 4 **Aves.** Birds are best-known for their feathers and wings. They are warm-blooded, and have a four-chambered heart. They lay eggs which are covered with a shell, and which they incubate with their bodies until the young are born. Their beaks are horny and do not contain teeth. Parts of their skeleton are hollow, thus assisting in their flight. Bird migration, which sometimes involves a flight of thousands of miles, is still an unexplained phenomenon.

The ostrich and the penguin are birds that cannot fly. Birds of prey like the hawk and owl have strong curved beaks. Ducks have a flattened beak for straining food from water. Woodpeckers have powerful beaks for drilling into trees for insects. Songbirds have slender beaks for catching insects. Birds constitute one of our strong defenses against the insect menace.

Class 5 **Mammalia.** Mammals are the most highly developed animals. They are warm-blooded and have a four-chambered heart. They breathe by lungs, and have a diaphragm which divides the body into two distinct cavities, the thorax or chest, and the abdomen. They are covered with hair. The young are born alive after developing in a special structure, the uterus, and are fed with milk secreted by mammary glands.

These characteristics apply to all mammals, including man, dog, whale, bat, and elephant. The duckbill platypus and the anteater, however, are exceptions that lay eggs which hatch outside of the mother's body; they then nurse the young on milk. The pouched mammals, such as the kangaroo and the opossum, give birth to premature young that they place in a pouch to complete their development and nourishment. Sea-living mammals such as the whale, porpoise and seal have limbs, modified as flippers. The blue whale, the largest animal on earth, grows more than 100 feet long and weighs 150 tons. Other mammals differ in their tooth structure: grinding teeth—hoofed mammals such as cow, deer, giraffe; large canine teeth—bear, dog, lion; long front incisor teeth for gnawing—rodents such as rat, beaver, squirrel.

The primates are mammals with the most highly developed brain, that walk more or less erect. Their limbs have a hand with an opposable thumb for grasping. Examples are: man, chimpanzee, gorilla, monkeys. The human, with the most highly developed brain of all, stands at the top of the animal kingdom, and is classified as *Homo sapiens*—man, the wise. All human beings today are members of this one genus and species.

CHAPTER REVIEW

Select the correct choice for each of the following statements.

1. Simple animals that contain pores leading to a hollow central cavity, are classified as (A) hydra (B) coelenterates (C) molluscs (D) sponges (E) echinoderms

2. Two warm-blooded groups of vertebrates are (A) mammals and sharks (B) birds and amphibia (C) mammals and birds (D) amphibia and reptiles (E) mammals and amphibia

3. The science dealing with the study of animals is known as (A) physiology (B) physiography (C) psychology (D) zoology (E) botany

4. An animal that is *not* an invertebrate is (A) jellyfish (B) starfish (C) goldfish (D) sponge (E) plasmodium

5. Hydra and coral are classified as members of the (A) Coelenterata (B) Porifera (C) Planaria (D) Infusoria (E) Echinodermata

6. Three phyla that include worms are (A) Annelida, Nematoda, and Flagellata (B) Annelida, Sarcodina, and Platyhelminthes (C) Annelida, Nematoda, and Platyhelminthes (D) Annelida, Flagellata, and Sarcodina (E) Annelida, Flagellata, and Platyhelminthes

7. All of the following are parasitic worms except (A) filaria (B) leech (C) liver fluke (D) hookworm (E) ringworm

8. Animals without backbones are classified as (A) invertebrates (B) vertebrates (C) protozoa (D) molluscs (E) rotifers

9. A phylum that is characterized by radial symmetry is (A) Chordata (B) Echinodermata (C) Pisces (D) Aves (E) Arthropoda

10. The animal that is *not* included in the same phylum as the others is (A) starfish (B) sea urchin (C) sea lion (D) sea cucumber (E) sea lily

11. The octopus, clam, and snail are alike in (A) having a backbone (B) having jointed legs (C) being classified as mollusks (D) being classified as vertebrates (E) having a three-chambered heart

12. Animals having a chitinous exoskeleton, and jointed legs are classified as (A) turtles (B) echinoderms (C) coelenterates (D) arthropoda (E) chordates

13. Of the following, the closest relative of the lobster is the (A) grasshopper (B) squid (C) sea horse (D) oyster (E) eel

14. Spiders may be distinguished from insects because they have (A) two antennae (B) jointed legs (C) eight legs (D) a separate head, thorax, and abdomen (E) biting mouth parts

15. Animals that possess a notochord are classified as (A) social insects (B) crustacea (C) chilopoda (D) notobrates (E) chordates

16. All of the following are classes of vertebrates except (A) Pisces (B) Amphibia (C) Reptilia (D) Arthropoda (E) Mammalia

17. An animal with a four-chambered heart is the (A) lungfish (B) whale (C) codfish (D) eel (E) shark

18. An animal that has a tadpole stage in its life history is (A) toad (B) turtle (C) tortoise (D) tiger moth (E) lizard

19. Although the duckbill platypus lays eggs, it is classified as (A) bird (B) reptile (C) rodent (D) mammal (E) primate

20. Of the following, the closest relative of the bat is the (A) robin (B) seal (C) crow (D) penguin (E) ostrich

21. Two classes of vertebrates that always breathe by lungs are (A) birds and amphibia (B) birds and insects (C) mammals and amphibia (D) mammals and insects (E) birds and mammals

22. The characteristic that makes a kangaroo a mammal is (A) internal fertilization (B) external development of its young (C) backbone (D) production of milk (E) presence of lungs

23. All of the following are primates except (A) chimpanzee (B) gorilla (C) giraffe (D) human (E) monkey

24. Animals that have a pupa stage in their development are (A) protozoa (B) insects (C) crustacea (D) segmented worms (E) frogs

25. The most numerous group of animals is the (A) amphibia (B) protozoa (C) worms (D) birds (E) insects

ANSWER KEY

1-D	6-C	11-C	16-D	21-E
2-C	7-E	12-D	17-B	22-D
3-D	8-A	13-A	18-A	23-C
4-C	9-B	14-C	19-D	24-B
5-A	10-C	15-E	20-B	25-E

38 The plant kingdom

BOTANY is the study of plants. About a third of a million species have been named and classified.

Phylum 1 **Bryophyta.** Bryophytes are small green land plants that include the mosses and liverworts. They are more complex than the thallophytes. They have simple stems with leaflike structures, without flowers or true roots. Their life history includes an alternation of generations, with both a spore stage (*sporophyte*) and a sex-cell stage (*gametophyte*).

Class 1 Hepaticae. The *liverworts,* such as Marchantia, have a small leaf-like structure that lies flat on the ground, with simple root-like projections. There are separate structures for producing sperm and egg cells. After fertilization, spores are produced, each of which may give rise to a new plant. Liverworts are found in moist shaded places. About 4,000 species have been identified.

Class 2 Musci. The mosses are somewhat better known than the liverworts, and a little more advanced in structure. They have a more erect appearance, with many leaf-like structures on their simple stem, and with larger root-like structures. They have a separate spore case at the end of a stalk.

Sphagnum is a moss that accumulates in swamps or bogs to form *peat.* Peat moss is used by gardeners to improve their soil. Peat eventually become partly carbonized, and can be used as fuel. About 15,000 species of moss have been classified.

Phylum 2 **Tracheophyta.** The tracheophytes are the large group of vascular plants, that is, plants with well-defined conducting systems. The sporophyte stage is prominent, with a tiny gametophyte stage. The plants have well-developed roots, stems and leaves.

Subphylum 1 **Lycopsida.** This class includes the club-moss and ground pine. In former geologic eras, these plants were as large as trees. Now they remain as low plants that grow close to the ground in the woods. They produce spores in such large numbers that at times they seem to be covered with a yellow dust. Lycopod powder consists of these spores. About 600 club mosses have been identified.

Subphylum 2 **Sphenopsida.** The horsetails are relatives of the ferns that grow in sandy places and along railroad tracks, and contain large amounts of minerals. For this reason, they were used to scour pots before the days of scouring powders. Instead of true leaves, horsetails have whorls of scalelike leaves along their stem. Spores are produced in a cone-like structure at the top of the stem.

Subphylum 3 **Pteropsida.** This group includes broad-leaved tracheophytes such as ferns, gymnosperms and angiosperms.

Class 1 Filicineae. This class contains the true ferns, such as the royal fern, Boston fern and the Christmas fern. They generally have an underground woody stem that sends up leaves. The undersurface of the leaves may be seen to contain tiny brown dots; these are the *sori* or the spore cases which produce spores. Other ferns produce spores on separate stems. Most ferns grow in wooded areas. About 4,400 ferns have been identified.

The sporophyte stage is well-developed, while the gametophyte is reduced in size to less than a half inch. About 250,000,000 years ago ferns were very prominent and grew to the size of trees. Over the ages, compression and decay of masses of these ancient ferns have formed our present coal deposits.

Seed Plants. The highest developed and most numerous group of plants are the seed-producing plants. About 175,000 different species have been identified. These are the plants all about us that we are most acquainted with. Some of the seed plants live for one year, and are known as *annuals* (wheat, bean, cotton, marigold). Others complete their life cycle in two years, and are called *biennials* (carrot, foxglove). Trees, most shrubs and many flowering plants are *perennials,* that is, they live on, year after year; some of them seem to die during the winter, and then come up again in the spring (phlox, aster, dandelion). Seed plants are classified in two classes, according to their manner of forming seeds:

Class 2 **Gymnospermae.** Pines, spruce, hemlock and other gymnosperms form "naked seeds" that are usually produced on the scales of a cone, and drop out when ripe. Many gymnosperms have leaves that are modified as needles and are evergreen. The giant sequoia redwood trees that are over two thousand years old, are also members of this class.

Class 3 **Angiospermae.** The angiosperms include flowering plants (rose, violet), trees with broad leaves (oak, maple), and plants used as crops (corn, bean). They all have covered seeds which develop in the part of a flower called the ovary. Most of them are *deciduous,* losing their leaves in the fall. Botanists distinguish between two subclasses of angiosperms:

Sub-class 1 **Dicotyledons.** The dicots have seeds that contain, in addition to the tiny embryo plant, two parts or cotyledons, such as occur in the bean or peanut. Their leaves have veins arranged in a network. They include many families of flowering plants, such as the legumes (peas, beans, clover), composites (daisy, dandelion, aster), rose (roses, strawberry, apple) and mustard (cabbage, cauliflower).

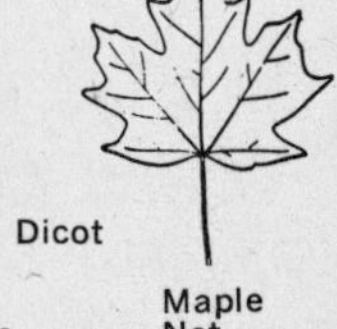

Sub-class 2 **Monocotyledons.** The monocots have seeds with one cotyledon. Their leaves have parallel veins. Examples: the grains (corn, wheat, rice and barley), grasses, onion, pineapple, lily, orchid.

CHAPTER REVIEW

Select the correct choice for each of the following statements.

1. The branch of biology that deals with the study of plants is (A) paleontology (B) genetics (C) zoology (D) botany (E) endocrinology

2. All of the following plants have a life history which includes an alternation of generations except (A) liverwort (B) sphagnum (C) moss (D) horsetail (E) spirogyra

3. Plants possessing leaves, stems and roots, but no seeds are classified as (A) Musci (B) Hepaticae (C) Filicinae (D) Bryophta (E) Angiosperms

4. The most highly developed plants are classified as (A) saprophytes (B) angiosperms (C) bryophytes (D) vertebrates (E) lycopods

5. An example of a biennial plant is (A) wheat (B) bean (C) cotton (D) marigold (E) carrot

6. An example of a gymnosperm is (A) jimson weed (B) pine (C) oak (D) maple (E) daisy

7. All of the following are deciduous plants except (A) spruce (B) oak (C) aster (D) willow (E) poplars

8. The corn plant (A) is a monocot (B) has leaves with veins arranged in a network (C) is a gymnosperm (D) has an embryo with several cotyledons (E) is a dicot

9. Plants that have covered seeds are classified as (A) gymnosperms (B) angiosperms (C) horsetails (D) ground pines (E) Marchantia

10. The closest relatives of mosses are (A) smuts (B) legumes (C) algae (D) liverworts (E) molds

ANSWER KEY

1-D	3-C	5-E	7-A	9-B
2-E	4-B	6-B	8-A	10-D

Part Two

PRACTICE College Board Achievement Tests IN BIOLOGY

THE COLLEGE BOARD ACHIEVEMENT TEST IN BIOLOGY

The College Board Achievement Test in Biology takes one hour. The total number of questions may vary in each test, along with the content covered, and the arrangement of the questions. There is an attempt made to include questions that range from easy to hard.

The test is planned to measure both (1) what you know about the subject matter generally taught in the biology course, and (2) how effectively you can use the knowledge you possess. The latter aspect is emphasized. It can best be described in terms of the following educational objectives:

EDUCATIONAL OBJECTIVES TESTED IN THE ACHIEVEMENT TEST IN BIOLOGY

1. **Knowledge:** It is expected that in mastering the content of biology, you will have learned much factual information about phenomena, characteristics, names, and important laws. However, more important than remembering facts is the ability to understand them and apply them. For example, it is desirable to name the parts of an organism; it is also important to understand how they work.

2. **Comprehension:** This objective calls for the ability to translate the information contained in the question into another form. For example, a law or principle may appear in the question in the form of words; the same information may then be described in the answer by a graph, an equation, a set of data, or an example, The opposite may also be true, in which the symbolic or numerical form of a question is to be recognized as a verbal statement of information.

3. **Application:** In addition to knowing a law, principle or concept, you are also expected to recognize how it applies to a particular situation. The situation may deal with familiar or unfamiliar material occurring in nature or the laboratory. Since one of the goals of biology education is to learn to apply what has been learned, application questions are viewed as being especially important.

The next three objectives are considered to be higher educational objectives.

4. **Analysis:** This calls for the ability to break a relatively extensive communication such as a passage, a graph or a chart into its various parts. It will be necessary to recognize the relationships among the parts of the selected presentation, and the way in which they are organized. The question may go beyond the specific sample being presented, and may ask about procedures that would provide information to strengthen or change the reasoning. The questions dealing with the analysis of a particular communication are usually presented in a series, all dealing with the same situation.

5. **Synthesis:** Questions testing this objective are based on the need to put together elements or parts in such a way as to result in a pattern or structure not clearly present up to now. The process of providing opportunities for synthesis in biology requires a good deal of time. Questions dealing with synthesis are likely to be time-consuming. Consequently, the achievement test generally does not contain many such questions.

6. **Evaluation:** The questions based on this objective call for the use of more than one criterion in selecting the correct answer. Thus, one of the criteria may involve the application of a science principle or concept; the other is usually a criterion of quality—i.e., selecting the fastest, the safest, the easiest, or the most accurate.

ORGANIZATION OF THE PRACTICE COLLEGE BOARD ACHIEVEMENT TESTS IN BIOLOGY

The ten practice achievement tests that follow are designed to give you practice in answering questions based on the various objectives being tested. In some cases, where there is a series of questions, there may be some overlap in the objectives being tested for. In general, with some variation in some of the tests, the questions are arranged as follows:

1. **Knowledge:** Part I—Types A, B, C, D and E
2. **Comprehension:** Part II
3. **Application:** Part III—Types A, B and C
4. **Analysis:** Part IV—Types A, B and C
5. **Synthesis:** Part V
6. **Evaluation:** Part VI

PRACTICE COLLEGE BOARD ACHIEVEMENT TEST IN BIOLOGY

NO. 1

PART I—TYPE A

Directions: **The accompanying diagram shows a generalized cell as seen under the electron microscope. Certain of the parts are labeled with numbers. Each of the questions is followed by five suggested labels or answers. For each question, select the one best answer. A label may be used once, more than once, or not at all.**

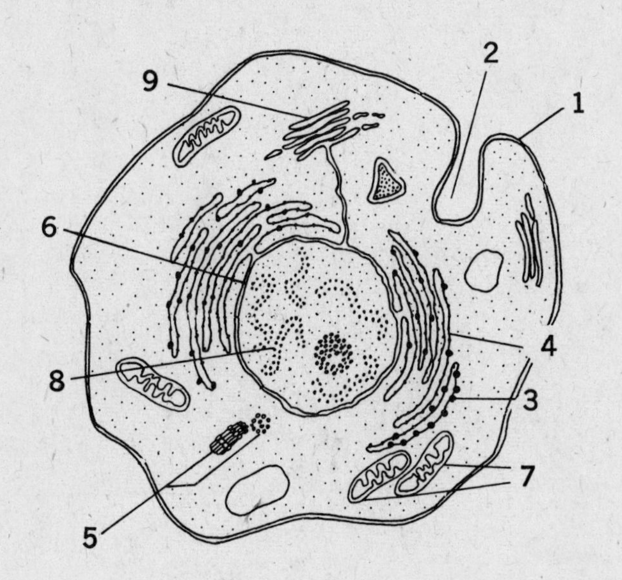

1. Serves as the boundary between the living cell and its environment:
(A) 2 (B) 1 (C) 6 (D) 4 (E) 5

2. Functions as the site of protein synthesis:
(A) 8 (B) 5 (C) 3 (D) 7 (E) 2

3. All are organelles except:
(A) 7 (B) 5 (C) 9 (D) 2 (E) 3

4. Is believed to function in intercellular transport:
(A) 3 (B) 7 (C) 8 (D) 4 (E) 5

5. Acts as a site for cellular respiration:
(A) 7 (B) 9 (C) 8 (D) 5 (E) 4

6. Contains molecules of DNA:
(A) 5 (B) 3 (C) 9 (D) 4 (E) 8

7. Is known as the powerhouse of the cell:
(A) 6 (B) 2 (C) 1 (D) 5 (E) 7

8. Becomes active only during mitosis:
(A) 3 (B) 9 (C) 5 (D) 4 (E) 2

9. Controls the transport of materials into and out of the nucleus:
(A) 1 (B) 2 (C) 9 (D) 4 (E) 6

10. Carries the genetic code:
(A) 4 (B) 5 (C) 9 (D) 8 (E) 6

1. A B C D E
2. A B C D E
3. A B C D E
4. A B C D E
5. A B C D E
6. A B C D E
7. A B C D E
8. A B C D E
9. A B C D E
10. A B C D E

TYPE B

Directions: **Each of the following statements is followed by five suggested answers. Select one answer which is best in each case.**

11. The Hardy-Weinberg principle applies to (A) the cell theory (B) the theory of spontaneous generation (C) experimental embryology (D) population genetics (E) germ theory of disease

11. A B C D E

12. Of the following diseases, the one caused by a worm is (A) sickle-cell anemia (B) Klinefelter's syndrome (C) ringworm (D) trichinosis (E) Down's syndrome

12. A B C D E

13. The scientist who first brought about the artificial production of mutations through the use of X-rays was (A) Mendel (B) Mendeleff (C) Metchnikoff (D) Morgan (E) Muller

13. A B C D E

14. In testing a substance for the presence of glucose, it should be heated with (A) Benedict's solution (B) Burrough's solution (C) iodine solution (D) iodine crystals (E) glycagon

14. A B C D E

15. The exchange of gases between a leaf and its environment takes place through special structures known as (A) cambium (B) lenticels (C) phloem (D) stomata (E) xylem

15. A B C D E

16. The biome in which an ecologist would expect to find the coldest year-round temperature is (A) taiga (B) coniferous forest (C) deciduous forest (D) tundra (E) grassland

16. A B C D E

TYPE C

Directions: **For each of the processes listed from 17 to 30, select the letter** (A), (B), (C), or (D), **depending on whether**

(A) it applies only to A
(B) it applies only to B
(C) it applies to both A and B
(D) it applies to neither A nor B

(A) *Higher green plants*	(C) *Both*
(B) *Mammals*	(D) *Neither*

17. aerobic respiration — 17. A B C D

18. ingestion — 18. A B C D

19. bioluminescence — 19. A B C D

20. active transport — 20. A B C D

21. assimilation — 21. A B C D

22. locomotion — 22. A B C D

23. digestion — 23. A B C D

24. extraterrestrial habitat — 24. A B C D

25. mutation — 25. A B C D

26. photoperiodism — 26. A B C D

27. transpiration — 27. A B C D

28. irritability — 28. A B C D

29. photosynthesis — 29. A B C D

30. evolution — 30. A B C D

TYPE D

Directions: **The following questions require two answers. One of the items in the left column** (A, B, or C) **is related to four of the items in the right column** (numbered 1–5). **Select the letter of the item which is so related; then select the number of the remaining item which does not belong.**

EXAMPLE

		A B C	1 2 3 4 5
(A) paramecium	1. cilia	▮ ‖ ‖	‖ ‖ ‖ ▮ ‖
(B) ameba	2. macronucleus		
(C) spirogyra	3. trichocyst		
	4. pyrenoid		
	5. micronucleus		

31. (A) carbohydrate
(B) lipid
(C) protein

1. carboxyl group
2. CH_2O basic structure
3. glycerol molecule
4. fatty acid molecules
5. OH group

31. A B C 1 2 3 4 5

32. (A) conjugation
(B) fertilization
(C) fission

1. egg
2. bud
3. testes
4. ovary
5. sperm

32. A B C 1 2 3 4 5

33. (A) reptile
(B) mammal
(C) fish

1. porpoise
2. shark
3. whale
4. walrus
5. seal

33. A B C 1 2 3 4 5

34. (A) gymnosperm
(B) filicinae
(C) angiosperm

1. maple
2. pine
3. hemlock
4. fir
5. spruce

34. A B C 1 2 3 4 5

35. (A) hormone
(B) vitamin
(C) enzyme

1. mucus
2. pepsin
3. erepsin
4. ptyalin
5. lipase

35. A B C 1 2 3 4 5

36. (A) nucleus
(B) cell membrane
(C) cell wall

1. chromosome
2. flagellum
3. gene
4. chromatin
5. DNA

36. A B C 1 2 3 4 5

37. (A) dominant
(B) blending
(C) recessive

1. albinism
2. blue eyes
3. Rh positive
4. blond hair
5. color blindness

37. A B C 1 2 3 4 5

38. (A) Lamarck
(B) Ehrlich
(C) Darwin

1. survival of fittest
2. germ theory
3. overproduction
4. natural selection
5. struggle for existence

38. A B C 1 2 3 4 5

TYPE E

Directions: **The five lettered headings below are followed by a list of numbered statements. For each numbered statement, choose the one lettered heading (A–E) which is most closely related to it.**

(A) Crop rotation
(B) Contour plowing
(C) Reforestation
(D) Erosion
(E) Insect control

39. A burned out forest in California is being planted with tree seedlings

39. A B C D E

40. A farmer plows in concentric circles around a hill

40. A B C D E

41. A windstorm carries so much dust that visibility is reduced almost to zero

41. A B C D E

42. A farmer plants clover on a field that had been used for corn

42. A B C D E

43. Some vanishing wildlife is beginning to reappear in previously heavily timbered areas

43. A B C D E

44. An airplane sprays a hillside with DDT

44. A B C D E

45. A crop of peanuts on a cotton plantation restores nitrogen to the soil

45. A B C D E

(A) Meiosis
(B) Mitosis
(C) Fertilization
(D) Mutation
(E) Crossing-over

46. How may the number of chromosomes be reduced?

46. A B C D E

47. How may the chromosome number be restored?

47. A B C D E

48. How may the chromosome number be kept constant in all the cells of an organism?

48. A B C D E

49. How may chromosomes exchange materials?

49. A B C D E

50. How may chromosomes change in composition?

50. A B C D E

51. How may the monoploid number of chromosomes be formed?

51. A B C D E

PART II

Directions: **A technique has been perfected by which the tip of a root can be cut off and placed in a sterile nutrient solution. Here, under controlled conditions of temperature, it will grow in length and produce secondary roots. A research scientist made daily measurements of a tomato root tip growing at 33°–35°C this way. The accompanying graph shows the results.**

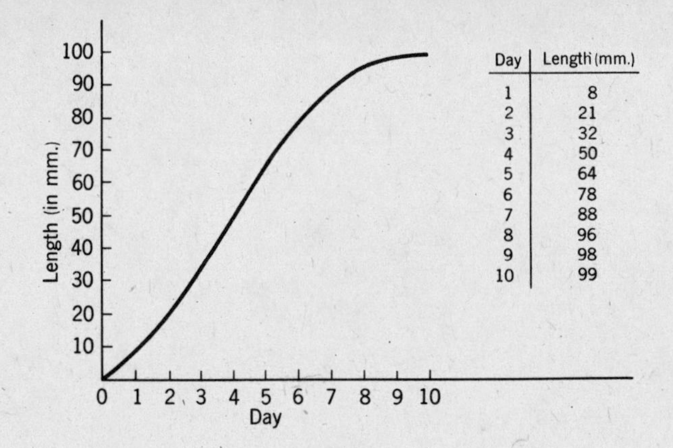

Day	Length (mm.)
1	8
2	21
3	32
4	50
5	64
6	78
7	88
8	96
9	98
10	99

Referring to the graph, select the best answer for each of the following questions.

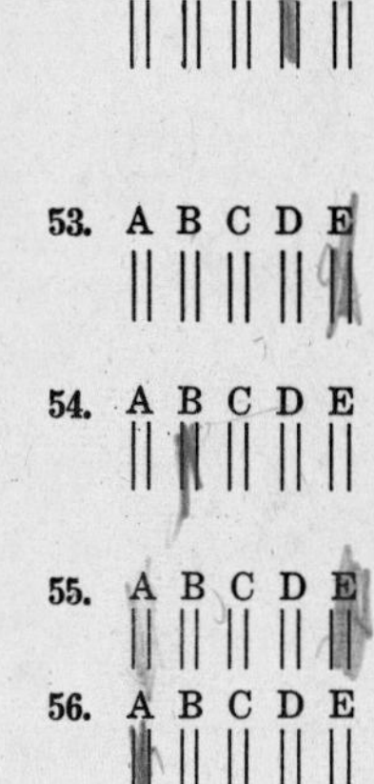

52. The time when the length of the root was increasing at the greatest rate was (A) during the first day (B) between the first and second day (C) between the second and third day (D) between the third and fourth day (E) between the sixth and seventh day

53. The root tip showed no growth (A) at the end of the first day (B) at the end of the 3rd day (C) at the end of the 8th day (D) at the end of the 10th day (E) at no time shown on the graph

54. The root tip had achieved more than half of the total length indicated on the graph within the first (A) 3 days (B) 4 days (C) 5 days (D) 6 days (E) 7 days

55. The smallest increment in growth was achieved during (A) the first 3 days (B) the last 3 days (C) the middle 3 days (D) the 3rd day (E) the 8th day

56. The S-shaped appearance of the growth curve indicates that (A) the rate of growth was greatest after a few days, and then it decreased (B) the root tip increased in length at a uniform rate (C) the root tip grew faster than the region of elongation of the root (D) the root tip grew more slowly than the region of elongation (E) mitotic division of the root tip cells occurred at intervals of thirty minutes

57. The graph would have appeared as a straight line (A) if the growth after the 8th day had been at the same rate as on the preceding day (B) if the root tip had increased in length at the same rate daily (C) if the temperature had been maintained at a constant level of 37°C. (D) if the growth increment of the first two days had been identical (E) if the nutrient solution had been re-enforced with root growth hormone

PART III—TYPE A

Directions: **Each item consists of an *assertion* (statement) in the left-hand column, and a *reason* on the right-hand side. Select**

(A) if both assertion and reason are true, and *are related* as cause and effect
(B) if both assertion and reason are true, but *are not related* as cause and effect
(C) if the assertion is true, but the reason is a false statement
(D) if the assertion is false, but the reason is a true statement
(E) if both assertion and reason are false statements

DIRECTIONS SUMMARIZED:

	(A)	(B)	(C)	(D)	(E)
Assertion:	True	True	True	False	False
Reason:	True	True	False	True	False
	Cause and effect	*Not* cause and effect			

	Assertion:		Reason:	
58.	An enzyme is considered to be an organic catalyst	BECAUSE	its molecules are huge compared to the molecules of the substrate	58. A B C D E
59.	Chromosomes do not occur in homologous pairs in a sperm cell	BECAUSE	mitosis takes place when the male primary sex cell divides	59. A B C D E
60.	During replication, a molecule of DNA will separate between adenine and uracil	BECAUSE	uracil is found only in DNA, not RNA	60. A B C D E
61.	The climax stage of a biotic succession remains stable until the environment changes	BECAUSE	its populations are in balance with each other and the environment	61. A B C D E
62.	When a person inhales, his lungs fill with air	BECAUSE	his diaphragm and his ribs are elevated	62. A B C D E
63.	A pollen grain that lands on a stigma germinates	BECAUSE	it forms a pollen tube	63. A B C D E
64.	Enzymes in the small intestine attack food particles	BECAUSE	some of them are changed from protein to glycerin	64. A B C D E
65.	The Japanese beetle has become more of a threat in this country than in Japan	BECAUSE	it goes through a life cycle of complete metamorphosis	65. A B C D E
66.	Plants exposed to high radiation show mutations	BECAUSE	genes are affected and changed	66. A B C D E
67.	There is less topsoil in the United States today than there was in the year 1650	BECAUSE	of excessive erosion	67. A B C D E
68.	We are certain that Neanderthal man lived during the time of the dinosaurs	BECAUSE	their remains have been found together	68. A B C D E
69.	Lamarck's Theory of Use and Disuse is acceptable to most scientists today	BECAUSE	there is little scientific evidence that acquired characteristics are inherited	69. A B C D E
70.	The arm of man and the flipper of a whale have similar bone structure	BECAUSE	their embryos have gill slits	70. A B C D E

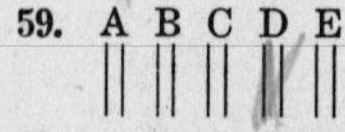

TYPE B

Directions: **Each of the following statements is followed by five suggested answers. Select the one answer which is best in each case.**

71. Which of the following would be most likely to occur in an ecosystem?
(A) As the number of prey decreases, the number of predators increases
(B) As the number of predators increases, the number of prey increases
(C) As the number of prey increases, the number of predators increases
(D) As the number of prey increases, the number of predators decreases
(E) As the number of predators decreases, the number of prey decreases

71. A B C D E

72. There is no pure race in the human species of today chiefly because of (A) inbreeding (B) isolation (C) cross-breeding (D) racial laws (E) linkage

72. A B C D E

73. The transpiration rate of a tree would most probably be increased by (A) increase in temperature only (B) increase in both humidity and temperature (C) increase in humidity and decrease in temperature (D) increase in both humidity and air movement (E) increase in both temperature and air movement

73. A B C D E

74. Astronauts are not likely to find active heterotroph aggregates in their exploration of the moon because (A) the force of gravity on the moon is too weak (B) the force of gravity on the moon is too strong (C) there are no X-rays, ultra-violet radiation, or cosmic rays on the surface of the moon (D) there is no atmosphere or water on the moon (E) it is now suspected that there are moonquakes on the surface

74. A B C D E

75. The scientific name of the dog is *Canis familiaris* and that of the coyote is *Canis latrans*. This indicates that both the dog and the coyote are members of the (A) same genus but different species (B) same species but different genera (C) same genus but different classes (D) same species but different classes (E) same class but different genera

75. A B C D E

76. If a family has four children, three boys and a girl, what are the chances that the next child will be a girl? (A) 0 (B) 1 in 2 (C) 1 in 3 (D) 1 in 4 (E) 1 in 8

76. A B C D E

TYPE C

***Directions:* In each of the following questions, there is a statement followed by four choices of which *one or more* are correct. Indicate the correct combination by choosing the following letters:**

(A) if only 1, 2, and 3 are correct
(B) if only 1 and 3 are correct
(C) if only 2 and 4 are correct
(D) if only 4 is correct
(E) if some other combination is correct

DIRECTIONS SUMMARIZED

(A)	(B)	(C)	(D)	(E)
1, 2, 3 only	1, 3 only	2, 4 only	4 only	some other combination

77. Carbon dioxide plays an important role in the metabolism of living things. It
1. is given off as a waste by animals during aerobic respiration
2. is given off as a waste by green plants during aerobic respiration
3. is taken in by green plants during photosynthesis
4. stimulates the breathing rate of human beings

77. A B C D E

78. The phenomenon of linkage helps to explain
1. the dominance of brown eyes
2. the predominance of hemophilia among males
3. why blending occurs in the Japanese four o'clock flower
4. why certain characteristics are inherited together

78. A B C D E

79. Iodine is needed by the body in small quantities. When it is absorbed, it is concentrated
1. largely in the thyroid gland
2. equally in all the cells of the body
3. mostly in the spleen
4. especially in freshly-produced ACTH

79. A B C D E

80. Insects are classified as arthropods because they
1. have jointed legs
2. have an exoskeleton
3. have segmented bodies
4. have a chitinous covering

80. A B C D E

81. Animal and plant cells are similar in having
1. nucleus
2. cytoplasm
3. cell membrane
4. mitochondria

81. A B C D E

82. Ganglia serve as
1. centers of thought
2. collections of cell bodies
3. connections between sensory and motor neurons
4. part of the autonomic nervous system

82. A B C D E

83. In forming a habit, one should
1. have a desire for it
2. experience a sense of satisfaction
3. practice it
4. replace the original stimulus with a new one

83. A B C D E

84. Radiocarbon is useful
1. in determining the age of early historic objects
2. in treating hyperthyroidism
3. because it has a half-life of 5,760 years
4. because it has the same number of neutrons as ordinary carbon

84. A B C D E

PART IV—TYPE A

Directions: **Refering to the following selection, choose the best answer for each of the questions that follows.**

*"The interior of the cell is distinguished from the outer world by the presence of very large and highly complex molecules. In fact, whenever such molecules turn up in the nonliving environment, one can be sure they are the remnants of dead cells. On the primitive earth, life must have had its origin in the spontaneous synthesis of complicated macromolecules at the expense of smaller molecules. Under present-day conditions, the capacity to synthesize large molecules from simpler substances remains one of the supremely distinguishing capacities of cells.

"Among these macromolecules are proteins. In addition to making up a major portion of the 'solid' substance of cells, many proteins (enzymes) have catalytic properties; that is, they are capable of greatly accelerating the speed of chemical reactions inside the cell, particularly those involved in the transformation of energy. The synthesis of proteins from the simpler units of the 20-odd amino acids goes forward under the regulation of deoxyribonucleic acid (DNA) and ribonucleic acid (RNA), by far, the most highly structured of all the macromolecules in the cell. In recent years and months investigators have shown that DNA, localized in the nucleus of the cell, presides at the synthesis of RNA, which is found in both the nucleus and the cytoplasm. The RNA in turn arranges the amino acids in proper sequence for linkage into protein chains. The DNA and the RNA may be compared to the architect and contractor who collaborate on the construction of a nice-looking house from a heap of bricks, stones and tiles."

85. All of the following statements are true *except* (A) macromolecules are very large and highly complex (B) macromolecules are largely present in nonliving things (C) macromolecules were probably formed spontaneously on the primitive earth (D) proteins are examples of macromolecules (E) macromolecules were formed from smaller molecules

85. A B C D E

86. The correct sequence of events is (A) arrangement of amino acids into proteins–synthesis of RNA–DNA (B) synthesis of RNA–DNA–arrange-

86. A B C D E

*From "The Living Cell" by John Brachet, Sept. 1961, by *Scientific American, Inc.* All rights reserved.

ment of amino acids into proteins (C) DNA–arrangement of amino acids into proteins–synthesis of RNA (D) RNA–synthesis of DNA–arrangement of amino acids into proteins (E) DNA–synthesis of RNA–arrangement of amino acids into proteins

87. The selection states that RNA is found in (A) nucleus only (B) cytoplasm only (C) both nucleus and cytoplasm (D) proteins (E) amino acids

87. A B C D E

88. All of the following statements are false except (A) all proteins are enzymes (B) all proteins have catalytic properties (C) enzymes speed up chemical reactions dealing with the transformation of energy (D) the 20-odd amino acids are synthesized from proteins (E) DNA and RNA are highly-structured proteins

88. A B C D E

TYPE B

Directions: **Each of the following problems is followed by five suggested answers. Select the one answer which is best in each case.**

89.

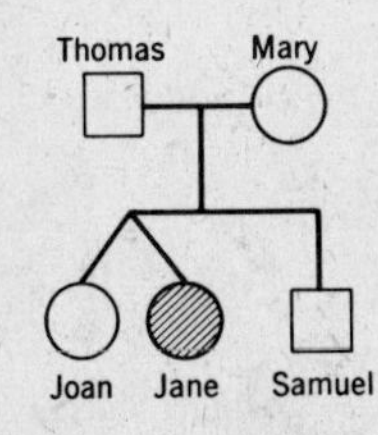

In this pedigree of a family, brown eyes are indicated as ○ and blue eyes as ●. Joan and Jane are twins. From this chart, it can be determined that: (A) Thomas and Mary are homozygous for brown eyes (B) Joan and Jane are identical twins (C) Jane is heterozygous for blue eyes (D) Jane is homozygous for blue eyes (E) Joan and Samuel are homozygous for brown eyes

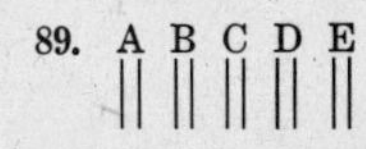

90. An animal breeder mated two heterozygous black guinea pigs. (Black is dominant.) It would be possible for the litter to contain: (A) only black guinea pigs (B) only white guinea pigs (C) some black and some white guinea pigs (D) 75% black, 25% white guinea pigs (E) all of the above

90. A B C D E

91. A plant breeder obtained 252 red, 235 white, and 503 pink flowers, when he crossed pink snapdragons (A) The phenotype of the hybrid is the same as its genotype (B) The phenotype of the pure type is different from its genotype (C) The phenotype of the hybrid is different from its genotype (D) The offspring show the phenotype of the dominant gene (E) All of the above statements are false

91. A B C D E

92. A color-blind man marries a normal homozygous woman. The chances of their having a color-blind son are (A) 25%–75% (B) 50%–50% (C) 25%–50%–25% (D) 100% (E) zero

92. A B C D E

TYPE C

Directions: **A colony of bacteria growing on a agar plate is studied. A slide of bacteria is prepared, using methylene blue stain. The steps in making such a slide are shown in the accompanying drawings:**

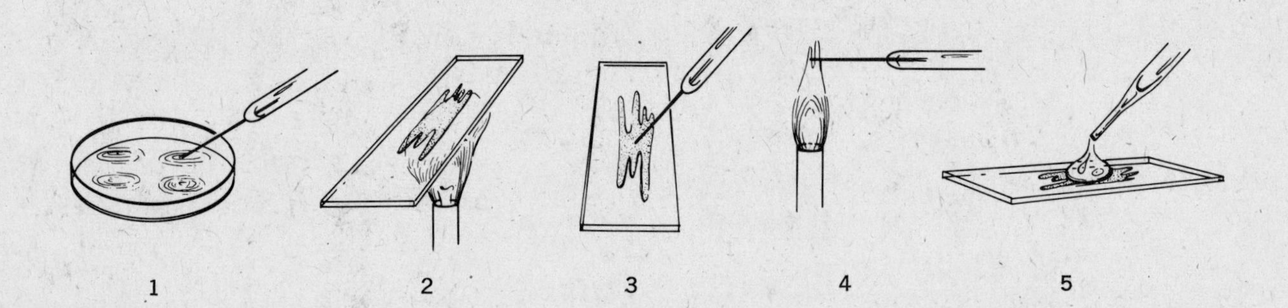

In each of the following questions, select the best answer.

93. The correct order of procedures in making the slide is (A) 1–2–3–4–5 (B) 1–2–5–4–3 (C) 2–4–3–5–1 (D) 3–1–2–4–5 (E) 4–1–3–2–5

93. A B C D E

94. The best way to distribute the bacteria on the slide with the needle: (A) keep the bacteria together (B) spread the bacteria apart (C) touch the needle to three or four spots on the slide (D) touch the needle to one spot on the slide (E) apply as much of the colony on the slide as possible

94. A B C D E

95. The bacteria are "fixed" on the slide in the procedure(s) numbered (A) 2 (B) 4 (C) 5 (D) 4 and 5 (E) 3 and 5

95. A B C D E

96. The needle is heated in the flame (A) both before and after using (B) only before being touched to the colony (C) in order to be warm enough to hold the bacteria (D) to sterilize the bacteria in the colony (E) for none of the reasons stated

96. A B C D E

97. Methylene blue stain is used in this case because (A) it stains the nucleus clearly (B) it stains only the cell wall of the bacteria (C) it enables the bacteria to be seen in the living condition (D) it stains the protoplasm of the bacteria (E) it is harmless to living bacteria

97. A B C D E

98. Each colony of bacteria (A) can continue to grow indefinitely in this Petri dish (B) contains bacteria that can be seen with the naked eye (C) contains many types of bacteria (D) originated from a single bacterium (E) reaches its largest size in about a month of incubation

98. A B C D E

99. In making a hanging drop slide from a liquid culture of bacteria, a loop would be used instead of the straight needle, and then the only procedure pictured above that would be followed is (A) 1 (B) 2 (C) 3 (D) 4 (E) 5

99. A B C D E

100. Of the following microorganisms, the only one that would appear rod-shaped is (A) *Bacillus anthracis* (B) *Endameba histolytica* (C) *Plasmodium falciparum* (D) *Spirochaeta pallida* (E) *Streptococcus hemolyticus*

100. A B C D E

PART V

Directions: **Each of the following questions has five suggested answers. Select the one answer that is best in each case.**

101. The range of random variation in the size of 1,000 lima beans is shown as a bell-shaped curve in the following graph:

101. A B C D E

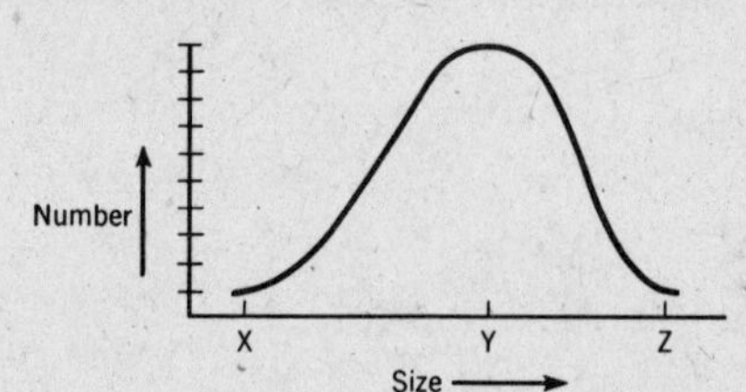

If a plant breeder selected only the longest seeds for planting each year for ten years, the range of variation in the size of 1,000 lima beans would most likely be as shown in:

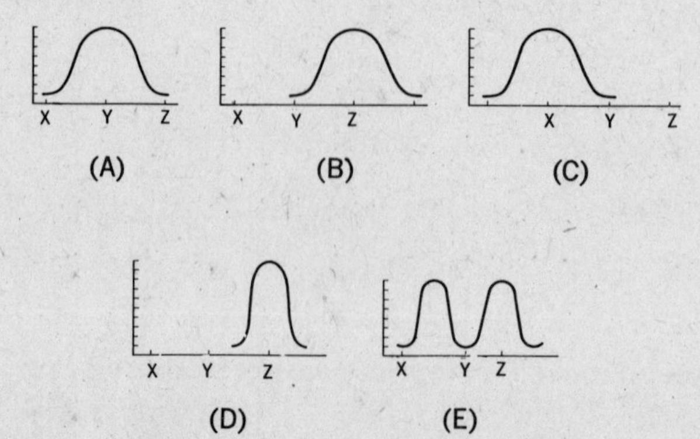

102. The following series of equations occurs in living things:

102. A B C D E

$$\text{(A)}\ C_6H_{12}O_6 + 6O_2 \xrightarrow[\text{enzymes}]{\text{energy}} 6CO_2 + 6H_2O + \text{energy}$$

$$\text{(B)}\ 6CO_2 + 12H_2O \xrightarrow[\text{chlorophyll enzymes}]{\text{energy}} C_6H_{12}O_6 + 6H_2O + 6O_2 + \text{energy}$$

Which of the following organisms is capable of performing the processes represented by both equation A and equation B (A) ameba (B) earthworm (C) bread mold (D) bean plant (E) none of these organisms

103. The relationship described in a food chain may also be expressed as a food pyramid. The major difference is that in the food pyramid, as depicted in the accompanying diagram, there is an indication of both the relative numbers of individuals and the amounts of energy involved at each level.

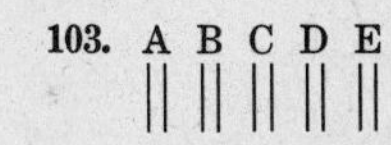

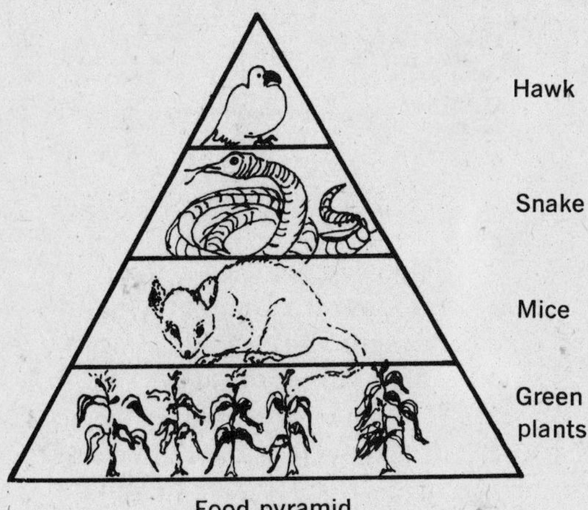

Food pyramid

This relationship may best be summarized as follows:
(A) When the top of the pyramid is reached, the number of individuals decreases but the amount of energy increases.
(B) When the top of the pyramid is reached, the number of individuals increases, and the amount of energy remains the same as at the other levels.
(C) At the bottom level of the pyramid, both the number of individuals and the amount of energy involved are greatest.
(D) At the bottom level of the pyramid, both the number of individuals and the amount of energy involved are lowest.
(E) At all the levels, the number of individuals and the amount of energy are the same.

104. It is known that the number of cricket chirps that can be counted in one minute varies with the temperature. The table shows the relationship:

104. A B C D E

Number of chirps/min.	Temperature (F)
47	49°
71	55°
150	75°

On the basis of this information, it can be predicted that, at 60°F, the number of cricket chirps would most probably be (A) 61 (B) 71 (C) 81 (D) 91 (E) 101

PART VI

Directions: **Each of the following questions has five suggested answers. Select the one answer that is best in each case.**

105. Which one of the following procedures would give the most convincing results in testing whether or not a certain new drug would be effective in curing cancer in white mice?
(A) Injecting a similar number of white mice and white rats with cancer, and comparing the number of mice that are cured of cancer with the number of rats that are cured.
(B) Injecting a dozen white mice with cancer and determining how many of them are cured of cancer.
(C) Injecting the drug into a large number of mice with cancer and comparing the number that are cured of cancer with the known number of mice that are cured without any treatment.
(D) Injecting the drug into a large number of white mice with cancer, and injecting the same amount of distilled water into another large number of white mice with cancer, and comparing the number of mice in each group that die of cancer.
(E) Injecting different doses of the drug into a large number of white mice with cancer and determining the number of white mice that are cured in each group.

105. A B C D E

106. In studying cyclosis in Elodea cells, the best way to focus on the chloroplasts as they are carried through the depth of the cells is to turn to the high power of the microscope,
(A) use the coarse adjustment and open the diaphragm wide
(B) use the fine adjustment and adjust the diaphragm for optimal light
(C) use the fine adjustment, and close the diaphragm all the way
(D) use the fine adjustment and focus the mirror in direct sunlight
(E) use the coarse adjustment and focus the mirror in strong indirect light

106. A B C D E

107. The energy requirements of different people vary. Which of the following people would most likely have a Calorie requirement of 4,700? (A) 15-year old girl weighing 115 pounds (B) clerk weighing 180 pounds (C) typist weighing 128 pounds (D) coal miner weighing 148 pounds (E) housewife weighing 138 pounds

107. A B C D E

108. Analysis of a certain complex compound shows that its molecule contains a phosphate molecule, a ribose molecule and a pyrimidine nitrogen base. Which of the following is the most informative statement that can properly be made about the compound on the basis of this information? (A) It is most likely ribonucleic acid (B) It is DNA (C) It is an inorganic compound (D) It contains thymine (E) It is a polypeptide

108. A B C D E

109. In comparing the effectiveness of enzymes at various pH concentrations, the accompanying graph was prepared to show the maximum rate of activity for maltase.

109. A B C D E

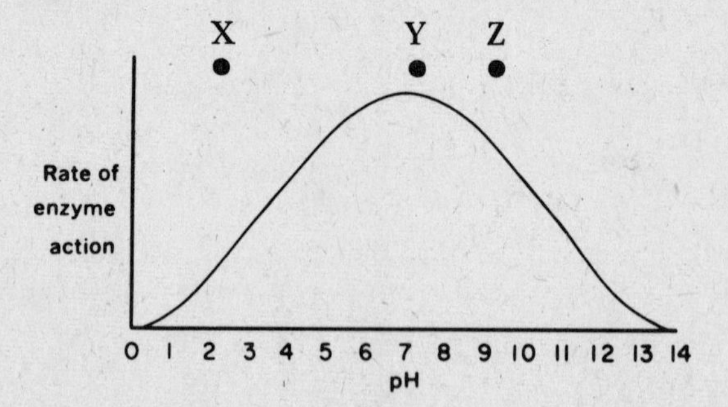

It was also determined that pepsin is most active at pH 1.5–2.2, and trypsin at pH 7.9–9.0. The graph shows that
(A) point X represents pepsin and point Y represents trypsin
(B) point X represents maltase and point Y represents pepsin

(C) point X represents trypsin and point Y represents pepsin
(D) point X represents pepsin and point Z represents maltase
(E) point X represents pepsin and point Y represents maltase

ANSWER KEY: TEST NO. 1

1. (B)	12. (D)	23. (C)	34. (A)1	45. (A)	56. (A)	67. (A)	78. (D)	89. (D)	100. (A)
2. (C)	13. (E)	24. (D)	35. (C)1	46. (A)	57. (B)	68. (E)	79. (E)	90. (E)	101. (D)
3. (D)	14. (A)	25. (C)	36. (A)2	47. (C)	58. (B)	69. (D)	80. (E)	91. (A)	102. (D)
4. (D)	15. (D)	26. (A)	37. (C)3	48. (B)	59. (C)	70. (B)	81. (E)	92. (E)	103. (C)
5. (A)	16. (D)	27. (A)	38. (C)2	49. (E)	60. (E)	71. (C)	82. (C)	93. (E)	104. (D)
6. (E)	17. (C)	28. (C)	39. (C)	50. (D)	61. (A)	72. (C)	83. (A)	94. (B)	105. (D)
7. (E)	18. (B)	29. (A)	40. (B)	51. (A)	62. (C)	73. (E)	84. (B)	95. (A)	106. (B)
8. (C)	19. (D)	30. (C)	41. (D)	52. (D)	63. (B)	74. (D)	85. (B)	96. (A)	107. (D)
9. (E)	20. (C)	31. (B)2	42. (A)	53. (E)	64. (C)	75. (A)	86. (E)	97. (D)	108. (A)
10. (D)	21. (C)	32. (B)2	43. (C)	54. (B)	65. (B)	76. (B)	87. (C)	98. (D)	109. (E)
11. (A)	22. (B)	33. (B)2	44. (E)	55. (B)	66. (A)	77. (E)	88. (C)	99. (D)	

ANSWERS EXPLAINED: TEST NO. 1

1. (B) cell membrane
2. (C) ribosome
3. (D) pinocytic vesicle
4. (D) endoplasmic reticulum
5. (A) mitochondrion
6. (E) chromosome
7. (E) mitochondrion
8. (C) centriole
9. (E) nuclear membrane
10. (D) chromosome

11. (A) The Hardy-Weinberg principle states that the gene pool of a population remains constant from generation to generation, if the population is large, there are random matings, and no new factors are introduced, such as mutations or migration.

12. (D) The trichinella worms are contained in raw pork, and enter the body alive if the meat is undercooked.

13. (E) Herman J. Muller produced mutations artificially in Drosphila flies in 1927.

14. (A) When glucose is heated with Benedict's solution, a green to brick-red color is obtained.

15. (D) Stomates are the tiny openings in the leaf epidermis.

16. (D) The tundra is the treeless region in the far north. It has an arctic climate of severe cold, with a long dark winter, and a mild summer having continuous daylight.

17. (C) By respiration, green plants and animals combine oxygen with food to obtain energy.

18. (B) Animals cannot make their own food, so they must obtain it and take it into their bodies.

19. (D) Bioluminescence refers to the ability of some lower organisms to give off light. i.e., certain bacteria, protozoa, fungi, deep-water fish, and the firefly.

20. (C) Active transport takes place when the cell uses energy to move molecules across the cell membrane, from a region of low concentration to a region of high concentration of the molecules.

21. (C) Assimilation is the process by which the protoplasm of all living things is formed.

22. (B) Practically all animals have the ability to move from one place to another.

23. (C) All living things change insoluble food into soluble form.

24. (D) Some scientists believe that life may exist away from the earth, in other solar systems.

25. (C) A mutation is a change in the gene that may suddenly appear in plants and animals.

26. (A) Photoperiodism is the flowering response of green plants to the amount of light they receive.

27. (A) Higher green plants give off excess water vapor from their leaves.

28. (C) The protoplasm of living things has the ability to respond to its environment.

29. (A) Only green plants have the ability to make carbohydrates from water and carbon dioxide in the presence of light.
30. (C) All living things have undergone changes over the ages; such changes are still going on.
31. (B)2 All are parts of a lipid molecule, except for the CH_2O structure, which is characteristic of carbohydrates.
32. (B)2 All the structures are involved in fertilization except a bud, which results from asexual reproduction in yeast and hydra.
33. (B)2 All are mammals except the shark, which is a fish.
34. (A)1 All are gymnosperms except maple, which is an angiosperm.
35. (C)1 All are enzymes except mucus, which is a lubricating fluid produced in the alimentary canal.
36. (A)2 All are in the nucleus except flagellum, which is made of cytoplasm.
37. (C)3 All are examples of recessive traits except Rh positive, which is dominant.
38. (C)2 Darwin's theory of evolution included all of these parts; germ theory deals with germs as the cause of disease.
39. (C) Reforestation is a conservation measure aimed at protecting our forests.
40. (B) This prevents rainwater from running downhill rapidly and producing gullies.
41. (D) When soil is not held together by roots of plants, wind carries it away and acts as an agent of erosion.
42. (A) The roots of clover plants have swellings called nodules in which nitrogen-fixing bacteria are located; they form nitrates, thereby enriching the soil which is depleted by a crop such as corn.
43. (C) Reforestation helps conserve forests which act as a home for wildlife.
44. (E) Harmful insects such as the gypsy moth are destroyed in this way.
45. (A) Peanuts are like clover in question 42, in enriching soil which has been depleted by continuous planting of cotton.
46. (A) During meiosis, gametes are formed with the monoploid number of chromosomes.
47. (C) When sperm and egg cells unite, the monoploid number of each contributes to the diploid number of the fertilized egg.
48. (B) During mitotic division, the resulting cells receive the same number of chromosomes as the original cell.
49. (E) When crossing-over occurs, parts of chromosomes become attached to other chromosomes.
50. (D) The genes, which make up chromosomes, are changed during a mutation.
51. (A) When the gametes are being formed, reduction-division occurs, reducing their chromosome number to half.
52. (D) The curve on the graph shows the steepest rise during this period of time.
53. (E) Growth occurred at all times.
54. (B) At that time, its length was 50 mm, of a total length of 99 mm.
55. (B) The curve began to level off during this period, indicating a slower rate of growth.
56. (A) Growth was slow at first, then rapid, then slow again.
57. (B) At that rate, there would have been a constant increase in length, leading to a straight line graph.
58. (B) An enzyme is capable of affecting the rate of a chemical reaction without itself being changed. It also has large molecules.
59. (C) During meiosis the homologous pairs of chromosomes in the primary sex cell split, giving the resulting sperm cells half the number of chromosomes.
60. (E) Uracil is present in RNA. In DNA, the base that pairs with adenine is thymine.
61. (A) The climax stage represents the final stage in succession, unless there is a change in environment.
62. (C) During inhalation, the diaphragm is flattened while the ribs are raised, thus enlarging the chest cavity and causing air to enter the lungs.
63. (B) The pollen tube is formed as the result of the germination of the pollen grain, which absorbs the sugar fluid on the stigma.
64. (C) Enzymes reduce proteins to amino acids.
65. (B) The Japanese beetle has been so successful in this country because of the absence of many of its natural enemies.
66. (A) Radiations cause genes to change, resulting in mutations.
67. (A) One third of our supply of topsoil has been lost by erosion.
68. (E) The dinosaurs died out during the Mesozoic Era, millions of years before man appeared on the scene.
69. (D) Most scientists do not accept Lamarck's theory because of the lack of evidence that acquired characteristics can be inherited.

70. (B) Both organisms had a common ancestor and still retain enough genes, despite the changes that have occurred, to have similar bone structure.

71. (C) The number of predators depends on their ability to catch prey. If there is more prey available, the predators will be more successful, and their numbers will increase.

72. (C) The extensive migrations of human beings over the ages has resulted in cross-breeding.

73. (E) With an increase in temperature there is an increase in the activities of the plant, including transpiration. Air movement removes moisture from the immediate vicinity of the stomates, making it possible for more water vapor of transpiration to evaporate.

74. (D) According to the heterotroph hypothesis, heterotroph aggregates are supposed to have formed in the ancient seas present on the earth's surface during its early history.

75. (A) *Canis* is the genus name for both animals; but each has a different species name, either *familiaris*, or *latrans*.

76. (B) Each time an egg is fertilized, it unites with a sperm that may contain either the X or the Y chromosome. Thus, there is one chance out of two that a sperm with an X chromosome will unite with the egg, resulting in a girl.

77. (E) All the statements are correct.

78. (D) Linkage refers to the inheritance together of genes on the same chromosome.

79. (E) The thyroid is the only gland that absorbs large amounts of iodine.

80. (E) All of the characteristics apply to insects.

81. (E) All cells contain these structures.

82. (C) Ganglia are collections of cell bodies lying outside of the spinal cord and brain.

83. (A) The first three items apply to habit formation; the fourth applies to conditioning.

84. (B) Radiocarbon has been used by Dr. Libby as an "atomic clock" in dating early objects.

85. (B) The selection opens with the statement that the interior of the cell is distinguished from the outside world by its content of macromolecules.

86. (E) DNA contains the genetic code in the nucleus. It transmits the coded information to messenger RNA which leaves the nucleus and passes into the cytoplasm where it becomes located in a ribosome. Here, transfer RNA molecules pick up amino acid molecules and line them up on the messenger RNA, to be linked together into proteins.

87. (C) RNA is formed in the nucleus from the pattern, or template, of DNA. It then passes into the cytoplasm.

88. (C) Enzymes are organic catalysts that can speed up the rate of a biochemical reaction without being changed.

89. (D) Blue eyes is a recessive trait, in which two genes are needed for the trait to appear. Thomas and Mary are heterozygous for brown eyes, as is seen from the appearance of blue-eyed Jane. Joan and Jane are fraternal twins, since they do not have the same eye color. Joan and Samuel could be heterozygous for brown eyes, since their parents also produced a blue-eyed child.

90. (E) When heterozygous black guinea pigs are mated, the possible ratio is 1:2:1; however this applies to very large numbers. In a single litter, all the possibilities indicated may appear.

91. (A) In incomplete dominance, the hybrid shows a blending of the colors of both parents, and therefore is seen to have one gene from each.

92. (E) By crossing $\underline{X}Y$ and XX, the results would be: all normal male and all normal heterozygous female.

93. (E) The needle is flamed (4); it is touched to a colony of bacteria (1); the bacteria are spread over the slide (3); the bacteria are fixed by passing the slide through the flame (2); the bacteria are stained with a dye (5).

94. (B) This prevents clumping of bacteria.

95. (A) This kills the bacteria and attaches them to the slide.

96. (A) It is heated to sterilize it and to destroy bacteria that may be on it before it is used, and prevent contamination after it is used.

97. (D) It is a common general stain that is easily applied.

98. (D) A single bacterium will multiply repeatedly, at half-hour intervals under favorable conditions, until it forms so many bacteria that they can be seen with the naked eye as a colony.

99. (D) It is important to sterilize the needle before and after use.

100. (A) A bacillus is rod-shaped bacterium.

101. (D) By selecting only the largest beans each year, the beans containing genes for smallness are gradually eliminated. The number of large beans keeps increasing until they are most numerous, with relatively little variation.

102. (D) Equation (A) represents the reactions involved in aerobic respiration. Equation (B) represents the reactions involved in photosynthesis. A bean plant carries on respiration all the time, and photosynthesis in the light.

103. (C) Green plants are the basis of the food chain. They are also the base of the food pyramid, where they occur in the largest number. Green plants serve as the source of energy for the other organisms involved. There is a decrease in the amount of energy as it is passed along from the producer to the consumers.

104. (D) When the temperature increases 6° from 49° to 55°, the number of chirps is increased 24 (71 − 47), or about 4 chirps per degree. As the table shows, this rate of increase in the number of chirps also takes place when the temperature is raised 20°, from 55° to 75°; the number of chirps is increased from 71 to 150, or by 79 chirps ($4 \times 20 = 80$, actually). Therefore, on this basis, when the temperature increases 5° from 55° to 60°, the number of chirps would be increased by 20 (5×4), or from 71 to 91.

105. (D) An experiment needs a control to show that the results would not have occurred anyway. Distilled water, which is used to dissolve the drug, is given to a control group of mice. If the group given the drug survives, while the group given the distilled water dies, it is an indication that the drug prevented the mice from dying of cancer.

106. (B) The fine adjustment allows the high power objective to be raised or lowered a very tiny distance, and to keep a sharp focus on the chloroplasts as they are being carried by the streaming cytoplasm up and down the sides of the cell. The diaphragm adjusts the amount of light reaching the slide and needs to be adjusted so that the best condition of illumination exists.

107. (D) A coal miner, being most active physically, requires the greatest number of calories to carry on his heavy labor.

108. (A) Ribonucleic acid, or RNA, is composed of nucleotides that contain a phosphate, ribose, which is a five-carbon sugar, and a nitrogen base, either a purine (adenine or guanine), or a pyrimidine (cytosine or uracil). DNA has a deoxyribose sugar and a pyrimidine base which is thymine rather than uracil.

109. (E) The maximum activity of pepsin (X) is shown on the graph at pH 1.5-2.2. The maximum activity of maltase (Y) is shown to take place at a pH of about 7.

PRACTICE COLLEGE BOARD ACHIEVEMENT TEST IN BIOLOGY

NO. 2

PART I—TYPE A

Directions: **The diagram on the next page shows parts of the human digestive system. Certain of its parts are labeled with numbers. Each of the statements is followed by five suggested labels or answers. A label may be used once, more than once, or not at all. For each statement, select the best answer.**

1. Serves as the passageway for food from the mouth to the stomach:
(A) 1 (B) 3 (C) 5 (D) 7 (E) 9

2. Produces a hormone that enables the cells to utilize glucose:
(A) 10 (B) 5 (C) 6 (D) 3 (E) 2

3. Stores glycogen:
(A) 4 (B) 8 (C) 3 (D) 5 (E) 2

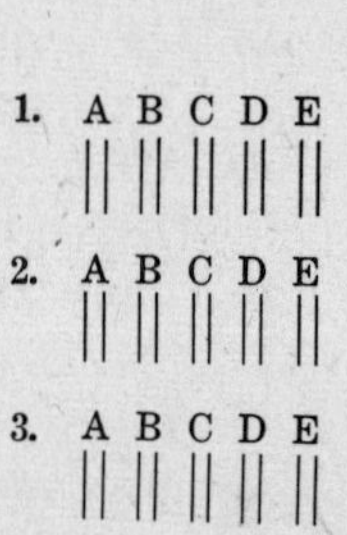

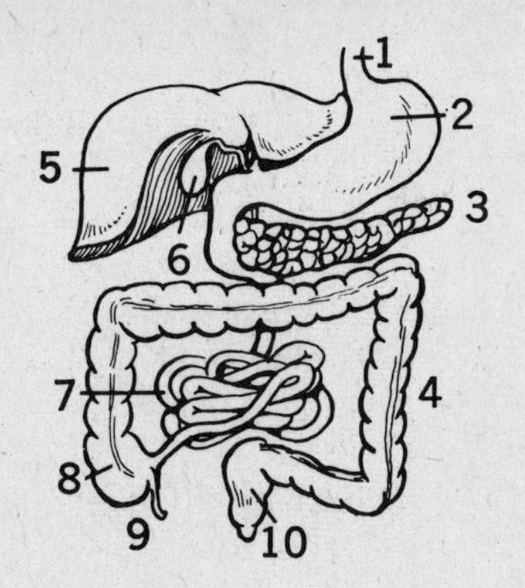

4. Stores bile:
(A) 7 (B) 6 (C) 3 (D) 9 (E) 1

4. A B C D E

5. Produces hydrochloric acid:
(A) 5 (B) 7 (C) 2 (D) 4 (E) 8

5. A B C D E

6. Serves no useful purpose to the body:
(A) 4 (B) 7 (C) 9 (D) 10 (E) 6

6. A B C D E

7. Absorbs digested fatty acids and glycerol:
(A) 7 (B) 8 (C) 10 (D) 2 (E) 3

7. A B C D E

8. Absorbs digested amino acids:
(A) 2 (B) 7 (C) 8 (D) 4 (E) 10

8. A B C D E

9. Receives indigestible remains of food:
(A) 7 (B) 9 (C) 2 (D) 1 (E) 4

9. A B C D E

10. Produces an enzyme that starts protein digestion:
(A) 5 (B) 7 (C) 3 (D) 8 (E) 2

10. A B C D E

TYPE B

Directions: **Each of the following statements is followed by five suggested answers. Select the one answer which is best in each case.**

11. The balance of nature has been upset most seriously by: (A) the effects of lightning in a forest (B) the spread of the English sparrow (C) man (D) spring floods (E) the spread of the gypsy moth

11. A B C D E

12. When two glucose ($C_6H_{12}O_6$) molecules are combined to form a molecule of maltose ($C_{12}H_{22}O_{11}$), the formula of the latter is not $C_{12}H_{24}O_{12}$ because: (A) hydrolysis takes place (B) dehydration synthesis takes place (C) transpiration takes place (D) polypeptides are formed (E) water is added

12. A B C D E

13. Colchicine is a chemical which has been used successfully on plants to (A) stimulate the formation of chromosomal mutations (B) stimulate the production of ADP (C) stimulate the evolution of CO_2 (D) prevent self-pollination (E) prevent fruit drop

13. A B C D E

14. Of the following terms, the one that includes all the others is (A) oxidation (B) respiration (C) excretion (D) metabolism (E) digestion

14. A B C D E

15. An animal that has been completely exterminated by man is (A) dinosaur (B) bison (C) passenger pigeon (D) whale (E) American eagle

15. A B C D E

16. The structure in mammals where urea is removed from the blood is the (A) alveolus (B) spleen (C) nephron tubule (D) gall baldder (E) villus

16. A B C D E

17. The type of cell which contains an axon is part of (A) bone tissue (B) adipose tissue (C) epithelial tissue (D) nerve tissue (E) connective tissue

17. A B C D E

18. All of the following reproduce by asexual reproduction except (A) paramecium (B) geranium (C) tulip (D) hydra (E) horse

18. A B C D E

19. In flowering plants, fertilization takes place in the (A) corolla (B) ovule (C) calyx (D) pollen tube (E) pollen grain

19. A B C D E

20. Choose the group of ecological terms which is in correct order, from simplest to most complex: (A) organism–population–community–ecosystem–biosphere (B) organism–population–community–biosphere–ecosystem (C) population–organism–community–ecosystem–biosphere (D) biosphere–ecosystem–organism–community–population (E) ecosystem–biosphere–organism–community–population

20. A B C D E

21. When an astronaut experiences acceleration during blast-off, there is an increase in (A) DNA (B) PGAL (C) RNA (D) G (E) RDP

21. A B C D E

22. The first living things on earth probably inhabited the (A) water (B) soil (C) air (D) rocks (E) sand

22. A B C D E

23. Of the following types of animals, the one that is believed to have appeared after reptiles is (A) amphibia (B) fish (C) birds (D) starfish (E) eels

23. A B C D E

24. Auxin serves to increase the rate of (A) digestion (B) circulation (C) reproduction (D) growth (E) variation

24. A B C D E

TYPE C

Directions: **For each of the scientists listed from 25 to 34, write the letter (A), (B), (C), or (D), depending on whether his name**

(A) applies only to (A)
(B) applies only to (B)
(C) applies to both (A) and (B)
(D) applies to neither (A) nor (B)

[(A)–genetics (C)–both
(B)–cell studies (D)–neither]

25. Watson — 25. A B C D

26. Morgan — 26. A B C D

27. Lysenko — 27. A B C D

28. Beadle — 28. A B C D

29. Lamarck — 29. A B C D

30. Leeuwenhoek — 30. A B C D

31. Pavlov — 31. A B C D

32. Crick — 32. A B C D

33. Mendel — 33. A B C D

34. Oparin — 34. A B C D

TYPE D

Directions: **The following questions require two answers. One of the items in the left column** (A, B, or C) **is related to four of the items in the right column** (numbered 1–5). **Select the letter of the item which is so related; then select the number of the remaining item which does not belong.**

EXAMPLE:

(A) paramecium
(B) ameba
(C) spirogyra

1. cilia
2. macronucleus
3. trichocyst
4. pyrenoid
5. micronucleus

35. (A) herbivore
(B) carnivore
(C) decomposer

1. lion
2. vulture
3. hawk
4. fungus
5. frog

35. A B C 1 2 3 4 5

36. (A) ameba
(B) yeast
(C) spirogyra

1. food vacuole
2. cellulose
3. contractile vacuole
4. cyst
5. pseudopod

36. A B C 1 2 3 4 5

37. (A) muscle tissue
(B) adipose tissue
(C) epithelial tissue

1. skeletal
2. cardiac
3. smooth
4. striated
5. ciliated

37. A B C 1 2 3 4 5

38. (A) vitamin
(B) antibody
(C) hormone

1. riboflavin
2. adrenin
3. insulin
4. thyroxin
5. cortisone

38. A B C 1 2 3 4 5

39. (A) excretion
(B) blood clotting
(C) enzyme formation

1. vitamin A
2. platelet
3. fibrinogen
4. thrombin
5. vitamin K

39. A B C 1 2 3 4 5

40. (A) assimilation
(B) digestion
(C) respiration

1. alveolus
2. trachea
3. retina
4. bronchus
5. diaphragm

40. A B C 1 2 3 4 5

41. (A) ecology
(B) atomic calendar
(C) drugs

1. lead
2. heroin
3. radiocarbon
4. half-life
5. uranium

41. A B C 1 2 3 4 5

TYPE E

Directions: **The five lettered headings below are followed by a list of numbered statements. For each numbered statement, choose the one lettered heading (A–E) which is most closely related to it.**

(A) Natural selection
(B) Passive immunity
(C) Heterotroph hypothesis
(D) Vestigial structure
(E) Use and disuse

42. Strains of mosquitoes have appeared that are not affected by DDT.

42. A B C D E

43. A steelworker develops large shoulder muscles.

43. A B C D E

44. Amino acids are formed from molecules acted on by various forms of energy.

44. A B C D E

45. The foot of the modern horse has two splints on either side of the large toe.

45. A B C D E

(A) Test cross
(B) Mutation
(C) Sex Linkage
(D) Independent assortment
(E) Segregation

46. How is hemophilia inherited?

46. A B C D E

47. How do tall pea plants produce short pea plants?

47. A B C D E

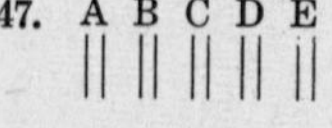

48. How is a gene affecting pigmentation changed to produce an albino?

48. A B C D E

49. How do dark-haired parents with brown eyes have a child with blond hair and brown eyes?

49. A B C D E

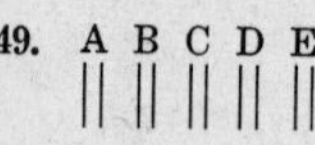

50. How can a black guinea pig be shown to be either homozygous black or heterozygous black?

50. A B C D E

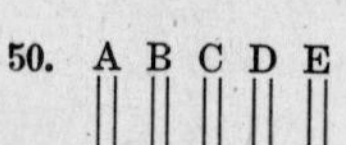

51. How did the seedless orange appear?

51. A B C D E

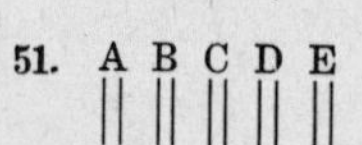

PART II

Directions: **A male hamster was allowed to go through a maze for a number of trials. Its progress was timed, and plotted on the accompanying graph. Referring to the graph, select the best answer for each of the following questions.**

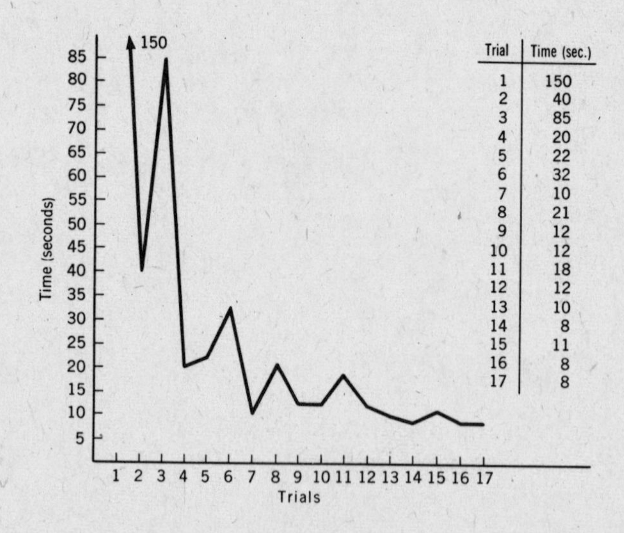

Trial	Time (sec.)
1	150
2	40
3	85
4	20
5	22
6	32
7	10
8	21
9	12
10	12
11	18
12	12
13	10
14	8
15	11
16	8
17	8

52. The hamster apparently learned the maze after trial number (A) 13 (B) 14 (C) 15 (D) 16 (E) 17

53. The hamster made its greatest improvement in trial number (A) 2 (B) 4 (C) 6 (D) 8 (E) 10

54. The least amount of improvement was in trials (A) 1–3 (B) 6–8 (C) 5–7 (D) 10–12 (E) 11–13

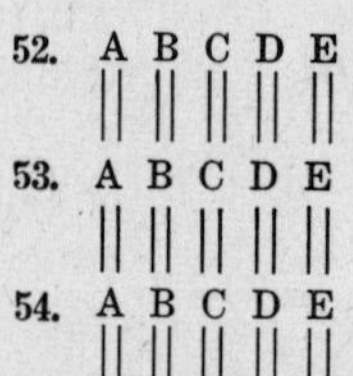

55. After it had learned the maze without error, the shortest time taken by the hamster in going through was (A) 8 seconds (B) 10 seconds (C) 12 seconds (D) 20 seconds (E) 30 seconds

55. A B C D E

56. The hamster went through the maze the first time by (A) instinct (B) insight (C) trial and error (D) reflex (E) habit

56. A B C D E

57. If the female mate of the hamster were permitted to watch him as he learned the maze, the amount of time needed by her to learn it could be predicted as (A) being less (B) being the same (C) being less only at the end (D) being less only at the beginning (E) being uncertain

57. A B C D E

58. If this maze were to be used for a white mouse, the rate of learning would be (A) identical (B) almost identical (C) slower (D) faster (E) unpredictable

58. A B C D E

PART III—TYPE A

Directions: **In each of the following questions there is a statement followed by four choices, of which *one* or *more* are correct. Indicate the correct combination by choosing the following letters:**

(A) if only 1, 2, and 3 are correct
(B) if only 1 and 3 are correct
(C) if only 2 and 4 are correct
(D) if only 4 is correct
(E) if some other combination is correct

DIRECTIONS SUMMARIZED:

(A)	(B)	(C)	(D)	(E)
1, 2, 3 only	1, 3 only	2, 4 only	4 only	some other combination

59. Mitochondria are considered to be the powerhouses of the cell because they
1. form DNA nucleotides
2. contain enzymes associated with energy release
3. are the site of protein synthesis
4. are the site of energy release

59. A B C D E

60. Toads are classified as amphibia because they have
1. microscopic eggs
2. a three-chambered heart
3. a warm-blooded metabolism
4. a tadpole stage

60. A B C D E

61. The fibrovascular bundles of a plant
1. conduct water upwards
2. contain supporting tissue
3. conduct food materials downwards
4. contain xylem cells

61. A B C D E

62. The nucleus of a cell
1. contains DNA
2. forms chromosomes
3. controls reproduction
4. contains mitochondria

62. A B C D E

63. The roots of a plant serve
1. to absorb water
2. to absorb dissolved minerals
3. to anchor the plant
4. to form carbohydrates

63. A B C D E

64. The left atrium of the human heart
1. receives blood from the vena cava

64. A B C D E

2. receives blood from the left ventricle
3. receives deoxygenated blood
4. receives oxygenated blood

65. The pituitary gland is important because it
1. regulates other endocrine glands
2. combines with vitamin C to prevent scurvy
3. controls growth of the skeleton
4. is both an endocrine and a digestive gland

65. A B C D E

66. In response to the stimulus of light, a green plant
1. carries on photosynthesis
2. shows positive phototropism
3. carries on respiration
4. stores food

66. A B C D E

67. The autonomic nervous system of man
1. directs internal reflexes
2. is responsible for automatic thoughts
3. carries on activities without our being aware of them
4. determines our conscious activities

67. A B C D E

TYPE B

Directions: **Each of the following statements is followed by five suggested answers. Select the one answer which is best in each case.**

68. Of the following organic compounds, the one that represents a protein is (A) $C_{12}H_{22}O_{11}$ (B) $C_6H_{12}O_6$ (C) $C_{17}H_{35}COOH$ (D) $(C_6H_{10}O_5)_n$ (E) $C_{708}H_{1130}O_{224}N_{180}S_4P_4$.

68. A B C D E

69. Near the equator, at 5°S latitude, the top of a seven-thousand foot mountain would have a biome that would most closely resemble (A) tropical rain forest (B) desert (C) tundra (D) coniferous forest (E) taiga

69. A B C D E

70. Bread will not spoil and become green as rapidly in a refrigerator as at room temperature because the cooling serves to (A) eliminate protein synthesis (B) inhibit the growth of mold (C) retain a normal moisture content (D) stimulate starch hydrolysis (E) check germination of yeast spores

70. A B C D E

71. Rivers become muddy because (A) of salmon migration upstream (B) of the use of check-dams to reduce erosion (C) beaver dams are built (D) strip-cropping loosens the soil (E) of erosion resulting from forest destruction

71. A B C D E

72. Cigarette smoking is causing concern among doctors because (A) they cannot agree on which is the best filter (B) tobacco is not as good as it used to be (C) of the increase in the incidence of lung cancer (D) the DDT used to spray the tobacco plants comes out in the smoke (E) advertisements are incorrect in claiming that cigarettes give a person a lift

72. A B C D E

73. That evolution has occurred is now accepted as a doctrine rather than a theory because (A) Congress passed a law to that effect in 1958 (B) Darwin received a posthumous Nobel Prize for his book on natural selection (C) it is now certain that the earth is at least four billion years old (D) there is ample evidence that species have changed over the ages (E) uranium-lead studies have identified the history of the horse

73. A B C D E

74. Ciliated epithelial cells in the nasal passages are useful because they (A) keep back dust and bacteria (B) provide the sense of smell (C) reduce the breathing rate when the air is impure (D) reduce the humidity of inhaled air (E) fan the air and cool it, especially in summer

74. A B C D E

75. Cro-Magnon man is thought to be a type of *Homo sapiens* because (A) he lived in caves (B) he knew how to use fire (C) he made weapons of stone (D) he lived together in small groups (E) his brain capacity was about the same as ours

75. A B C D E

76. Plasma is used in most transfusions rather than whole blood because (A) it can be made artificially (B) it only requires the addition of dehydrated red blood cells (C) it does not require typing (D) type O blood has been added to it (E) it does not contain figrinogen for clotting

76. A B C D E

77. Legume plants enrich the soil because (A) they remove selenium, a deadly poison (B) they encourage the breeding of earthworms (C) their roots contain nitrogen-fixing bacteria (D) the underground stems of potatoes are considered to be tubers (E) their root systems penetrate six feet below the surface

77. A B C D E

78. The Drosophila fly is useful in studies of heredity because (A) its eggs hatch every other month (B) it is immune to Texas fever (C) it may breed by spontaneous generation (D) it carries on conjugation during unfavorable conditions (E) it is easily bred

78. A B C D E

TYPE C

Directions: **Each item consists of an *assertion* (statement) in the left-hand column and a *reason* on the right-hand side. Select**

(A) if both assertion and reason are true, and *are related* as cause and effect
(B) if both assertion and reason are true, but *are not related* as cause and effect
(C) if the assertion is true, but the reason is a false statement
(D) if the assertion is false, but the reason is a true statement
(E) if both assertion and reason are false statements

DIRECTIONS SUMMARIZED:

	(A)	(B)	(C)	(D)	(E)
Assertion:	True	True	True	False	False
Reason:	True	True	False	True	False
	Cause and effect	*Not* cause and effect			

Assertion:	Reason:	
79. Messenger RNA is important in protein synthesis	BECAUSE it carries the code from DNA to the ribosomes	79. A B C D E
80. A spider is not considered to be a member of the insect class	BECAUSE it is an arthropod	80. A B C D E
81. A man in a rocket may black-out during the take-off stages	BECAUSE of weightlessness	81. A B C D E
82. A radioactive isotope may be detected with a microscope	BECAUSE it gives off radiations	82. A B C D E
83. The whooping crane is slowly beginning to increase in number	BECAUSE game laws protect it	83. A B C D E
84. Contour planting is practiced on a hillside	BECAUSE crop rotation enriches the soil	84. A B C D E
85. Nordics are a pure race	BECAUSE they tend to have blond hair	85. A B C D E
86. Fossils are found in igneous rock	BECAUSE they are quickly formed there	86. A B C D E
87. In the struggle for existence, the fittest usually survive	BECAUSE all variations are inherited in the next generation	87. A B C D E
88. The eyes of an albino animal appear pink	BECAUSE they have no pigment	88. A B C D E
89. Grafting is a way of producing new types of plants	BECAUSE it combines traits of the stock and the scion	89. A B C D E

PART IV—TYPE A

Directions: **Referring to the following selection, choose the best answer for each of the questions that follows:**

*"From the centers of respiration and photosynthesis the same well-defined molecule—adenosine triphosphate (ATP)—carries the free energy extracted from foodstuffs or from sunlight to all the energy-expending processes of the cell. ATP, which was first isolated from muscle by K. Lohmann of the University of Heidelberg some 30 years ago, contains three phosphate groups linked together. In the test tube the terminal group can be detached from the molecule by the drastic, one-step reaction of hydrolysis to yield adenosine diphosphate (ADP) and simple phosphate. As this reaction proceeds, the free energy of the ATP molecule appears as heat and entropy, in accordance with the second law of thermodynamics. In the cell, however, the terminal phosphate group is not merely detached by hydrolysis but is transferred to a specific acceptor molecule. The free energy of the ATP molecule is largely conserved by 'phosphorylation' of the acceptor molecule, the energy content of which is now raised so that it can participate in an energy-requiring process such as biosynthesis or muscle contraction. Left over from this 'coupled reaction' is ADP. In the thermodynamics of the cell ATP may be considered as the energy-rich, or 'charged' form of the energy carrier and ADP as the energy-poor, or 'discharged' form."

90. The best title of this selection is: (A) The 2nd Law of Thermodynamics (B) Phosphorylation (C) Free Energy of ATP (D) Superior Energy Content of ADP (E) Lohmann's Phosphate Fantasy

90. A B C D E

91. The number of phosphate groups linked together in ATP is (A) 1 (B) 2 (C) 3 (D) 4 (E) 0

91. A B C D E

92. According to the selection, ATP carries free energy from (A) food and sunlight (B) only food (C) only sunlight (D) respiration (E) photosynthesis

92. A B C D E

93. The terminal phosphate group may be detached from ATP artificially in the test tube and naturally in the cell. The difference is that in the cell, (A) the terminal group yields ADP (B) the free energy appears as heat and entropy (C) the terminal group is detached by hydrolysis (D) the terminal group is transferred to an acceptor molecule (E) the free energy is stored for photosynthesis

93. A B C D E

94. The ATP → ADP relationship may be summed up as (A) energy-rich → charged (B) energy-poor → discharged (C) energy rich → discharged (D) energy-poor → charged (E) discharged → charged

94. A B C D E

95. ATP was first isolated from (A) food (B) sunlight (C) photosynthesis (D) muscle (E) adenosine diphosphate

95. A B C D E

TYPE B

Directions: **Each of the following problems is followed by five suggested answers. Select the one answer which is best in each case.**

96. The half-life of carbon-14 is 5,760 years; at the end of this time its radioactivity is reduced by 50%. A sample of charcoal from a cave containing remains of primitive man was estimated to be 17,280 years old because its radioactivity had been reduced to (A) one-half (B) one-quarter (C) one-eighth (D) three-quarters (E) two-thirds

97. A man who is normal for color vision marries a normal heterozygous woman. The chances of their sons being color blind is (A) zero (B) 100% (C) 50–50 (D) 25–50–25 (E) 25–75

97. A B C D E

*From "How Cells Transform Energy" by Albert L. Lehninger, Sept. 1961, by *Scientific American, Inc.*

98. Japanese four-o'clock flowers were crossed, and produced only offspring with pink flowers. The parent flowers were (A) pink and pink (B) red and white (C) pink and red (D) pink and white (E) red and blue

98. A B C D E

99. Two coins are tossed simultaneously a hundred times, and the results are tallied as: tails-tails (TT), heads-tails (HT), heads-heads (HH). According to the laws of probability, the most likely results would be (A) 50TT; 50HH (B) 50TT; 50TH (C) 25TT, 50HH, 25TH (D) 25TT; 50TH; 25HH (E) 25HT, 50TT, 25HH

99. A B C D E

100. In some of Mendel's experiments, three-quarters of the offspring showed the dominant trait. The parents were (A) both recessive (B) both heterozygous (C) both dominant (D) one heterozygous, the other homozygous dominant (E) one recessive, the other homozygous dominant

100. A B C D E

PART V

Directions: **Each of the following questions has five suggested answers. Select the one answer that is best in each case.**

101. Assume that John's father has type *A* blood and his mother has type *O* blood. John's blood type will most likely be (A) *A* or *B* (B) *A* or *O* (C) *A* or *AB* (D) *B* or *O* (E) *AB* or *O*

101. A B C D E

102. Study the accompanying diagram. It shows the results of Englemann's experiment, in which he found clusters of bacteria gathered in different amounts along a green alga filament which was being illuminated with the spectrum of light from a prism.

102. A B C D E

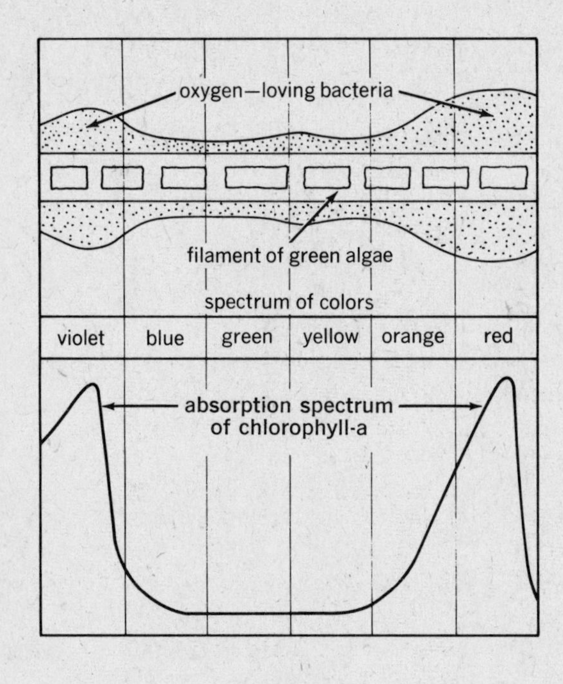

He noted that (A) the bacteria concentrated where oxygen production by photosynthesis was the greatest, in violet and blue light (B) the bacteria concentrated where oxygen production by photosynthesis was the greatest, in violet and red light (C) the bacteria concentrated where oxygen production by photosynthesis was the greatest, in green light (D) the filament of green alga prevented the bacteria from absorbing light in the green zone (E) the bacteria produced the greatest amounts of oxygen in the red zone

103. In a rare disease known as malignant hyperpyrexia, the muscles become extremely rigid. Research on the muscles indicates that there is an abnormality of their calcium metabolism, causing them to lock in contraction.

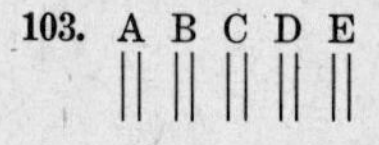
103. A B C D E

In seeking a treatment, doctors would try
(A) X-ray treatment, because the radiations may kill some of the cells
(B) ultra-violet treatment, because it sterilizes the air and kills bacteria
(C) cobalt-60 radiation which inhibits the growth of cancer cells
(D) a drug called procaine, because it is known to block some of calcium's effect on muscle physiology
(E) vitamin D, which helps in proper bone formation

104. Some plants, such as asters and chrysanthemums, are considered to be short-day plants because they flower in the fall, when there are less than 12 hours of daylight. To get them to flower in July, when there are more than 12 hours of sunlight, a botanist would
(A) place a transparent plastic cover over them every night
(B) place a transparent plastic cover over them every morning
(C) place them in a darkroom after 11 hours of daylight each afternoon, and move them out of doors again early each morning
(D) place them in a darkroom at 11 P.M. each night and return them out of doors early each morning
(E) cover them with an opaque box every night

104. A B C D E

PART VI

Directions: **Each of the following questions has five suggested answers. Select the one answer that is best in each case.**

105. When a chewed cracker is tested for the presence of simple sugar, a suitable control would be (A) to test the saliva with iodine (B) to test the unchewed cracker for the presence of simple sugar (C) to test the unchewed cracker with iodine (D) to heat the chewed cracker slowly in alcohol (E) to use Benedict's solution without heating

105. A B C D E

106. The best method of determining the relation of leaf-surface area to transpiration rate would be to compare the transpiration rates of
(A) two identical geranium plants exposed to different wind velocities
(B) two identical geranium plants exposed to different temperatures
(C) a normal geranium plant and one whose stomates have been closed by spreading vaseline over the lower epidermis of the leaves
(D) two identical geranium plants, one with the leaves removed
(E) a geranium plant placed in the light, and one placed in the dark

106. A B C D E

107. In examining the tail of a living goldfish with a microscope, the capillaries can best be found by looking for (A) the blood vessels in which the red blood corpuscles pass through in single file (B) the small blood vessels in which there is a pulsating flow of blood (C) the blood vessels which do not have valves (D) the blood vessels which flow toward the end of the tail (E) the blood vessels which flow toward the head

107. A B C D E

108. In an experiment to determine whether germinating bean seeds carry on respiration, the liquid indicator used would be (A) hydrogen peroxide, because of its oxygen content (B) hydrochloric acid because it liberates carbon dioxide when added to marble chips (C) lime water because it turns cloudy in the presence of carbon dioxide (D) glucose because it furnishes energy (E) auxin because it is the growth-promoting substance in plants

108. A B C D E

109. After watching the behavior of earthworms in the ground, a biologist suggested that the penetration of air into the soil promotes root development of plants. He then set up the following experiment:

109. A B C D E

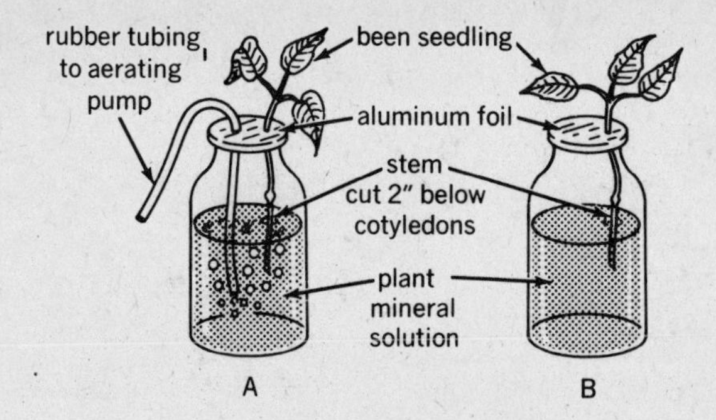

The important data to be recorded in the experiment will come from the observation of the increase in (A) leaf size (B) stem size (C) number of leaves (D) number of roots (E) number of buds

ANSWER KEY: TEST NO. 2

1. (A)	**12.** (B)	**23.** (C)	**34.** (D)	**45.** (D)	**56.** (C)	**67.** (B)	**78.** (E)	**89.** (E)	**100.** (B)
2. (D)	**13.** (A)	**24.** (D)	**35.** (B)4	**46.** (C)	**57.** (E)	**68.** (E)	**79.** (A)	**90.** (C)	**101.** (B)
3. (D)	**14.** (D)	**25.** (C)	**36.** (A)2	**47.** (E)	**58.** (E)	**69.** (C)	**80.** (B)	**91.** (C)	**102.** (B)
4. (B)	**15.** (C)	**26.** (C)	**37.** (A)5	**48.** (B)	**59.** (C)	**70.** (B)	**81.** (C)	**92.** (A)	**103.** (D)
5. (C)	**16.** (C)	**27.** (D)	**38.** (C)1	**49.** (D)	**60.** (C)	**71.** (E)	**82.** (D)	**93.** (D)	**104.** (C)
6. (C)	**17.** (D)	**28.** (A)	**39.** (B)1	**50.** (A)	**61.** (E)	**72.** (C)	**83.** (A)	**94.** (C)	**105.** (B)
7. (A)	**18.** (E)	**29.** (D)	**40.** (C)3	**51.** (B)	**62.** (A)	**73.** (D)	**84.** (B)	**95.** (D)	**106.** (D)
8. (B)	**19.** (B)	**30.** (B)	**41.** (B)2	**52.** (C)	**63.** (A)	**74.** (A)	**85.** (D)	**96.** (C)	**107.** (A)
9. (E)	**20.** (A)	**31.** (D)	**42.** (A)	**53.** (A)	**64.** (D)	**75.** (E)	**86.** (E)	**97.** (C)	**108.** (C)
10. (E)	**21.** (D)	**32.** (C)	**43.** (E)	**54.** (D)	**65.** (B)	**76.** (C)	**87.** (C)	**98.** (B)	**109.** (D)
11. (C)	**22.** (A)	**33.** (A)	**44.** (C)	**55.** (A)	**66.** (E)	**77.** (C)	**88.** (A)	**99.** (D)	

ANSWERS EXPLAINED: TEST NO. 2

1. (A) esophagus
2. (D) pancreas
3. (D) liver
4. (B) gall bladder
5. (C) stomach
6. (C) appendix
7. (A) small intestine
8. (B) small intestine
9. (E) large intestine
10. (E) stomach
11. (C) The spread of man's civilization has interfered with wildlife, forests, and soil.
12. (B) In dehydration synthesis, large organic molecules are built up from smaller building blocks, with the release of water. When two glucose molecules are linked to form maltose, a hydrogen atom (H) is removed from one glucose and a hydroxyl group (OH) is removed from the other, to form a molecule of water.
13. (A) When the chemical colchicine is applied to the rapidly growing parts of a plant, such as the roots or the bud tips, or to seedlings, the chromosome number in the cell is doubled, resulting in a condition called polyploidy.
14. (D) Metabolism refers to all of the life activities.
15. (C) The last passenger pigeon died in 1914.
16. (C) As blood flows through each of the many dense networks of capillaries in the kidney called the glomerulus, a liquid containing urea diffuses into part of a urinary tubule associated with it.
17. (D) The axon is the long extension of a neuron that may be two or three feet in length, in some cases.
18. (E) The paramecium reproduces by binary fission; the geranium can be reproduced from a cutting; the tulip develops from a bulb; hydra may form buds. The horse reproduces sexually, producing egg and sperm cells.
19. (B) The egg nucleus is contained in the embryo sac located within the ovule. When the pollen tube grows into the

ovule, a sperm nucleus enters and fertilizes the egg nucleus.

20. (A) An organism is either a single-celled or a many-celled living thing. Large numbers of organisms of one species make up a population. A community consists of all the plant and animal populations interacting in a given environment. An ecosystem is a self-sustaining living community in relation to the physical environment. The biosphere is that portion of the earth in which ecosystems operate, and includes the biologically inhabited soil, water and air.

21. (D) The force of gravity (G) acting on the astronaut is very great as his space vehicle gathers ever-increasing speed.

22. (A) It is believed that the original simplest forms of life developed in the water where conditions for existence were suitable.

23. (C) All of the other forms were already present when reptiles appeared on the scene.

24. (D) Auxin is a plant growth hormone that causes the cells to increase in growth.

25. (C) James D. Watson was one of the scientists who first described the structure of the DNA molecule, and received the Nobel Prize in 1962.

26. (C) Morgan received the Nobel Prize for his studies of heredity in Drosophila.

27. (D) Lysenko is a Russian scientist who believes that acquired characteristics can be inherited.

28. (A) Beadle provided evidence, from his research on the mold *Neurospora*, that a single gene governs the synthesis of a single enzyme in the cell.

29. (D) Lamarck proposed a theory of evolution based on the inheritance of acquired characteristics.

30. (B) Leeuwenhoek was among the earliest microscopists to see the cells of living things.

31. (D) Pavlov studied conditioned reflexes in dogs.

32. (C) Francis H. C. Crick worked with Watson in describing the structure of DNA, and shared the 1962 Nobel Prize with him.

33. (A) Mendel is considered to be the father of heredity.

34. (D) Alexander I. Oparin proposed the heterotroph hypothesis to explain the origin of simple life on earth.

35. (B)4 All of the animals are meat-eating, or carnivores, except for the fungus which is generally a decomposer.

36. (A)2 All are structures associated with ameba, except cellulose, which is found only in plant cells.

37. (A)5 All are types of muscle tissue except ciliated, which refers to a type of epithelial tissue containing cilia.

38. (C)1 All are hormones except riboflavin, which is vitamin B_2.

39. (B)1 All play a role in blood clotting except vitamin A, which is needed for healthy eye condition.

40. (C)3 All deal with respiration except retina, which is the part of the eye that receives light impulses.

41. (B)2 Because radioactive elements decompose into other elements at a fixed rate, they can be used as an atomic calendar in determining the age of various remains. Heroin is a habit-forming drug.

42. (A) By natural selection, mosquitoes that were unaffected by DDT survived, and produced a new variety of mosquitoes.

43. (E) Exercise of his muscles causes them to increase in size.

44. (C) According to the heterotroph hypothesis, amino acids were bonded together in the "hot thin soup" of primitive oceans from molecules originating in the gases of the atmosphere, under the influence of such forms of energy as solar radiation, cosmic rays, lightning and the earth's radioactivity.

45. (D) The ancestor of the horse, Eohippus, had four toes on its front legs. One of these became the foot of the modern horse, while the splints represent the remains of two of the other toes.

46. (C) Hemophilia is inherited as a sex-linked recessive trait.

47. (E) Heterozygous tall pea plants contain recessive genes for shortness which segregate to produce short plants.

48. (B) The change in a normal gene for pigment to the albino condition is inherited and is known as a mutation.

49. (D) These traits are inherited independently of each other.

50. (A) It is crossed with a white guinea pig possessing two recessive genes. If one or more white offspring are produced, the black guinea pig is heterozygous; if only black offspring are produced, it is probably homozygous black.

51. (B) The seedless orange appeared as a mutation on an orange tree that had only oranges with seeds.

52. (C) After trial number 15, the time required to go through the maze was at a minimum, indicating that the hamster was able to go directly to the end without any mistakes.

53. (A) In the second trial, the time was reduced from 150 seconds to 40 seconds.

54. (D) During trials 10, 11, and 12, the times were 12, 18, and 12 seconds, representing the least amount of improvement.

55. (A) The hamster went through the maze without an error, when it took only 8 seconds.

56. (C) The hamster entered many blind alleys as it went through the maze. It went through the correct alleys by chance. Its progress was determined by trial and error.

57. (E) There is no certainty that the female would learn by the experience of the male in going through the maze.

58. (E) Another animal, such as a white mouse, would learn the maze after its own series of trials and errors. It is not possible to predict how much time would be needed.

59. (C) Mitochondria are rod-like structures with inner, folded walls, that are located in the cytoplasm. They contain ATP, and act as sites of cellular respiration.

60. (C) Toads have eggs that can easily be seen with the naked eye; they are also cold-blooded.

61. (E) All choices are correct.

62. (A) Mitochondria are found in the cytoplasm outside of the nucleus.

63. (A) Carbohydrates are formed in green leaves during photosynthesis.

64. (D) The left atrium receives oxygenated blood coming from the lungs via the pulmonary veins.

65. (B) The pituitary gland is also known as the master gland.

66. (E) Choices 3 and 4 are carried on independently of the presence of light.

67. (B) The autonomic nervous system is outside of the central nervous system and does not function in thought or in conscious activities.

68. (E) A protein contains not only carbon (C), hydrogen (H) and oxygen (O), but also nitrogen (N) and sometimes other elements such as sulfur (S) and phosphorus (P).

69. (C) At high altitudes, the climate is as cold as it is in the north.

70. (B) At the low temperature of a refrigerator, metabolism of living things is reduced, and growth is inhibited.

71. (E) When a forest is destroyed, the trees can no longer bind the soil with their roots. Rains loosen the soil and carry it down to the rivers, which become muddy in appearance.

72. (C) The great increase in the number of cases of lung cancer has been linked to the increase in cigarette smoking in recent years.

73. (D) Evidence for evolution is taken from many fields, including paleontology, comparative anatomy, embryology, vestigial structures, physiology, geographic distribution, classification, and heredity.

74. (A) The cilia beat dust and bacteria outward, thus keeping them out of the respiratory passages.

75. (E) Of all the types of primitive man, Cro-Magnon resembles modern man the most in brain capacity and size. His drawings on the walls of caves are evidence of an advanced type of intelligence.

76. (C) When plasma is separated from whole blood, the red blood cells are removed. This avoids the problem of agglutination during a transfusion.

77. (C) Nitrogen-fixing bacteria are located in the nodules of legume plants. They convert nitrogen, which cannot be used directly by plants, into nitrates, which can be used.

78. (E) The Drosophila fly can be grown in half-pint bottles and stored on a shelf in large numbers. Its life cycle is completed within two weeks.

79. (A) Protein synthesis takes place in the ribosomes, where messenger RNA brings the DNA code from the nucleus.

80. (B) A spider is a member of the arthropod phylum, but it is not classified as an insect, because it has eight legs and only two main body parts.

81. (C) During take-off, the acceleration force, or G, may be so great that blood does not reach the brain fast enough, causing blackout. Weightlessness does not produce this effect.

82. (D) A radioactive isotope is detected by an instrument such as a Geiger counter. The radiations enter the tube and cause a tiny amount of current to flow, producing a clicking sound.

83. (A) The whooping crane was on the verge of extinction in 1939, when it numbered only 14 individuals. As the result of game laws protecting it, its numbers have increased to 38 or more.

84. (B) Contour plowing on a hillside helps to conserve the soil. The furrows are plowed around the hill, rather than up and down, thus preventing the water from rushing down during a rainstorm and carving away the topsoil. Crop rotation is unrelated to this method of conserving the soil. Instead, it is a way of enriching the soil with nitrogen compounds, when legume plants are planted in a field that is used at other times for growing corn.

85. (D) There are no pure races of mankind because of extensive interbreeding among human beings. The Nordic type is found in northern Europe, and is characterized by blond hair and blue eyes.

86. (E) Fossils are found in sedimentary rock, where they are formed over long periods of time as the sediment changes to rock. Any organism that came in contact with molten lava that was cooling into igneous rock would be burned to a crisp.

87. (C) Darwin pointed out in his theory of natural selection that there is a struggle for existence among living things. Those with the most favorable variations are the fittest, and will survive. However, it has been shown that variations caused by the environment are not inherited.

88. (A) An albino is lacking in pigment. An animal such as a white mouse therefore has pink eyes because of the presence of blood vessels in the eyes; there is no pigment to give the eyes another color.

89. (E) In grafting, the genetic qualities of the scion and the stock remain unchanged. The scion receives only nutritional materials from the stock.

90. (C) The paragraph refers several times to the fact that ATP carries the free energy extracted from foodstuffs.

91. (C) ATP is adenosine *tri*phosphate, which indicates that there are three phosphate groups in the molecule.

92. (A) The opening statement of the selection indicates that ATP carries the free energy extracted from foodstuffs or from sunlight.

93. (D) In the cell, the terminal phosphate is not detached merely by hydrolysis, but is transferred by a specific acceptor molecule.

94. (C) In the thermodynamics of the cell, ATP may be considered as the energy-rich, or "charged" form of the energy carrier and ADP as the energy-poor, or "discharged" form.

95. (D) ATP was first isolated from muscle by K. Lohmann of the University of Heidelberg some 30 years ago.

96. (C) After 5,760 years, its radioactivity is reduced in half; after another 5,760 years it is reduced to a quarter; and after another 5,760 years, it is reduced to an eighth.

97. (C) By crossing XY and X$\underline{X}$,

	X	Y
X	XX	XY
$\underline{X}$	$\underline{X}$X	$\underline{X}$Y

The possible results are: Male: 50% normal, 50% color-blind; Female: 100% normal (50% heterozygous)

98. (B) When red and white four-o'clock flowers are crossed, the results are 100% pink; each offspring receives one gene for white and one gene for red, and is pink. This illustrates incomplete dominance, or blending

	W	W
R	RW	RW
R	RW	RW

99. (D) In a large number of tosses, by chance, the coins would sort out in a 1:2:1 ratio. If they were tossed only a few times, this ratio would not apply.

100. (B) When hybrids are crossed, the genes segregate into the gametes, and then recombine in the fertilized eggs in a ratio of 25% homozygous dominant, 50% heterozygous dominant, and 25% recessive. The total percentage of dominant offspring is therefore 75%, or three-quarters of all.

101. (B) Red blood cells may contain either *A* or *B* proteins for either type *A* or type *B* blood. These proteins are determined by genes *A* and *B*. Both of these genes are dominant to gene *O*, which does not cause either protein to be produced, and results in blood type *O*. When the genes *A* and *B* are both present, they produce both proteins, resulting in blood type *AB*. The following diagram shows that John's father, with type *A* blood can have either *AA* or *AO* genes, while his mother, with *O* blood can have only *OO* genes.

	A	A	Results:
O	AO	AO	100% AO–Type *A*
O	AO	AO	

	A	O	
O	AO	OO	50% AO–Type *A*
O	AO	OO	50% OO–Type *O*

102. (B) Chlorophyll absorbs certain wavelengths of light more effectively than others. Most absorption takes place at the blue-violet, and orange-red ends of the spectrum. Here the rate of photosynthesis is highest, and the most oxygen is liberated.

103. (D) Since calcium metabolism is abnormal during the disease, the muscles contract. The drug procaine appears to block some of calcium's effects on the muscles, and may be useful in overcoming the effects of the disease.

104. (C) This is an example of photoperiodism. If the plants are given only 11 hours of light daily, they will be stimulated to produce flowers during the summer.

105. (B) The cracker should be tested first to show that it does not contain any simple sugar. Then, it can be safely inferred that sugar in the chewed cracker was produced by the digestion of the starch into sugar.

106. (D) In this way, the transpiration rate of the plant with leaves can be compared to that of the plant without leaves. The result will indicate the relation of leaf-surface to transpiration.

107. (A) Capillaries are the smallest of the blood vessels. Their walls are only a cell in thickness. Their opening is so small that the red blood corpuscles are spread out in passing through, one at a time, or in single file.

108. (C) Germinating seeds carry on respiration, and give off carbon dioxide. In the presence of this gas, clear lime water turns a milky color.

109. (D) The biologist was interested in determining whether aeration promotes root development. He would therefore find the answer by determining whether there was an increase in the number of roots in the experimental plant.

PRACTICE COLLEGE BOARD ACHIEVEMENT TEST IN BIOLOGY NO. 3

PART I—TYPE A

Directions: **The following diagram shows stages in fertilization and development of an animal embryo. Certain of the stages or parts are labeled with numbers. Each of the statements is followed by five suggested labels or answers. A label may be used once, more than once, or not at all. For each statement select the best answer.**

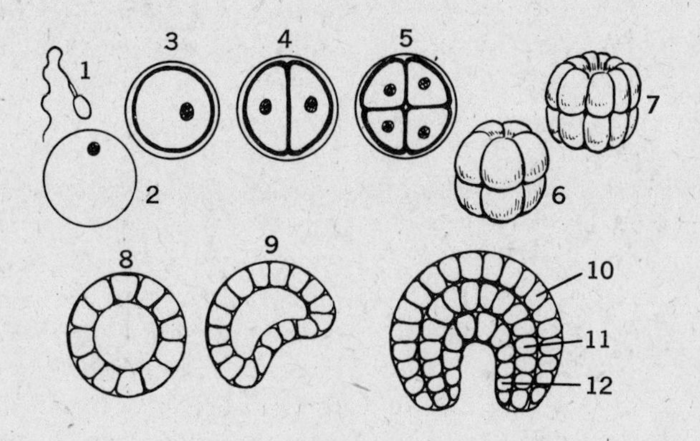

1. Enables sperm to move:
   (A) 1 (B) 2 (C) 8 (D) 10 (E) 11

1. A B C D E

2. First cell to contain the diploid number of chromosomes:
   (A) 2 (B) 3 (C) 4 (D) 6 (E) 9

2. A B C D E

3. Embryo appears as a hollow ball of cells
   (A) 4 (B) 6 (C) 8 (D) 10 (E) 12

3. A B C D E

4. Layer differentiates into the nervous system
   (A) 6 (B) 6 (C) 8 (D) 9 (E) 10

4. A B C D E

5. Gamete containing stored food
   (A)1 (B) 2 (C) 3 (D) 8 (E) 9

5. A B C D E

6. Layer differentiates into the lining of the alimentary canal
   (A) 6 (B) 7 (C) 10 (D) 11 (E) 12

6. A B C D E

7. Female gamete containing the monoploid (haploid) number of chromosomes
   (A) 2 (B) 3 (C) 4 (D) 5 (E) 8

7. A B C D E

8. First cell to undergo mitotic division
   (A) 3 (B) 4 (C) 5 (D) 6 (E) 8

8. A B C D E

9. Layer differentiates into skeleton
   (A) 6 (B) 7 (C) 9 (D) 10 (E) 11

9. A B C D E

10. Layer differentiates into epidermis of the skin
    (A) 8 (B) 9 (C) 10 (D) 11 (E) 12

10. A B C D E

TYPE B

Directions: **Each of the following statements is followed by five suggested answers. Select the one answer that is best in each case.**

11. The best description of an enzyme is that it (A) becomes hydrolyzed during chemical reactions (B) becomes dehydrated during chemical synthesis (C) speeds up the rate of chemical reactions (D) serves as an inorganic catalyst (E) is the source of energy for a chemical reaction.

11. A B C D E

12. If gastric juice is tested with a pH meter, its pH would most likely be about (A) 2 (B) 6 (C) 7 (D) 8 (E) 14

12. A B C D E

13. According to the heterotroph hypothesis, the correct sequence of events was (A) autotroph—heterotroph—organic molecules—aggregates of molecules (B) autotraph—organic molecules—aggregates of molecules—heterotroph (C) heterotroph—aggregates of molecules—autotroph—organic molecules (D) organic molecules—aggregates of molecules—heterotroph—autotroph (E) aggregates of molecules—organic molecules—heterotroph—autotroph

13. A B C D E

14. Of the following, the only one to contain chlorophyll is (A) sponge (B) blue-green mold (C) bread mold (D) slime mold (E) pleurococcus

14. A B C D E

15. In its early development, the embryo of man contains (A) gill slits (B) scales (C) fins (D) beak (E) six limbs

15. A B C D E

16. The cerebellum controls (A) reasoning (B) heart beat (C) muscular co-ordination (D) voluntary activity (E) sensations

16. A B C D E

17. The region where two neurons are in close contact is known as (A) mycelium (B) synapse (C) sori (D) sinus (E) pellicle

17. A B C D E

18. The skeleton of man resembles most closely the skeleton of a (A) cat (B) garter snake (C) chicken (D) goldfish (E) frog

18. A B C D E

19. The scientist who discovered insulin was (A) Banting (B) Trudeau (C) Pasteur (D) Goldberger (E) Eijkman

19. A B C D E

20. It is necessary for an astronaut to sip semi-solid food through a straw because of the effects of (A) radiation (B) acceleration (C) deceleration (D) psychological isolation (E) weightlessness

20. A B C D E

21. Eating undercooked pork may lead to (A) hookworm disease (B) tuberculosis (C) trichinosis (D) ringworm (E) filariasis

21. A B C D E

22. The autonomic nervous system controls (A) sight (B) thinking (C) digestion (D) hearing (E) memory

22. A B C D E

23. Nongreen plants that live on dead organic matter are known as (A) symbionts (B) parasites (C) scavengers (D) saprophytes (E) legumes

23. A B C D E

24. The part of a plant that develops into a fruit is the (A) ovary (B) ovule (C) stigma (D) anther (E) style

24. A B C D E

TYPE C

Directions: **For each of the organisms listed from 25–36, select the letter** (A), (B), (C), **or** (D), **depending on whether it**

(A) applies only to (A)
(B) applies only to (B)
(C) applies to both (A) and (B)
(D) applies to neither (A) nor (B)

[(A) vertebrate (C) both
(B) invertebrate (D) neither]

25. Lobster — 25. A B C D

26. Brontosaurus — 26. A B C D

27. Perch — 27. A B C D

28. Pencillium — 28. A B C D

29. Vorticella — 29. A B C D

30. Jellyfish — 30. A B C D

31. Horsetail — 31. A B C D

32. Sparrow — 32. A B C D

33. Turtle — 33. A B C D

34. Squid — 34. A B C D

35. Starfish — 35. A B C D

36. Codfish — 36. A B C D

TYPE D

Directions: **The following questions require two answers. One of the items in the left column** (A, B, or C) **is related to four of the items in the right column** (numbered 1–5). **Select the letter of the item which is so related; then select the number of the remaining item which does not belong.**

EXAMPLE:

(A) paramecium	1. cilia
(B) ameba	2. macronucleus
(C) spirogyra	3. trichocyst
	4. pyrenoid
	5. micronucleus

A B C 1 2 3 4 5 (A and 4 marked)

37.

(A) paramecium	1. pellicle
(B) yeast	2. cilia
(C) virus	3. cell wall
	4. oral groove
	5. contractile vacuole

37. A B C 1 2 3 4 5

38. (A) stem
(B) fruit
(C) flower

1. tendril
2. anther
3. ovule
4. stamen
5. pistil

38. A B C 1 2 3 4 5

39. (A) Pavlov
(B) Mendel
(C) Leeuwenhoek

1. dominant
2. segregation
3. recessive
4. opsonin
5. independent assortment

39. A B C 1 2 3 4 5

40. (A) Archaeozoic Era
(B) Mesozoic Era
(C) Cenozoic Era

1. *Trilobite*
2. *Pterodactyl*
3. *Brontosaurus*
4. *Steogosaurus*
5. *Tricerotops*

40. A B C 1 2 3 4 5

41. (A) embryological evidence
(B) vestigial evidence
(C) physiological evidence

1. iris diaphragm
2. appendix
3. ear muscles
4. coccyx
5. third eyelid

41. A B C 1 2 3 4 5

42. (A) tuber
(B) root
(C) leaf

1. guard cells
2. palisade layer
3. pith
4. spongy layer
5. stomate

42. A B C 1 2 3 4 5

43. (A) mold
(B) moss
(C) algae

1. prothallium
2. mycelium
3. hypha
4. sporangium
5. rhizoid

43. A B C 1 2 3 4 5

44. (A) ingestion
(B) digestion
(C) excretion

1. skin
2. lung
3. bladder
4. kidney
5. appendix

44. A B C 1 2 3 4 5

TYPE E

Directions: **The five lettered headings below are followed by a list of numbered statements. For each numbered statement, choose the one lettered heading (A–E) which is most closely related to it.**

(A) Habit (C) Tropism
(B) Instinct (D) Reflex
(E) Conditioned response

45. How does the stem of a plant grow when light comes from one side?

45. A B C D E

46. How does a person's eye react when a cinder blows in it?

46. A B C D E

47. How does a secretary take shorthand notes?

47. A B C D E

48. How does a bee build a hive?

48. A B C D E

49. How does a horse respond to the command of "whoa"?

49. A B C D E

50. How does a pianist play the "Emperor Concerto"?

50. A B C D E

PART II

The crystals of phenylthiocarbamide (PTC) appear bitter to some people but tasteless to others. When 855 people were tested for their ability to taste various concentrations of PTC, the accompanying graph was obtained. Solution 1 contained 1.3 grams of PTC per 1,000 cc.; each numbered solution has a concentration half as strong as the one preceding it.

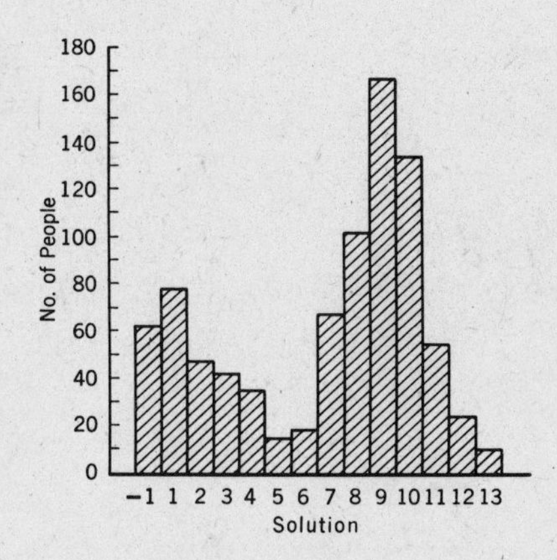

51. The ability to taste PTC was greatest for the solutions numbered (A) −1 and 1 (B) 1 and 9 (C) 5 and 13 (D) −1 and 13 (E) 2 and 10

51. A B C D E

52. The smallest number of people able to taste PTC was for the solution numbered (A) 3 (B) 6 (C) 9 (D) 12 (E) 13

52. A B C D E

53. The greatest number of people tasted the solution having a concentration of (A) 5% (B) 0.5% (C) 0.05% (D) 0.005% (E) 0.0005%

53. A B C D E

54. The most concentrated solution of PTC was numbered (A) −1 (B) 1 (C) 5 (D) 9 (E) 13

54. A B C D E

55. The largest number of people who could taste all the concentrations of PTC, from the highest to the lowest, was (A) 10 (B) 20 (C) 40 (D) 80 (E) 160

55. A B C D E

56. The ability to taste PTC is present in two-thirds of Americans; genetically speaking, therefore, it appears to be a (A) recessive trait (B) dominant trait (C) blended trait (D) linked trait (E) sex-linked trait

56. A B C D E

PART III—TYPE A

Directions: **In each of the following questions, there is a statement followed by four choices, of which *one* or *more* are correct. Indicate the correct combination by choosing the following letters:**

(A) if only 1, 2, and 3 are correct
(B) if only 1 and 3 are correct
(C) if only 2 and 4 are correct
(D) if only 4 is correct
(E) if some other combination is correct

DIRECTIONS SUMMARIZED:				
(A)	(B)	(C)	(D)	(E)
1, 2, 3	1, 3	2, 4	4	some other
only	only	only	only	combination

57. Among the most likely results of repeated spraying by airplane of a large area with DDT would be
1. an increase in the proportion of DDT-resistant to DDT-susceptible insects in the area
2. a reduction in the number of pine seeds in the area
3. the lessening of the food supply of some birds in the area
4. the elimination of weeds in that area

57. A B C D E

58. When scientific expeditions study the living things on isolated oceanic islands, they generally find many unique species among them. Among the causative factors involved in this situation would be
1. mutation
2. natural selection
3. isolation resulting in failure to interbreed with organisms on other islands
4. extensive competition

58. A B C D E

59. When one strand of a DNA molecule is arranged as adenine-guanine-thymine-cytosine, the other strand is arranged as
1. adenine-guanine-thymine-cytosine
2. cytosine-thymine-guanine-adenine
3. guanine-adenine-cytosine-thymine
4. thymine-cytosine-adenine-cytosine

59. A B C D E

60. The digestion of proteins takes place in the
1. stomach
2. esophagus
3. small intestine
4. large intestine

60. A B C D E

61. The dinosaurs became extinct
1. at the end of the Mesozoic Era
2. after man appeared
3. before man appeared
4. at the end of the Paleozoic Era

61. A B C D E

62. The whooping crane is close to extinction because of
1. the Japanese beetle
2. the spread of civilization
3. pollution of waterways
4. threats to its nesting sites

62. A B C D E

63. Radioactive iodine is useful as a tracer
1. in locating metastases of the thyroid gland
2. because it is absorbed by thyroid tissue
3. in controlling intermediate stages of leukemia
4. because it activates the Geiger counter

63. A B C D E

64. During the process of fermentation, yeast cells
1. liberate oxygen
2. give off carbon dioxide
3. manufacture glucose
4. produce alcohol

64. A B C D E

65. Ameba and elodea are alike in
1. possessing a nucleus
2. carrying on respiration
3. using carbohydrates for energy
4. excreting carbon dioxide as a waste

65. A B C D E

66. Sunshine is important to the body because
1. it stimulates the skin to form vitamin D
2. its ultraviolet rays kill bacteria
3. it promotes the formation of healthy bones
4. it is needed for the formation of red blood cells

66. A B C D E

67. In the process of photosynthesis
1. green plants eliminate carbon dioxide
2. chloroplasts produce carbohydrates
3. saprophytes liberate oxygen
4. green plants take in carbon dioxide

67. A B C D E

68. The right ventricle of the human heart
1. receives blood from right atrium
2. sends blood to the lungs
3. contains deoxygenated blood
4. sends blood into the aorta

68. A B C D E

69. In cross-pollination of flowers
1. pollen is carried to the stigma of the same flower
2. pollen is carried to the stigma of another flower
3. the petals serve to prevent self-pollination
4. bees may actively transport pollen

69. A B C D E

TYPE B

Directions: **Each of the following statements is followed by five suggested answers. Select the one answer which is best in each case.**

70. All of the following organic compounds are examples of carbohydrates except (A) lactose ($C_{12}H_{22}O_{11}$) (B) glucose ($C_6H_{12}O_6$) (C) galactose ($C_6H_{12}O_6$) (D) glycerol ($C_3H_5(OH)_3$) (E) glycogen ($(C_6H_{10}O_5)_n$)

70. A B C D E

71. The driver of a car should not drink liquor because (A) his coordination is reduced (B) he becomes oversensitive to horn-blowing (C) his vision becomes myopic (D) he becomes overly anxious about accidents (E) he cannot ask for directions properly

71. A B C D E

72. ACTH is useful in treating some kinds of arthritis because (A) it stimulates the production of thyroxin (B) it strengthens the heartbeat (C) it speeds up the rate of blood clotting (D) it stimulates the production of cortisone (E) it increases the breathing rate

72. A B C D E

73. Humus on the floor of a forest is valuable because (A) it allows rainwater to run off quickly (B) it serves as food for deer and woodchucks (C) it absorbs a great deal of water (D) it provides a home for termites (E) it encourages burrowing by moles

73. A B C D E

74. Identical twins are alike because (A) they come from the same uterus (B) they were nourished by the same placenta (C) they possess identical genes (D) they originated from one egg fertilized by two sperms (E) they have the same number of chromosomes

74. A B C D E

75. A man in space is in danger of damage from radiation because (A) he is nearer the sun (B) he is not shielded by the atmosphere (C) it is not possible to shield him against uranium atoms in space (D) a Geiger counter ceases to function at such high altitudes (E) the intense cold of outer space lowers his resistance to roentgen exposure

75. A B C D E

76. Fossils of the horse have not been found in rocks of the Paleozoic Era because (A) such rocks are too old to study (B) bones of the horse were too fragile to be embedded in rock (C) the horse had not yet appeared on earth (D) primitive man broke up the bones to eat the marrow (E) some regions of the earth, such as the Antarctic, have not been thoroughly studied

76. A B C D E

77. Valves in the veins aid in circulation because (A) the spurting action is too uneven (B) the heart sends blood into the veins with great pressure (C)

77. A B C D E

they separate oxygenated from deoxygenated blood (D) they filter out blood clots (E) they keep the blood from flowing backward

78. A person with insufficient iron in his diet may become anemic because (A) not enough oxyhemoglobin is formed in the liver (B) the spleen removes too many leucocytes from the blood (C) the clotting reaction time is reduced (D) iron is used in building up hemoglobin (E) iron is used in making fibrinogen

78. A B C D E

79. The production of apples depends on bees because (A) they destroy the worms that infest apples (B) they make it possible for apple flowers to carry on self-pollination (C) they cross-pollinate apple flowers (D) their honey adds to the flavor of apples (E) they keep animals away by means of their stingers

79. A B C D E

80. The trachea of man retains a strong, rigid structure because (A) it contains rings of cartilage (B) air pressure keeps it inflated (C) it is supported by the diaphragm (D) it is located near the heart (E) it is next to the esophagus, which lends it support

80. A B C D E

TYPE C

Directions: **Each item consists of an *assertion* (statement) in the left-hand column, and a *reason* on the right-hand side. Select**

(A) if both assertion and reason are true, and *are related* as cause and effect
(B) if both assertion and reason are true, but *are not related* as cause and effect
(C) if the assertion is true, but the reason is a false statement
(D) if the assertion is false, but the reason is a true statement
(E) if both assertion and reason are false statements

DIRECTIONS SUMMARIZED:

	(A)	(B)	(C)	(D)	(E)
Assertion:	True	True	True	False	False
Reason:	True	True	False	True	False
	Cause and effect	*Not* cause and effect			

	Assertion:		Reason:
81.	New combinations of original genes may occur	BECAUSE	crossing-over of chromosomes takes place
82.	The rabbit became a serious pest when it was introduced into Australia	BECAUSE	there were very few natural enemies there
83.	Uranium eventually changes into lead	BECAUSE	its rays may be detected with a Geiger counter
84.	The pituitary gland is known as the master gland	BECAUSE	it interacts with other ductless glands
85.	A dog can live without its medulla	BECAUSE	the nerves of the spinal cord connect the brain with other parts of the body
86.	Human inheritance is difficult to study	BECAUSE	there is a small number of children in a family
87.	Pedigreed animals are usually not bred with each other	BECAUSE	hybrid vigor always results when different strains are mated
88.	The mental ability of Java man is well known	BECAUSE	his fossils have been found on the island of Java

81. A B C D E

82. A B C D E

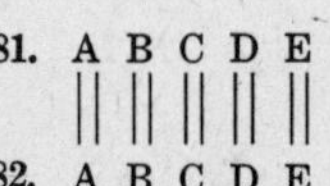

83. A B C D E

84. A B C D E

85. A B C D E

86. A B C D E

87. A B C D E

88. A B C D E

PART IV—TYPE A

In showing the requirement of light for photosynthesis, a demonstration is conducted in which the steps shown in the accompanying diagram are followed:

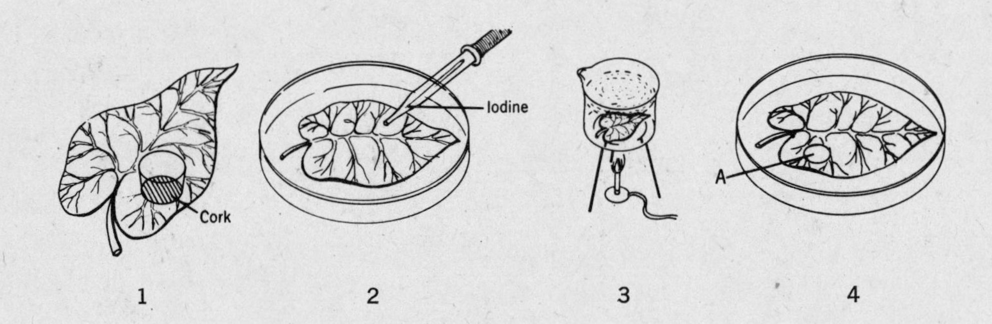

89. The correct sequence of steps is in performing this demonstration: (A) 1–2–3–4 (B) 1–2–4–3 (C) 1–3–2–4 (D) 1–4–3–2 (E) 1–4–2–3

89. A B C D E

90. In Step 3, the leaf is heated in alcohol (A) in order to remove its chlorophyll (B) to show that heat is not one of the factors in photosynthesis (C) to keep the chloroplasts from streaming (D) to bring photosynthesis to a halt (E) to coagulate the protoplasm in the chloroplasts

90. A B C D E

91. In step 4, the circular area marked A (A) is shown to contain starch (B) is shown to contain glucose (C) did not carry on photosynthesis (D) is stained blue-black by the iodine (E) was bleached by the iodine

91. A B C D E

92. The cork was placed over part of the leaf in step 1 (A) to protect the underlying leaf from the ultraviolet rays (B) to keep light from reaching the underlying cells (C) to allow the leaf to float in alcohol (D) to vary the conditions under which photosynthesis occurs (E) to demonstrate the relationship between photosynthesis and respiration

92. A B C D E

93. Before iodine is added to the leaf the color of the leaf is (A) green (B) blue-black (C) brown (D) white (E) all green except for the part that was covered by the cork

93. A B C D E

94. The control in this demonstration is (A) Part A in step 4 (B) the part of the leaf not covered by the cork (C) the part of the leaf covered by the cork (D) the leaf in step 2 (E) not shown in the accompanying drawings

94. A B C D E

95. If the cork in A is removed after twenty-four hours, it will be observed that (A) the entire leaf begins to wilt (B) the entire leaf turns a pale yellow (C) the underlying part of the leaf has turned white (D) the underlying part of the leaf looks as normal as the rest of the leaf (E) the underlying part of the leaf is dead

95. A B C D E

96. Iodine is added (A) to preserve the leaf (B) to act as a disinfectant (C) to stimulate the action of the chloroplasts (D) to supply a necessary element for photosynthesis (E) to determine whether starch is present in the part of the leaf that was covered

96. A B C D E

TYPE B

Directions: **Each of the following problems is followed by five suggested answers. Select the one answer which is best in each case.**

97. At rest, a man had a pulse beat of 76, and a breathing rate of 14. He was allowed to inhale air in which the composition of carbon dioxide was changed from 0.03% to 0.3%. As a result, (A) his pulse beat fell to 14 (B) his breathing rate could be estimated by dividing by 10% (C) his pulse

97. A B C D E

beat could be determined by using the equation $\frac{76}{14}=\frac{0.3\%}{0.03\%}$ (D) his breathing rate remained the same (E) his breathing rate increased

98. In a large litter of guinea pigs, one-half of the offspring were white (black is dominant). The parents were probably (A) Bb × bb (B) Bb × Bb (C) Bb × BB (D) BB × bb (E) bb × bb

98. A B C D E

99. A color-blind son will be produced from: (Key $\underline{X}$ = chromosome with gene for color-blindness; X = chromosome with gene for normal vision) (A) $\underline{X}\underline{Y} \times XX$ (B) $XY \times XX$ (C) $\underline{X}Y \times XX$ (D) $X\underline{Y} \times XX$ (E) $XY \times X\underline{X}$

99. A B C D E

100. Richard — Alice; Charles [?] — Joan; Sylvia, Joseph

In this pedigree of a family, brown eyes are indicated as ○, blue eyes as ◍. The eye color of Charles is not given. From this chart, it can be determined that (A) Richard and Alice are both homozygous for brown eyes (B) Charles is probably homozygous for brown eyes (C) Charles is probably heterozygous for brown eyes (D) Charles is homozygous for blue eyes (E) Sylvia's eyes could not be brown since her mother is blue-eyed

100. A B C D E

101. When Mendel self-pollinated many pea plants that were heterozygous for seed color, he obtained about 8,000 seeds in the next generation (yellow seed is dominant, green seed is recessive). The number that were yellow and the number green was as follows (A) 4,000–4,000 (B) 5,000–3,000 (C) 3,000–5,000 (D) 6,000–2,000 (E) 2,000–6,000

101. A B C D E

102. To calculate the amount of protein actually used by the body, the urine is chemically analyzed for its nitrogen content. One gram of nitrogen shows that 5.94 liters of oxygen were used up by protein, and that 26.51 calories of heat were produced. The caloric value of oxygen in the oxidation of protein is (A) 5.94 calories per liter (B) 26.51 calories per liter (C) 4.47 calories per liter (D) 20.67 calories per liter (E) 31.45 calories per liter

102. A B C D E

PART V

Directions: **Each of the following questions has five suggested answers. Select the one best answer in each case.**

103. A taxonomist member of an expedition in the Amazon jungle came across a new organism. When he attempted to classify it, he realized that it had never been discovered before. He observed these characteristics: It measured 24 mm. across; it was multinucleate; its naked protoplasm showed cyclosis and spread slowly over a decaying log; it produced spores that formed flagellate cells which then fused into an ameboid form. He pondered whether to classify it in the protist, plant or animal kingdom. He finally decided on the protist. The chief consideration that led him to do so was that

(A) it could not move and therefore was not an animal
(B) it did not contain chlorophyll and therefore was not a plant
(C) it seemed to resemble the algae
(D) it had the characteristics of a slime mold
(E) as a zoologist, he had never seen an animal like this

103. A B C D E

104. The Biloxi soybean flowers in September. In order to find out if the time of flowering is determined by maturation time (the interval of growth and development required for plants to reach the flowering stage), a scientist planted seeds on three different occasions—in early spring, in June, and in July. By the end of the summer, the three different groups of plants had reached different sizes, depending on when they had first been planted. However, in September, they all bloomed at the same time. The scientist concluded that

(A) summer temperature determined the time of flowering

104. A B C D E

(B) maturation time determined the time of flowering
(C) plant seedlings go through a process like that of puberty in children before flowers are produced
(D) the date of seed germination dictated the time of flowering
(E) the date of seed germination did not dictate the time of flowering

105. A research team at the Yerkes Primate Research Center found that chimpanzees can match objects by touch to photographs of identical objects. To do this, the brain must perform an operation called cross-modal integration. The process of reading also requires cross-modal integration—that is, there is an association of sounds with written words, and the association of written words with their meanings. Using the techniques developed in their research with chimpanzees, the investigators are now studying children with dyslexia, a reading disability, for a further understanding of this disorder. They believe that (A) chimpanzees can be taught to read by dyslexia (B) dyslexia may be due to a child's not being able to perform certain cross-modal integrations (C) dyslexia is a difficult form of cross-modal integration (D) children do not go through the process of cross-modal integration (E) the chimpanzee's vision is too sharp for it to have dyslexia

105. A B C D E

PART VI

106. A certain human enzyme is shown to be active in a test tube placed in a brightly lighted room at a pH of 7.2 and at a temperature of 37°C. Which would most likely speed up the rate of the action of the enzyme? (A) reducing the amount of light (B) eliminating any amount of light (C) increasing the amount of enzyme present (D) decreasing the amount of substrate present (E) raising the temperature to 100°C.

106. A B C D E

107. A group of 24 frogs was separated into two equal groups. Group *A* was placed in an environment in which the temperature was a constant 35°F. Group *B* was placed in a similar environment, except that the temperature was kept at a constant 65°F. Both groups were given equal amounts of food at the start of the experiment and every 24 hours thereafter. Immediately before each daily feeding, the excess food from the previous feeding was removed and measured. Thus, it was possible to determine the daily consumption of food by each group. Each day, the frogs in each group were checked for heartbeat and breathing rate. At the end of the experiment, the accompanying bar graphs were prepared;

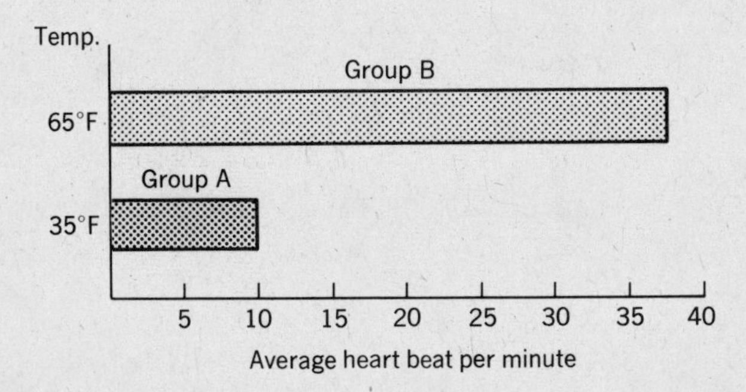

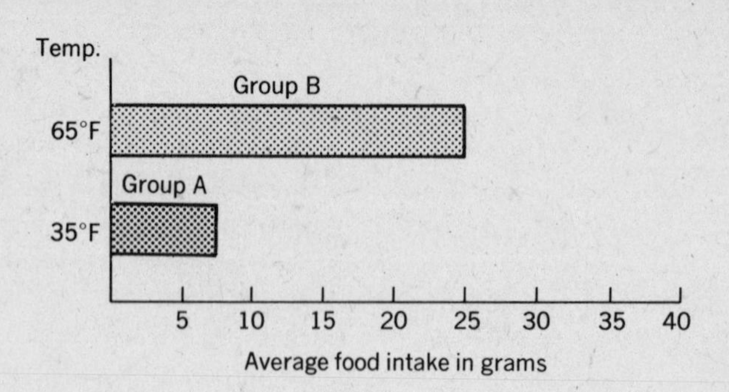

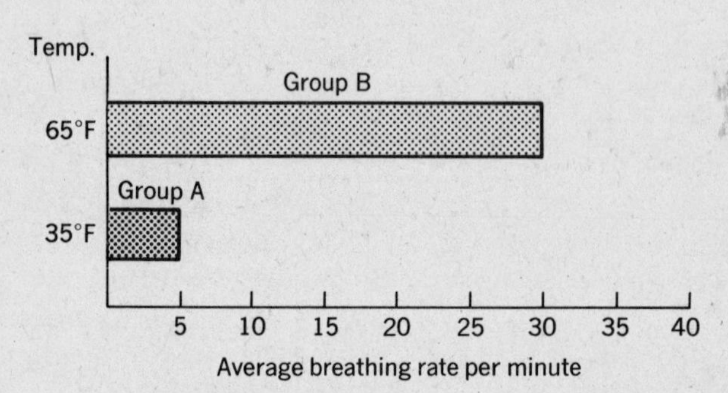

The primary question which the scientist was probably studying in this experiment was (A) how do living things adapt to a change in temperature (B) in what ways do frogs adapt to a change in temperature (C) how much oxygen does a frog need when the temperature changes (D) how much blood is circulated through the heart of a frog when the temperature changes (E) how much food does a frog need

107. A B C D E

108. Referring to the previous question, one could reasonably assume after an examination of the three graphs that at low temperatures the frogs (A) became more active (B) produced less carbon dioxide (C) required more food (D) required more oxygen (E) produced more carbon dioxide

108. A B C D E

109. The experimental variable in the experiment described in the two preceding questions was the (A) heartbeat rate (B) breathing rate (C) amount of food (D) temperature (E) number of experimental animals

109. A B C D E

ANSWER KEY: TEST NO. 3

1. (A)	**12.** (A)	**23.** (D)	**34.** (B)	**45.** (C)	**56.** (B)	**67.** (C)	**78.** (D)	**89.** (C)	**100.** (C)
2. (B)	**13.** (D)	**24.** (A)	**35.** (B)	**46.** (D)	**57.** (B)	**68.** (A)	**79.** (C)	**90.** (A)	**101.** (D)
3. (C)	**14.** (E)	**25.** (B)	**36.** (A)	**47.** (A)	**58.** (A)	**69.** (C)	**80.** (A)	**91.** (C)	**102.** (C)
4. (E)	**15.** (A)	**26.** (A)	**37.** (A)3	**48.** (B)	**59.** (E)	**70.** (D)	**81.** (A)	**92.** (B)	**103.** (D)
5. (B)	**16.** (C)	**27.** (A)	**38.** (C)1	**49.** (E)	**60.** (B)	**71.** (A)	**82.** (A)	**93.** (D)	**104.** (E)
6. (E)	**17.** (B)	**28.** (D)	**39.** (B)4	**50.** (A)	**61.** (B)	**72.** (D)	**83.** (B)	**94.** (B)	**105.** (B)
7. (A)	**18.** (A)	**29.** (B)	**40.** (B)1	**51.** (B)	**62.** (E)	**73.** (C)	**84.** (A)	**95.** (D)	**106.** (C)
8. (A)	**19.** (A)	**30.** (B)	**41.** (B)1	**52.** (E)	**63.** (E)	**74.** (C)	**85.** (D)	**96.** (E)	**107.** (B)
9. (E)	**20.** (E)	**31.** (D)	**42.** (C)3	**53.** (E)	**64.** (C)	**75.** (B)	**86.** (A)	**97.** (E)	**108.** (B)
10. (C)	**21.** (C)	**32.** (A)	**43.** (A)1	**54.** (A)	**65.** (E)	**76.** (C)	**87.** (E)	**98.** (A)	**109.** (D)
11. (C)	**22.** (C)	**33.** (A)	**44.** (C)5	**55.** (A)	**66.** (A)	**77.** (E)	**88.** (D)	**99.** (E)	

ANSWERS EXPLAINED: TEST NO. 3

1. (A) tail
2. (B) fertilized egg
3. (C) blastula
4. (E) ectoderm
5. (B) ovum
6. (E) endoderm
7. (A) ovum
8. (A) fertilized egg
9. (E) mesoderm
10. (C) ectoderm

11. (C) An enzyme is an organic catalyst. It affects the rate of a chemical reaction without being changed.
12. (A) Gastric juice in the stomach gives an acid response to a pH meter because it contains hydrochloric acid.
13. (D) According to the heterotroph hypothesis, life on earth evolved through a sequence of stages. The atmosphere of the primitive earth is supposed to have contained hydrogen, methane (CH_4) ammonia (NH_3) and water vapor. As these gases were washed down by heavy rains into the early seas, they were acted on by ultraviolet radiation, cosmic rays, the earth's heat and radioactivity. Molecules were bonded together into large organic molecules. Some of these gathered into more complex aggregates of molecules. Since these aggregates used molecules present in the ocean as food, they acted as heterotrophs. Some of them developed a method of using carbon dioxide to manufacture their own food, and became autotrophs.
14. (E) Pleurococcus is a green alga; sponge is an animal, and the next three organisms are examples of fungi.
15. (A) This is taken as evidence that man is descended from a distant vertebrate that breathed by gills. He still retains enough of the original genes to form gill slits in the embryonic stage, which disappear shortly afterward.
16. (C) All of the other activities are centered in the cerebrum, except for heartbeat, which is a function of the medulla.
17. (B) In a synapse, the nerve impulse passes from one neuron to another; there is no direct connection between the two neurons at this point.
18. (A) Both man and cat are descended from a mammal ancestor. Although they have become quite different from each other, they still retain many similarities.
19. (A) Banting and his co-workers announced the discovery of insulin in 1922. He later received the Nobel Prize for his work.
20. (E) During weightlessness, semisolid food resembling baby food is taken in from a plastic squeeze tube with a straw. More solid food might not enter the esophagus properly without the effect of gravity to bring it to the point of being acted on by peristalsis. Also, water does not pour without the effect of gravity.
21. (C) Undercooked pork may contain the tiny trichinella worm, which can cause trichinosis.
22. (C) Involuntary, automatic and internal activities, such as digestion and heartbeat, are controlled by the autonomic nervous system.
23. (D) Plants such as mushrooms and molds are examples of fungi which obtain their nutrition in this way.
24. (A) After pollination, the ovary of a flower matures and becomes a fruit. The pea pod is a fruit, since it originated from the ovary of the pea flower. The ovules within the ovary become the seeds.
25. (B) A lobster is an arthropod, and does not have a backbone.
26. (A) Brontosaurus was a reptile, and had a backbone.
27. (A) A perch is a fish, and has a backbone.
28. (D) Penicillium is a blue-green mold. It is a fungus, and is a member of the protist kingdom.
29. (B) Vorticella is a one-celled animal, and is classified as a protozoan.
30. (B) A jellyfish is a simple invertebrate, among the first to have three layers of cells. It is classified as a coelenterate.
31. (D) Pleurococcus is an example of the algae, the simplest green protists.
32. (A) A sparrow is a bird, and has a backbone.
33. (A) A turtle is a reptile, and has a backbone.
34. (B) A squid is a member of the mollusk phylum, and does not have a backbone.
35. (B) A starfish is classified as a member of the echinodermata phylum; it does not have a backbone.
36. (A) A codfish is a fish, and has a backbone.

37. (A)3 All are structures found in paramecium except cell wall, which is present only in plants.

38. (C)1 All are structures found in a flower except tendril, which is a modified part of a stem used by climbing plants such as the grape.

39. (B)4 All terms deal with Mendel's discoveries of the laws of heredity, except opsonin, which is an antibody that helps white blood corpuscles engulf bacteria.

40. (B)1 All are examples of dinosaurs which lived during the Mesozoic Era, except the trilobite, which was dominant during the Paleozoic Era.

41. (B)1 All are examples of vestigial structures in man, except the iris, which gives color to the eye and which controls the amount of light entering the eye.

42. (C)3 All are structures of a leaf, except pith, which is found in a stem.

43. (A)1 All are parts of a mold, except prothallus, which develops from a germinating spore of a moss.

44. (C)5 All function as organs of excretion, except the appendix, which is a vestigial structure without any apparent function.

45. (C) The stem of a plant grows toward the light; this is an example of positive phototropism.

46. (D) A reflex action occurs; tears are stimulated to flow, and the eye closes.

47. (A) By habit formation, the secretary has learned to take notes automatically and quickly.

48. (B) A bee builds a hive by instinct, one reflex leading to another, until the chain of reflexes is complete.

49. (E) The horse has learned to associate the sound of the command with a pull on the reins, and has developed a conditioned response to stop at the command.

50. (A) Playing the piano is an example of a habit that has been developed by so much practice that the responses are automatic.

51. (B) The two high points of the graph were for solution 1 (78 people), and solution 9 (166 people).

52. (E) Only ten people tasted solution B.

53. (E) Solution 1 contained 1.3 grams per 1000 cc., giving a concentration of 0.13%. Each numbered solution had a concentration of half as much. The answer is obtained by dividing each concentration in half until solution 9 is reached.

54. (A) Each solution following solution 1 had half the concentration of the preceding one. Solution −1 therefore had twice the concentration of solution 1.

55. (A) As shown on the graph, ten people could taste solution 13, and more than ten people could taste all the others.

56. (B) It is a dominant trait because most people possess this ability. The number of people who are not tasters is relatively small, and they show the recessive trait.

57. (B) At first, most of the insects would be susceptible to DDT, and would die. However, a few would resist the chemical. As they reproduced, their offspring would continue to be resistant and in time, there would be more of these than the susceptible ones. Birds that depended on these insects for their food supply would find fewer of the susceptible type available for food.

58. (A) Mutations would change the characteristics of the birds. By natural selection, the fittest would survive. Since they were isolated on the islands, they would not be able to interbreed with other birds and their characteristics would become different from those of other birds elsewhere.

59. (E) All of the choices are incorrect. The correct answer is: thymine-cytosine-adenine-guanine.

60. (B) The digestion of proteins starts in the stomach, and is completed in the small intestine.

61. (B) The dinosaurs became extinct during the Mesozoic Era, which ended about 60 million years ago; primitive man first appeared on earth less than a million years ago.

62. (E) The whooping crane is threatened by all of the factors except the first.

63. (E) All of the items are correct except number 3.

64. (C) When yeast cells ferment sugar, they give off carbon dioxide and alcohol.

65. (E) All four answers are correct.

66. (A) Upon exposure to direct sunlight, the substance dehydrocholesterol in the skin is converted into vitamin D. This vitamin is needed for the proper formation of the bones. The ultraviolet rays of sunshine destroy bacteria. However, sunshine has no effect on the formation of red blood cells.

67. (C) During photosynthesis, green plants take in carbon dioxide and water, and produce carbohydrates; oxygen is eliminated as a by-product.

68. (A) The first three items apply to the right ventricle; the fourth applies to the left ventricle.

69. (C) Bees play an important part in cross-pollination by picking up pollen on their bodies and depositing it on the stigma of another flower as they visit it to obtain nectar.

70. (D) Glycerol is a lipid. It does not contain hydrogen and oxygen in the 2:1 ratio, as in water.

71. (A) Even a small concentration of alcohol affects the brain, reducing muscular co-ordination. A driver may not realize this, and will not react as effectively in an emergency as he normally would.

72. (D) It has recently been found that injections of cortisone, which is produced by the adrenal cortex, relieve the painful symptoms of arthritis. The production of cortisone is stimulated by the hormone ACTH, which is produced by the pituitary gland.

73. (C) Humus is composed of decayed organic matter, and acts as a sponge in absorbing water. During a rain, it prevents water run-off, reducing the danger of erosion.

74. (C) Identical twins originated from one fertilized egg. During cleavage, the developing mass of cells split into two separate groups, each of which became a twin. Since they possess the same genes, they are identical in their characteristics.

75. (B) The earth's atmosphere helps to shield us from the radiations present in space. An astronaut is not protected in this way, and may be affected by primary cosmic rays, as well as by other types of radiation that can penetrate his space vehicle.

76. (C) Mammals did not appear on earth until the end of the Mesozoic Era. The Paleozoic Era came to a close about 125 million years earlier.

77. (E) The blood is kept flowing in a forward direction by the pulsating pressure in the arteries. The flow in the veins is smooth, and may be against the force of gravity, as in the legs. The valves in the veins keep the blood moving toward the heart by preventing a back flow.

78. (D) An anemic person has a lack of red blood corpuscles. Hemoglobin is a major component of these cells, and contains iron.

79. (C) Apple flowers require cross-pollination if they are to develop into apples.

80. (A) The rings of cartilage keep the passageway of the trachea from collapsing.

81. (A) Crossing-over takes place during meiosis, when homologous chromosomes may twist about each other and exchange adjacent parts. When they separate, the chromosomes now have a portion that originally came from the other chromosome of the pair. This leads to new combinations of the genes located on the chromosomes.

82. (A) The rabbit population is normally kept in check by its natural enemies, such as the fox. In the absence of natural enemies, the rabbits in Australia multiplied so rapidly that they have become a menace to agriculture.

83. (B) Uranium is a radioactive element that is constantly breaking down. One of its stages is radium; lead is the final product.

84. (A) The pituitary gland, located at the base of the brain, interacts with: the adrenal cortex by producing ACTH; the reproductive organs; the thyroid gland; etc.

85. (D) The medulla is a part of the brain that controls breathing and heart beat. If it is removed, death results.

86. (A) Comparatively little is known about human heredity because of the small number of children produced in a family, the length of a generation, and the difficulty of obtaining data.

87. (E) Pedigreed animals of a certain breed are always mated with each other to retain their valuable characteristics, and to prevent the introduction of other genes.

88. (D) Little is known about Java man because of the limited fossil remains that have been unearthed.

89. (C) The leaf is covered with a cork (1); after exposure to light, the leaf is boiled in alcohol to remove the chlorophyll (3); the bleached leaf is treated with iodine (2); the exposed part of the leaf turns blue-black (4).

90. (A) Chlorophyll is soluble in alcohol. It is removed from the leaf so that the effect of iodine on the starch in the leaf may be observed.

91. (C) This area contained no starch, because it had been covered with the cork.

92. (B) By keeping light from reaching a part of the leaf, it was possible to show that starch would not be made in this area.

93. (D) When the leaf was boiled in alcohol, all of the chlorophyll was removed, and the leaf turned white.

94. (B) Since a part of the leaf was covered, and could not carry on photosynthesis, the rest of the leaf served as the control and showed that photosynthesis takes place in the light.

95. (D) There is sufficient food present in the covered area to keep the cells alive for twenty-four hours.

96. (E) Iodine gives a blue-black color in the presence of starch. It is used to determine whether the covered part of the leaf carried on photosynthesis and produced starch.

97. (E) An increase in the amount of carbon dioxide in the blood stimulates the breathing center in the medulla to speed up the rate of breathing.

98. (A) The results of crossing Bb and bb are 50% Bb (heterozygous black), and 50% bb (white):

	B	b
b	Bb	bb
b	Bb	bb

99. (E) The possible results of crossing XY and $X\underline{X}$ are

	X	Y
X	XX	XY
$\underline{X}$	$X\underline{X}$	$\underline{X}Y$

Male: 50% color-blind, 50% normal
Female: all normal (50% carriers)

100. (C) Charles is heterozygous for brown eyes, and contains a gene for blue eyes; his wife, Joan, is blue-eyed. Their daughter Sylvia, who is blue-eyed, received a gene from each. Since Charles has a recessive gene for blue eyes, either one or both of his parents, Richard and Alice, are heterozygous for brown eyes. Sylvia might have had brown eyes if she had received a gene for brown eyes from Charles.

101. (D) Mendel learned that hybrids produce dominant offspring in a ratio of 3 : 1. Since yellow is dominant over green, there were three times as many yellow as green seeds.

102. (C) If 5.94 liters of oxygen were used up in producing 26.51 calories of heat, then the caloric value of one liter of oxygen can be calculated by dividing 26.51 (total calories of heat) by 5.94 (total liters of oxygen).

103. (D) The characteristics described by the taxonomist are those of a slime mold. Slime molds are classified in the protist kingdom.

104. (E) The date of seed germination will determine when the new plants will start to grow. However, the time of flowering is determined by another factor, namely the amount of light received by the plants. This is an example of photoperiodism.

105. (B) Through this research, the scientists learned that cross-modal integration takes place in learning. It also occurs in reading. It is possible that dyslexia, a reading disability, results because of a failure to perform the necessary cross-modal integrations.

106. (C) The rate of enzyme action is influenced by several factors, including pH and temperature. Many human enzyme reactions take place at a pH of around 7. The temperature of the human body is 37°C. By increasing the amount of enzyme, the rate of action would be increased.

107. (B) Since other factors were kept constant, the experiment obviously deals with reactions to different temperatures. The scientist was specifically examining the adaptation of the frog.

108. (B) As seen from the bar graph, the breathing rate of the frogs kept at 35°F was much lower than that of the frogs kept at the higher temperature. Since breathing rate is directly controlled by the amount of the carbon dioxide in the bloodstream (i.e., the higher the carbon dioxide content, the faster the breathing rate), it is safe to assume that the frogs at the lower temperature produced less carbon dioxide. Also, since these frogs breathed more slowly, they undoubtedly took in less oxygen, and produced less carbon dioxide.

109. (D) An experimental variable is a condition of the experiment that is changed to see the effect of the change on different aspects of the organism involved. In this case, the temperature was varied in an attempt to study the effect of this change on heart beat, breathing rate, and amount of food consumed.

PRACTICE COLLEGE BOARD ACHIEVEMENT TEST IN BIOLOGY NO. 4

PART I—TYPE A

Directions: **The following diagram shows the parts of a paramecium. Certain of the stages or parts are labeled with numbers. Each of the statements is followed by five suggested labels or answers. A label may be used once, more than once, or not at all. For each statement select the best answer.**

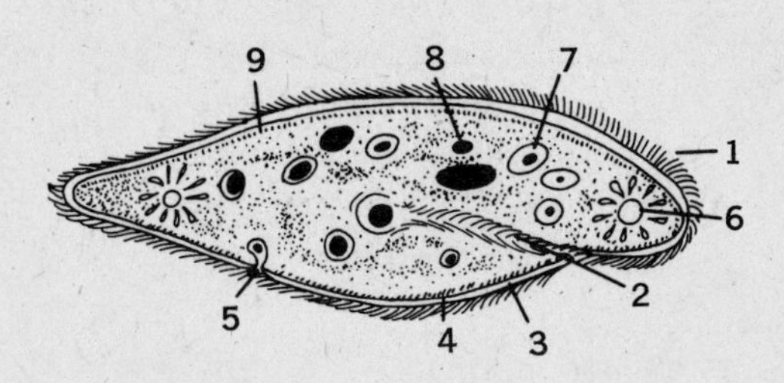

1. Is composed primarily of lipid and protein
(A) 1 (B) 3 (C) 4 (D) 6 (E) 9

1. A B C D E

2. Keeps the paramecium moving
(A) 1 (B) 2 (C) 5 (D) 7 (E) 8

2. A B C D E

3. Serves to excrete solid wastes
(A) 1 (B) 3 (C) 5 (D) 7 (E) 9

3. A B C D E

4. Maintains its shape
(A) 1 (B) 3 (C) 4 (D) 5 (E) 9

4. A B C D E

5. Serves to digest food
(A) 2 (B) 4 (C) 5 (D) 6 (E) 7

5. A B C D E

6. Serves to take in food
(A) 2 (B) 3 (C) 4 (D) 5 (E) 6

6. A B C D E

7. Eliminates excess water
(A) 2 (B) 5 (C) 6 (D) 7 (E) 8

7. A B C D E

8. Possibly serves a protective weapon
(A) 2 (B) 3 (C) 4 (D) 8 (E) 9

8. A B C D E

9. Contains the chromosomes
(A) 3 (B) 5 (C) 6 (D) 7 (E) 8

9. A B C D E

10. Permits diffusion of dissolved gases
(A) 3 (B) 4 (C) 5 (D) 6 (E) 8

10. A B C D E

11. Controls reproduction
(A) 4 (B) 5 (C) 6 (D) 7 (E) 8

11. A B C D E

TYPE B

Directions: **Each of the following statements is followed by five suggested answers. Select the one answer which is best in each case.**

12. The adjoining structural formula represents (A) an atom (B) an ion (C) a molecule (D) an element (E) a mixture

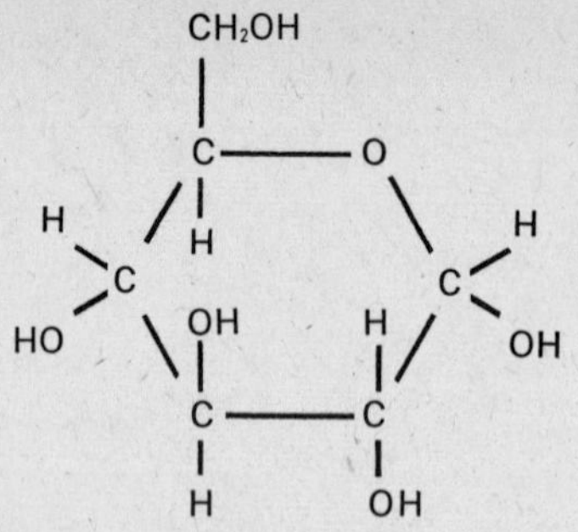

```
               CH2OH
                 |
                 C———————O
          H    / |        \      H
           \  /  H         \    /
            C                C
           /                  \
        HO  \  OH       H  /   OH
             \ |        | /
               C————————C
               |        |
               H        OH
```

Glucose

13. The production of glucose, a six-carbon sugar, is most closely associated with the chemical reactions that occur during (A) aerobic respiration (B) anaerobic respiration (C) fermentation (D) the dark phase of photosynthesis (E) the light phase of photosynthesis

13. A B C D E

14. The mutation theory was first proposed by (A) Dujardin (B) De Vries (C) Funk (D) Fleming (E) Gorgas

14. A B C D E

15. An atom of radioactive carbon is similar to an atom of ordinary carbon in having the same (A) number of neutrons (B) number of protons (C) half-life (D) atomic weight (E) mass

15. A B C D E

16. Coronary thrombosis refers to a blood clot in the (A) heart (B) brain (C) lung (D) leg (E) abdomen

16. A B C D E

17. The gray matter of the brain contains (A) pseudopodia (B) axons (C) ganglia (D) macronuclei (E) dendrites

17. A B C D E

18. Of the following, the most advanced type of man was (A) Java man (B) Peking man (C) Cro-Magnon man (D) Neanderthal man (E) Heidelberg man

18. A B C D E

19. A characteristic of organisms that excrete uric acid as their main nitrogenous waste is that they (A) usually live in salt water (B) usually live in fresh water (C) are usually land dwellers (D) can carry on only anaerobic respiration (E) cannot metabolize proteins

19. A B C D E

20. Of the following chemical substances, the only one related to the nervous system is (A) gibberellic acid (B) acetylcholine (C) insulin (D) deoxyribonucleic acid (E) opsonin

20. A B C D E

21. The dominant form of life during the Mesozoic Era were the (A) trilobites (B) amphibia (C) dinosaurs (D) birds (E) mammals

21. A B C D E

22. In the take-off of an astronaut, the increase in speed is known as (A) deceleration (B) gravitation (C) weightlessness (D) acceleration (E) shock wave

22. A B C D E

23. The seed develops from the part of a plant called the (A) hilum (B) pollen tube (C) anther (D) oviduct (E) ovule

23. A B C D E

24. Of the following, the term that includes all the others is (A) mitosis (B) reduction-division (C) oogenesis (D) meiosis (E) spermatogenesis

24. A B C D E

25. The end product of protein digestion is (A) amino acid (B) proteoses (C) peptide (D) peptone (E) glycerin

25. A B C D E

26. Red corpuscles are to hemoglobin as chloroplasts are to (A) guard cells (B) palisade cells (C) chlorophyll (D) photosynthesis (E) cytoplasm

26. A B C D E

TYPE C

Directions: **For each of the organisms listed from 27 to 38, select the letter** (A), (B), (C), or (D), **depending on whether it**

(A) applies only to (A)
(B) applies only to (B)
(C) applies to both (A) and (B)
(D) applies to neither (A) nor (B)

[(A) Invertebrate (C) Both
(B) Mammal (D) Neither]

27. mouse 27. A B C D
28. housefly 28. A B C D
29. canary 29. A B C D
30. euglena 30. A B C D
31. whale 31. A B C D
32. python 32. A B C D
33. starfish 33. A B C D
34. sea lily 34. A B C D
35. bat 35. A B C D
36. bread mold 36. A B C D
37. spider 37. A B C D
38. kangaroo 38. A B C D

TYPE D

Directions: **The following questions require two answers. One of the items in the left column** (A, B, or C) **is related to four of the items in the right column** (numbered 1–5). **Select the letter of the item which is so related; then select the number of the remaining item which does not belong.**

EXAMPLE:

(A) paramecium
(B) ameba
(C) spirogyra

1. cilia
2. macronucleus
3. trichocyst
4. pyrenoid
5. micronucleus

A B C 1 2 3 4 5

39. (A) transpiration
(B) protein synthesis
(C) photosynthesis

1. water
2. sunlight
3. carbon dioxide
4. sodium chloride
5. chlorophyll

39. A B C 1 2 3 4 5

40. (A) parasite
(B) herbivore
(C) saprophyte

1. mushroom
2. wheat rust
3. yeast
4. bacteria of decay
5. bread mold

40. A B C 1 2 3 4 5

41. (A) igneous rock
(B) sedimentary rock
(C) metamorphic rock

1. sandstone
2. limestone
3. shale
4. conglomerate
5. granite

41. A B C 1 2 3 4 5

42. (A) endocrine deficiency
(B) vitamin deficiency
(C) enzyme deficiency

1. diabetes
2. pernicious anemia
3. cretinism
4. myxedema
5. dwarfism

42. A B C 1 2 3 4 5

43. (A) epithelium
(B) neuron
(C) muscle

1. axon
2. pleura
3. dendrite
4. cyton
5. terminal branch

43. A B C 1 2 3 4 5

44. (A) spleen
(B) liver
(C) kidney

1. ureter
2. renal artery
3. urethra
4. villus
5. tubule

44. A B C 1 2 3 4 5

TYPE E

Directions: **The five lettered headings below are followed by a list of numbered statements. For each numbered statement, choose the one lettered heading (A–E) which is most closely related to it.**

(A) Conjugation (C) Parthenogenesis
(B) Diffusion (D) Metamorphosis
(E) Differentiation

45. How were fatherless rabbits obtained?

45. A B C D E

46. How does a tadpole develop into a frog?

46. A B C D E

47. How is the developing young mouse embryo nourished?

47. A B C D E

48. How does a hen's egg develop feathers, beak, etc.?

48. A B C D E

49. How does paramecium reproduce sexually?

49. A B C D E

50. How does a frog's egg develop without fertilization?

50. A B C D E

PART II

In an ecological study of vegetation in a cultivated garden bed that had been neglected for eight weeks, a record was made of the number of weeds that were found in an area measuring a square foot. Twenty-five random trials were made and the findings summarized in Table I. By multiplying the totals by four, the percentage occurrence of a hundred trial areas is obtained. The species were then arranged in five groups according to frequency of occurrence, as shown in Table II,

TABLE I

SPECIES	TOTAL OCCURRENCE IN 25 TRIALS
Poa	25
Stellaria	25
Capsella	24
Senecio	10
Lamium	1
Veronica	2

TABLE II

Group I	0.20 per cent
Group II	21–40 per cent
Group III	41–60 per cent
Group IV	61–80 per cent
Group V	81–100 per cent

51. The number of species present in Group II is (A) 0 (B) 1 (C) 2 (D) 3 (E) 4

52. The least common species classified in Group I is (A) Stellaria (B) Capsella (C) Lamium (D) Senecio (E) Veronica

53. Of the six species studied, the number classified in Group V is (A) 1 (B) 2 (C) 3 (D) 4 (E) none

54. Thirty-three per cent of the species observed were in Group (A) I (B) II (C) III (D) IV (E) V

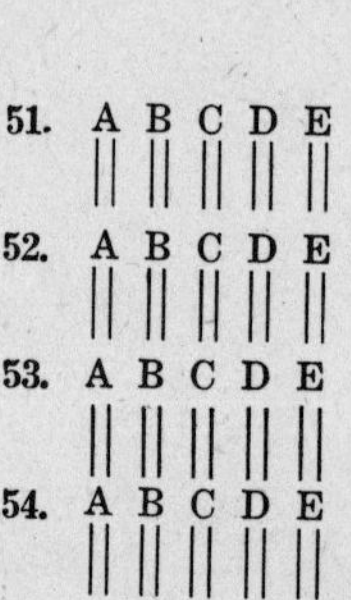

55. There were no species represented in Groups (A) I and II (B) II and III (C) III and IV (D) II and IV (E) IV and V

56. From this study it may be concluded that the two most common weeds to be found in this entire neglected garden bed will probably be (A) Senecio and Capsella (B) Senecio and Lamium (C) Poa and Veronica (D) Poa and Lamium (E) Poa and Stellaria

55. A B C D E

56. A B C D E

PART III—TYPE A

Directions: **In each of the following questions, there is a statement followed by four choices of which *one or more* are correct. Indicate the correct combination by choosing the following letters:**

(A) if only 1, 2 and 3 are correct
(B) if only 1 and 3 are correct
(C) if only 2 and 4 are correct
(D) if only 4 is correct
(E) if some other combination is correct

DIRECTIONS SUMMARIZED:

(A)	(B)	(C)	(D)	(E)
1, 2, 3	1, 3	2, 4	4	some other
only	only	only	only	combination

57. Egg cells of a mammal
1. are fertilized in the ovary
2. contain the monoploid number of chromosomes
3. carry the Y chromosome
4. are microscopic

57. A B C D E

58. The formation of healthy bones is aided by
1. vitamin D
2. calcium
3. exposure to sunlight
4. drinking milk

58. A B C D E

59. Enzymes
1. are present in all cells
2. help digest nutrients
3. serve in the oxidation of carbohydrates
4. are present in bile

59. A B C D E

60. Among the hazards confronting man in space are
1. inability to communicate with earth
2. intense radiation
3. vitamin deficiency
4. psychological isolation

60. A B C D E

61. Charles Darwin made important contributions
1. in the field of evolution
2. to our understanding of the struggle for existence
3. on the effects of natural selection
4. in the study of physiology

61. A B C D E

62. Conservation of our soil resources is
1. assisted by terracing
2. threatened by gullying
3. promoted by rotation of crops
4. unimportant now that atomic energy is available

62. A B C D E

63. The recent discoveries about DNA
1. are making the cure of TB possible
2. explain how hormones affect the pituitary

63. A B C D E

3. challenge the germ theory
4. are useful in understanding the gene

64. A neuron is different from other animal cells
1. in possessing a nucleus
2. because it can transmit impulses
3. because it does not carry on respiration
4. in containing dendrites

64. A B C D E

TYPE B

Directions: **Each of the following statements is followed by five suggested answers. Select the one answer which is best in each case.**

65. Foods have to be digested before they can be used by the body because (A) the stomach is the center of all digestion (B) the villi digest only those nutrients they can absorb (C) only insoluble materials can pass through membranes (D) only soluble materials can pass through membranes (E) assimilation always takes place before digestion

65. A B C D E

66. Weeds are harmful because (A) their seeds serve as food for birds (B) they interfere with the growth of useful plants (C) they prevent the cross-pollination of useful plants (D) they poison the soil (E) their roots do not bind the soil

66. A B C D E

67. A germinating young seed plant does not require food from the soil because (A) it can do without any food for the first 100 hours (B) it obtains its energy by respiration (C) the parent flowering plant nourishes the young plant (D) it can carry on conjugation if there is enough moisture (E) it lives on the stored food in the seed

67. A B C D E

68. Gametes are pure for each character because they contain (A) one chromosome of each pair (B) pairs of chromosomes (C) two pairs of chromosomes (D) the diploid number of chromosomes (E) the sex chromosomes

68. A B C D E

69. Uranium is useful in determining the age of rock because (A) U-235 is more effective than U-238 in giving off gamma rays (B) it changes into lead at a fixed rate (C) it first changes to radiocarbon which has a half-life of 5,760 years (D) sedimentary rock always contains uranium (E) it is possible to observe its disintegration into lead in the microscale laboratory

69. A B C D E

70. The arm of man, the wing of a bat, and the flipper of a whale have the same basic structure because (A) they are used for the same purpose (B) these animals had a common ancestor (C) these animals have identical genes (D) these animals all have a backbone (E) these animals are descended from each other

70. A B C D E

71. The sperm cell determines the sex of the offspring because it contains (A) two X chromosomes (B) two Y chromosomes (C) both an X and a Y chromosome (D) either an X or a Y chromosome (E) more X chromosomes than the egg

71. A B C D E

72. Young people should not take narcotics because they (A) are expensive (B) are sold by "pushers" (C) lead to vitamin failure (D) are painkillers (E) are habit-forming

72. A B C D E

73. The photometer in the accompanying illustration is an instrument for measuring the transpiration rate of plants. A branch is cut from a geranium plant and is inserted and sealed into rubber tubing. All parts of the apparatus, including the funnel, are completely filled with water. As water is lost from the leaves by transpiration, the level of the water in a calibrated capillary tube descends.

The distance per minute that the water moves down the capillary tube measures the (A) rate of photosynthesis (B) rate of water loss from leaves (C) amount of water absorbed by the leaves (D) cohesiveness of the water molecule (E) amount of water used in aerobic respiration

73. A B C D E

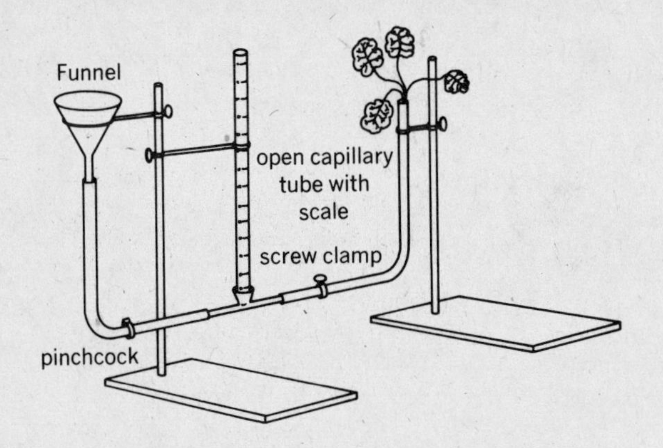

74. The garden pea was well suited for Mendel's experiments on heredity because (A) it is normally cross-pollinated by bees (B) he wanted a plant in which tallness was dominant (C) there is a new generation every two weeks (D) all the peas in a pod are alike (E) it possesses simple contrasting hereditary traits

74. A B C D E

75. The theory of spontaneous generation was disproved because (A) parthenogenesis has been made to occur artificially (B) Pasteur showed that flies arise from maggots (C) cleavage has been shown to occur among all vertebrates (D) living things have been shown to arise only from other living things (E) living organisms multiply by binary fission

75. A B C D E

76. The bean is classified as a dicot because (A) its leaves have parallel veins (B) its flower parts are in groups of three (C) when it germinates, the plumule remains underground (D) it has two cotyledons in its seed (E) its seed has the diploid number of chromosomes

76. A B C D E

77. Blood clots because (A) fibrinogen is transformed into fibrin (B) the red blood cells form fibrinogen (C) the antibodies react with the red blood cells (D) the platelets trap the red blood cells (E) the white blood cells engulf the platelets

77. A B C D E

78. Milk is considered to be an almost perfect food because (A) it is a natural product (B) it contains all of the vitamins (C) it has practically no calories (D) it contains most of the nutrients (E) babies cannot live without it

78. A B C D E

TYPE C

Directions: **Each item consists of an *assertion* (statement) in the left-hand column, and a *reason* on the right-hand side. Select**

(A) if both assertion and reason are true, and *are related* as cause and effect
(B) if both assertion and reason are true, but *are not related* as cause and effect
(C) if the assertion is true, but the reason is a false statement
(D) if the assertion is false, but the reason is a true statement
(E) if both assertion and reason are false statements

DIRECTIONS SUMMARIZED:

	(A)	(B)	(C)	(D)	(E)
Assertion:	True	True	True	False	False
Reason:	True	True	False	True	False
	Cause and effect	*Not* cause and effect			

	Assertion:	Reason:	
79.	A person with Type A blood can receive only Type A blood during a transfusion	BECAUSE his blood would be agglutinated by Type B blood	79. A B C D E
80.	A person paralyzed by polio can breathe in an "iron lung"	BECAUSE the apparatus sends pure air into the lungs through the nostrils	80. A B C D E
81.	The large artery leading into the kidneys carries more wastes than the large vein leading out of them	BECAUSE the kidneys are connected to the bladder by the ureters	81. A B C D E
82.	The liver converts glucose into glycogen	BECAUSE adrenin permits the cells to make use of glucose	82. A B C D E
83.	A habit is like a reflex	BECAUSE the response is automatic in both cases	83. A B C D E
84.	Although it lives in the water, a shark is considered to be a mammal	BECAUSE it has a small amount of hair distributed over its body	84. A B C D E
85.	An animal breeder is not interested in the appearance of mutations	BECAUSE most mutations are harmful	85. A B C D E
86.	The swellings on the roots of legume plants such as clover are harmful	BECAUSE they contain bacteria	86. A B C D E
87.	Eohippus is considered to be an ancestor of the horse	BECAUSE it had four toes	87. A B C D E
88.	The pulmonary artery carries deoxygenated blood	BECAUSE it sends blood to the lungs	88. A B C D E

PART IV—TYPE A

Directions: **Referring to the following selection, choose the best answer for each of the questions that follows:**

*"There is a transmission of genetic information from the DNA in the chromosomes to the sites of protein synthesis. Textbooks often show this flow as an arrow from the nucleus to the cytoplasm. There is no corresponding arrow from the cytoplasm back to the nucleus. Absence of the return arrow might suggest to a biologist that there is something essential missing in the scheme, for in all biological systems that have been carefully studied there is a feedback. In the cell there is indeed evidence for a feedback control directed from the cytoplasm to the chromosomes. In some cases, the feedback comes quickly and lasts for only a short time. In the pancreas, for example, when the cells are stimulated so that their cytoplasm synthesizes digestive enzymes, tracer experiments show that within a few minutes there is a rise in the uptake of amino acids into proteins in the chromosomes. There are also less immediate and more enduring cytoplasmic influences on chromosomes. Among these are the profound changes associated with cell differentiation.

"But although protein synthesis in chromosomes has been shown to be subject to feedback control, there is at present no evidence that the sequence of bases in DNA can be altered by feedback. The chromosomes of germ cells have changed in the course of evolution so that they carry genetic information that is effective in adapting an organism to its environment. The DNA of a germ cell has been shaped by evolution so that it can determine the synthesis of enzymes and other proteins that make for a viable organism. The important point here is that the changes that have taken place during the course of evolution in the DNA molecules of the germ cells of an organism are not the direct result of a feedback from the cytoplasm to the chromosomes in the nucleus. Changes in DNA itself,

*From "How Cells Make Molecules" by V. G. Allfrey and A. E. Mirsky, Sept. 1961, by *Scientific American, Inc.*

according to the generally held views of biologists today, are due to mutation and selection."

89. According to current thought, changes in DNA result from (A) feedback from cytoplasm to the chromosomes (B) feedback from the chromosomes to the cytoplasm (C) mutation and selection (D) protein synthesis (E) enzyme synthesis

89. A B C D E

90. There is a transmission of genetic information from DNA to (A) amino acids (B) digestive enzymes (C) chromosomes (D) proteins (E) the sites of protein synthesis

90. A B C D E

91. All of the following statements are true except (A) when pancreas cells synthesize digestive enzymes, there is an increase in the uptake of amino acids into proteins of the chromosomes (B) there are cytoplasmic influences on the chromosomes (C) there is evidence for a feedback control from the cytoplasm to the chromosomes (D) there is no evidence for a feedback control from the cytoplasm to the chromosomes (E) the profound changes associated with cell differentiation indicate that there are cytoplasmic influences on the chromosomes

91. A B C D E

92. According to the selection, the direction of flow of information from DNA in the chromosomes may be represented most accurately as (A) nucleus → cytoplasm (B) cytoplasm → nucleus (C) cytoplasm ← nucleus (D) nucleus ← cytoplasm (E) nucleus ⇄ cytoplasm

92. A B C D E

93. All of the following statements are false except (A) the sequence of bases of the DNA molecule is not altered by feedback (B) the chromosomes have not been affected by evolution (C) the adaptation of an organism to the environment is independent of the genetic information carried in the germ cells (D) the DNA of a germ cell is unaffected by evolution (E) the DNA of a germ cell has no effect on protein synthesis

93. A B C D E

Directions: **Referring to the following selection, choose the best answer for each of the questions that follows:**

*"All cells synthesize proteins, some continuously, others for only a part of their life cycle. The proteins they make are enormously varied, differing in size, shape, over-all chemical composition and physical properties. But whatever their function, and regardless of their size, shape, solubility or enzyme activity, all proteins have an underlying similarity in constitution: they are all made up of the relatively simple molecular units of amino acid. The synthesis of a protein from these smaller units is conceptually a simple process, involving the joining of the individual amino acids to form long chains. The length of the chain and the sequence of the amino acids vary, of course, from one protein to another. But the essential unit of structure, the link that prolongs the chain, is ubiquitous; it is the peptide bond, the chemical union between the carboxyl group (COOH) of one amino acid and the amino group (NH_2) of the next amino acid in the chain."

94. Proteins are (A) all alike (B) synthesized in all cells (C) alike in size and shape (D) formed from other proteins (E) made of DNA

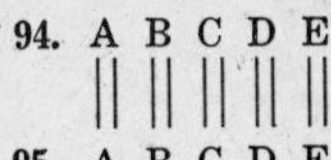

95. During protein synthesis, individual amino acids are (A) joined to form long chains (B) separated out of long chains (C) digested into their separate elements (D) hydrolyzed into their molecular units (E) combined with RNA nucleotides

96. The peptide bond in proteins (A) unites with the carboxyl group of amino acids (B) unites the amino groups of amino acids (C) unites a carboxyl group and an amino group (D) unites a COOH group with a COOH group (E) unites a NH_2 group with a NH_2 group

96. A B C D E

*From "How Cells Make Molecules" by V. G. Allfrey and A. E. Mirsky, Sept. 1961, by *Scientific American, Inc.*

TYPE B

Directions: **Each of the following problems is followed by five suggested answers. Select the one answer which is best in each case.**

97. The chirping of a cricket has been found to be related to the temperature of the air. If the temperature is between 45°–80°F, count the number of chirps in 15 seconds, and add 37; the resulting number gives the air temperature. From this it can be determined that if the number of chirps is timed at 22 in 15 seconds, the temperature is (A) 22°F (B) 37°F (C) 45°F (D) 59°F (E) 80°F

97. A B C D E

98. Of the following, the largest ovum is produced by a (A) mouse (B) elephant (C) whale (D) giraffe (E) sparrow

98. A B C D E

99. A color-blind man married a normal woman who is heterozygous for color vision. The chances of their two daughters being color blind are (A) zero (B) 100 (C) 25–50–25 (D) 50–50 (E) 25–75

99. A B C D E

100. The basal metabolism of a normal adult man was determined to proceed at such a rate as to liberate 39 calories an hour for each square meter of skin surface. In a case of myxedema, the basal metabolism rate of a patient was measured at being (A) the same (B) 40% below normal (C) 100% above normal (D) 40 calories (E) 100 calories

100. A B C D E

101. In a large litter of guinea pigs, three-quarters of the offspring were black (black is dominant). The parents were most likely (A) BB × ww (B) Bw × ww (C) Bw × Bw (D) Bw × BB (E) BB × BB

101. A B C D E

102. A cross between two black guinea pigs produces some black offspring and some white offspring. The gene makeup of the two parents is most likely (A) homozygous black (B) heterozygous black (C) heterozygous white (D) homozygous white (E) recessive white

102. A B C D E

PART V

Directions: **Each of the following statements is followed by five suggested answers. Select the one answer which is best in each case.**

103. These 24-hour clocks show the effects of different periods of light and dark on chrysanthemum and black-eyed Susan. In I, the long night produces flowering in chrysanthemum only. In II, the short night produces flowering only in black-eyed Susan.

103. A B C D E

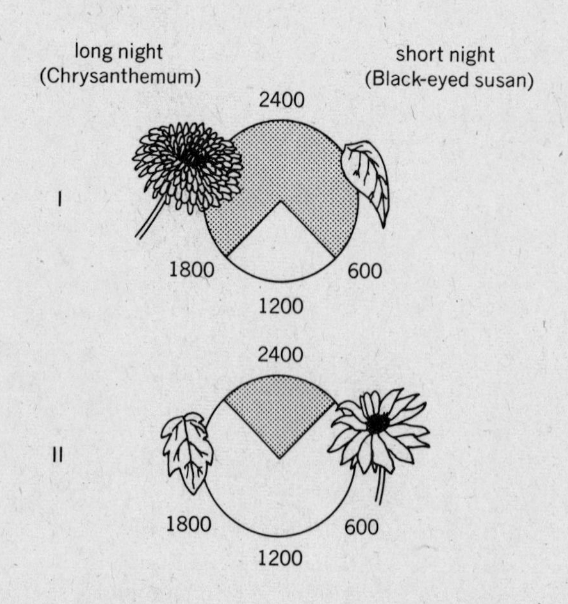

From these observations, it is concluded that
(A) chrysanthemums bloom in the summer, when nights are short.
(B) chrysanthemums bloom in the fall when the nights are long.
(C) chrysanthemums and black-eyed Susan bloom at the same time.
(D) black-eyed Susan blooms when the nights are long.
(E) it is not possible to predict under what conditions black-eyed Susan will bloom.

104. Which is indicated by the information on the accompanying graph?

104. A B C D E

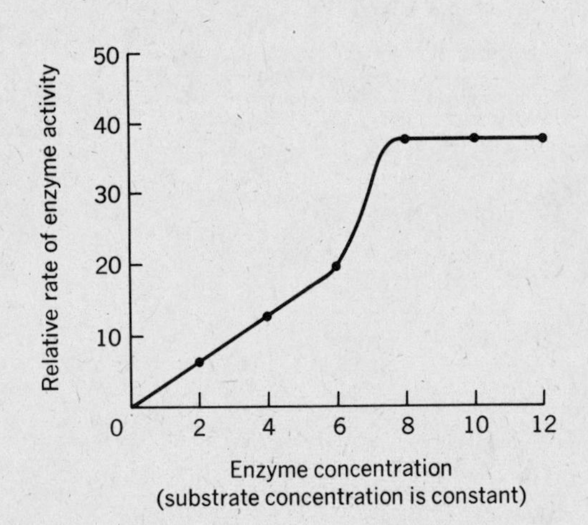

(A) The rate of enzyme action is directly dependent upon the substrate concentration.
(B) The rate of enzyme action increases constantly with an increase in enzyme concentration.
(C) The rate of enzyme action becomes stabilized when a certain enzyme concentration is reached.
(D) Enzyme concentration has no effect upon the rate of enzyme action.
(E) When the substrate concentration is increased, the enzyme concentration is decreased.

105. A new electro-optical device, the Optisat, has been developed by scientists at the National Institute of Health, for measuring blood oxygen levels during use of the artificial lung. The new method of measuring blood oxygen saturation depends on the fact that hemoglobin changes its optical absorption properties when it combines with oxygen to form oxyhemoglobin. Hemoglobin absorbs less visible red light than does oxyhemoglobin. The light passing through the blood is picked up by a photoelectric cell and converted into an electric current which is used to compute saturation. The usefulness of the Optisat depends on the fact that

105. A B C D E

(A) hemoglobin contains more oxygen than oxyhemoglobin
(B) hemoglobin has the same optical absorption properties as oxyhemoglobin
(C) hemoglobin absorbs more visible red light than oxyhemoglobin
(D) the resulting electrical signal is proportional to blood oxygen saturation
(E) the resulting electrical signal speeds up the conversion of hemoglobin to oxyhemoglobin

PART VI

106. A nutrient solution containing bacteria which normally have a red-colored pigment was placed under ultra-violet rays for several days. The colonies which developed were all red-colored, with the exception of one colony which was white. At this point, the investigator should

(A) draw no conclusions
(B) conclude that ultra-violet rays mutated the genes of the bacteria in the white colony
(C) conclude that the white color is a genetic change
(D) conclude that the white color is an environmental change
(E) expect the white colony to turn red again if there is no further exposure to ultra-violet rays

107. Which of the following statements is considered scientific evidence that the size of man's brain may have increased?
(A) modern man is capable of performing on a higher intellectual than the cave man
(B) the size of the brain of man in more primitive societies is smaller than that of the Neanderthal man
(C) fossils indicate that the capacities of the cranium (skull cavity) of man was once less than it is now
(D) Java man and Peking man did not walk erect
(E) man's intelligence has continuously increased as higher intelligence has become increasingly essential for survival

107. A B C D E

108. In order to test a substance for the presence of glucose, the *minimum* apparatus necessary in addition to the substance to be tested, Benedict's solution, and a test tube would be (A) iodine solution (B) nitric acid (C) nitric acid and ammonia (D) a Bunsen burner (E) a Bunsen burner and iodine solution

108. A B C D E

109. The value of using a green geranium leaf with a white edge in a demonstration rather than an all-green geranium leaf is that it shows
(A) an example of positive phototropism
(B) the relationship between cambium and epidermis
(C) that chlorophyll is needed for photosynthesis
(D) the rate of transpiration
(E) the cycle of photoperiodism

109. A B C D E

ANSWER KEY: TEST NO. 4

1. (C)	**12.** (C)	**23.** (E)	**34.** (A)	**45.** (C)	**56.** (E)	**67.** (E)	**78.** (D)	**89.** (C)	**100.** (B)
2. (A)	**13.** (D)	**24.** (D)	**35.** (B)	**46.** (D)	**57.** (C)	**68.** (A)	**79.** (D)	**90.** (E)	**101.** (C)
3. (C)	**14.** (B)	**25.** (A)	**36.** (D)	**47.** (B)	**58.** (E)	**69.** (B)	**80.** (C)	**91.** (D)	**102.** (B)
4. (B)	**15.** (B)	**26.** (C)	**37.** (A)	**48.** (E)	**59.** (A)	**70.** (B)	**81.** (B)	**92.** (E)	**103.** (B)
5. (E)	**16.** (A)	**27.** (B)	**38.** (B)	**49.** (A)	**60.** (C)	**71.** (D)	**82.** (C)	**93.** (A)	**104.** (C)
6. (A)	**17.** (E)	**28.** (A)	**39.** (C)4	**50.** (C)	**61.** (A)	**72.** (E)	**83.** (A)	**94.** (B)	**105.** (D)
7. (C)	**18.** (C)	**29.** (D)	**40.** (C)2	**51.** (B)	**62.** (A)	**73.** (B)	**84.** (E)	**95.** (A)	**106.** (A)
8. (E)	**19.** (C)	**30.** (A)	**41.** (B)5	**52.** (C)	**63.** (D)	**74.** (E)	**85.** (D)	**96.** (C)	**107.** (C)
9. (E)	**20.** (B)	**31.** (B)	**42.** (A)2	**53.** (C)	**64.** (C)	**75.** (D)	**86.** (D)	**97.** (D)	**108.** (D)
10. (B)	**21.** (C)	**32.** (D)	**43.** (B)2	**54.** (A)	**65.** (D)	**76.** (D)	**87.** (B)	**98.** (E)	**109.** (C)
11. (E)	**22.** (D)	**33.** (A)	**44.** (C)4	**55.** (C)	**66.** (B)	**77.** (A)	**88.** (A)	**99.** (D)	

ANSWERS EXPLAINED: TEST NO. 4

1. (C) cell membrane
2. (A) cilia
3. (C) anal spot
4. (B) pellicle
5. (E) food vacuole
6. (A) oral groove
7. (C) contractile vacuole
8. (E) trichocyst
9. (E) micronucleus
10. (B) cell membrane
11. (E) micronucleus

12. (C) The diagram refers to the structural formula of the glucose molecule. A molecule is the smallest part of a substance that still has the properties of the substance.

13. (D) Carbon fixation takes place during the dark phase of photosynthesis. A number of intermediate products are formed when carbon atoms are combined to form C—C bonds, and when hydrogen unites with carbon to form C—H bonds. The first stable product is PGAL (phosphoglyceraldehyde), which is then used in the synthesis of glucose.

14. (B) Hugo de Vries studied mutations in the evening primrose plant, and proposed the theory that genes change.

15. (B) Both forms of carbon (or isotopes) have six protons. The nucleus of an ordinary carbon atom has, in addition, six neutrons, giving it an atomic weight of 12. Radioactive carbon-14 has 8 neutrons in its nucleus, and has an atomic weight of 14.

16. (A) The coronary arteries supply the cells of the heart with nourishment. A blood clot blocks the flow of blood through them, causing the condition known as coronary thrombosis.

17. (E) The gray matter of the brain takes its color from the dendrites and cytons present in it. The white matter contains axons which are covered with a white axon sheath.

18. (C) Cro-Magnon man had a brain capacity somewhat like that of modern man, and larger than that of the other types of primitive man.

19. (C) In the grasshopper, nitrogenous wastes are largely excreted as uric acid. Since this substance is very insoluble, little water is used to carry it in solution. Instead, it is removed in solid crystal form by the Malpighian tubules which open into the digestive system. The wastes are then eliminated through the anus along with undigested food materials. This method of excretion serves as a water-conservation mechanism for certain land-dwelling organisms.

20. (B) Acetylcholine is produced along the length of a neuron and plays a part in the transmission of a nervous impulse. It is also formed in a synapse.

21. (C) The reptiles reached their greatest development in the dinosaurs during this era.

22. (D) A space capsule achieves speeds up to 17,500 miles an hour when it leaves the launching pad. The increase in speed needed to accomplish this terrific rate of travel in a short time is referred to as acceleration. The astronaut is affected by gravitational forces measured in G during this period.

23. (E) When double fertilization occurs within the ovule, one part develops into an embryo, and the rest becomes the food and covering. The matured ovule becomes the seed.

24. (D) Meiosis may occur in the male (spermatogenesis) or in the female (oögenesis). Reduction-division takes place, resulting in the monoploid number of chromosomes; and then mitosis, increasing the number of cells.

25. (A) The digestion of proteins is started in the stomach by the enzyme pepsin. Intermediate products of digestion, peptones and proteoses, are formed. The end product, amino acids, result from further digestion in the small intestine by the enzymes erepsin and trypsin.

26. (C) Red blood cells receive their color from the hemoglobin in them; chloroplasts are green because they contain chlorophyll.

27. (B) A mouse is a mammal because it is covered with hair and feeds its young on milk.

28. (A) A housefly is an insect, and is a member of the arthropod phylum.

29. (D) A canary is a vertebrate, since it has a backbone, but it is not a mammal.

30. (A) Euglena is often classified in the phylum protozoa, despite its chlorophyll content.

31. (B) The whale is the largest of the mammals.

32. (D) The python is a snake, and a vertebrate; it is not a mammal.

33. (A) A starfish is an echinoderm, and an invertebrate. It does not have a backbone.

34. (A) The sea lily is a type of echinoderm. It is an invertebrate.

35. (B) A bat is a flying mammal.

36. (D) A bread mold is a fungus, and a member of the protist kingdom.

37. (A) A spider is an arthropod, and an invertebrate.

38. (B) A kangaroo is a mammal that gives birth to undeveloped young, which it places into its pouch, where it completes its development.

39. (C)4 Photosynthesis takes place in green plants that contain chlorophyll, in the presence of light, with water and carbon dioxide serving as raw materials. Sodium chloride is not used.

40. (C)2 All are saprophytes and live on dead organic matter, except wheat rust, which is a parasite that destroys live wheat plants.

41. (B)5 All are examples of sedimentary rock and are formed from sediment except granite, which is a type of igneous rock that cooled from the molten form.

42. (A)2 All are conditions brought about by a lack of endocrines (diabetes—insulin; cretinism and myxedema—thyroxin; dwarfism—pituitary growth hormone) except pernicious anemia, which appears to be caused by a lack of vitamin B_{12}.

43. (B)2 All are structures present in a neuron except pleura, which is the lining covering the lungs and the inside of the thorax.

44. (C)4 All are related to the kidney and its function of excretion, except the villus, which is a microscopic structure in the small intestine that absorbs digested food.

45. (C) Gregory Pincus removed the ova of rabbits, treated them with chemical solutions, and stimulated them to undergo cleavage, even though they were not fertilized by sperm cells.

46. (D) A developing tadpole loses its gills and tail, and forms lungs and legs as it turns into a frog.

47. (B) The embryo receives its food from the mother's bloodstream. The nutrients diffuse into its own bloodstream in the placenta.

48. (E) The different layers of the gastrula differentiate into the various tissues and organs, i.e., the ectoderm into the skin covering, the mesoderm into muscles, the endoderm into the lining of the alimentary tract, etc.

49. (A) During conjugation, two paramecia lie side by side and exchange parts of their micronuclei.

50. (C) A frog's egg can be stimulated to develop by treatment with chemical solutions, electricity, and mechanical pricking.

51. (B) The species is Senecio. Ten multiplied by 4 gives a percentage occurrence of 40 per cent, for group II.

52. (C) Only one example of Lamium was found in twenty-five trials.

53. (C) The three in group V are Poa (100%), Stellaria (100%), and Capsella (96%).

54. (A) One-third, or two out of six of the species are in group I—Lamium and Veronica.

55. (C) No species occurred in 41–60%, or 61–80% of the cases.

56. (E) Poa and Stellaria were found to occur in every one of the twenty-five trials, making them the most numerous.

57. (C) Mammal egg cells contain practically no stored food, and are therefore microscopic; like other eggs, they have gone through reduction-division and have the monoploid number of chromosomes.

58. (E) All of the items are correct. Milk contains calcium which is used in building bones; exposure to direct sunlight converts a substance in the skin into vitamin D, which enables the body to utilize calcium.

59. (A) Bile does not contain enzymes. It serves to emulsify fats, thereby enabling fat-digesting enzymes in the small intestine to attack the broken-down fat particles more readily.

60. (C) Without the shielding effect of the atmosphere, an astronaut is exposed to the effect of primary cosmic rays. Isolation may lead to undesirable psychological effects such as weariness and hallucinations.

61. (A) Following his voyage on the *Beagle*, Darwin collected much data to show that evolution has taken place. He also proposed his theory of natural selection to explain how organisms have changed as a result of the survival of the fittest after a struggle for existence.

62. (A) Terracing prevents the erosion of soil on a hillside. Gullying on an unprotected slope results in the loss of topsoil. Rotation of standard crops, such as corn, with legume plants enriches the soil with nitrates.

63. (D) It has been shown that deoxyribonucleic acid (DNA) is the chemical component in the chromosomes that makes up the genes. DNA is a complex molecule whose atoms are arranged in two spiral chains.

64. (C) A neuron has the property of irritability, which enables it to receive and transmit impulses. It contains specialized structures such as axon and dendrites for performing these functions.

65. (D) During digestion, the nutrients of food are digested to simpler, soluble form, and are able to diffuse into the bloodstream to be carried to the cells of the body.

66. (B) Weeds crowd out more desirable plants, and keep them from growing properly.
67. (E) The seed contains an embryo and stored food which nourishes the young plant until it can start obtaining the raw materials needed for food-making.
68. (A) During reduction-division, the members of a pair of chromosomes separate, so that each gamete has one set of genes.
69. (B) Uranium is an element that keeps decaying into other elements in the radioactive series. One of them is radium. The end-product is lead, which is not radioactive. By computing the ratio of uranium to lead, the age of the rock is determined.
70. (B) These animals have changed since the time of their common ancestor, but they still retain enough of the same genes for their forelimbs to be similar in structure.
71. (D) When a sperm cell containing an X chromosome fertilizes an egg cell, the fertilized egg will contain the XX chromosomes and will become a female; fertilization by a sperm containing the Y chromosome results in a fertilized egg containing the XY chromosomes, which will become a male.
72. (E) Even a small dose of narcotics may lead to drug addiction. The habit is difficult to break, and leads to painful withdrawal symptoms.
73. (B) Water evaporates from the leaf by the process of transpiration. As it is drawn up by the stem, it takes along with it water from the capillary tube. The faster water evaporates, the more rapidly will it move down the capillary tube.
74. (E) Mendel used seven contrasting traits in his experiments.
75. (D) Redi was the first scientist to show that flies arise from previously existing flies.
76. (D) Its leaves also have a network of veins.
77. (A) A series of reactions takes place, concluding with the formation of fibrin threads.
78. (D) However, it contains little or no vitamin C after it has been pasteurized, requiring that babies receive an additional source of this vitamin.
79. (D) A person with type A blood may receive either type A, or type O, the universal donor type.
80. (C) The iron lung alternately increases and releases the pressure around the chest, causing air to leave and to enter the lungs.
81. (B) As the blood flows through the kidneys, wastes are removed from it. The renal artery brings blood to the kidneys. The purified blood leaving the kidneys is collected into the renal vein.
82. (C) Under the influence of insulin, glucose is converted into glycogen for storage in the liver. The cells of the body can use glucose for energy when insulin is present.
83. (A) A reflex is inborn, while a habit is acquired. In both, however, the response is automatic and requires no thought.
84. (E) A shark is a fish; it breathes by gills, and has no hair.
85. (D) Breeders are always on the lookout for the rare useful mutations. The hornless condition in cattle originated as a mutation.
86. (D) The nodules on legume roots are useful because they contain nitrogen-fixing bacteria which change nitrogen into nitrates for use by plants.
87. (B) Eohippus is connected with the present-day horse by a series of intermediate fossils which show the gradual change from a four-toed foot to a foot with one large toe, having the remains of two side toes as splints.
88. (A) The pulmonary artery is the only artery carrying deoxygenated blood; it goes from the right ventricle to the lungs.
89. (C) From an evolutionary point of view, DNA is quite stable. Any changes in it would most likely be due to an alteration of its structure.
90. (E) The introductory statement of the selection indicates that genetic information is transmitted from the DNA to the sites of protein synthesis.
91. (D) The contrary of this is stated in the middle of the first paragraph.
92. (E) The main point of the first paragraph indicates that the flow of information is in both directions.
93. (A) The first sentence of the second paragraph states that the sequence of bases in DNA is not altered by feedback.
94. (B) The opening words of the selection indicate that all cells synthesize proteins, some continuously, others for only a part of their life cycle.
95. (A) Proteins are large polymers of many repeating amino acid units.
96. (C) When amino acids are bonded together, a C—N bond forms between the carboxyl group of one amino acid and the amino group of the next amino acid. The C—N bond is known as a peptide bond. A chain

of amino acids bonded in this way is called a polypeptide.

97. (D) Add 22 to 37.

98. (E) All the animals are mammals, with microscopic ova, having practically no stored food, except the sparrow, which has relatively large eggs containing a large amount of stored food.

99. (D) The possible results are:

	X	Y
X	XX	XY
X	XX	XY

Female: 50% normal heterozygous
50% color-blind
Male: 50% normal
50% color-blind

100. (B) In myxedema, there is an underproduction of thyroxin, resulting in a reduced basal metabolism rate.

101. (C) When hybrids are crossed, the results are:

	B	b
B	BB	Bb
b	Bb	bb

25% homozygous black }
50% heterozygous black } 75% black
25% white

102. (B) In guinea pigs, black coat color is dominant to white. Black guinea pigs that produce white offspring must carry a recessive gene and must be heterozygous black. This can be shown as follows:

	B	b
B	BB	Bb
b	Bb	bb

Possible results
25% BB—homozygous black
50% Bb—heterozygous black
25% bb—white

103. (B) This is an example of photoperiodism. Chrysanthemums require a long period of darkness in order to flower. In the fall, when the hours of daylight are less than 12–13, such short-day flowers will bloom.

104. (C) The graph shows that the relative rate of enzyme activity rises as the enzyme concentration increases until the concentration is about 7. After that point, the relative rate of enzyme activity levels off at about 35, and remains the same.

105. (D) The blood oxygen level depends on the amount of oxygen that has been absorbed by hemoglobin to form oxyhemoglobin. Since hemoglobin absorbs less visible red light than does oxyhemoglobin, the amount of light passing through blood will vary with the relative amounts of hemoglobin and oxyhemoglobin that are present. As a result, the electric signal produced will vary with the blood oxygen saturation.

106. (A) There is insufficient evidence to draw any conclusion. There was no control. It is possible that this result could have been obtained without the ultraviolet treatment. Before a mutation or a gene change is accepted, there would have to be continued culturing of the bacteria, to determine whether the characteristic was inherited in the next generation.

107. (C) By measuring the capacity of the cranium, or skull capacity, it is possible to determine the size of the brain. The brain capacity of primitive man was as follows: Java man—940 cc.; Peking man—1,000 cc. Modern man's brain capacity is about 1,500 cc.

108. (D) The substance to be tested must be heated in Benedict's solution before the proper color reaction can be obtained.

109. (C) The white edge will be seen not to contain any starch, while the green part does contain starch. Since there is no chlorophyll present in the edge, starch was not produced there.

PRACTICE COLLEGE BOARD ACHIEVEMENT TEST IN BIOLOGY

NO. 5

PART I—TYPE A

Directions: **The following diagram shows structures involved in flower fertilization. Certain of the parts are labeled with numbers. Each of the statements is followed by five suggested labels or answers. A label may be used once, more than once, or not at all. For each statement, select the best answer.**

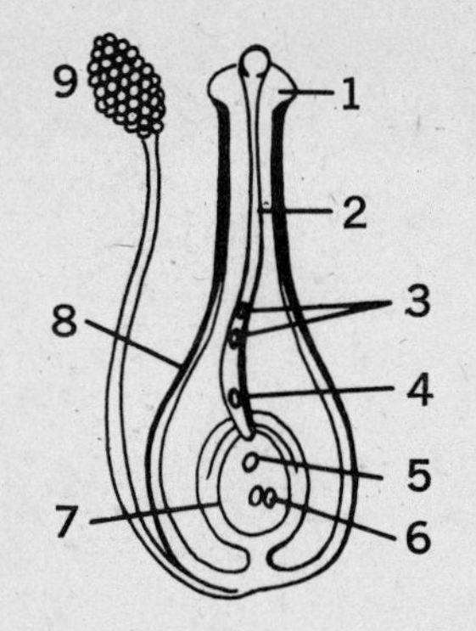

1. Forms pollen grains
   (A) 1 (B) 3 (C) 4 (D) 6 (E) 9

2. Will develop into the fruit
   (A) 1 (B) 5 (C) 7 (D) 8 (E) 9

3. Grows toward the ovule
   (A) 2 (B) 5 (C) 6 (D) 7 (E) 8

4. Will develop into the seed
   (A) 1 (B) 2 (C) 7 (D) 8 (E) 9

5. Receives pollen during pollination
   (A) 1 (B) 2 (C) 3 (D) 4 (E) 7

6. Controls the growth of the pollen tube
   (A) 2 (B) 4 (C) 5 (D) 6 (E) 8

7. After fertilization, forms the embryo of the seed
   (A) 3 (B) 4 (C) 5 (D) 7 (E) 8

8. Fertilizes the nuclei in the ovule
   (A) 2 (B) 3 (C) 4 (D) 8 (E) 9

9. Serves as a passageway for the sperm nuclei
   (A) 1 (B) 2 (C) 3 (D) 5 (E) 6

10. After fertilization, forms the food of the seed
    (A) 4 (B) 5 (C) 6 (D) 7 (E) 8

1. A B C D E
2. A B C D E
3. A B C D E
4. A B C D E
5. A B C D E
6. A B C D E
7. A B C D E
8. A B C D E
9. A B C D E
10. A B C D E

TYPE B

Directions: **Each of the following statements is followed by five suggested answers. Select the one answer that is best in each case.**

11. Neurohumors may best be described as
    (A) chemicals produced by the endocrine glands
    (B) secretions that take the place of enzymes in muscle activity
    (C) secretions that slow down muscle activity
    (D) chemicals involved in the passage of an impulse across a synapse
    (E) chemicals carried in the plasma to the organs they stimulate

12. The appearance of new insect populations that are resistant to chemical insecticides is a current illustration of (A) inheritance of acquired characteristics (B) biological control (C) natural selection (D) intraspecific control (E) ecological succession

13. In which biome is the greatest amount of food per acre produced? (A) marine (B) tropical rain forest (C) tundra (D) grassland (E) taiga

11. A B C D E
12. A B C D E
13. A B C D E

14. Development of an organism from an unfertilized egg is known as (A) fertilization (B) conjugation (C) maturation (D) parthenogenesis (E) oögenesis

14. A B C D E

15. The part of the cell in which proteins are synthesized is the (A) ribosome (B) nucleus (C) centriole (D) mitochondrion (E) Golgi body

15. A B C D E

16. Oxygen is carried throughout the body by the (A) platelets (B) plasma (C) veins (D) red blood cells (E) lymphocytes

16. A B C D E

17. Digested fats in the small intestine are absorbed by the (A) capillaries (B) lacteals (C) venules (D) arterioles (E) sinusoids

17. A B C D E

18. The temperature of the body is regulated by the excretion of (A) urea (B) carbon dioxide (C) sweat (D) urine (E) bile salts

18. A B C D E

19. A synapse is (A) part of an axon (B) part of a neuron (C) a gap between neurons (D) a chemical impulse (E) an electrical impulse

19. A B C D E

20. The process by which a lobster grows a new claw is known as (A) regeneration (B) rheotropism (C) budding (D) reclamation (E) the anaphase stage

20. A B C D E

21. Soil erosion is caused by (A) terracing (B) contour plowing (C) strip cropping (D) rotation of crops (E) overgrazing

21. A B C D E

22. The changing of nutrients into protoplasm is known as (A) absorption (B) digestion (C) oxidation (D) assimilation (E) circulation

22. A B C D E

23. Cancer may be cured by all of the following methods except (A) surgery (B) immunization (C) X-ray treatment (D) radium treatment (E) use of radioisotopes

23. A B C D E

24. The adrenal gland is located next to the (A) liver (B) brain (C) bladder (D) kidney (E) neck

24. A B C D E

TYPE C

Directions: **For each of the structures listed from 25 to 36, select the letter (A), (B), (C), or (D), depending on whether it**

(A) applies only to (A)
(B) applies only to (B)
(C) applies to both (A) and (B)
(D) applies to neither (A) nor (B)

[(A) Euglena (C) Both
(B) Ameba (D) Neither]

25. cell membrane — 25. A B C D

26. food vacuole — 26. A B C D

27. pseudopodia — 27. A B C D

28. chloroplast — 28. A B C D

29. cilia — 29. A B C D

30. nucleus — 30. A B C D

31. trichocyst — 31. A B C D

32. flagellum — 32. A B C D

33. eyespot — 33. A B C D

34. cytoplasm — 34. A B C D

35. oral groove — 35. A B C D

36. chromatin — 36. A B C D

TYPE D

Directions: **The following questions require two answers. One of the items in the left column** (A), (B), **or** (C) **is related to four of the items in the right column** (numbered 1–5). **Select the letter of the item which is so related; then select the number of the remaining item which does not belong.**

EXAMPLE:

(A) paramecium	1. cilia	A B C 1 2 3 4 5
(B) ameba	2. macronucleus	
(C) spirogyra	3. trichocyst	
	4. pyrenoid	
	5. micronucleus	

37. (A) aerobic respiration	1. lactic acid	37. A B C 1 2 3 4 5
(B) anaerobic respiration	2. oxygen	
(C) transpiration	3. alcohol	
	4. 2 ATP	
	5. fermentation	

38. (A) protein	1. fish	38. A B C 1 2 3 4 5
(B) fat	2. spaghetti	
(C) carbohydrate	3. cake	
	4. bread	
	5. candy	

39. (A) thyroxin	1. goiter	39. A B C 1 2 3 4 5
(B) insulin	2. hemophilia	
(C) estrogen	3. cretinism	
	4. basal metabolism	
	5. myxedema	

40. (A) evolution	1. Lamarck	40. A B C 1 2 3 4 5
(B) cell doctrine	2. De Vries	
(C) disease	3. Darwin	
	4. Banting	
	5. Wallace	

41. (A) inborn behavior	1. reflex	41. A B C 1 2 3 4 5
(B) conditioned behavior	2. negative geotropism	
(C) intelligent behavior	3. instinct	
	4. positive hydrotropism	
	5. habit	

42. (A) blending	1. hemophilia	42. A B C 1 2 3 4 5
(B) dominant human traits	2. normal clotting of blood	
(C) recessive human traits	3. brown eyes	
	4. dark hair	
	5. Rh positive	

43. (A) igneous rock	1. amber	43. A B C 1 2 3 4 5
(B) Paleozoic Era	2. appendix	
(C) fossil	3. mammoth	
	4. tar pit	
	5. imprint	

TYPE E

Directions: **The five lettered headings below are followed by a list of numbered statements. For each numbered statement, choose the one lettered heading (A)–(E) which is most closely related to it.**

(A) Grafting (C) Germination
(B) Pollination (D) Cutting
(E) Runner

44. How do bees help in the production of apples?

44. A B C D E

45. How does a gardener obtain more geranium plants?

45. A B C D E

46. How can more seedless oranges be obtained?

46. A B C D E

47. How does a strawberry plant spread?

47. A B C D E

48. How can a pear tree grow apples?

48. A B C D E

49. How does a bean seed form a new bean plant?

49. A B C D E

PART II

A project was conducted to determine the factors involved in frog hibernation. The effects of reduced temperature on the breathing rate were studied. Graph I was prepared to show the results. Graph II shows the results when the experiment was repeated. During the first experiment, the frog attempted to hibernate at 40°F. During the second experiment it attempted to hibernate at 55°, 53°, 50°, 46°, 41°, 36°, and 34°. The attempts to hibernate included these activities: The frog closed its eyes and expelled air from its lungs; it attempted to dig in at the bottom of the jar as though there were mud there.

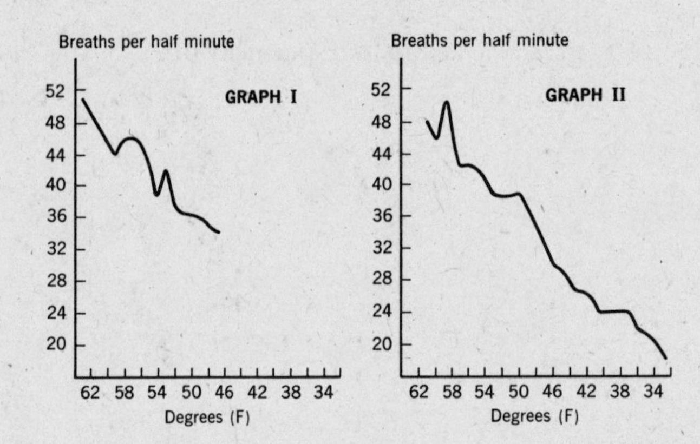

50. From this project, it can be concluded that (A) frogs hibernate when the temperature is reduced (B) as the temperature is lowered, the heart beat is reduced (C) frog hibernation is not related to temperature (D) the breathing rate is reduced as the temperature is lowered (E) frogs stop breathing when they attempt to hibernate

50. A B C D E

51. The results of the second graph indicate that (A) the frog resisted hibernating in the second experiment (B) the frog had learned to hibernate from its previous experience (C) the frog showed identical responses to the same temperatures (D) the frog showed varying responses to the same temperatures (E) the frog showed signs of fatigue

51. A B C D E

52. The experiment was done a second time for all of the following reasons except (A) to improve on the techniques of the first experiment (B) to check on the first set of results (C) to find the rate of respiration from 48° to 43°F (D) to locate the brain center that controls hibernation (E) to see whether hibernation could be induced artificially

52. A B C D E

53. At 59°F, the breathing rate (A) was higher than at 60° in Graph I (B) was lower than at 60° in Graph II (C) was at its highest point in Graph II (D) was at its lowest point in Graph I (E) showed an increase in Graph I

53. A B C D E

54. When the frog attempted to hibernate, it did so (A) from habit (B) from instinct (C) to avoid any further experimentation (D) because hibernation is a tropism (E) as an example of trial and error learning

54. A B C D E

55. While a frog is hibernating during the winter, it (A) uses up very little energy (B) uses up a great deal of energy (C) wakes up only long enough to catch a few insects (D) wakes up only to clear a breathing tunnel through the snow (E) is at the end of its life history, lays its eggs, and gradually dies

55. A B C D E

PART III—TYPE A

Directions: **In each of the following questions, there is a statement followed by four choices, of which *one or more* are correct. Indicate the correct combination by choosing the following letters:**

(A) if only 1, 2, and 3 are correct
(B) if only 1 and 3 are correct
(C) if only 2 and 4 are correct
(D) if only 4 is correct
(E) if some other combination is correct

DIRECTIONS SUMMARIZED:

(A)	(B)	(C)	(D)	(E)
1, 2, 3	1, 3	2, 4	4	some other
only	only	only	only	combination

56. According to the "lock and key" concept of enzyme action, the specificity of enzymes is based on
1. relative amounts of lipid present in the enzyme
2. amount of water present
3. amount of substrate present
4. arrangement of the atoms at the active site of the enzyme

56. A B C D E

57. The nutrient glucose is useful to the body
1. in preventing goiter
2. as a source of energy
3. to build protoplasm
4. during the process of respiration

57. A B C D E

58. Protozoa are considered to be
1. invertebrates
2. chordates
3. one-celled protists
4. bryophytes

58. A B C D E

59. The cambium layer of a plant
1. serves primarily for food storage
2. carries on secondary photosynthesis
3. conducts liquids up the stem
4. contains actively dividing cells

59. A B C D E

60. The diaphragm in man
1. separates the chest and abdominal cavities
2. is composed of skeletal muscle
3. is useful in breathing
4. causes the bends at high altitudes

60. A B C D E

61. A turtle is considered to be a vertebrate because
1. of its external skeleton
2. it has scales
3. it lays eggs
4. it has a backbone

61. A B C D E

62. A person who has type A blood
1. can safely receive type O blood
2. can safely receive type AB blood
3. can be given type A blood without danger
4. can safely be given a transfusion of type B blood

62. A B C D E

63. Nitrogenous wastes
1. are excreted through the kidneys
2. are eliminated from the lungs
3. result from the breakdown of proteins
4. contain urea

63. A B C D E

64. The pituitary is known as the master gland
1. because it is located close to the brain
2. because it controls the growth of the skeleton
3. or the gland of emergency
4. because it controls most of the other endocrine glands

64. A B C D E

65. The axon of a neuron
1. stimulates red blood cell formation
2. may be over a foot in length
3. transmits impulses
4. is covered with a fatty sheath

65. A B C D E

66. In forming a habit
1. the autonomic ganglia are retained
2. inborn responses are developed
3. one sensory and one motor neuron are involved
4. one learns through conscious effort

66. A B C D E

TYPE B

Directions: **Each of the following statements if followed by five suggested answers. Select the one answer which is best in each case.**

67. The Mesozoic Era is called the Age of Reptiles because (A) reptiles first appeared then (B) reptiles reached their greatest development (C) the dinosaurs exterminated all other types of reptiles (D) reptiles successfully resisted man (E) the dinosaurs became extinct

67. A B C D E

68. Darwin found that finches on the various Galapagos Islands were somewhat different from each other because (A) of different food (B) of geographic isolation (C) of different natural enemies (D) of different climatic variations (E) of the inheritance of acquired characteristics

68. A B C D E

69. In diabetes, excess sugar is found in the urine because (A) the basal metabolism is decreased (B) the kidneys are in a state of hypertension (C) the cells cannot utilize glucose (D) an excess of insulin is produced (E) insufficient thyroxin is formed

69. A B C D E

70. Paramecium is able to avoid obstacles in its path because (A) its sense organs receive responses (B) its contractile vacuoles burst (C) its flagellum moves it away (D) its protoplasm has the property of irritability (E) its conditioned responses are inborn

70. A B C D E

71. You can't drown a grasshopper by keeping its head under water because (A) it breathes through openings in its abdomen (B) the spiracles in its head filter out oxygen (C) it can hold its breath for hours (D) it is cold-blooded (E) the maxillae and mandibles serve as an airtight valve

71. A B C D E

72. During conditions of great acceleration, a space man may black out because (A) the human body cannot stand speeds of 17,500 miles an hour (B) the G force is lower than the weight of his body (C) gravity is overcome (D) the blood is prevented from reaching his head (E) weightlessness is suddenly achieved

72. A B C D E

73. Chestnut trees are becoming extinct in the eastern part of the United States because (A) of overpicking of the chestnut crop (B) the Japanese beetle eats the leaves and roots (C) the tent caterpillar defoliates the trees (D) they are susceptible to the white pine blister rust (E) a fungus disease, the blight, is killing the trees

73. A B C D E

74. A blood clot in the coronary artery may be fatal because (A) it leads to atherosclerosis (B) varicose veins invariably result as a by-product (C) the portal circulation is interfered with (D) the flow of blood to the heart muscle is blocked (E) it leads to the coagulation of lymph

74. A B C D E

75. The whale is classified as a mammal chiefly because (A) it breathes air (B) it has a backbone (C) it has a four-chambered heart (D) it bears live offspring (E) it produces milk

75. A B C D E

76. The fact that, under certain conditions, stomatal openings become smaller enables the plants to avoid excessive loss of (A) carbon dioxide (B) oxygen (C) water (D) essential plant hormones (E) chlorophyll

76. A B C D E

77. Down's syndrome (mongolism) is a human defect caused by the presence of an extra chromosome; the affected individual has 47 chromosomes in the body cells instead of 46. This condition results from (A) linkage (B) non-linkage (C) nondisjunction (D) sex linkage (E) chromosomal replication

77. A B C D E

78. To the species involved, a major advantage of sexual reproduction over asexual reproduction is that sexual reproduction results in (A) greater number of offspring (B) greater variety of offspring (C) greater size of offspring (D) more rapid development of offspring (E) less vulnerability of offspring

78. A B C D E

79. The phase-contrast microscope has aided our understanding of the cell by permitting (A) detailed observations of unstained living cells (B) magnification of cells up to 100,000 times their normal size (C) observation of objects a centimeter or more in thickness (D) chemical analysis of parts of the cell (E) ultra-violet light penetration of cells

79. A B C D E

TYPE C

Directions: **Each item consists of an *assertion* (statement) in the left-hand column, and a *reason* on the right-hand side. Select**

(A) if both assertion and reason are true, but *are related* as cause and effect
(B) if both assertion and reason are true, but *are not related* as cause and effect
(C) if the assertion is true, but the reason is a false statement
(D) if the assertion is false, but the reason is a true statement
(E) if both assertion and reason are false statements

DIRECTIONS SUMMARIZED:

	(A)	(B)	(C)	(D)	(E)
Assertion:	True	True	True	False	False
Reason:	True	True	False	True	False
	Cause and effect	*Not* cause and effect			

Assertion:		Reason:
80. Fats are used by the body for growth and repair	BECAUSE	they contain nitrogen which is needed for making protoplasm
81. Twins of the opposite sex are fraternal twins	BECAUSE	they contain the same genetic make-up
82. Sperm and egg cells have the diploid number of chromosomes	BECAUSE	they undergo reduction-division
83. Mutations may be induced artificially by the use of radioactive materials	BECAUSE	the genes can be changed by radiation

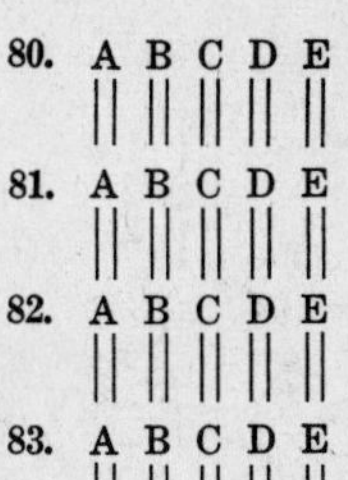

84. Plants, but not animals, may be improved by selection and inbreeding BECAUSE they can be made to breed true by vegetative propagation

84. A B C D E

85. In the early stages, the embryo of man and bird are similar BECAUSE they are land-living animals

85. A B C D E

86. Anthropologists define members of the Latin race as being short, with dark features BECAUSE they live in the Mediterranean area

86. A B C D E

87. The decay of leaves into humus is useful BECAUSE it reduces the number of bacteria in the soil

87. A B C D E

88. Yeast cells obtain energy by fermentation BECAUSE they give off carbon dioxide and alcohol

88. A B C D E

PART IV—TYPE A

Directions: **Referring to the following selection, choose the best answer for each of the questions that follows:**

*"Much of our knowledge of the nerve cell has been obtained from the giant axon of the squid, which is nearly a millimeter in diameter. It is fairly easy to probe this useful fiber with microelectrodes and to follow the movement of radioactively labeled substances into it and out of it. The axon membrane separates two aqueous solutions that are almost equally electroconductive and that contain approximately the same number of electrically charged particles, or ions. But the chemical composition of the two solutions is quite different. In the external solution more than 90 per cent of the charged particles are sodium ions (positively charged) and chloride ions (negatively charged). Inside the cell these ions together account for less than 10 per cent of the solutes; there the principal positive ion is potassium and the negative ions are a variety of organic particles (doubtless synthesized within the cell itself) that are too large to diffuse easily through the axon membrane. Therefore the concentration of sodium is about 10 times higher *outside* the axon, and the concentration of potassium is about 30 times higher *inside* the axon. Although the permeability of the membrane to ions is low, it is not indiscriminate; potassium and chloride ions can move through the membrane much more easily than sodium and the large organic ions can. This gives rise to a voltage drop of some 60 to 90 millivolts across the membrane, with the inside of the cell being negative with respect to the outside."

89. The axon membrane separates two aqueous solutions that (A) contain approximately the same number of electrically charged particles, are almost equally electroconductive and have the same chemical composition (B) contain approximately the same number of electrically charged particles, are unequally electroconductive and have the same chemical composition (C) contain approximately the same number of electrically charged particles, are unequally electroconductive and have different chemical compositions (D) contain approximately the same number of electrically charged particles, are almost equally electroconductive and have different chemical compositions (E) contain widely varying numbers of electrically charged particles, are almost equally electroconductive and have the same chemical composition.

90. In the external solution of the axon membrane (A) more than 90 per cent of the charged particles are positively charged sodium ions, negatively charged chloride ions and positively charged potassium ions (B) more than 90 per cent of the charged ions are positively charged sodium ions and negatively charged chloride ions (C) more than 90 per cent of the charged ions are negatively charged sodium ions and positively charged chloride ions (D) more than 90 per cent of the charged ions are positively charged

90. A B C D E

potassium ions (E) more than 90 per cent of the charged ions are positively charged potassium ions and negatively charged chloride ions

91. The concentration of sodium outside the axon and the concentration of potassium inside the axon (A) is the same (B) varies as 10 is to 30 (C) varies as 30 is to 10 (D) is equal to the sum of 10 and 30 (E) is equal to the difference between 10 and 30

91. A B C D E

92. All of the following statements about the giant axon of the squid are true except (A) it has given us much of our knowledge of the nerve cell (B) it can be probed with microelectrodes (C) like other axons, it is microscopic (D) the movement of radioactive substances into and out of it can be traced (E) it is almost a millimeter in diameter

92. A B C D E

93. The electrical charge on the inside and the outside of the cell can be described as follows: (A) negative on the inside, positive on the outside (B) positive on the inside, negative on the outside (C) negative on both the inside and the outside (D) positive on both inside and outside (E) negative on the inside, neutral on the outside

93. A B C D E

TYPE B

Directions: **Each of the following problems is followed by five suggested answers. Select the one answer which is best in each case.**

94. The volume of a cell varies as the cube of its diameter, and the surface as the square. Compared to a full-grown cell, two cells that have just been formed by binary fission have a ratio of surface to volume that is (A) increased (B) decreased (C) unchanged (D) determined by its metabolic rate (E) determined by the chromosome number

94. A B C D E

95. A man who is normal for blood clotting ability marries a normal heterozygous woman who came from a family having a history of hemophilia. The chances of their daughters being normal heterozygous is (A) 50–25–25 (B) zero (C) 100 (D) 50–50 (E) 25–75

95. A B C D E

96. In Project Ozma, a radio telescope listens for signs of extraterrestrial life. At times, it is directed at Alpha Centauri, the star nearest our solar system, only about four light years away. Radio waves travel at the speed of light so that if we send a signal, the number of years it would take for us to receive a reply would be at least (A) 4 light years (B) 8 light years (C) 12 light years (D) 93 million years (E) 6,000 billion years

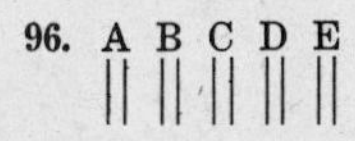

97. Of the following the smallest ovum is produced by a(n) (A) robin (B) salmon (C) painted turtle (D) elephant (E) ostrich

98. There are 25 yellow and 25 green peas in each of two jars. A student simultaneously removes one bean from each jar at random and keeps a tally of the color of the pairs of beans obtained. According to the law of probability, the final result would most likely be (A) 50 yellow-yellow; 50 green-green (B) 25 yellow-yellow; 50 yellow-green; 25 green-green (C) 100 yellow-green (D) 25 yellow-yellow; 25 green-green (E) 50 yellow-green

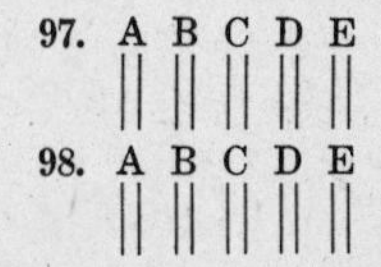

TYPE C

You are assigned a project to determine whether the rate of photosynthesis in Elodea, an aquatic plant, is affected by the addition of small amounts of carbon dioxide to the battery jar in which the plants are growing. The plants appear to be giving off bubbles of gas.

Directions: **Select the best answers in each of the following questions.**

99. You can collect the gas by (A) keeping the Elodea in a beaker under a bell jar (B) placing a funnel over the Elodea, with a test tube at its end (C) inserting two electrodes in the water, covered with test tubes (D) covering

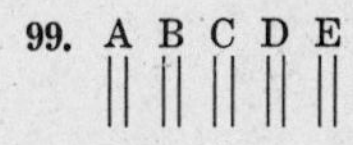

the plants with a cone of filter paper (E) using a thistle tube in a collecting bottle

100. The gas can be shown to be oxygen by (A) testing it with limewater (B) burning it as it is generated (C) inserting a glowing splint into a test tube of the gas (D) passing it over potassium chlorate (E) bringing it into contact with moist litmus paper

101. Additional amounts of carbon dioxide may be added by all the following methods except (A) exhaling into the water through a straw (B) keeping a fish in the water (C) adding hydrochloric acid to marble chips (D) adding sulphuric acid to zinc (E) adding some club soda to the water

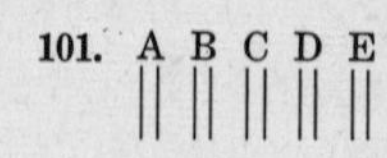

102. The effect of added amounts of carbon dioxide will probably be (A) change of leaf color to blue-black color (B) increased production of bubbles of oxygen (C) greater phototropic response (D) observable increase in leaf size (E) an increase in the rate of respiration

102. A B C D E

103. Besides carbon dioxide, another factor which must be supplied if photosynthesis is to take place is (A) glucose (B) starch (C) solution of chlorophyll (D) oxygen (E) light

103. A B C D E

104. A control in this project would be to (A) place a similar set-up in the dark (B) compare the production of bubbles when no additional carbon dioxide is added (C) compare the production of bubbles when the temperature of the water is raised (D) use a smaller battery jar (E) generate oxygen into the battery jar

104. A B C D E

105. If carbon dioxide containing radioactive carbon-14 were used (A) the newly produced starch would contain carbon-14 (B) the production of oxygen would be increased (C) the production of oxygen would be reduced (D) the oxygen being released would be radioactive (E) half of the bubbles of oxygen would stimulate a Geiger counter

105. A B C D E

106. Before you could conclude that additional amounts of carbon dioxide stimulate the rate of photosynthesis in all aquatic plants, it would be necessary to repeat the experiment (A) with more Elodea plants (B) with the same amount of Elodea plants (C) with variable amounts of Elodea plants (D) with exactly the same materials (E) with many other aquatic plants

106. A B C D E

PART VI

Directions: **Each of the following statements is followed by five suggested answers. Select the one answer which is best in each case.**

107. An experiment was conducted to determine which of five mouthwashes was most effective. The experimenter inoculated nutrient agar plates with bacteria from his mouth. Five blotting paper discs, each of which had been soaked in a different mouthwash, were placed on the center of each agar surface. Sterile techniques were used throughout. The plates were placed in an incubator. In actual use, the immediate effectiveness of mouthwashes may differ from the results obtained in the experiment because

(A) none is strong enough to be an effective germ killer
(B) the most effective ones are likely to have an unpleasant taste
(C) bacteria may develop immunity to their effects
(D) some bacteria may be anaerobic
(E) a mouthwash does not remain in the mouth long enough to be fully effective

107. A B C D E

108. The accompanying graph compares the calorie requirements of boys and girls. If there is a direct relationship between physical growth and calorie intake, the graph indicates that most girls will have completed their physical growth about (A) the same time as boys (B) one or two years later than boys (C) one or two years sooner than boys (D) four years sooner than boys (E) six years sooner than boys

108. A B C D E

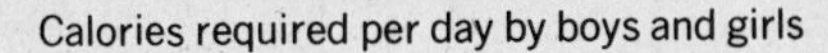

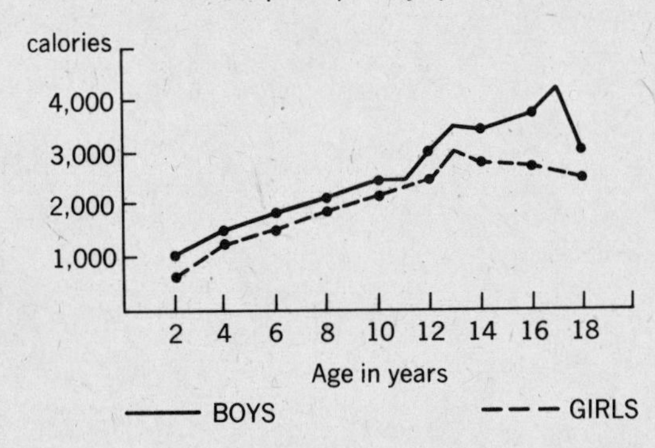

109. The Chinese medical practice of acupuncture involves the technique of inserting needles along certain nerve pathways to deaden pain and to make it possible to operate without the use of anesthesia. Claims are also made for its value in curing afflications ranging from headaches to tuberculosis. American doctors who recently visited China have observed heart and lung surgery in which the patients were fully conscious and ate oranges during the operations. Nevertheless, American doctors remain skeptical of the technique's values. The chief reasons for their doubts is

(A) they are not able to read Chinese
(B) they feel that Chinese medicine is backward compared to American medicine
(C) the Chinese doctors themselves do not understand how acupuncture works
(D) there have been insufficient scientific studies to validate the claims
(E) Chinese doctors and their patients are subject to thought control by the Communist government of the country

109. A B C D E

ANSWER KEY: TEST NO. 5

1. (E)	**12.** (C)	**23.** (B)	**34.** (C)	**45.** (D)	**56.** (D)	**67.** (B)	**78.** (B)	**89.** (D)	**100.** (C)
2. (D)	**13.** (A)	**24.** (D)	**35.** (D)	**46.** (A)	**57.** (C)	**68.** (B)	**79.** (A)	**90.** (B)	**101.** (D)
3. (A)	**14.** (D)	**25.** (C)	**36.** (C)	**47.** (E)	**58.** (B)	**69.** (C)	**80.** (E)	**91.** (B)	**102.** (B)
4. (C)	**15.** (A)	**26.** (B)	**37.** (B)2	**48.** (A)	**59.** (D)	**70.** (D)	**81.** (C)	**92.** (C)	**103.** (E)
5. (A)	**16.** (D)	**27.** (B)	**38.** (C)1	**49.** (C)	**60.** (A)	**71.** (A)	**82.** (D)	**93.** (A)	**104.** (B)
6. (B)	**17.** (B)	**28.** (A)	**39.** (A)2	**50.** (D)	**61.** (D)	**72.** (D)	**83.** (A)	**94.** (A)	**105.** (A)
7. (C)	**18.** (C)	**29.** (D)	**40.** (A)4	**51.** (D)	**62.** (B)	**73.** (E)	**84.** (D)	**95.** (D)	**106.** (E)
8. (B)	**19.** (C)	**30.** (C)	**41.** (A)5	**52.** (D)	**63.** (E)	**74.** (D)	**85.** (B)	**96.** (B)	**107.** (E)
9. (B)	**20.** (A)	**31.** (D)	**42.** (B)1	**53.** (C)	**64.** (D)	**75.** (E)	**86.** (D)	**97.** (D)	**108.** (D)
10. (C)	**21.** (E)	**32.** (A)	**43.** (C)2	**54.** (B)	**65.** (E)	**76.** (C)	**87.** (C)	**98.** (B)	**109.** (D)
11. (D)	**22.** (D)	**33.** (A)	**44.** (B)	**55.** (A)	**66.** (D)	**77.** (C)	**88.** (B)	**99.** (B)	

ANSWERS EXPLAINED: TEST NO. 5

1. (E) anther
2. (D) ovary
3. (A) pollen tube
4. (C) ovule
5. (A) stigma
6. (B) tube nucleus
7. (C) egg nucleus
8. (B) sperm nuclei
9. (B) pollen tube
10. (C) endosperm nuclei

11. (D) Neurohumors are chemical stimulants that are secreted by neurons. Acetylcholine is an example of a neurohumor that is secreted by the end brush of an axon during the passage of an impulse across a synapse.

12. (C) Most of the insects would be killed by the insecticides. A few, however, may have a genetic makeup that makes them immune. By natural selection, they are thus able to survive and to produce offspring that are resistant to the chemical insecticides.

13. (A) More than 70 percent of the earth's surface is covered by water. The greatest amount of food production on earth takes place in the ocean along the edges of the land masses. The algae serve as the basis of the plankton, the teeming mass of microscopic life that serves as the beginning of the food chains of the oceans.

14. (D) During the summer the eggs of plant aphids develop without fertilization. Frog eggs can be made to develop artificially by pricking their membrane, treating them with chemicals, etc.

15. (A) A ribosome is a tiny granule located on the endoplasmic reticulum of the cytoplasm. It is the place where amino acids are bonded together to synthesize proteins.

16. (D) The red blood cells contain hemoglobin, which combines readily with oxygen to form oxyhemoglobin.

17. (B) The villi contain capillaries, which absorb glucose and amino acids, and lacteals, which absorb fatty acids. The lacteals combine to form lymphatics, which eventually lead to the thoracic duct, a large lymph vessel which passes its contents into the bloodstream in the neck region.

18. (C) As sweat evaporates, it cools the body. In hot weather, the blood vessels in the skin dilate, and permit more liquid to be withdrawn from them by the sweat glands.

19. (C) Neurons are not directly connected with each other. Impulses pass from the end brush of the axon of one neuron to the dendrites of another across the area of a synapse.

20. (A) Many lower animals have the power to grow back a lost part.

21. (E) All of the items are examples of practices leading to soil conservation, except the last. By overgrazing, plants are killed and their roots stop binding the soil. This leads to erosion by water and wind.

22. (D) Protoplasm has the ability to make more of itself chiefly by using amino acids.

23. (B) Immunization is a method of protecting people against diseases caused by germs.

24. (D) The adrenal gland is located at the top of the kidneys.

25. (C) All cells have a cell membrane as the outer layer of their cytoplasm.

26. (B) The food of an ameba is digested in its food vacuole.

27. (B) An ameba moves and engulfs food by the flowing action of its pseudopodia, or false feet.

28. (A) Euglena has the ability to carry on photosynthesis in its chloroplasts.

29. (D) Cilia are found in other protozoa such as paramecium and vorticella.

30. (C) Practically all cells have a nucleus.

31. (D) Trichocysts are found in paramecium, and are thought to be useful in protecting it.

32. (A) Euglena moves by whipping its hairlike flagellum back and forth.

33. (A) Euglena's eyespot is sensitive to light, and enables it to move to a location where it carries on photosynthesis.

34. (C) All cells contain cytoplasm.

35. (D) Paramecium takes in food through its oral groove.

36. (C) Chromatin is found in the nucleus of cells.

37. (B)2 During anaerobic respiration, carbon dioxide and alcohol (or lactic acid) are produced. The net yield of energy for each molecule of glucose that is broken down is 2 ATP's. Anaerobic respiration is also referred to as fermentation. Oxygen is not used. When it is, aerobic respiration occurs.

38. (C)1 All the foods are rich in carbohydrates except fish, which is a good source of proteins.

39. (A)2 All are related to the thyroid gland, except hemophilia, which is bleeder's disease brought about through the inheritance of a recessive, sex-linked gene.

40. (A)4 All the scientists proposed theories of evolution except Banting, who discovered the hormone insulin.

41. (A)5 All relate to inborn, unlearned, automatic behavior except habit, which is an acquired type of behavior.

42. (B)1 All are examples of dominant traits in humans except hemophilia, which is a recessive trait.

43. (C)2 All are related to fossil formation except appendix, which is a vestigial structure in man.

44. (B) Bees cross-pollinate apple flowers by carrying pollen from one flower to another. After pollination, the ovary of a flower matures into the apple fruit.

45. (D) He cuts stems from the plant and keeps them in water or moist sand until the roots develop. Then, when planted, each cutting develops into a new plant.

46. (A) A branch of a seedless orange (scion) is grafted onto a tree that normally bears oranges with seeds (stock). The scions grow and produce more seedless oranges.

47. (E) A strawberry plant sends out horizontal stems called runners, which develop roots and leaves to form new plants.

48. (A) Apple branches (scions) can be grafted onto the pear tree. They will obtain water and minerals through the pear tree, but will yield only apples.

49. (C) When a seed is planted it absorbs moisture, and germinates to form a new plant. The embryo lives on the stored food of the seed until it has formed roots and leaves and can make its own food.

50. (D) The graphs show that the number of breaths per minute decreased with the reduction in temperature.

51. (D) When the experiment was repeated, it was seen that the responses differed. Thus, in Graph I, there were 44 breaths per minute at 60°F. In Graph II at the same temperature, there were 46 breaths per minute.

52. (D) All of the items except (D) are instances of good scientific procedure.

53. (C) At this temperature, the breathing rate in Graph II was 50 breaths per minute. In Graph I, at 60°, it was only 44. In Graph II, it was 46 at 60°.

54. (B) This is an example of an inborn, unlearned and automatic type of behavior, involving a series of reflexes.

55. (A) It is able to survive because its basal metabolism rate is at such a low point that it conserves its energy.

56. (D) Although an enzyme molecule is comparatively huge, only a small portion of the enzyme functions when it is active. The localized region is called the active site of the enzyme. Here, the arrangement of the atoms permits them to interact with the atoms of the substrate molecule.

57. (C) Cells obtain their energy from the oxidation of nutrients, including glucose. Oxygen is used, and carbon dioxide and water given off as wastes. The process is known as respiration.

58. (B) Protozoa include such one-celled protists as paramecium and ameba.

59. (D) The growth of a stem takes place because of the activity of the cambium layer of cells. These cells are actively dividing, and form xylem, phloem, and more cambium cells, enlarging the stem.

60. (A) The diaphragm is a horizontal sheet of skeletal muscle below the thorax. It contracts and relaxes in response to stimuli from the medulla of the brain which controls the breathing rate.

61. (D) The last item applies because it is the only one true of all vertebrates.

62. (B) Type O blood is the universal type and can be given to people who have the other types, without causing the red blood cells to clump together.

63. (E) All of the items are correct except the second.

64. (D) The pituitary gland produces many hormones, some of which interact with the gonads, the thyroid, and the adrenal cortex.

65. (E) All of the items are correct except the first.

66. (D) A habit is an acquired form of behavior involving desire, practice, and a feeling of satisfaction. After it has been learned, it is automatic and requires no thought.

67. (B) The dinosaurs were giant reptiles that dominated the earth during the Mesozoic Era.

68. (B) As the result of being isolated, any changes that occurred among the finches of one island remained concentrated there. These birds did not interbreed with those from other islands, and remained distinct from them.

69. (C) Insulin permits the cells to use glucose for energy. In diabetes, where there is a shortage of insulin, the glucose remains unused and is excreted as a waste.

70. (D) Paramecium moves forward until it bumps into an object. Then it reverses the movement of its cilia, backs up, and moves forward again at a different angle. In this way, it can avoid obstacles in its way. At other times, it simply reverses its movement and goes off in a completely different direction.

71. (A) The grasshopper breathes by a system of tubes that connect with the openings, or spiracles, in its abdomen. Air is pumped in and out of the tubes by the contractions and expansions of the abdomen.

72. (D) Starting from a stationary position, the astronaut achieves a speed of about 17,500 miles an hour in only a few minutes. The great acceleration subjects his body to 6 or more G's. This gravitational pull may be great enough to keep the blood from reaching his brain fast enough, and he may become unconscious.

73. (E) The chestnut blight is a parasitic fungus that was first seen destroying chestnut trees at the beginning of this century. By now there are practically no chestnut trees left in the Eastern United States.

74. (D) The coronary arteries supply the cells of the heart with nourishment. A clot blocks the flow of blood through them, and causes a heart attack. This condition is known as coronary thrombosis.

75. (E) Of all the characteristics mentioned, the one that applies only to mammals is the last one. The first three also apply to birds, and the fourth is true of some fish and snakes.

76. (C) Stomates are the tiny openings in the epidermis of a leaf. The size of the stomatal openings is controlled by pairs of guard cells. Water is constantly given off through these openings during transpiration. During conditions of dryness, the guard cells reduce the size of the openings thereby reducing the excessive loss of water.

77. (C) In nondisjunction, a pair of homologous chromosomes fails to separate during meiosis. This results in gametes having one chromosome more or less than the monoploid number. After fertilization, the new individual has more or less than the normal 2n number of chromosomes.

78. (B) Many unlike gametes are involved, each having its own genetic make-up. When gametes are brought together, during fertilization, there will be many different combinations of genes, resulting in a great deal of variety in the offspring.

79. (A) The phase microscope uses different arrangements of light waves that are out of phase with each other, resulting in variations in light intensity. This makes possible the examination of normal living cells without the necessity of staining them.

80. (E) Both statements apply to proteins, not fats. Fats contain only the elements carbon, hydrogen, and oxygen.

81. (C) Fraternal twins originate from two separate eggs that were fertilized at the same time. If the sperm that unites with one egg has an X chromosome, while the other sperm has a Y chromosome, a girl and a boy will develop. They will have different sets of genes.

82. (D) When reduction-division occurs, the members of a pair of chromosomes separate, so that the cells receive one of each pair. This results in their having the monoploid number.

83. (A) Mutations are caused by changes in the genes. Such changes can be brought about by the action of radiation particles that strike a gene and alter its molecular make-up.

84. (D) Animals may be made to breed true by inbreeding them, that is, by crossing them with members of the same litter which have the same desirable genes. Also, by selection, they are bred only with animals of the same type.

85. (B) They are derived from a common ancestor which changed during the ages to form both man and birds. However, they still retain some of the same genes, and have certain similarities.

86. (D) Anthropologists do not classify humans on the basis of the language they use. The term Latin race is used loosely by nonscientists to refer to peoples who speak the Latin languages, such as Italian, French, and Spanish.

87. (C) Humus is soil with a high organic content. It absorbs water freely, and is useful in preventing run-off of water and gullying on slopes.

88. (B) They obtain their energy from sugar without adding oxygen; they give off alcohol in addition to carbon dioxide.

89. (D) The description of the two aqueous solutions is contained in the third and fourth sentences of the selection.

90. (B) Sodium ions are positively charged (Na^+) and chloride ions are negatively charged (Cl^-).

91. (B) The concentration of sodium is about 10 times higher outside the axon, and the concentration of potassium is about 30 times higher inside the axon.

92. (C) The giant axon of the squid is unusual because it is nearly a millimeter in diameter. Practically all other axons are microscopic in diameter.

93. (A) Potassium and chloride ions can move through the axon membrance much more easily than sodium. This gives rise to a voltage drop across the membrane. The inside is negative with respect to the outside.

94. (A) As a cell grows, its volume increases more rapidly than its surface. If it did not divide when it reaches its maximum size, food could not diffuse into it rapidly enough, and wastes could not diffuse out rapidly enough to keep it alive. When it divides in half, the surface is increased compared to the volume which is reduced.

95. (D) The possible genetic combinations would be:

	X	Y
X	XX	XY
X	XX	XY

Female: 50% normal homozygous
50% normal heterozygous
Male: 50% normal
50% color-blind

96. (B) It would take four light years for our signal to be received, and four more light years for the signal to be returned to us.

97. (D) Mammal ova contain practically no stored food, and are microscopic. The eggs of birds and turtles are large because of the large amount of stored food they contain.

98. (B) By chance, a 1:2:1 ratio is generally obtained when there are large numbers involved. With smaller numbers, the results are not predictable because they may tend to concentrate in one particular area.

99. (B) As the bubbles of gas are given off, they collect under the funnel, and then float up through it into the test tube at its narrow end. They displace the water in the test tube.

100. (C) The glowing splint will burst into flames if the gas in the test tube is oxygen.

101. (D) All of the methods except the fourth liberate carbon dioxide. Hydrogen is produced by the addition of sulfuric acid to zinc.

102. (B) The supply of carbon dioxide in the water is not used up, and the plant will continue to carry on photosynthesis, liberating more oxygen.

103. (E) Photosynthesis takes place only in the presence of light, which furnishes the energy needed for this process.

104. (B) If few or no bubbles of oxygen are given off, it may be concluded that the additional supply of carbon dioxide was responsible for the increased production of oxygen bubbles.

105. (A) The molecules of carbon are combined in the formation of starch.

106. (E) Other aquatic plants might react differently. A scientific conclusion should therefore be based on many cases.

107. (E) In the experiment, the mouthwash was allowed to remain in contact with the bacteria for hours at a time. In actual use, however, the mouthwash would be in contact with the bacteria for only a very short time. Thus, the practical effectiveness of the mouthwash may not be indicated. An experiment could be devised in which the mouthwash is poured onto the plate and then poured off after a short time interval. The plate could then be examined for the effects of the mouthwash.

108. (D) Assuming that increase in calorie intake is directly related to physical growth, the growth of girls seems to level off at about age 13. Boys, however, as seen by the graph, apparently complete their physical growth at age 17, or four years later than girls.

109. (D) In view of the doubts that remain, along with many unanswered questions, much further scientific investigation is needed before the validity of the claims can be finally resolved.

PRACTICE COLLEGE BOARD ACHIEVEMENT TEST IN BIOLOGY

NO. 6

PART I—TYPE A

Directions: **The following diagram shows parts of the human heart. Certain of the parts are labeled with letters. Next to each of the statements below, write the letter of the specific part most closely related to it. A label may be used once, more than once, or not at all.**

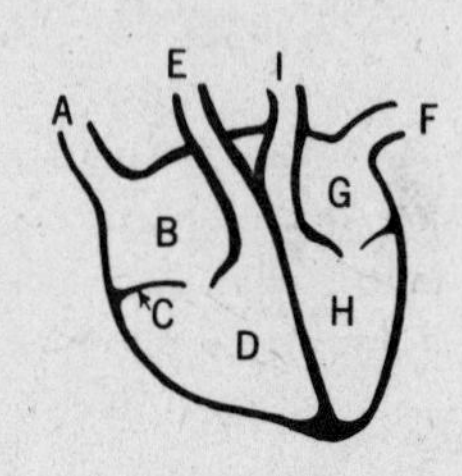

1. Chamber of the heart which receives blood from the lungs 1. ....
2. Chamber of the heart from which blood is sent to the lungs 2. ....
3. Prevents the blood from flowing backward 3. ....
4. Chamber of the heart which receives blood from the body and the head 4. ....
5. Chamber of the heart from which blood is sent to the body and the head 5. ....
6. The largest artery of the body 6. ....
7. The chamber of the heart having the thickest walls 7. ....
8. The blood vessel which carries blood to the lungs 8. ....
9. The blood vessel which carries blood from the lungs to the heart 9. ....
10. The blood vessel through which blood travels from the heart to most of the body 10. ....

TYPE B

Directions: **Each of the following statements is followed by five suggested answers. Select the one answer which is best in each case.**

11. A limitation of the electron microscope is that it (A) suffers loss of resolving power above a magnification of 1000x (B) cannot be used to study living things (C) cannot magnify above 1000x (D) uses short wave electrons (E) requires the use of balsam oil, which is becoming scarce

    11. A B C D E

12. Which statement concerning ribosomes is true? (A) They function in cell division (B) They contain DNA (C) They are the sites of carbohydrate synthesis (D) They are the sites of protein synthesis (E) They permit cells to contract

    12. A B C D E

13. As the number of neutrons in the nucleus of an atom increases, the atomic number of the atom (A) increases (B) decreases (C) remains the same (D) varies with the form of the isotope (E) depends on the number of electron shells

    13. A B C D E

14. Because urea is a nitrogen compound, it cannot be derived from the metabolism of (A) peptides (B) polypeptides (C) polysaccharides (D) proteins (E) amino acids

    14. A B C D E

15. During the process of respiration, energy is transferred from glucose molecules to molecules of (A) ACTH (B) DNA (C) RNA (D) ADP (E) BCG

15. A B C D E

16. The species number of chromosomes is restored in a new individual by the process of (A) regeneration (B) fertilization (C) reduction-division (D) parthenogenesis (E) metamorphosis

16. A B C D E

17. In humans, large amounts of carbohydrates are stored in the (A) liver (B) gall bladder (C) spleen (D) coccyx (E) pancreas

17. A B C D E

18. In examining a slide of blood with the high power of the microscope (A) the mirror should be horizontal (B) the light from the sun should be utilized (C) the slide should first be dipped in cedar oil (D) it is preferable to use the coarse adjustment (E) it is preferable to use the fine adjustment

18. A B C D E

19. An organism that exists in a mutually helpful relationship with another organism is called (A) photosynthetic (B) independent (C) saprophytic (D) parasitic (E) symbiotic

19. A B C D E

20. The tube that brings urine from the kidneys to the bladder is called (A) urethra (B) ureter (C) uterus (D) Eustachian tube (E) bile duct

20. A B C D E

21. All of the following diseases are caused by nutritional deficiency except (A) scurvy (B) anemia (C) rickets (D) xerophthalmia (E) rabies

21. A B C D E

22. A nerve cell that receives stimuli from the outside is known as a(n) (A) motor neuron (B) sensory neuron (C) efferent neuron (D) associative neuron (E) ganglion

22. A B C D E

23. The three-layered, cuplike stage in the development of many animals is called (A) blastula (B) zygote (C) zygospore (D) gastrula (E) ectoderm

23. A B C D E

24. A disease that may be inherited is (A) malaria (B) trichinosis (C) hemophilia (D) pellagra (E) typhus

24. A B C D E

25. The period of 4 billion years represents (A) the estimated age of the earth (B) the duration of the age of dinosaurs (C) how long ago living things originated on earth (D) the speed of light in years (E) how long ago primitive man first appeared

25. A B C D E

26. Inbreeding is practiced by a breeder to (A) insure fertilization (B) maintain a desired type (C) obtain a new type (D) produce a hybrid (E) encourage mutations

26. A B C D E

27. The only pure race of man is (A) Nordic (B) Aryan (C) Alpine (D) Caucasoid (E) none of these

27. A B C D E

TYPE C

Directions: **The following questions require two answers. One of the items in the left column** (A), (B), or (C) **is related to four of the items in the right column** (numbered 1–5). **Select the letter of the item which is so related; then select the number of the remaining item which does not belong.**

EXAMPLE:

(A) paramecium	1. cilia
(B) ameba	2. macronucleus
(C) spirogyra	3. trichocyst
	4. pyrenoid
	5. micronucleus

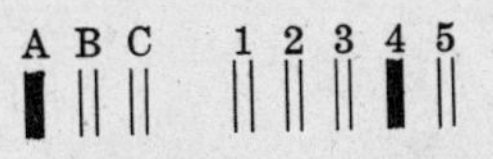

28. (A) internal fertilization	1. rabbit
(B) external fertilization	2. bat
(C) external development	3. whale
	4. frog
	5. human

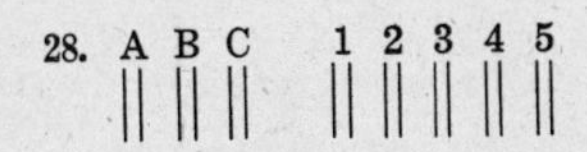

29. (A) fruit
(B) root
(C) stem

1. radish
2. cucumber
3. coconut
4. acorn
5. tomato

29. A B C 1 2 3 4 5

30. (A) insect
(B) spider
(C) coelenterate

1. millipede
2. termite
3. ant
4. bee
5. beetle

30. A B C 1 2 3 4 5

31. (A) flower
(B) conjugation
(C) vegetative propagation

1. tuber
2. bulb
3. zygospore
4. runner
5. cutting

31. A B C 1 2 3 4 5

32. (A) dicot
(B) monocot
(C) bryophyte

1. bean
2. pea
3. pumpkin
4. wheat
5. peanut

32. A B C 1 2 3 4 5

33. (A) parthenogenesis
(B) pollination
(C) phototropism

1. anther
2. stigma
3. pollen
4. stamen
5. phloem

33. A B C 1 2 3 4 5

34. (A) pancreas
(B) spleen
(C) pituitary

1. ACTH
2. growth
3. stimulates gonads
4. milk production
5. digests starch

34. A B C 1 2 3 4 5

35. (A) vaccination
(B) clotting
(C) peristalsis

1. fibrinogen
2. platelet
3. secretin
4. thrombin
5. fibrin

35. A B C 1 2 3 4 5

TYPE D

Directions: **The five lettered headings below are followed by a list of numbered statements. For each numbered statement, choose the one lettered heading (A–E) which is most closely related to it.**

(A) Tropism
(B) Simple reflex
(C) Instinct
(D) Conditioned response
(E) Habit

36. A secretary takes dictation in shorthand

37. A plant that was blown over on its side begins to grow upward

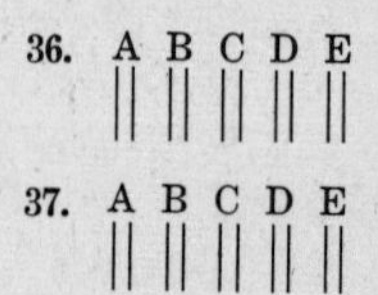

38. A robin builds a nest

39. A man jumps at the sound of an automobile horn behind him

40. A boy blinks as a basketball is thrown at his face

41. The roots of a poplar tree grow toward an underground spring

42. A dog's mouth waters when his master picks up his feeding dish

38. A B C D E
39. A B C D E
40. A B C D E
41. A B C D E
42. A B C D E

(A) Petrification
(B) Sedimentation
(C) Mutation
(D) Radiation from carbon-14
(E) Refrigeration

43. How were the strata of rock in the Grand Canyon formed?

44. How was the woolly mammoth kept intact up to the present time?

45. How was the bone of a dinosaur preserved?

46. How did the horse change through the ages?

47. How was the wood of a tree turned to stone?

48. How is the age of the remains of primitive man determined?

43. A B C D E
44. A B C D E
45. A B C D E
46. A B C D E
47. A B C D E
48. A B C D E

PART II

It has been found that the flowering rate of plants is related to the length of daylight. This phenomenon is known as photoperiodism. Some plants are short-day plants; they flower only when they are exposed to short periods of light. Other plants are long-day plants; they produce flowers only when they are exposed to long periods of light.

Thus, the ragweed plant starts making flowers when the day is just about 14.5 hours long. At Washington, D.C., this takes place about July 1, so the plant produces its flowers, sheds its pollen, and scatters its seeds in August. In northern Maine, however, the long summer days do not shorten to 14.5 hours until after August 1. By the time flowers could begin to form and produce seeds, frost generally kills the plants; so northern Maine has little or no ragweed.

On the other hand, the plant *Sedum telephium*, which needs a day of 16 hours or more, flowers in Vermont; however, it does not flower in Virginia.

The graph on page 268 shows the annual change in the length of day throughout the year at four latitudes: Miami (26°), San Francisco (37°), Chicago (42°), and Winnipeg (50°).

49. The cocklebur plant requires 9 hours or more of darkness (or 15 hours or less of light), in order to flower. It therefore will flower immediately when it is ripe to do so in (A) Miami (B) Chicago (C) Winnipeg (D) all cities on March 21 (E) none of these cities

50. The cocklebur will begin to form flower buds in Winnipeg about (A) March 3 (B) April 3 (C) May 3 (D) August 3 (E) January 3

49. A B C D E
50. A B C D E

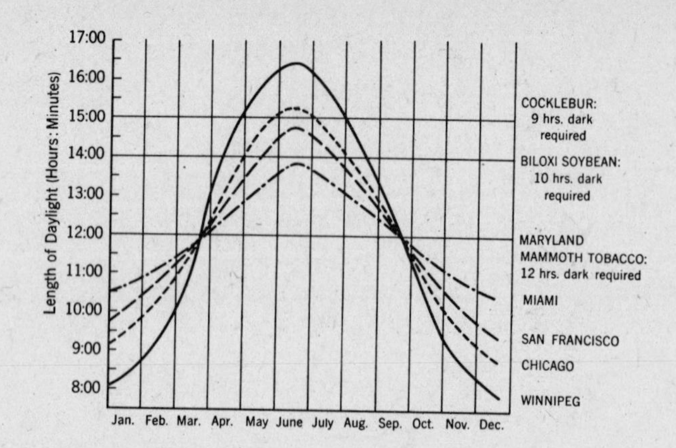

51. In Chicago, the flower buds of cocklebur begin to appear (A) before they do in Miami (B) before they do in San Francisco (C) before they do in Winnipeg (D) before they do in all of the other cities (E) before March 21

51. A B C D E

52. The Maryland Mammoth tobacco is a short-day plant, forming flowers when the days are 10–12 hours long; it will therefore flower (A) only in Winnipeg in the middle of the summer (B) in all the cities except Winnipeg in the middle of the summer (C) very early in the summer in all the cities (D) in the middle of the summer in all the cities (E) very late in the summer in all the cities

52. A B C D E

53. The Biloxi soybean plant requires 10 hours of darkness (14 hours of light) in order to produce flowers; in San Francisco it will begin to produce flowers (A) before July 15 (B) after July 15 (C) after it does in Chicago (D) before it does in Miami (E) at no time during the summer

53. A B C D E

54. Of the four cities shown, the one having the longest day length is (A) Miami (B) San Francisco (C) Chicago (D) Winnipeg (E) none, because they all have the same day length

54. A B C D E

PART III—TYPE A

Directions: **Each item consists of an *assertion* (statement) in the left-hand column, and a *reason* on the right-hand side. Select**

(A) if both assertion and reason are true, and *are related* as cause and effect
(B) if both assertion and reason are true, but *are not related* as cause and effect
(C) if the assertion is true, but the reason is a false statement
(D) if the assertion is false, but the reason is a true statement
(E) if both assertion and reason are false statements

DIRECTIONS SUMMARIZED:

	(A)	(B)	(C)	(D)	(E)
Assertion:	True	True	True	False	False
Reason:	True	True	False	True	False
	Cause and effect	*Not* cause and effect			

Assertion: | Reason:

55. Lichens help form soil from rock BECAUSE they are made up of a combination of an alga and a fungus

56. The passenger pigeon became extinct in this country BECAUSE the telegraph replaced this bird as a method of sending messages

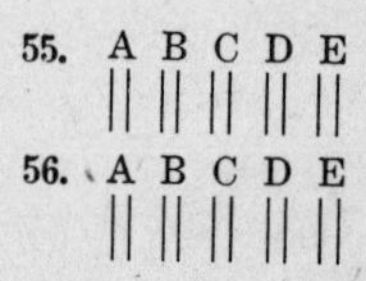
55. A B C D E
56. A B C D E

57. Uranium is useful as an "atomic clock" in determining the age of rock BECAUSE it is one of the elements in the radioactive series that eventually turn to lead

57. A B C D E

58. Cro-Magnon man is considered to be an ancestor of modern *Homo sapiens* BECAUSE he lived in caves

58. A B C D E

59. Kangaroos and other pouched mammals were originally found only in Australia BECAUSE they were prevented from migrating by the ocean barrier

59. A B C D E

60. A mammal embryo is able to develop within its mother's uterus BECAUSE it is born alive

60. A B C D E

61. Double fertilization takes place within the ovule of a flower BECAUSE there are two sperm nuclei in the pollen tube

61. A B C D E

62. The cerebrum of mammals controls breathing rate BECAUSE it is the center of voluntary activities

62. A B C D E

63. Respiration in animals and plants is a similar process BECAUSE living protoplasm requires energy

63. A B C D E

TYPE B

Directions: **Each of the following statements is followed by five suggested answers. Select the one answer which is best in each case.**

64. In exhaling, air leaves the lungs because (A) the diaphragm flattens out (B) the ribs are raised (C) the chest cavity becomes larger (D) the chest cavity becomes smaller (E) the lungs contract of their own accord

64. A B C D E

65. Green plants are added to an aquarium because (A) they liberate carbon dioxide (B) they liberate oxygen (C) they balance its appearance (D) they supply the fish with carbohydrates (E) they are a good source of ATP for young tropical fish

65. A B C D E

66. In certain communities many cases of simple goiter have been found to occur because (A) there was an excess of fluorine in the water (B) there was prejudice against the use of codliver oil (C) the soil was lacking in iron (D) there was insufficient iodine in the diet (E) radioactive fall-out affected the adolescents

66. A B C D E

67. Earthworms are useful because (A) they scare away grubs of the Japanese beetle (B) they break down proteins in the soil into nitrates (C) their castings provide shelter for useful underground insects (D) they improve soil drainage (E) they convert ammonia, resulting from decay, into nitrites

67. A B C D E

68. Salmon swim up rivers from the ocean to (A) spawn (B) seek food (C) escape enemies (D) winter over (E) follow the gulf stream

68. A B C D E

69. A drug addict has difficulty breaking the habit because (A) he is afraid to break the law (B) he knows his friends rely on him for protection (C) doctors insist on an operation (D) hospitalization is too expensive (E) he suffers from withdrawal symptoms

69. A B C D E

70. Some molds are useful because (A) they help make beer (B) they help make bread (C) they form antibiotics (D) they give buttermilk its taste (E) they synthesize sulfa drugs

70. A B C D E

71. The cells of a frog embryo have the same number of chromosomes as the fertilized egg because (A) mitosis takes place (B) reduction-division takes place (C) meiosis takes place (D) the tadpole undergoes metamorphosis (E) the frog is an amphibian

71. A B C D E

72. Most cells lacking a cell wall would also lack (A) mitochondria (B) chloroplasts (C) cell membrane (D) vacuole (E) Golgi apparatus

72. A B C D E

73. Plankton is the name given to the algae, protozoa and other forms of microscopic life that abound on the surface of the ocean. Which best describes the

73. A B C D E

most important role of the plankton in nature? (A) It serves as the basis of the food chain for marine life (B) It is useful as human food (C) It is symbiotic to land forms of microscopic life (D) It serves directly as the basic diet of most fish (E) It is the basis of autotrophic nutrition

74. In an area populated by all of the following animals, which would probably have the largest population? (A) foxes (B) field mice (C) grasshoppers (D) skunks (E) moles

74. A B C D E

TYPE C

Directions: **In each of the following questions, there is a statement followed by four choices, of which *one or more* are correct. Indicate the correct combination by choosing the following letters:**

(A) if only 1, 2, and 3 are correct
(B) if only 1 and 3 are correct
(C) if only 2 and 4 are correct
(D) if only 4 is correct
(E) if some other combination is correct

DIRECTIONS SUMMARIZED:

(A)	(B)	(C)	(D)	(E)
1, 2, 3	1, 3	2, 4	4	some other
only	only	only	only	combination

75. According to Linnaeus' system of classification
1. plants and animals are grouped within the same phylum
2. organisms have a genus and a species name
3. organisms are classified mainly according to habitat
4. man is designated as *Homo sapiens*

75. A B C D E

76. In human beings, starch is digested chiefly
1. in the mouth, and small intestine
2. in the mouth and large intestine
3. in the stomach and small intestine
4. in the stomach and large intestine

76. A B C D E

77. During take-off, a human being in a space capsule would
1. have a sensation of weightlessness
2. be subject to a force of several G
3. experience anoxia
4. show the effects of acceleration

77. A B C D E

78. When first introduced into Australia, the rabbit
1. multiplied rapidly
2. was kept in check by foxes and weasels
3. was affected by few natural enemies
4. had difficulty in finding enough food to survive

78. A B C D E

79. Forests are useful
1. in providing a shelter for wildlife
2. in preventing floods
3. in preventing erosion
4. in providing man with natural products

79. A B C D E

80. Man's existence on earth
1. began during the latter part of the Cenozoic Era
2. has been studied by the discovery of fossil remains
3. has been marked by the appearance of primitive types that have disappeared
4. began during the age of dinosaurs

80. A B C D E

81. Darwin's theory of evolution has been brought up to date by
1. the application of Lamarck's theory
2. studies of mutations
3. the concept of use and disuse
4. work in population genetics

81. A B C D E

82. The plant breeder has advantages over the animal breeder because
1. he can maintain the gene type by vegetative propagation
2. animals do not breed true
3. there are more offspring
4. he can hybridize by grafting

82. A B C D E

83. During fission
1. two cells of the same size result
2. the nucleus divides by mitosis
3. animal cells divide equally, but plant cells divide unequally
4. one of the cells receives the haploid number of chromosomes, and the other receives the diploid number

83. A B C D E

84. The thyroid gland
1. is known as the gland of emergency
2. stimulates the liver to store glycogen
3. controls the production of ACTH
4. regulates the rate of metabolism

84. A B C D E

PART IV—TYPE A

Directions: **Referring to the following selection, choose the best answer for each of the questions that follows:**

*"The energy of sunlight comes in packets called photons, or quanta; light of different colors or wavelengths is characterized by different energy content. When light strikes and is absorbed by certain metallic surfaces, the energy of the impinging photons is transferred to electrons of the metal. This 'photoelectric' effect can be measured by the resulting flow of electric current. In the green-plant cell, solar energy of a particular range of wavelengths is absorbed by the green pigment chlorophyll. The absorbed energy raises an electron from its normal energy level to a higher level in the bond structure of this complex molecule. Such 'excited' electrons tend to fall back to their normal and stable level, and when they do they give up the energy they have absorbed. In a pure preparation of chlorophyll, isolated from the cell, the absorbed energy is re-emitted in the form of visible light, as it is from other phosphorescent or fluorescent organic and inorganic compounds.

"This chlorophyll itself in the test tube cannot store or usefully harness the energy of light; the energy escapes quickly, as though by short circuit. In the cell, however, chlorophyll is so connected spatially with other specific molecules that when it is excited by the absorption of light, the 'hot' or energy-rich, electrons do not simply fall back to their normal positions. Instead these electrons are led away from the chlorophyll molecule by associated 'electron carrier' molecules and handed from one to the other around a circular chain of reactions. As they traverse this external path the excited electrons give up their energy bit by bit and return to their original positions in the chlorophyll, which is now ready to absorb another photon. The energy given up by the electrons has meanwhile gone into the formation of ATP from ADP and phosphate; that is, into recharging the ATP system of the photosynthetic cell."

*From "How Cells Transform Energy" by Albert L. Lehninger, Sept. 1961, by *Scientific American, Inc.*

85. All of the following statements about the energy of sunlight are true except (A) different colors of light have different energy content (B) it occurs as photons (C) when light strikes certain metals its energy is transferred to the electrons of the metals (D) all wavelengths of light have the same energy content (E) the photoelectric effect occurs when light energy is absorbed by certain metallic surfaces

85. A B C D E

86. When solar energy reaches a green plant cell (A) it is absorbed (B) all of its wavelengths are reflected by the green chlorophyll (C) an electron is created (D) an electron drops from its normal energy value to a lower level (E) an electron becomes "excited" and is destroyed

86. A B C D E

87. Under the influence of light, a green plant cell forms (A) electrons (B) quanta (C) photons (D) ADP (E) ATP

87. A B C D E

88. In pure form, chlorophyll (A) does not absorb light energy (B) gives up absorbed light energy as visible light (C) is an inorganic compound (D) is free of electrons (E) has a normal and stable photon

88. A B C D E

89. In the green plant cell, the "hot" electrons (A) quickly fall back to their normal position (B) quickly give up all their absorbed energy (C) are passed along by "electron carriers" (D) are converted to photons (E) never return to their original position in the chlorophyll molecule

89. A B C D E

TYPE B

Directions: **Each of the following problems is followed by five suggested answers. Select the one answer which is best in each case.**

90. A study has been made of the daily activities of the average farmer's wife in New York State, weighing 62 kilograms (136 pounds). The following table shows a day's activities and the output of energy in Calories per kilogram (2.2 pounds) per hour.

90. A B C D E

ACTIVITY	HOURS	OUTPUT OF ENERGY IN CALORIES PER KG. HOUR
Sleeping	8	1.0
Sitting quietly	2	0.4
Eating	1½	0.4
Laundering	1½	1.3
Sweeping	1	1.4
Dishwashing	2	1.0
Driving car	2	0.9
Going upstairs	1	14.7
Going downstairs	1	4.1
Walking	4	2.0

Her total energy requirements in Calories for such a day are (A) 2,000–2,200 (B) 2,200–2,400 (C) 2,400–2,600 (D) 2,600–2,800 (E) 2,800–3,000

91. In a large litter of guinea pigs, one-quarter of the offspring were white (black is dominant). The parents most likely were (A) bb × bb (B) bb × Bb (C) bb × BB (D) Bb × Bb (E) Bb × BB

91. A B C D E

92. A color-blind man marries a normal woman who is heterozygous for color vision. The chances of their two sons being color-blind are (A) 100 (B) zero (C) 50–50 (D) 25–75 (E) 25–50–25

92. A B C D E

93. The number of different gene combinations possible in the gametes of a trihybrid pea plant, TtYySs, is (A) 2 (B) 4 (C) 6 (D) 8 (E) 10

93. A B C D E

94. In this pedigree of a family, brown eyes are indicated as ○ and blue eyes as ◍. From this chart it can be determined that (A) all individuals are

94. A B C D E

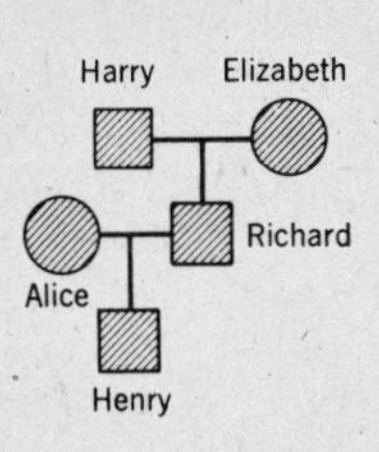

heterozygous for blue eyes (B) if there were other children born to Harry and Elizabeth, they might have had brown eyes (C) if other children were born to Alice and Richard, some of them might have brown eyes (D) there is a greater chance of males in this family having brown eyes in the next generation (E) Alice and Richard could continue to have only blue-eyed children

95. If a bacillus divides every half hour, under ideal conditions of food and temperature, after four hours, the number of bacilli would be (A) 4 (B) 8 (C) 24 (D) 128 (E) 256

95. A B C D E

TYPE C

A blood slide is to be prepared and stained with Wright's stain. The steps in making such a slide are shown in the accompanying drawings:

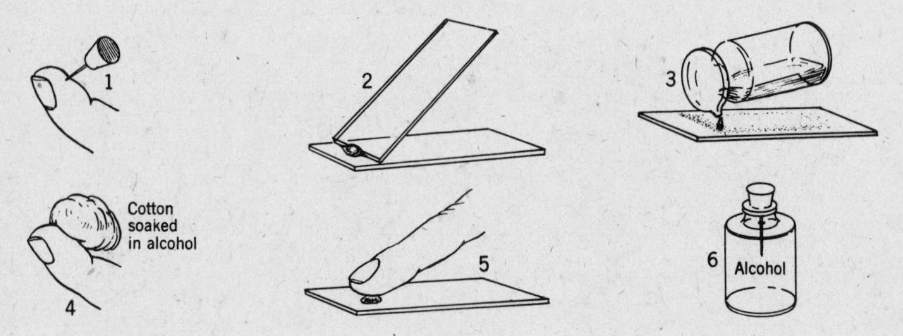

96. The correct sequence of steps in making the slide is (A) 4–6–1–5–2–3 (B) 4–6–1–5–3–2 (C) 4–6–1–3–5–2 (D) 4–1–6–3–5–2 (E) 4–1–6–3–2–5

96. A B C D E

97. The purpose of step number 2 is (A) to concentrate the red blood cells at one end (B) to clear the slide of dust (C) to spread the blood evenly over the slide (D) to keep the blood from drying up (E) to keep the blood from clotting

97. A B C D E

98. The purpose of step 6 is (A) to keep the needle from rusting (B) to mix the alcohol with the blood cells (C) to make it easier to draw blood (D) to sterilize the needle (E) to comply with the state laws dealing with the use of alcohol by minors

98. A B C D E

99. Ordinarily, the number of drops placed on the slide (A) varies with the age of the person (B) varies with the weight of the person (C) depends on the person's health (D) is one (E) is six

99. A B C D E

100. The purpose of step 4 is (A) to shrink the blood cells (B) to bring more blood to the surface of the skin (C) to prevent clotting (D) to keep the blood cells from plasmolyzing (E) to sterilize the skin

100. A B C D E

101. After the application of Wright's stain (A) the nucleus of red blood cells is clearly stained (B) the nucleus of white blood cells is clearly stained (C) the white blood cells can be seen ingesting bacteria (D) the blood types can be determined (E) Rh negative blood remains colorless

101. A B C D E

102. The most numerous component of the blood is seen to be (A) white blood cells (B) red blood cells (C) blood platelets (D) plasmagenes (E) phagocytes

102. A B C D E

103. Bunching of blood cells together on the slide may be caused by (A) the presence of blood type AB (B) the presence of blood type O (C) failure to spread the blood film evenly (D) allowing the slide to dry in air (E) using a concentrated solution of Wright's stain

103. A B C D E

PART V

Directions: **Each of the following situations has five suggested answers. Select the one answer that is best in each case.**

104. When Charles Darwin visited the Galapagos Islands as a naturalist on H.M.S. *Beagle*, he was impressed by the differences he saw in some of the native animals. Among the birds, he noticed a number of different species of finch. Some finches were insect-eating and had narrow beaks; others were seed-eating and had stubby bills. Yet all the finches were very similar in most of their other characteristics. Putting all his observations together, he concluded that (A) the different species of finch had been imported to control the insect population (B) the different species of finch had developed as a natural enemy to eat the seeds of the weeds (C) since the species of finch were different from each other, they illustrated the basic idea of variation (D) the different species of finch had evolved from one ancestor and had developed their specific characteristics in isolation from each other (E) overproduction was taking place among the finches, because of the absence of other vertebrates on the islands

104. A B C D E

105. Cholera is one of mankind's most ancient diseases. In recent years, it spread on a massive scale among Bengali refugees from East Pakistan. It is a terrifying disease because it strikes suddenly and sweeps through entire villages and cities. The victim becomes giddy, may vomit, and has extended diarrhea, excreting great quantities of liquid waste from his rectum. Death may occur within several days and is due chiefly to the effects of severe dehydration. The obvious treatment is (A) to examine the patient's discharge immediately for evidence of the germ, *Vibrio cholerae* (B) to inject into the patient a solution that matches in amount and composition what is lost in vomiting and diarrhea (C) strict quarantine of the patient so that the nurse and doctor do not have excessive contact with him (D) immunization against similar disease carried in polluted water, such as typhoid (E) to drink a cup of hot tea every half hour

105. A B C D E

PART VI

Directions: **Each of the following questions has five suggested answers. Select the one answer that is best in each case.**

106. In an experiment, a few drops each of concentrated $MnSO_4$ (manganese sulfate) and NaOH (sodium hydroxide) added to a solution containing O_2 (oxygen) will cause a brown precipitate. The greater the O_2 content, the deeper the brown color of the precipitate. Three tubes, marked *A*, *B*, and *C* respectively, were filled with water. A sprig of Elodea was placed in tubes *A* and *B*. Tubes *A* and *C* were placed in the sunlight. Tube *B* was placed in the dark. After three hours the Elodea was removed and three drops of each testing substance were added to all three tubes. The precipitate of darkest color was found in tube *A*. Which is most clearly illustrated by this experiment? (A) Only green plants carry on photosynthesis (B) Green plants carry on respiration (C) Green plants give off O_2 during photosynthesis (D) Sunlight alters the O_2 content of water (E) Green plants use up O_2 in the dark

106. A B C D E

107. In the preceding question, for an additional control, the experimenter should have (A) used another type of water plant (B) used another test for oxygen (C) used bromthymol blue (D) set up a tube in the dark with no plant (E) added oxygen to three drops of the testing substance

107. A B C D E

108. A flock of prize poultry became infected with a strange disease of unknown cause. Half of the flock died. Scientists were called in to try to determine the exact cause of the disease. Which group of materials would be most

108. A B C D E

useful in this research? (A) sterilizer, microscope, penicillin, filter paper, petri dishes (B) thistle tubes, agar nutrient, penicillin, filter paper, micrometer (C) sterilizer, agar nutrient, microscope, incubator, petri dishes (D) petri dishes, filter paper, microscope, tripod magnifier, sterilizer (E) incubator, penicillin discs, agar nutrient, filter paper, micrometer

109. In the situation described in the previous question, at which point is it reasonable to believe that the research problem is being solved? (A) when the suspected germ has been recovered from the tissues of birds made experimentally ill by inoculation (B) when all the infected birds begin to recover from the disease, instead of dying (C) when half the infected birds still alive begin to recover from the disease (D) when there is evidence that the remaining healthy birds have developed a natural immunity to the disease (E) when there is no further incidence of the disease among the members of the flock

109. A B C D E

ANSWER KEY: TEST NO. 6

1. (G)	12. (D)	23. (D)	34. (C)5	45. (A)	56. (C)	67. (D)	78. (B)	89. (C)	100. (E)
2. (D)	13. (C)	24. (C)	35. (B)3	46. (C)	57. (A)	68. (A)	79. (E)	90. (D)	101. (B)
3. (C)	14. (C)	25. (A)	36. (E)	47. (A)	58. (B)	69. (E)	80. (A)	91. (D)	102. (B)
4. (B)	15. (D)	26. (B)	37. (A)	48. (D)	59. (A)	70. (C)	81. (C)	92. (C)	103. (C)
5. (H)	16. (B)	27. (E)	38. (C)	49. (A)	60. (B)	71. (A)	82. (B)	93. (D)	104. (D)
6. (I)	17. (A)	28. (A)4	39. (B)	50. (D)	61. (A)	72. (B)	83. (E)	94. (E)	105. (B)
7. (H)	18. (E)	29. (A)1	40. (B)	51. (C)	62. (D)	73. (A)	84. (D)	95. (E)	106. (C)
8. (E)	19. (E)	30. (A)1	41. (A)	52. (E)	63. (A)	74. (C)	85. (D)	96. (A)	107. (D)
9. (F)	20. (B)	31. (C)3	42. (D)	53. (B)	64. (D)	75. (C)	86. (A)	97. (C)	108. (C)
10. (I)	21. (E)	32. (A)4	43. (B)	54. (D)	65. (B)	76. (E)	87. (E)	98. (D)	109. (A)
11. (B)	22. (B)	33. (B)5	44. (E)	55. (B)	66. (D)	77. (C)	88. (B)	99. (D)	

ANSWERS EXPLAINED: TEST NO. 6

1. (G) left atrium
2. (D) right ventricle
3. (C) valves
4. (B) right atrium
5. (H) left ventricle
6. (I) aorta
7. (H) left ventricle
8. (E) pulmonary artery
9. (F) pulmonary vein
10. (I) aorta

11. (B) The electron microscope uses a stream of electrons and produces an image of the object on a screen or a photographic plate. The specimen is placed in a vacuum chamber because any air in the tube would interfere with the motion of electrons. Living cells could not exist in the vacuum.

12. (D) Ribosomes are tiny granules found in the endoplasmic reticulum of the cytoplasm. They contain most of the RNA of the cell, and are the site of protein synthesis, including enzyme formation.

13. (C) The atomic number is determined by the number of protons in the nucleus. An increase in the number of neutrons creates a new isotope, but does not change the atomic number of the element.

14. (C) Urea is a nitrogenous waste resulting from the breakdown, or deaminization, of amino acids in the liver. Polysaccharides are large carbohydrate molecules consisting of many repeating glucose molecules bonded together; the only elements they contain are carbon, hydrogen and oxygen.

15. (D) During cellular respiration, energy is released from glucose by a series of complex steps. When glucose is broken down, some of its hydrogen is removed, releasing energy which is transferred to ADP with the addition of a high energy phosphate molecule to convert it to ATP.

16. (B) The sperm and egg cells each have the monoploid number of chromosomes. At fertilization they unite to form the diploid number of chromosomes. All the cells of the organism contain this number, which is typical of the species.

17. (A) Glucose is stored in the liver as glycogen, or animal starch.

18. (E) In using the high power, the fine adjustment should be used, because of the small working distance between the end of the objective and the slide.

19. (E) Symbiotic organisms live together in an intimate relationship. For example, a lichen is composed of an alga and a fungus. The alga makes food by photosynthesis, while the fungus supplies water and minerals. This type of symbiosis in which both organisms benefit is called mutualism.

20. (B) Each kidney is connected with the bladder by a ureter.

21. (E) Rabies is caused by a virus. The nutrients missing in the other diseases are: scurvy—vitamin C; anemia—iron; rickets—vitamin D; xerophthalmia—vitamin A.

22. (B) A sensory or afferent neuron connects a sense organ with the brain or the spinal cord.

23. (D) The gastrula is formed after the blastula, or hollow-ball stage.

24. (C) Hemophilia, or bleeder's disease, is caused by the presence of a recessive, sex-linked gene.

25. (A) Most scientists now believe that the earth is at least 4 billion years old.

26. (B) By inbreeding, he prevents other types of genes from being introduced, and so he can maintain the desired type.

27. (E) There are no pure races of man because of intermarriages through the centuries, which have produced mixed sets of genes in all of the racial types.

28. (A)4 All are examples of mammals, with internal fertilization, except for the frog; it has external fertilization.

29. (A)1 All are examples of fruits having developed from the ovary of a flower, except radish, which is an enlarged root containing stored food.

30. (A)1 All are examples of insects, and have six legs, while a millipede has many legs, two per segment.

31. (C)3 All are examples of vegetative propagation except zygospore, which is formed from the union of two gametes, as in spirogyra or bread mold.

32. (A)4 All are examples of dicots except wheat, which is a monocot.

33. (B)5 All are involved in pollination except phloem, which is part of a fibro-vascular bundle, and is used for the circulation of food in a stem or root.

34. (C)5 All are related to the activity of the pituitary gland except the last item, which refers to the activity of the enzyme ptyalin.

35. (B)3 All are involved in the clotting of blood except secretin, which is a hormone produced in the small intestine and which stimulates the flow of pancreatic juice.

36. (E) This is an example of a learned automatic activity that was acquired by having a desire to learn shorthand, accompanied by a feeling of satisfaction, and after much practice.

37. (A) Turning responses of plants are known as tropisms. In this case, the plant shows negative geotropism and grows upward, away from the force of gravity. It may also be stimulated to grow toward light, showing positive phototropism.

38. (C) This is an example of an instinct, which is an inborn, unlearned series of reflexes.

39. (B) There is no thought involved in the immediate reflex until the cerebrum receives additional impulses.

40. (B) There is an example of an inborn, unlearned and automatic response.

41. (A) This is an example of positive hydrotropism.

42. (D) There is an association between the original reflex of saliva production in the presence of food, and the feeding dish which contains the food. This is a learned response.

43. (B) They originated in the layers of sediment that were deposited by rivers. Under the pressure of the water over a long period of time, these layers, or strata, became solidified into rock.

44. (E) Its body was frozen, and prevented from decaying for thousands of years.

45. (A) Under the pressure of the water, the particles of bone were gradually replaced by mineral matter, finally resulting in petrified bone.

46. (C) By a series of mutations, the earliest type of horse (Eohippus) changed from a small animal with four toes on its forelegs to the present animal standing on a foot composed of one toe, having two small side splints which are the remains of toes.

47. (A) Under water, the organic structure of

the tree was gradually replaced by the fine mineral matter which covered it, turning it into petrified wood.

48. (D) This is done by computing the rate of decay of the carbon-14 present in these remains. It is known that the half-life of carbon-14 is 5,760 years.

49. (A) Since the day length in Miami is less than fourteen hours throughout the summer, the cocklebur plant has more than nine hours of darkness daily, and so can flower as soon as it is ripe to do so. In other parts of the country it will begin to flower when the days become short enough for this amount of darkness to set in daily.

50. (D) At that time of the year, the day length becomes reduced to fifteen hours, and then proceeds to get shorter. This exposes the cocklebur plants to nine or more hours of darkness, and they begin to form flower buds.

51. (C) In Chicago, the day length begins to become reduced to fifteen hours or less around July 1. This exposes the cocklebur plants to nine or more hours of darkness and they begin to form flower buds after that date. In Miami and in San Francisco, with their shorter day lengths, flowering takes place at an earlier date, whenever the buds are formed.

52. (E) Toward the end of September, all of the cities begin to have twelve hours or less of daylight. The plant therefore begins to produce flowers after that date.

53. (B) After July 15, the day length in San Francisco is reduced to fourteen hours. From then on, the plant will begin to form flowers.

54. (D) Winnipeg has a day length of sixteen and a half hours in the middle of June.

55. (B) They break up the surface of the rock very slowly as they gain a foothold there, and hold the particles together.

56. (C) The spread of civilization destroyed its food and its nesting places, and unrestricted hunting reduced its numbers.

57. (A) As it gives off radiations, it decays into other radioactive elements, until finally nonradioactive lead is formed. The ratio of both elements in the rock is used as a measure of the length of time that has elapsed since this breakdown started.

58. (B) His similar physical features, his similar brain capacity, and the evidence of his culture indicate that he was a form of present-day man.

59. (A) They changed over the ages in a different way from the mammals on the other side of the barrier. Since they could not interbreed, they retained their own characteristics.

60. (B) It is nourished on food and oxygen supplied in the mother's bloodstream, and diffused into its own bloodstream in the placenta.

61. (A) One sperm nucleus unites with the egg nucleus; the other unites with the endosperm nucleus.

62. (D) The medulla of the brain controls the breathing rate. It is connected with the diaphragm by nerves, and stimulates it to contract and expand more rapidly when the concentration of carbon dioxide in the blood passing through it is high; this results in a faster breathing rate.

63. (A) During respiration, nutrients are combined with oxygen, to release energy.

64. (D) When we exhale, the diaphragm is raised and ribs are lowered, making the chest cavity smaller. This compresses the lungs and forces the air out of them.

65. (B) During photosynthesis, green plants give off oxygen. The carbon dioxide given off by the fish is used by the plants.

66. (D) This was especially true in the Midwest, where sea food, with its iodine content, was not available as food. Now iodized salt is used there to avoid goiter.

67. (D) As they burrow through the soil, earthworms loosen the soil.

68. (A) As part of their reproductive cycle, salmon leave the ocean and swim up rivers to fresh water to lay their eggs.

69. (E) Once the habit is established, the body craves the drug. Painful symptoms occur when the drug is not supplied. Although he may wish to break the habit, the addict usually cannot stand the pain, and will go back to using the drug.

70. (C) Some antibiotics obtained from molds are penicillin, streptomycin, and aureomycin.

71. (A) Each time the cells divide, they receive the same number of chromosomes as the fertilized egg had. During mitosis, the chromosomes split.

72. (B) Plants have cells with cell walls made of cellulose. Green plants would also contain chloroplasts.

73. (A) Plankton is the basis of all life in the ocean. The microscopic algae and protozoa serve as food for the slightly larger forms of life, which in turn are eaten by other, large animals, including small fish. These are then preyed upon by larger fish.

74. (C) The grasshopper is the most numerous consumer listed. It is then eaten by secondary consumers which are present in smaller numbers.

75. (C) Linnaeus classified living things on the basis of structural similarities. He gave organisms two names, i.e. *Homo* is the genus, and *sapiens* is the species name.

76 (E) Only the first item is correct.

77. (C) When a space capsule takes off from the launching pad, it reaches a speed of 17,500 miles an hour in only a few minutes. This terrific acceleration subjects an astronaut to intensive gravitational pull, or G.

78. (B) Without the natural enemies that are present in other countries, and that keep its numbers in check, the rabbits multiplied at an enormous rate in Australia, and became a serious menace to agriculture.

79. (E) All of the items are correct.

80. (A) All of the items are correct except the last one. Dinosaurs had become extinct long before man appeared. Some primitive types of man that died out were: Java man, Peking man, and Neanderthal man.

81. (C) Since Darwin's time, additional research has given us greater insight into the causes of variations. Today, variations are explained as being caused by recombination of genes, and by mutations. New understanding of how species change has also come through the study of population genetics. This deals with the distribution of genes in a population, rather than in a single individual.

82. (B) In reproduction by vegetative propagation, no new genes are introduced, so that the original genetic make-up is maintained in the offspring.

83. (E) The first two items are correct.

84. (D) Thyroxin regulates the rate of oxidation in the body. It helps determine the basal metabolism rate of the individual.

85. (D) Sunlight is composed of different colors or wavelengths, with varying energy content.

86. (A) In the green plant cell, solar energy of a particular range of wavelengths is absorbed by the green pigment chlorophyll.

87. (E) When light is absorbed by chlorophyll, the energy-rich electrons are passed along to associated "electron carriers." As they traverse this path, the excited electrons give up their energy bit by bit. This energy goes into the formation of ATP from ADP and phosphate.

88. (B) In a pure preparation of chlorophyll, isolated from the cell, the absorbed energy of sunlight is re-emitted in the form of visible light, as it is from other phosphorescent or fluorescent organic and inorganic compounds.

89. (C) In the green plant cell, the "hot" or energy-rich electrons do not simply fall back to their normal position as they do in a pure preparation of chlorophyll in a test tube. Instead, the electrons are passed along by a number of "electron carriers," and give up their energy, bit by bit. As a consequence, ATP is formed.

90. (D) The output of energy for each activity is computed as follows:

ACTIVITY	HOURS	OUTPUT PER KG.		OUTPUT FOR 62 KG.
Sleeping	8	1.0 calories	× 62	496.0 calories
Sitting quietly	2	0.4 calories	× 62	49.6 calories
Eating	1½	0.4 calories	× 62	37.2 calories
Laundering, etc.	1½, etc.	1.3 calories, etc.	× 62	120.9 calories, etc.
		Total		2,718.7 calories

91. (D) When hybrids are crossed, the possible results are as follows:

	B	b
B	BB	Bb
b	Bb	bb

25% homozygous black
50% heterozygous black
25% white

92. (C) The possible results are arrived at in this way:

	X	Y
X	XX	XY
X	XX	XY

Male: 50% normal
50% color-blind
Female: 50% normal carrier
50% color-blind

93. (D) The possible types of gene combinations

in the gamets are: TYs, TYs, TyS, Tys, tYS, tYs, tyS, tys.

94. (E) The gene for blue eyes is recessive. Therefore Alice and Richard, who are blue-eyed, have only genes for blue eyes, and their children will continue to have only blue eyes.

95. (E) In four hours, the bacilli would have divided eight times. After the first division there would be two; after the second division, there would be four, and so on, until 256 would be formed.

96. (A) The finger is wiped with alcohol to sterilize it (4); then a needle is used which was dipped in alcohol to sterilize it (6); the finger is punctured with the needle (1); a drop of blood is deposited at one end of a clean slide (5); a thin film of the blood is smeared over the slide with the aid of another slide (2); the stain is added and left for one to two minutes (3)—after that, water is added for two to three minutes and then the slide is rinsed off.

97. (C) A thin layer of cells can easily be studied, especially if they are spread out and not clumped.

98. (D) This is important in avoiding an infection.

99. (D) One drop is sufficient for making a thin film.

100. (E) Otherwise an infection may result.

101. (B) The nucleus takes the stain readily.

102. (B) There are about 5 million red blood cells per cubic millimeter. Blood appears red because of the presence of a large number of these cells.

103. (C) An even amount of pressure and smoothness of movement are needed to make a thin blood film.

104. (D) When members of a population are isolated into smaller groups, they are prevented from breeding with each other. In such small populations, changes in the gene pool take place more readily. Any mutations give rise to distinct characteristics, leading to the development of different species.

105. (B) Since the victims of cholera become dehydrated, because of the loss of great quantities of water from their bodies, it is obviously important to replace the lost liquid as quickly as possible.

106. (C) The dark brown color in tube *A* indicates that more O_2 was formed in that tube. We may assume that photosynthesis carried on by the Elodea sprig was responsible for the O_2. There was little color in tube *C* because there was no Elodea sprig to produce O_2. The Elodea in tube *B* could not photosynthesize due to lack of light.

107. (D) The experimenter could see that there was less O_2 in the tube with the Elodea placed in the dark than in the light. He could not, however, compare O_2 content in a tube with Elodea that was placed in the dark, with a tube without Elodea, also in the dark.

108. (C) In determining the cause of the disease, germs would be studied with the aid of the materials indicated. A sterilizer would permit the agar nutrient to be prepared in the petri dishes in sterile condition, so that there would be no germs to start with. The germs would be grown on the agar nutrient. The microscope would reveal the type and structure of the germs. The petri dishes containing the germs would be placed in the incubator where the germs would be able to grow at the body temperature of the poultry.

109. (A) In attempting to determine the cause of the disease, the steps known as Koch's postulates would be followed: (1) the same germ must be found in all cases of the disease; (2) this germ must be isolated in pure culture; (3) when this germ is inoculated into healthy animals, they must develop the original disease; (4) the same germ must then be obtained from the diseased animals.

PRACTICE COLLEGE BOARD ACHIEVEMENT TEST IN BIOLOGY

NO. 7

PART I—TYPE A

Directions: **The following diagram shows structure of the head. Certain of the parts are labeled with letters. Next to each of the statements below, write the letter of the specific part most closely related to it. A label may be used once, more than once, or not at all.**

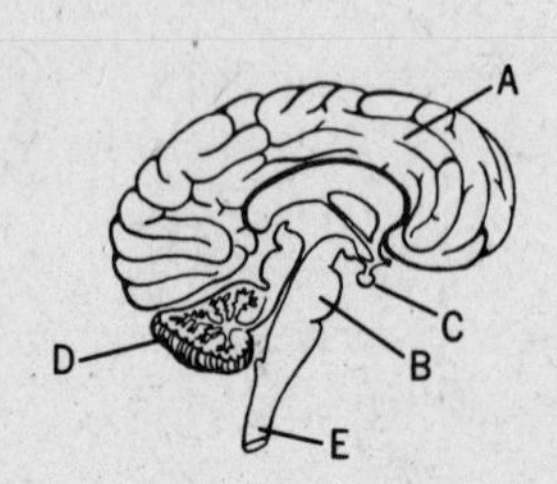

1. Serves as the center of memory 1. ....

2. Serves as the reflex center of the lower part of the body 2. ....

3. Regulates the rate of breathing 3. ....

4. Controls the use of voluntary muscles 4. ....

5. Serves as center of muscular coordination 5. ....

6. Controls balance 6. ....

7. Serves as center of sight 7. ....

8. Controls the growth of the skeleton 8. ....

9. Serves as the center of heartbeat 9. ....

10. Produces ACTH 10. ....

TYPE B

Directions: **Each of the following statements is followed by five suggested answers. Select the one answer which is best in each case.**

11. If the ocular of a microscope is marked 10x and the high power objective is marked 44x, the total magnification is (A) 100 (B) 144 (C) 440 (D) 1044 (E) 1936 — 11. A B C D E

12. A neuron does not contain (A) a cyton (B) an axon (C) dendrites (D) a synapse (E) a terminal branch — 12. A B C D E

13. A person may become poisoned if he inhales a small amount of carbon monoxide because (A) carbon monoxide dissolves delicate lung tissue (B) carbon monoxide combines with hemoglobin more readily than does oxygen (C) carbon monoxide stimulates the heart excessively (D) carbon monoxide reduces the vital capacity of the lungs (E) even a small reduction of oxygen in the inhaled air is dangerous — 13. A B C D E

14. Cooked egg white is tested with nitric acid (A) to demonstrate the presence of protein (B) to neutralize the albumen (C) to observe chemical digestion (D) to demonstrate absorption (E) to see whether it contains vitamin C — 14. A B C D E

15. A recent method of determining the age of prehistoric bones is by measuring their content of radioactive (A) calcium (B) phosphorus (C) nitrogen (D) carbon (E) potassium — 15. A B C D E

16. The innermost chamber of the respiratory system into which air can be drawn is the (A) bronchiole (B) bronchus (C) air sac (D) bronchial tube (E) sinus

16. A B C D E

17. An animal that develops a placenta is a (A) goldfish (B) tropical fish (C) lizard (D) salamander (E) cow

17. A B C D E

18. In aerobic respiration, the final hydrogen acceptor is (A) chlorophyll (B) carbon dioxide (C) water (D) ATP (E) molecular oxygen

18. A B C D E

19. The splints in a horse's foot are the remains of (A) toes (B) teeth (C) flippers (D) coccyx (E) gill slits

19. A B C D E

20. Fossils of insects have been preserved in resinous material called (A) tar-pit asphalt (B) granite (C) amber (D) lignite (E) shale

20. A B C D E

21. Antibodies are chemicals that are (A) non-specific (B) produced by the body in response to an antigen (C) synthesized from carbohydrates (D) synthesized from glycogen (E) transported by red blood cells

21. A B C D E

22. Eohippus is considered to be an early ancestor of the (A) camel (B) bison (C) horse (D) elephant (E) giraffe

22. A B C D E

23. Hormones are distributed through the body by (A) blood plasma (B) ducts (C) lacteals (D) endocrines (E) enzymes

23. A B C D E

24. All of the following are reflexes except (A) blinking (B) sneezing (C) swallowing (D) typewriting (E) coughing

24. A B C D E

25. The turning of a plant toward the light is an example of (A) instinct (B) conditioned response (C) acquired reflex (D) tropism (E) habit

25. A B C D E

26. The idea of natural selection was an important part of the theory proposed by (A) Redi (B) Lamarck (C) Weismann (D) Langerhans (E) Darwin

26. A B C D E

27. The largest brain capacity is observed in (A) chimpanzee (B) gorilla (C) Java man (D) Heidelberg man (E) Cro-Magnon man

27. A B C D E

28. Chromosomes do not normally occur as homologous pairs in (A) somatic cells (B) fertilized eggs (C) gametes (D) zygotes (E) nerve cells

28. A B C D E

TYPE C

Directions: **The following questions require two answers. One of the items in the left column** (A), (B), **or** (C) **is related to four of the items in the right column** (numbered 1–5). **Select the letter of the item which is so related; then select the number of the remaining item which does not belong.**

EXAMPLE:

(A) paramecium	1. cilia
(B) ameba	2. macronucleus
(C) spirogyra	3. trichocyst
	4. pyrenoid
	5. micronucleus

A B C 1 2 3 4 5

29.

(A) carbohydrate	1. polypeptide
(B) lipid	2. amino acid
(C) protein	3. COOH group
	4. glucose
	5. NH_2 group

29. A B C 1 2 3 4 5

30.

(A) pneumonia	1. Schick test
(B) influenza	2. chest X-ray
(C) tuberculosis	3. pneumothorax
	4. sputum examination
	5. patch test

30. A B C 1 2 3 4 5

31. (A) filicineae (B) seed plant (C) bryophyte

1. gymnosperm
2. angiosperm
3. monocot
4. dicot
5. horsetail

31. A B C 1 2 3 4 5

32. (A) fern (B) moss (C) liverwort

1. spore
2. gametes
3. prothallium
4. hypha
5. sporangium

32. A B C 1 2 3 4 5

33. (A) bird (B) mammal (C) reptile

1. warm-blooded
2. egg with much food
3. hair
4. milk
5. microscopic egg

33. A B C 1 2 3 4 5

34. (A) liver (B) stomach (C) pancreas

1. islets of Langerhans
2. diabetes
3. ptyalin
4. insulin
5. amylase

34. A B C 1 2 3 4 5

35. (A) fat digestion (B) carbohydrate digestion (C) protein digestion

1. lipase
2. pepsin
3. trypsin
4. hydrochloric acid
5. erepsin

35. A B C 1 2 3 4 5

36. (A) four-chambered heart (B) three-chambered heart (C) two-chambered heart

1. chicken
2. dog
3. horse
4. frog
5. human

36. A B C 1 2 3 4 5

TYPE D

Directions: **The five lettered headings below are followed by a list of numbered statements. For each numbered statement, choose the one lettered heading (A–E) which is most closely related to it.**

(A) Blending
(B) Back-cross
(C) Dominance
(D) Independent Assortment
(E) Linkage

37. Some people have blond hair and blue eyes, while others have brown hair and blue eyes

37. A B C D E

38. When red and white snapdragons are cross-pollinated, the next generation of flowers will be pink

38. A B C D E

39. Most human beings have brown eyes

39. A B C D E

40. When tall yellow-seeded dihybrid pea plants are cross-pollinated, short yellow-seeded offspring appear

40. A B C D E

41. Geneticists find that Drosophila flies with black bodies always seem to have short wings

41. A B C D E

42. A plant breeder can tell whether a tall pea plant is homozygous tall or heterozygous tall by crossing it with a short one

42. A B C D E

43. You can predict what the offspring will be when pure tall plants are crossed with short plants

43. A B C D E

(A) Respiration
(B) Photosynthesis
(C) Protein synthesis
(D) Assimilation
(E) Diffusion

44. How does oxygen enter a green-plant cell?

44. A B C D E

45. How is oxygen used by a green-plant cell?

45. A B C D E

46. How is carbon dioxide used by a green-plant cell?

46. A B C D E

47. How is glucose obtained?

47. A B C D E

48. How are amino acids obtained?

48. A B C D E

49. How is more protoplasm made?

49. A B C D E

PART II

A young scientist studied the response of plant lice, or aphids, to various wave lengths of light. These small insects are parthenogenetic, and give rise to living young. They may be either wingless or winged, depending on various environmental conditions, such as temperature, humidity, length of day, intensity of light, abundance of food, and quality or color of light. They were grown on nasturtium plants that were covered with light filters of red, yellow, blue or gray (the control). The appearance of winged aphids and the reproduction rate are shown in the accompanying graphs

GRAPH 1: PER CENT OF WINGED APHIDS

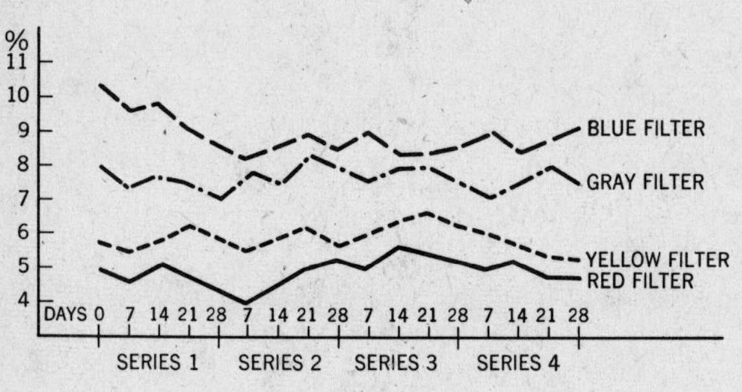

GRAPH 2: NUMBER OF OFFSPRING

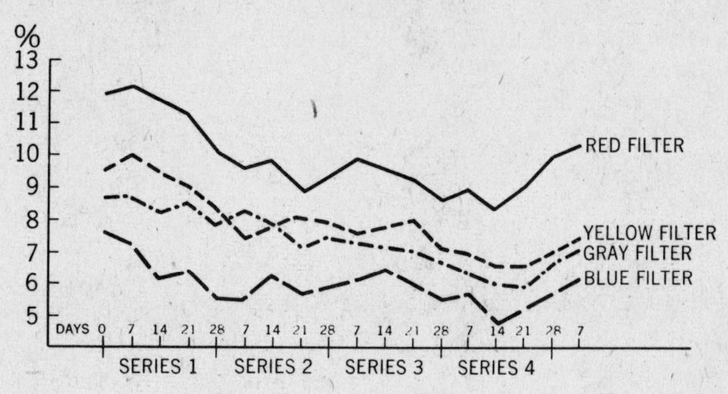

50. The greatest number of winged aphids were produced (A) under the blue filter (B) under the gray filter (C) under the yellow filter (D) under the red filter (E) under the control filter

50. A B C D E

51. Under the yellow filter (A) there were fewer winged aphids than under the red filter (B) there were more winged aphids than under the gray filter (C) there were more winged aphids than under the blue filter (D) there

51. A B C D E

were more winged aphids than under the red filter (E) there were more winged aphids than under the control filter

52. Under the red filter (A) there were the least winged aphids and the highest reproduction rate (B) there were the least winged aphids and lowest reproduction rate (C) there were the most winged aphids, and the highest reproduction rate (D) there were the least wingless aphids and the highest reproduction rate (E) there were the most wingless aphids and the lowest reproduction rate

52. A B C D E

53. Reproduction is inhibited most (A) under the red filter (B) under the yellow filter (C) under the gray filter (D) under the blue filter (E) under ordinary light

53. A B C D E

54. The production of wingless aphids is stimulated most by (A) both blue and gray light (B) both gray and yellow light (C) both yellow and red light (D) both red and blue light (E) both yellow and blue light

54. A B C D E

55. The highest reproduction rate is stimulated most by (A) both blue and gray light (B) both gray and yellow light (C) both yellow and red light (D) both red and blue light (E) both yellow and blue light

55. A B C D E

PART III—TYPE A

Directions: **Each item consists of an *assertion* (statement) in the left-hand column, and a *reason* on the right-hand side. Select**

(A) if both assertion and reason are true, and *are related* as cause and effect
(B) if both assertion and reason are true, but *are not related* as cause and effect
(C) if the assertion is true, but the reason is a false statement
(D) if the assertion is false, but the reason is a true statement
(E) if both assertion and reason are false statements

DIRECTIONS SUMMARIZED:

	(A)	(B)	(C)	(D)	(E)
Assertion:	True	True	True	False	False
Reason:	True	True	False	True	False
	Cause and effect	*Not* cause and effect			

Assertion:		Reason:	
56. The sperm is smaller than the ovum	BECAUSE	it has the power of locomotion	56. A B C D E
57. A man in space will experience a sensation of weightlessness	BECAUSE	acceleration produces a force of many G	57. A B C D E
58. Mendel is considered to be the father of genetics	BECAUSE	he developed the theory of the gene	58. A B C D E
59. *Archaeopteryx* is considered to be a link between reptiles and mammals	BECAUSE	it had teeth in its jaw	59. A B C D E
60. The navel orange is propagated by grafting	BECAUSE	it is seedless	60. A B C D E
61. The pollen grain grows in size after it reaches the stigma	BECAUSE	it sends a pollen tube to the anther	61. A B C D E
62. The American Indian has not been classified as a member of *Homo sapiens*	BECAUSE	natural barriers kept him from migrating and interbreeding	62. A B C D E
63. A forest fire may be a cause of floods	BECAUSE	the roots of trees absorb excess water	63. A B C D E

64. Strip-cropping is a factor leading to erosion, if practiced on hillsides BECAUSE close-growing strip crops such as grass slow down the runoff of rain water

64. A B C D E

65. The American bison was threatened with extinction BECAUSE Texas cattle fever was extremely contagious

65. A B C D E

TYPE B

Directions: **Each of the following statements is followed by five suggested answers. Select the one answer which is best in each case.**

66. Insects and spiders are classified as arthropods because (A) they are carnivorous (B) they have jointed legs (C) they have a backbone in the embryonic stages (D) they are land-dwellers (E) their eyes are simple

66. A B C D E

67. The chloroplasts in an Elodea green-plant cell move because (A) they are equipped with cilia (B) they have flagella (C) they are carried around by the streaming protoplasm (D) they form pseudopodia (E) they carry on plasmolysis

67. A B C D E

68. Living yeast cells are used in the baking industry because (A) they give off carbon dioxide (B) they multiply so rapidly that they make the dough expand (C) they produce alcohol (D) their buds form antibiotics (E) their spores give bread its flavor

68. A B C D E

69. A gardener plants lily bulbs rather than lily seeds because (A) they are easier to handle (B) seeds are often eaten by birds (C) they are usually planted in the spring (D) they breed true (E) the seeds require a special spreader

69. A B C D E

70. It is better to breathe through the nose than the mouth because (A) this is a shorter route to the lungs (B) the air becomes warmed and moistened in the nose (C) the number of breaths per day is reduced (D) silicosis is prevented in this way (E) experiments with the spirometer show a greater lung capacity as a result

70. A B C D E

71. The excretion of sweat by the skin is useful because (A) it tends to keep the pores from contracting (B) it eliminates salt from the body (C) it helps keep the sebaceous glands clean (D) its evaporation helps keep the body cool (E) its formation indicates healthy kidneys

71. A B C D E

72. The body often has extra strength in an emergency because (A) the storage of glucose is stimulated (B) the activity of the digestive tract is stimulated (C) the production of glycogen is stimulated (D) the rate of peristalsis is stimulated (E) the secretion of adrenin is stimulated

72. A B C D E

73. Eating undercooked pork is dangerous because (A) it may contain the live worms of trichinosis (B) its nutrients are not ready for digestion (C) it contains too high a fat content (D) it may contain the rickettsia of typhus (E) it should be sterilized to prevent rabies

73. A B C D E

74. Viruses do not fit into our concept of the cell because they (A) lack cytoplasm (B) cannot multiply (C) contain DNA (D) are so small (E) lack chlorophyll

74. A B C D E

75. A lack of certain vitamins in the diet results in muscular weakness because these vitamins (A) are oxidized to yield energy (B) have a high calorie content (C) are a part of certain respiratory enzymes (D) contain high-energy phosphate bonds (E) can only carry on anaerobic respiration

75. A B C D E

TYPE C

Directions: **In each of the following questions, there is a statement followed by four choices, of which *one or more* are correct. Indicate the correct combination by choosing the following letters:**

(A) if only 1, 2, and 3 are correct

(B) if only 1 and 3 are correct
(C) if only 2 and 4 are correct
(D) if only 4 is correct
(E) if some other combination is correct

DIRECTIONS SUMMARIZED:

(A)	(B)	(C)	(D)	(E)
1, 2, 3 only	1, 3 only	2, 4 only	4 only	some other combination

76. In the treatment of cancer, doctors use
1. radium treatment
2. surgery
3. X-ray treatment
4. toxin-antitoxin

76. A B C D E

77. One reason that algae are classified as protists is that
1. they are one-celled animals
2. their cells differ from those of higher organisms
3. they contain a nucleus
4. they contain cellulose

77. A B C D E

78. Bile is
1. useful in emulsifying fats
2. stored in the gall bladder
3. secreted by the liver
4. an enzyme

78. A B C D E

79. Carbohydrates are compounds that are
1. composed of carbon, hydrogen, and oxygen
2. used for assimilation
3. used by the body for energy
4. present largely in meats

79. A B C D E

80. Oxygen enters the bloodstream in the
1. trachea
2. sinuses
3. bronchi
4. alveoli

80. A B C D E

81. Calcium is important to the body for
1. blood clotting
2. muscle tone
3. proper formation of bones and teeth
4. hemoglobin production

81. A B C D E

82. Study of a stained drop of human blood under the microscope reveals the
1. cellular structure of white blood cells
2. nucleus of red blood cells
3. appearance of blood platelets
4. dissolved nutrients of the plasma

82. A B C D E

83. A daily diet of the Basic Four
1. supplies the essential nutrients
2. must be supplemented with vitamin pills
3. is complete except for the eighth nutrient
4. is lacking in roughage

83. A B C D E

84. The spinning of a web by a spider
1. is acquired
2. is inborn
3. is an example of conditioned behavior
4. is an example of instinct behavior

84. A B C D E

PART IV—TYPE A

Directions: **Referring to the following selection, choose the best answer for each of the questions that follows:**

*"It is now well known that in plant and animal cells the actual replication of the genetic material—the doubling of DNA—takes place only between divisions. This can best be shown by experiments in which a population of cells is fed for a brief interval with some radioactively labeled substance (usually thymidine) that is built into the newly formed DNA. The newly synthesized DNA is found only in the nuclei of cells that are in the interphase—never in cells that are going through mitosis. Refinements of such experiments show that DNA synthesis occupies only a certain part of the period between divisions.

"If a given cell is not destined to divide again, as is the case with the cells of many specialized organs (muscles and brain, for example), DNA synthesis does not begin. If it does begin, the rule is that it goes to completion; that is, the original amount of DNA is doubled. A less rigid rule is: If a cell does undertake DNA synthesis, not only is the doubling completed but also the cell will usually enter division. Studies of the intestinal cells of the rat . . . have shown that every cell makes a crucial decision within the first few hours after division; either it enters DNA synthesis and will divide again or it adopts the career of a differentiated cell and will never divide again."

85. The replication of DNA takes place during the stage of mitosis known as the (A) interphase (B) prophase (C) metaphase (D) anaphase (E) telophase

86. In mitosis research, thymidine is (A) part of the centrosome (B) a component of the spindle (C) a fibril in the centriole (D) the membranous structure of the cell plate (E) a radioactively labeled substance

87. The replication of DNA takes place (A) only in plant cells (B) only in animal cells (C) only in differentiated tissue (D) only in embryonic tissue (E) in both plant and animal cells

88. When the genetic material replicates, (A) DNA remains the same (B) DNA doubles (C) DNA triples (D) DNA is fertilized (E) DNA is halved

89. All of the following statements are true except: (A) if a cell is not to divide again, DNA synthesis does not begin (B) muscle and brain cells do not divide (C) DNA synthesis generally does not take place in differentiated cells (D) doubling of DNA takes place during the time when a cell is dividing (E) cell division has been studied in the intestinal cells of the rat

85. A B C D E
86. A B C D E
87. A B C D E
88. A B C D E
89. A B C D E

TYPE B

Directions: **Each of the following problems is followed by five suggested answers. Select the one answer which is best in each case.**

90. In the gene pool of a given population of rabbits, 80 per cent of all the gametes carry the dominant allele for gray coat and 20 per cent carry the recessive allele for white coat. From this information, it can be predicted that the per cent of rabbits which would be homozygous for gray coat would be (A) 20% (B) 40% (C) 64% (D) 80% (E) 96%

91. A color-blind daughter will be produced from: (Key: X = chromosome with gene for normal vision; $\underline{X}$ = chromosome with gene for color-blindness)
(A) $XX \times \underline{X}Y$
(B) $X\underline{X} \times XY$
(C) $XX \times X\underline{Y}$
(D) $X\underline{X} \times \underline{X}Y$
(E) $\underline{X}\underline{X} \times XY$

90. A B C D E
91. A B C D E

*From "How Cells Divide" by Daniel Mazia, Sept. 1961, by *Scientific American, Inc.*

92. A slice of white bread weighing 25 grams is burned in a calorimeter in which the central container is surrounded by 5 kilograms of water. The temperature of the water is raised from 12°C to 26°C. The number of Calories contained in the bread is (A) 14 (B) 25 (C) 50 (D) 70 (E) 125

92. A B C D E

93. Of the following, the largest ovum is produced by a (A) hippopotamus (B) elephant (C) ostrich (D) whale (E) chicken

93. A B C D E

94. Carbohydrates are organic compounds containing carbon, hydrogen, and oxygen, in which the hydrogen and oxygen occur in the same ratio as in water. Of the following compounds, the carbohydrate is (A) stearin, $C_{57}H_{110}O_6$ (B) thiamin, $C_{12}H_{18}N_4O_2S$ (C) palmatin, $C_{51}H_{98}O_6$ (D) cellulose, $(C_6H_{10}O_5)_n$ (E) riboflavin, $C_{17}H_{20}N_4O_6$

94. A B C D E

95. In comparison with the number of gametes produced from one primary sex cell during oögenesis, the number of gametes produced from one primary sex cell during spermatogenesis is (A) the same (B) half (C) twice as many (D) three times as many (E) four times as many

95. A B C D E

TYPE C

A student made a study of the structure of a lamb's heart he obtained at a butcher shop. He observed the exterior appearance first. Then he made a longitudinal cut from the upper end to the bottom of the heart, and laid it open so that he could see the inside.

96. The blood vessels he observed on the outside of the heart were the (A) coronary arteries (B) portal veins (C) renal veins (D) carotid arteries (E) brachial arteries

96. A B C D E

97. He knew that the large blood vessel leading to one side of the heart was a vein because (A) it was connected with a ventricle (B) it had thin walls and no valves (C) it had thick walls and valves (D) it had thick walls and no valves (E) it had thin walls with valves

97. A B C D E

98. He noticed that the thickest walls of the heart were located in the (A) right atrium (B) left atrium (C) right ventricle (D) left ventricle (E) valves of the heart

98. A B C D E

99. The small chambers he noticed at the upper part of the heart were the (A) atria (B) ventricles (C) entrance to the aorta (D) entrance to the pulmonary arteries (E) entrance to the pulmonary veins

99. A B C D E

100. He observed the valves of the heart and noticed that they were made of (A) epidermis (B) connective tissue (C) epithelial tissue (D) cartilage (E) dermis

100. A B C D E

101. He made a slide of a small piece of the heart wall, and under the microscope could see the (A) skeletal muscle (B) smooth muscle (C) striations (D) nuclei of the red blood cells (E) ciliated cells

101. A B C D E

102. He noticed that the heart was about the size of (A) his head (B) his eye (C) his fist (D) a chicken's heart (E) a frog's heart

102. A B C D E

103. He observed that the color of the heart was (A) white (B) blue (C) reddish brown (D) colorless (E) blue-black

103. A B C D E

PART V

Directions: **Each of the following situations has five suggested answers. Select the one answer that is best in each case.**

104. The heterotroph hypothesis proposed by Oparin and others, assumes that the early atmosphere of the earth lacked oxygen, carbon dioxide and nitrogen. The gases present were probably water vapor, hydrogen, methane and ammonia. Heavy rains washed these gases into the oceans where they mixed with dissolved materials. In the "hot thin soup" of the seas, the molecules of these substances were acted on by various forms of energy such as solar radiation, cosmic rays, lightning and radioactivity in the earth's surface. Organic molecules were formed, resulting in the first

104. A B C D E

proteins and carbohydrates. A scientist attempting to duplicate the formation of such organic compounds artificially in a "hot thin soup" would most likely: (A) set off electric sparks in a flask containing water vapor, hydrogen, ammonia and methane (B) introduce gamma rays into a mixture of carbon monoxide and carbon dioxide (C) conclude that it is not possible to duplicate conditions existing billions of years ago (D) inject DNA into a mixture of methane, ammonia, oxygen and carbon dioxide (E) use a concentration of cosmic rays on a mixture of simple organic compounds present as a soupy mixture in a TV tube

105. In the grassland plains of the Serengeti National Park between Kenya and Tanzania in East Africa, herds of zebras, wildebeests and gazelles form an ecological succession of migration. It has been found that the large zebra feeds on the upper parts of grasses and herbs. The somewhat smaller wildebeest follows and feeds on the midparts of the plants and the small gazelle comes along later to feed on the lower parts. The earlier members of the succession apparently prepare the structure of the vegetation for the members that follow them. An important conclusion from these observations is that (A) a reduction in the number of one species by disease or hunting could lead to a reduction in the numbers of another species (B) an increase in the numbers of one species could lead to a reduction in the numbers of another species (C) an increase in hunting of wildebeests would indirectly cause an increase in the numbers of gazelles (D) a shortage of grass would cause zebras to turn on the smaller gazelle for food (E) the wildebeest would drive out the gazelles if there were a shortage of grass

105. A B C D E

106. Dentistry students at Columbia University are now taught that tooth cavities are practically 100% preventable. This depends on a program that includes: proper brushing of the teeth to remove plaque (the invisible bacterial substance that sticks to teeth and initiates decay); taking in fluorides (in toothpastes or in drinking water); limiting the intake of sugar which interacts with the bacteria in plaque to form acids that attack tooth enamel; and regular visits to the dentist. The success of this program depends on the (A) ability of the teeth to absorb fluorides from toothpaste or drinking water (B) the use of the proper toothpaste by the average person (C) the willingness of the average person to practice the proper procedures regularly (D) the willingness of the average person to purchase the best type of toothbrush (E) the daily use of mouthwash to destroy the bacteria that form plaque

106. A B C D E

PART VI

Directions: **Each of the following questions has five suggested answers. Select the one answer that is best in each case.**

107. A dog was placed in a special room free of unrelated stimuli. On repeated trials a tone was sounded for 5 seconds; approximately 2 seconds later the dog was given powdered food. Trials 1, 10, 20, 30, 40 and 50 were test trials; that is, the tone was sounded for 30 seconds and no food powder was given. The following data were collected

107. A B C D E

Test Trial No.	*Drops of Saliva Secreted*	*No. Seconds Between Onset of Tone and Salivation*
1	0	...
10	6	18
20	20	9
30	60	2
40	62	1
50	59	2

Which part of the nervous system would be of *least* importance in explaining the behavior of the dog? (A) cerebellum (B) motor neuron (C) auditory nerve (D) spinal cord (E) salivary nerve

108. A student examined a microscope slide of cells that had been removed from a root and noted these three types: Cell No. 1—very thin walls; Cell No. 2—thick heavy walls, cells dead; Cell No. 3—mitotic division taking place. He would be correct in identifying the three cells as follows:

108. A B C D E

	Cell No. 1	*Cell No. 2*	*Cell No. 3*
(A)	cell in growing region	palisade cell	root cap cell
(B)	root hair cell	root cap cell	cell in growing region
(C)	root cap cell	root hair cell	cell in growing region
(D)	palisade cell	root hair cell	cell in growing region
(E)	tap root cell	cell in growing region	root hair cell

109. The following results were obtained in various crosses of squash plants which were round, oval and long:

109. A B C D E

(a) Round squash × oval squash produced 103 round squash, 105 oval squash
(b) Long squash × oval squash produced 94 long squash, 92 oval squash
(c) Oval squash × oval squash produced 119 round squash, 240 oval squash, 120 long squash

In order to obtain 100% long squash, the best cross to follow would be:
(A) round squash × oval squash
(B) oval squash × long squash
(C) oval squash × oval squash
(D) long squash × long squash
(E) round squash × long squash

ANSWER KEY: TEST NO. 7

1. (A)	**12.** (D)	**23.** (A)	**34.** (C)3	**45.** (A)	**56.** (B)	**67.** (C)	**78.** (A)	**89.** (D)	**100.** (B)
2. (E)	**13.** (B)	**24.** (D)	**35.** (C)1	**46.** (B)	**57.** (B)	**68.** (A)	**79.** (B)	**90.** (C)	**101.** (C)
3. (B)	**14.** (A)	**25.** (D)	**36.** (A)4	**47.** (B)	**58.** (C)	**69.** (D)	**80.** (D)	**91.** (D)	**102.** (C)
4. (A)	**15.** (D)	**26.** (E)	**37.** (D)	**48.** (C)	**59.** (D)	**70.** (B)	**81.** (A)	**92.** (D)	**103.** (C)
5. (D)	**16.** (C)	**27.** (E)	**38.** (A)	**49.** (D)	**60.** (A)	**71.** (D)	**82.** (B)	**93.** (C)	**104.** (A)
6. (D)	**17.** (E)	**28.** (C)	**39.** (C)	**50.** (A)	**61.** (E)	**72.** (E)	**83.** (E)	**94.** (D)	**105.** (A)
7. (A)	**18.** (E)	**29.** (C)4	**40.** (D)	**51.** (D)	**62.** (E)	**73.** (A)	**84.** (C)	**95.** (E)	**106.** (C)
8. (C)	**19.** (A)	**30.** (C)1	**41.** (E)	**52.** (A)	**63.** (C)	**74.** (A)	**85.** (A)	**96.** (A)	**107.** (A)
9. (B)	**20.** (C)	**31.** (B)5	**42.** (B)	**53.** (D)	**64.** (D)	**75.** (C)	**86.** (E)	**97.** (E)	**108.** (B)
10. (C)	**21.** (B)	**32.** (A)4	**43.** (C)	**54.** (C)	**65.** (C)	**76.** (A)	**87.** (E)	**98.** (D)	**109.** (D)
11. (C)	**22.** (C)	**33.** (B)2	**44.** (E)	**55.** (C)	**66.** (B)	**77.** (E)	**88.** (B)	**99.** (A)	

ANSWERS EXPLAINED: TEST NO. 7

1. (A) cerebrum
2. (E) spinal cord
3. (B) medulla
4. (A) cerebrum
5. (D) cerebellum
6. (D) cerebellum
7. (A) cerebrum
8. (C) pituitary gland
9. (B) medulla
10. (C) pituitary gland

11. (C) The magnification of the ocular (10) is multiplied by the magnification of the high power objective (44), to give a total magnification of 440.

12. (D) The synapse is the area where two neurons are very close together, and where the impulse travels from one neuron to the other.

13. (B) Carbon monoxide is extremely dangerous because it is odorless, and so can be inhaled without the person being aware of it. The hemoglobin of the red blood corpuscles combines more readily with it than with oxygen, and carries it to the cells of the body, where it acts as a poison.

14. (A) Egg white consists of a protein known as albumen. In testing for its presence, it is heated with nitric acid. A yellow color results. When ammonium hydroxide is added to the albumen, an orange color is obtained.

15. (D) Radioactive carbon, or carbon-14, has a half-life of 5,760 years. By measuring the amount of it left in bone, the age of the bone is determined.

16. (C) After entering the nasal passages, air passes through the trachea into the bronchi, the bronchial tubes, and finally into the air sacs where its oxygen enters the blood and where it picks up carbon dioxide given off by the blood.

17. (E) All mammals, with the exception of the duck-bill platypus and the anteater, have a placenta in which food and oxygen diffuse from the bloodstream of the mother into the bloodstream of the embryo.

18. (E) In the Krebs cycle of aerobic respiration, hydrogen atoms are passed from one carrier to another, releasing energy along the way. The final acceptor is oxygen which combines with the hydrogen to form water.

19. (A) The ancestor of the horse, Eohippus, had four toes on its forelegs. With the passage of millions of years, one toe became enlarged, and two others were reduced to small bones or splints high up on the horse's foot. The fourth toe disappeared altogether.

20. (C) Ants and wasps were trapped in the sticky resin of trees. In time, this resin hardened into amber, with the insects still present within it.

21. (B) When foreign substances (antigens) like germs enter the body, antibodies are produced in the blood tissue to counteract their effect.

22. (C) Eohippus lived 60 million years ago, was the size of a cat, had simple teeth, and walked on front feet that had four toes, and hind feet with three toes.

23. (A) Hormones are produced by the ductless glands and are circulated in the plasma.

24. (D) This is a habit, a learned automatic activity; reflexes are inborn, unlearned and automatic.

25. (D) A tropism is a turning response of plants; this is a case of positive phototropism.

26. (E) Darwin's theory of natural selection was based on the idea that there is variation among the many offspring produced by organisms, and that in the struggle for existence, the fittest survive.

27. (E) Cro-Magnon's brain capacity was a little larger than ours.

28. (C) Chromosomes usually occur in homologous pairs—that is, they are alike in size, shape and the kind of genes they carry. One originated in the sperm and the other in the egg. At fertilization, they became a homologous pair. However, when gametes are formed, each member of the pair of chromosomes separates into a different cell, with the result that gametes have the monoploid number of chromosomes; they now have a single member of each homologous pair of chromosomes.

29. (C)4 Proteins contain building blocks of amino acids, which contain the amine group (NH_2) and carboxyl group (COOH). A chain of amino acids is called a polypeptide. Glucose is an example of a carbohydrate.

30. (C)1 All of these are related to the control of tuberculosis except the Schick test, which indicates whether a person is immune to diphtheria.

31. (B)5 All are classified as spermatophytes except horsetail which is in the sphenopsida.

32. (A)4 All the structures are found in a fern except hypha, which is a stemlike structure present in molds.

33. (B)2 All of these are characteristic of mammals except egg with much food, which is true of the other vertebrates.

34. (C)3 All of these deal with the pancreas except ptyalin, which is produced by the salivary glands.

35. (C)1 All of these function in the digestion of proteins except lipase, which is a fat-digesting enzyme.

36. (A)4 Mammals and birds have a four-chambered heart; the frog has a three-chambered heart.

37. (D) Each characteristic is inherited independently of the other.

38. (A) In this example of blending inheritance, or incomplete dominance, neither red nor white is dominant, but both appear in the petals to produce a pink effect.

39. (C) In humans, the gene for brown eyes is dominant over the gene for blue eyes.

40. (D) Each characteristic is inherited independently of the other.

41. (E) The genes for both characteristics are located on the same chromosome, and so they are inherited together.

42. (B) If the offspring are only tall, the tall plant is probably homozygous tall, because the result of crossing homozygous tall and short plants is 100% heterozygous tall; if at least one of the offspring is short, the tall plant must be heterozygous, because a short plant would receive a gene for shortness from each parent.

43. (C) According to Mendel's Law of Dominance, the offspring of such pea plants would be hybrid tall, and would contain the recessive gene for shortness.

44. (E) Dissolved gases and minerals diffuse through the cell membrane of a cell.

45. (A) Green plants, like animals, carry on respiration. They release energy from food nutrients in the presence of oxygen, and give off carbon dioxide and water as wastes.

46. (B) Carbon dioxide, which is normally a waste product resulting from respiration, is taken in by green plants in the light during photosynthesis to make carbohydrates.

47. (B) Glucose is formed during the process of photosynthesis. Through a series of complex steps, involving both a light reaction and a dark reaction, oxygen is liberated from water, and the remaining hydrogen is combined with carbon dioxide to form certain intermediate products, leading to the formation of glucose.

48. (C) After a green plant has made carbohydrates, it combines them with nitrogen and other elements, to produce amino acids, the building blocks of proteins.

49. (D) Protoplasm is produced from amino acids; in this way more of the living material is made from nonliving nutrients.

50. (A) Under the blue filter, the percentage of winged aphids varied from 8 to 10%.

51. (D) Under the yellow filter, there were 5 to 6+% winged aphids, as compared with the red filter where there were only 4 to 5+% winged aphids.

52. (A) In graph 1, it is seen that the lowest percentage of winged aphids occurred under the red filter. In graph 2, it is seen that the highest number of offspring occurred under the red filter.

53. (D) The smallest number of offspring was produced under the blue filter, as shown in graph 2.

54. (C) Under these filters, the percentage of wingless aphids was highest, being in inverse proportion to the percentage of winged aphids.

55. (C) As shown in graph 2, the greatest number of offspring were produced under the red and yellow filters.

56. (B) It has no stored food.

57. (B) After feeling the stress of acceleration, he will enter a phase where there is equilibrium between the force of gravity and the forward motion of the space vehicle. At this point, he is weightless.

58. (C) Mendel first discovered laws of heredity in his work with pea plants, namely, the Law of Dominance, Law of Segregation, and Law of Independent Assortment.

59. (D) It is a link between reptiles and birds, possessing some characteristics of both groups, such as feathers, wings, teeth and a tail.

60. (A) A branch (scion) of the navel orange will bear more navel oranges after it is grafted onto a stock that produces oranges with seeds, because its genetic make-up is still the same.

61. (E) The pollen grain germinates, and a pollen tube grows from it down to the ovule.

62. (E) He is a member of the species, and he has migrated and interbred with other racial types.

63. (C) The roots of trees bind the topsoil, which is porous and absorbs water. After a fire kills the trees, the soil is no longer held together, and is washed away, leading to run-off of the water and consequent floods.

64. (D) It is a measure designed to avoid erosion on hillsides by preventing gullying, which is caused by the run-off of water on soil which is not bound by roots of plants such as legumes and grass.

65. (C) The spread of civilization, and unrestricted hunting reduced its numbers to the vanishing point, until it was protected by law.

66. (B) Insects have six legs; spiders have eight legs; they also have an exoskeleton.

67. (C) The streaming nature of protoplasm is illustrated in Elodea cells.

68. (A) Yeast ferments sugar and in the process gives off carbon dioxide, which swells and puffs up the dough used in baking bread and cake.

69. (D) The use of bulbs is an example of vegetative propagation; since no new genes are introduced, the plants will breed true to the original type.

70. (B) As air enters through the nose it is warmed and moistened by the membranes of the nasal passages. Dust and bacteria are caught by the hairs and cilia and are prevented from entering the lungs. Air that enters through the mouth is dry and cool, and remains largely unfiltered.

71. (D) Evaporation of a liquid results in cooling; in this case, as sweat evaporates, the skin is cooled. In hot weather, the body excretes more sweat than in cold weather.

72. (E) Adrenin is produced by the adrenal glands which are also known as the glands of emergency. Among its effects, it stimulates the liver to convert glycogen to glucose; the heart is caused to beat faster, and more glucose is brought to the muscles and brain; there is an increase in breathing rate, providing more oxygen. Under these conditions, the muscles have greater energy, and the brain can think more clearly.

73. (A) These worms may be taken into the body and cause the painful effects of trichinosis, resulting from their boring into skeletal muscles.

74. (A) Viruses contain a central core of nucleic acid (DNA or RNA) surrounded by a protein covering.

75. (C) In the process of cellular respiration, a number of enzymes are active in the release of energy from glucose. During the sequence of stages in the breakdown of glucose, vitamins such as thiamin and riboflavin help to form some of the enzymes that regulate the reactions.

76. (A) These methods can be used to cure cancer in the early stages.

77. (E) Only the second item is correct.

78. (A) Bile does not contain enzymes. By breaking fats down to small particles, it makes it easier for the enzyme lipase in the small intestine to digest them.

79. (B) In carbohydrates, hydrogen and oxygen occur in the same ratio as in water, there being twice as many atoms of hydrogen as oxygen, i.e., glucose – $C_6H_{12}O_6$.

80. (D) The walls of the alveoli, or air sacs, are one cell thick, permitting diffusion of oxygen to take place through them into the bloodstream.

81. (A) The element iron is needed for hemoglobin formation.

82. (B) A slide of blood is usually stained with Wright's stain.

83. (E) Only the first item is correct.

84. (C) In an instinct there is a series of reflexes producing a rather complex form of behavior.

85. (A) The actual replication of the genetic material – the doubling of DNA takes place only between divisions. Newly synthesized DNA is found only in the nuclei of cells that are in the interphase.

86. (E) Thymidine is a radioactively labeled substance that is fed into a population of cells and then traced in the nuclei of cells that are in the interphase stage of mitosis.

87. (E) The doubling of DNA takes place in both plant and animal cells shortly after they have divided, and before they will once again divide.

88. (B) As is indicated in the opening statement of the selection, the replication of DNA refers to its doubling.

89. (D) Doubling of DNA takes place between divisions, and not during the time when the cells are actually dividing.

90. (C) Rabbits that are homozygous for gray coat have two dominant alleles, *GG*. One of these alleles came from an egg and the other from a sperm. Since 80 per cent of the eggs and 80 per cent of the sperm carried *G*, when they are united, the per cent of the resulting zygotes having *GG* alleles would be 64 ($0.80 \times 0.80 = 0.64$).

91. (D) The possible results are:

	$\underline{X}$	Y
X	X$\underline{X}$	XY
$\underline{X}$	$\underline{X}\underline{X}$	$\underline{X}$Y

Female: 50% normal carrier
50% color-blind

Male: 50% normal
50% color-blind

92. (D) A Calorie is the amount of heat needed to raise the temperature of one kilogram of water one degree centigrade. In this case, the temperature of 5 kg. of water was increased 14°C; $5 \times 14 = 70$ Calories.

93. (C) The mammals have microscopic ova with practically no stored food in them. The ostrich egg is the largest egg because of the large amount of food it contains.

94. (D) There is twice as much hydrogen as oxygen.

95. (E) Four sperm cells are produced from each primary sex cell during spermatogenesis. In oögenesis, one egg and three polar bodies are formed.

96. (A) These blood vessels bring nourishment to the cells of the heart.

97. (E) The vena cava leads to the right atrium.

98. (D) The left ventricle pumps blood through the aorta to most of the body, compared to the right ventricle, which pumps blood only to the lungs.

99. (A) The atria receive blood from the large veins which lead into the heart, and send it to the adjoining ventricles through the valves.

100. (B) The valves contain tough fibers of connect tissue which permit them to shut tight when the ventricles contract, thus preventing the blood from flowing back into the atria.

101. (C) Cardiac muscle has microscopic striations, or striped areas.

102. (C) The heart of a lamb is roughly the same size as a man's heart.

103. (C) The muscle tissue of the heart is a reddish-brown color, similar to that of beef muscle.

104. (A) These are like the conditions present on the earth in its primitive stages. The experiments of Dr. Stanley Miller produced a number of organic molecules, including amino acids, using a similar procedure.

105. (A) The earlier members of the ecological succession prepare the structure of the vegetation for the members that follow them. A reduction in the numbers of the earlier members will reduce the amount of vegetation that can be prepared for the later arrivals. This in turn will cause a reduction in the numbers of the later species.

106. (C) The new program can practically prevent the formation of tooth cavities. All of the steps, however, must be followed, if success is to be achieved. If the average person is willing to practice the recommended procedure every day, there is a better chance that cavities can be avoided.

107. (A) The cerebellum is the seat of balance and muscular coordination. It would therefore be least involved in a conditioned reflex involving the production of saliva at the sound of a tone.

108. (B) A root hair cell is a delicate projection from the root epidermis, having very thin walls. Diffusion of water and minerals takes place through its walls from the soil. This cell is in close contact with the soil particles. Root cap cells are located at the end of a root and protect it as the root pushes its way through the soil. Its cells are constantly being worn away and replaced by new cells. The cells of the growing region of a root are constantly undergoing cell division. This region is in the root tip, just behind the root cap. At any time, various stages of mitotic division may be observed among these cells.

109. (D) The results illustrate incomplete dominance. The pure types are either round squash (RR) or long squash (LL). When they are crossed, the oval form appears (RL). By crossing only long squash, no other genes are introduced and only long squash are produced.
Key: LL – long squash

	L	L
L	LL	LL
L	LL	LL

Possible results:
100% LL – long squash

PRACTICE COLLEGE BOARD ACHIEVEMENT TEST IN BIOLOGY

NO. 8

PART I—TYPE A

Directions: **The following diagram shows a microscopic view of a drop of blood. Certain of the parts are labeled with letters. Next to each of the statements below, select the letter of the specific part most closely related to it. A label may be used once, more than once, or not at all.**

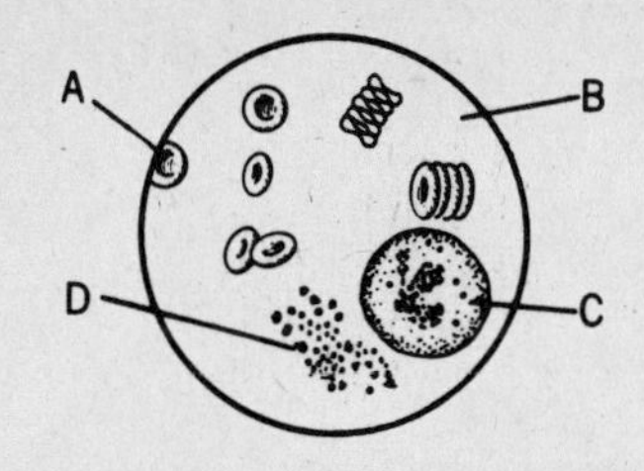

1. Carries oxygen — 1. A B C D

2. Engulfs bacteria — 2. A B C D

3. Carries endocrines — 3. A B C D

4. Requires iron for its structure — 4. A B C D

5. Makes the blood clot — 5. A B C D

6. Its numbers increase enormously in leukemia — 6. A B C D

7. Contains the protein hemoglobin — 7. A B C D

8. May be dried and used for transfusions — 8. A B C D

9. Contains dissolved fibrinogen — 9. A B C D

10. Is clumped together in the presence of the wrong blood type — 10. A B C D

11. Contains agglutinogens in Type AB blood — 11. A B C D

TYPE B

Directions: **Each of the following statements is followed by five suggested answers. Select the one answer which is best in each case.**

12. The statement that is true about the bond formed between two bonds of carbon is (A) It involves the sharing of protons (B) It involves the sharing of electrons (C) It involves the sharing of neutrons (D) It contains ADP (E) It cannot be broken — 12. A B C D E

13. Special structures for absorption in the small intestine are called (A) alveoli (B) tubules (C) mucus glands (D) villi (E) peristalsis — 13. A B C D E

14. All of the following statements about populations are true except (A) A population is made up of individuals of different species (B) Populations respond to favorable conditions by increasing in number (C) Conditions favorable for one population may be unfavorable for another (D) Individuals of a population interact with each other (D) Individuals of a population can interbreed — 14. A B C D E

15. An American animal that was once threatened with extinction, but is now increasing in numbers, is the (A) bison (B) passenger pigeon (C) dodo (D) heath hen (E) mule

15. A B C D E

16. The lichen is an example of a (A) dicot (B) monocot (C) moss (D) fungus and an alga (E) deciduous and an evergreen plant

16. A B C D E

17. The arthropod phylum contains all of the following animals except (A) butterfly (B) spider (C) snail (D) crab (E) centipede

17. A B C D E

18. The endocrine gland known as the gland of emergency is the (A) thyroid (B) parathyroid (C) pancreas (D) pituitary (E) adrenal

18. A B C D E

19. The system that produces acetylcholine is the (A) digestive (B) respiratory (C) excretory (D) reproductive (E) nervous

19. A B C D E

20. Pollen is formed in the (A) ovule (B) pistil (C) aster (D) anther (E) calyx

20. A B C D E

21. The scientist who gave us the binomial system of classification was (A) Metchnikoff (B) Linnaeus (C) Mendel (D) Mendeleeff (E) Ehrlich

21. A B C D E

22. The adult frog is different from the tadpole in that it has (A) a two-chambered heart (B) tail (C) lungs (D) internal gills (E) external gills

22. A B C D E

23. Cross-pollination is effective in all of the following plants except (A) wheat (B) apple (C) pear (D) corn (E) fern

23. A B C D E

24. Color-blindness in humans is a (A) sex-linked trait (B) dominant trait (C) blended trait (D) hybrid trait (E) pure trait

24. A B C D E

25. Which factor best explains the various species of finches found on the Galapagos Islands? (A) competition for one type of food (B) isolation (C) interbreeding (D) climate on the island (E) resistance to natural enemies

25. A B C D E

26. In sickle-cell anemia, the abnormal hemoglobin differs from normal hemoglobin by (A) A single amino acid (B) the amount of coenzymes present (C) the number of iron atoms (D) the number of magnesium atoms (E) the number of genes present

26. A B C D E

27. Which sequence is correctly arranged in order of decreasing average temperature? (A) desert, grassland, tundra, taiga (B) tropical forest, deciduous forest, tundra, taiga (C) deciduous forest, tropical forest, taiga, tundra (D) tropical forest, grassland, taiga, tundra (E) desert, tropical forest, tundra, taiga

27. A B C D E

TYPE C

Directions: **For each of the processes listed from 28–36, select the letter (A), (B), (C), or (D), depending on whether it**

(A) applies only to A
(B) applies only to B
(C) applies both to A and B
(D) applies to neither A nor B

(A) Paramecium	(C) Both
(B) Human	(D) Neither

28. Ingestion — 28. A B C D

29. Fission — 29. A B C D

30. Photosynthesis — 30. A B C D

31. Mitosis — 31. A B C D

32. Irritability — 32. A B C D

33. Conjugation — 33. A B C D

34. Fertilization — 34. A B C D

35. Transpiration — 35. A B C D

36. Meiosis — 36. A B C D

TYPE D

Directions: **The following questions require two answers. One of the items in the left column (A), (B), or (C) is related to four of the items in the right column (numbered 1–5). Select the letter of the item which is so related; then select the number of the remaining item which does not belong.**

EXAMPLE:

(A) paramecium	1. cilia	A B C 1 2 3 4 5
(B) ameba	2. macronucleus	
(C) spirogyra	3. trichocyst	
	4. pyrenoid	
	5. micronucleus	

37. (A) absorption
(B) photosynthesis
(C) food storage

1. nitrogen-fixing bacteria
2. stomate
3. spongy layer
4. guard cells
5. palisade layer

37. A B C 1 2 3 4 5

38. (A) small intestine
(B) spleen
(C) mouth

1. bile
2. villi
3. pancreatic juice
4. gastric juice
5. absorption

38. A B C 1 2 3 4 5

39. (A) hormone
(B) antibiotic
(C) vitamin

1. carotene
2. niacin
3. ascorbic acid
4. aureomycin
5. thiamin

39. A B C 1 2 3 4 5

40. (A) gonads
(B) goiter
(C) ganglia

1. synapse
2. sperm
3. ovary
4. ovum
5. testis

40. A B C 1 2 3 4 5

41. (A) blending inheritance
(B) plant breeding
(C) linkage

1. selection
2. salivary chromosome
3. hybridization
4. using mutations
5. grafting

41. A B C 1 2 3 4 5

TYPE E

Directions: **The five lettered headings below are followed by a list of numbered statements. For each numbered statement, choose the one lettered heading (A) to (E) which is most closely related to it.**

(A) Variation
(B) Overproduction
(C) Survival of the fittest
(D) Vestigial structure
(E) Mutation

42. "As alike as two peas in a pod" is not a true statement

42. A B C D E

43. Organisms with new traits suddenly appear

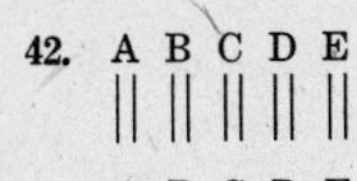
43. A B C D E

44. The skeleton of a whale contains hip bones

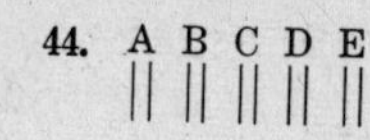

45. There are two splint bones on the foot of the horse

45. A B C D E

46. Our eye contains a "third eyelid"

46. A B C D E

(A) Peristalsis (C) Absorption
(B) Digestion (D) Circulation
(E) Emulsification

47. How does dissolved food enter the bloodstream?

47. A B C D E

48. How is food moved along the alimentary tract?

48. A B C D E

49. How does dissolved food reach all parts of the body?

49. A B C D E

50. How are fat particles made smaller in size?

50. A B C D E

51. How is food made soluble?

51. A B C D E

PART II

In a demonstration of trial-and-error learning, a pupil was given a puzzle consisting of four pieces of cardboard. He was told to arrange them in the form of a letter L. At a signal from another pupil who acted as timekeeper, he worked at the puzzle until the pieces fitted together, as shown in the accompanying diagram.

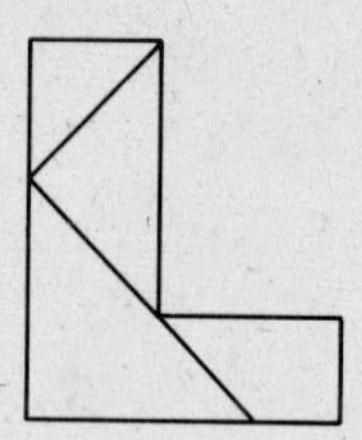

As soon as he finished, the time was recorded. The pieces were then scrambled and he attempted to put them together again. This was done for a total of ten times. Graph A was then constructed

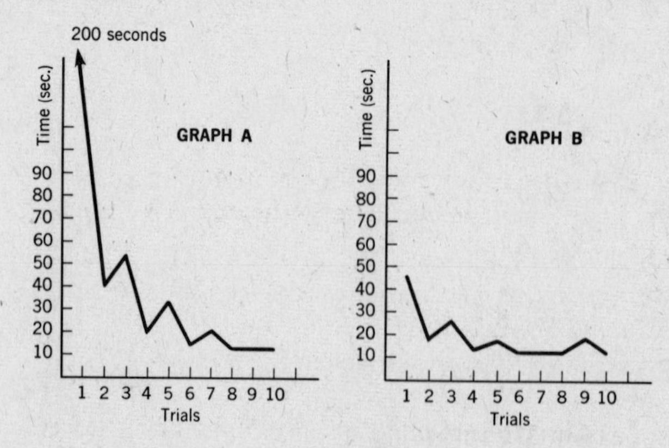

52. The greatest improvement was shown (A) in the first trial (B) in the second trial (C) in the fourth trial (D) in the sixth trial (E) in the tenth trial

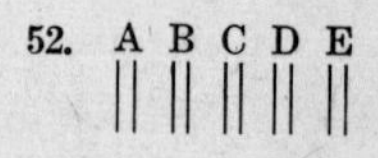

53. The least amount of improvement was shown (A) in the third trial (B) in the fifth trial (C) in the seventh trial (D) in the eighth trial (E) in the ninth trial

53. A B C D E

54. The peaks and valleys in the graph indicate that (A) learning is a rapidly progressive process (B) learning is a gradually progressive process (C) learning takes place all at one time (D) the student lost his confidence after the second trial (E) the student's concentration on the problem was uneven

54. A B C D E

55. The straight line after the eighth trial indicates all of the following except (A) he was putting the pieces together as quickly as possible (B) he could not improve any further (C) he had learned the puzzle perfectly (D) he gave up and made no more efforts to learn the puzzle (E) he made no more errors in solving it

55. A B C D E

56. The timekeeper, who had been watching the pupil, then attempted to put the puzzle together. According to Graph B, (A) he did it perfectly the first time (B) he learned the puzzle in less time than the first student (C) the fifth trial took less than the fourth (D) he did the puzzle perfectly after the sixth trial (E) he did not time himself accurately

56. A B C D E

57. If another student who had been watching were now to do the puzzle (A) his time for doing the puzzle perfectly would be better than the timekeeper's (B) he would need as much time as the timekeeper to do it perfectly (C) he would need less time for the first trial (D) after he had learned to do the puzzle, he would make no more errors (E) his results could not be predicted accurately

57. A B C D E

PART III—TYPE A

***Directions:* Each item consists of an *assertion* (statement) in the left-hand column, and a *reason* on the right-hand side. Select**

(A) if both assertion and reason are true, and *are related* as cause and effect
(B) if both assertion and reason are true, but *are not related* as cause and effect
(C) if the assertion is true, but the reason is a false statement
(D) if the assertion is false, but the reason is a true statement
(E) if both assertion and reason are false statements

DIRECTIONS SUMMARIZED:

	(A)	(B)	(C)	(D)	(E)
Assertion:	True	True	True	False	False
Reason:	True	True	False	True	False
	Cause and effect	*Not* cause and effect			

Assertion:	Reason:
58. The sperm determines the sex of a new individual	BECAUSE it carries either the X or Y chromosome
59. Pavlov is famous for his research on conditioned reflexes in dogs	BECAUSE he showed that reflexes can be changed
60. A beech-maple forest is a climax community	BECAUSE it is eventually replaced by a tall oak and pine tree forest
61. Corn is considered an American crop	BECAUSE it grows so well in this country
62. Fossils are always found in metamorphic rock	BECAUSE it is formed from sedimentary rock
63. Members of the Caucasoid stock always have the same blood group	BECAUSE all blood is either A, B, AB, or O

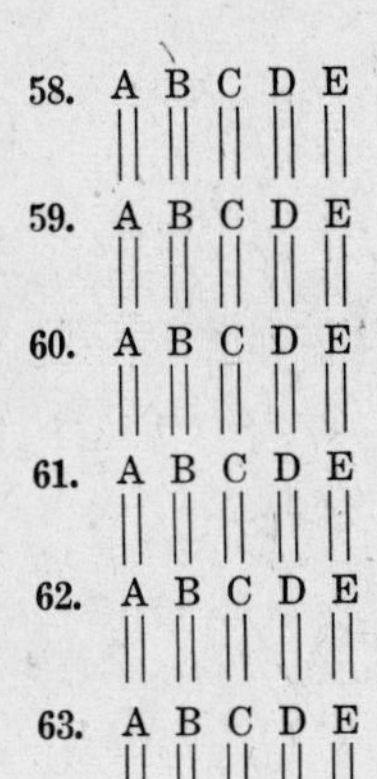

64. Dinosaurs became extinct shortly after primitive man appeared BECAUSE he sought out their breeding places and destroyed their eggs

64. A B C D E

65. The whale is considered to be a mammal and not a fish BECAUSE it has a backbone

65. A B C D E

66. The autonomic nervous system is not under our direct voluntary control BECAUSE it controls the internal organs

66. A B C D E

67. As the result of an underproduction of insulin, a child becomes a cretin BECAUSE ACTH controls the production of insulin

67. A B C D E

TYPE B

Directions: **Each of the following statements is followed by five suggested answers. Select the one answer which is best in each case.**

68. In hot weather, a person sweats a good deal (A) because he drinks more water (B) in order to cool off (C) to keep his pores open (D) because he loses salt (E) because his sweat glands expand

68. A B C D E

69. A person who is exercising breathes more rapidly because (A) his lungs use up more energy (B) his calorie production goes down (C) he gives off more carbon dioxide which stimulates the medulla (D) glycogen in the liver is converted to glucose (E) glycogen in the muscles is converted to glucose

69. A B C D E

70. A green plant grows toward the light because (A) light is needed for photosynthesis (B) this response is an instinct (C) otherwise it could not carry on carbohydrate synthesis (D) auxin collects on the illuminated side (E) auxin collects on the shaded side

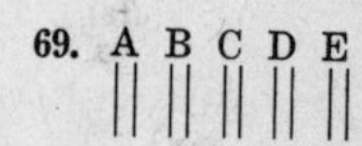

70. A B C D E

71. A blow on the head may cause loss of memory because (A) all reflexes are centered in the brain (B) the cerebrum is the center of memory (C) the cerebellum is the center of memory (D) the medulla is connected with the autonomic nervous system (E) the brain controls all involuntary activities

71. A B C D E

72. DDT is not as effective against certain insects as it used to be because (A) they now hide on the undersurface of leaves when it is sprayed (B) some of them stop breathing while the DDT is in the air (C) the noise of the airplane used in spraying scares them away (D) they close their spiracles until the DDT settles (E) resistant strains have arisen as mutations

72. A B C D E

73. A goldfish cannot live in water that has been boiled because (A) its mineral content has been reduced (B) it contains little oxygen (C) its symbiotic bacteria have been killed (D) its atmosphere pressure has been altered (E) the gills of the fish become coated with tiny air bubbles

73. A B C D E

74. Bread that is kept in a damp place may become covered with mold because (A) the carbohydrates of the bread become sublimated (B) dust forms cobwebs in dampness (C) mold spores germinate in dampness (D) the yeast in the bread begins to reproduce (E) the bacteria in the bread undergo binary fission

74. A B C D E

75. The gene pool of a population tends to remain stable if (A) mating is at random (B) there is extensive migration (C) the populations involved are small (D) there are frequent mutations (E) there are selected matings

75. A B C D E

76. Most marine fish cannot survive in fresh water because of their (A) inability to maintain homeostatic water balance (B) the lower oxygen concentration in fresh water (C) the higher carbon dioxide concentration in fresh water (D) the lack of food plants in fresh water (E) the great amount of pollution in fresh water

76. A B C D E

77. Multiple births may occur in humans when (A) there is an excess of sperm (B) multiple ovulation occurs (C) the corpus luteum appears (D) the pituitary oversecretes (E) menstruation occurs

77. A B C D E

TYPE C

Directions: **In each of the following questions, there is a statement followed by four choices, of which *one or more* are correct. Indicate the correct combination by choosing the following letters:**

(A) if only 1, 2 and 3 are correct
(B) if only 1 and 3 are correct
(C) if only 2 and 4 are correct
(D) if only 4 is correct
(E) if some other combination is correct

DIRECTIONS SUMMARIZED:

(A)	(B)	(C)	(D)	(E)
1, 2, 3 only	1, 3 only	2, 4 only	4 only	some other combination

78. A breeder ties a paper bag over a flower to
1. keep stray wind-blown pollen away
2. protect the flower from injury while it is developing
3. prevent insects from cross-pollinating it
4. identify a desirable breed

78. A B C D E

79. The discovery of DNA
1. explains the mechanism of mitosis
2. disproves the usefulness of the spindle fibers
3. indicates the values of centrosomes
4. is useful in studies of the gene

79. A B C D E

80. The fossil remains of *Eohippus*
1. show a relationship between *Archaeopteryx* and mammals
2. indicate that the horse once had four toes
3. explain the disappearance of the dinosaurs
4. prove that the ancestor of the horse was about the size of a cat

80. A B C D E

81. The presence of vestigial structures
1. proves that man descended from monkeys
2. supports the Lamarckian theory of evolution
3. was used by Darwin to explain struggle for existence
4. indicates descent with modification from a distant ancestor

81. A B C D E

82. Radiation sickness following exposure to radiation
1. is always fatal
2. can always be cured
3. is treated by contact with a Geiger counter
4. results in destruction of red blood cells

82. A B C D E

83. A man in a space capsule
1. experiences an increase of gravitational force during acceleration
2. requires less food because of his weightless condition
3. may black out if he is subjected to a force of too many G's
4. is free of radiation as soon as he passes beyond the earth's atmosphere

83. A B C D E

84. During development, the mammalian embryo
1. develops in the uterus
2. is nourished through the placenta
3. depends on the parent bloodstream for food and oxygen
4. undergoes spermatogenesis

84. A B C D E

85. Insect carriers of disease germs include the
1. housefly for typhus fever
2. Anopheles mosquito for malaria
3. Aedes mosquito for typhoid fever
4. flea for bubonic plague

85. A B C D E

PART IV—TYPE A

Directions: **Referring to the following selection, choose the best answer for each of the questions that follows:**

*"Photosynthesis and vision do not exhaust the potential of the luminous environment. Both plants and animals have evolved mechanisms to respond to the changing daily cycle of light and dark. It is this photoperiodism that provides the seasonal schedule for, among other things, the flowering of plants, the pupation of insects and the nesting of birds.... In all three processes light acts through absorption of a small, colored molecule—a chromophore—that is associated with a large molecule of protein.

"Chrysanthemums and many other plants flower in response to the increasing length of nights as fall approaches. If the long nights are experimentally interrupted by exposing the plants to short periods of light near midnight, the plants will not flower. Red light with an absorption maximum at a wavelength of 660 nanometers is most effective in preventing flower formation. Thus we anticipate that the light-receiving pigment in the plant is blue—the complementary color to the absorbed red. If shortly after exposure to red light the plants are exposed to light near the limit of vision in the far red (730 nanometers), they will flower."

86. Photoperiodism (A) takes place during photosynthesis (B) takes place during vision (C) is the result of photosynthesis (D) takes place only during the dark (E) is the response to the daily cycle of light and dark

86. A B C D E

87. Photoperiodism applies to all of the following except (A) the pupation of insects (B) the seasonal schedule of living things (C) photosynthesis (D) the nesting of birds (E) the flowering of plants

87. A B C D E

88. A chromophore is present (A) in photosynthesis, vision and photoperiodism (B) only in photosynthesis and vision (C) only in vision and photoperiodism (D) only in photosynthesis and photoperiodism (E) in none of the processes named

88. A B C D E

89. All of the following statements about the flowering of chrysanthemums are false except (A) they flower as the nights grow shorter (B) they flower if exposed to short periods of light near midnight (C) red light is most effective in preventing flowering (D) the light-receiving pigment in the plant is red (E) the wavelength of 660 nanometers is most effective for flower formation

89. A B C D E

TYPE B

Directions: **Each of the following problems is followed by five suggested answers. Select the one answer which is best in each case.**

90. If we know the carbohydrate, fat, and protein content of bread, we can estimate the total available calories contained in it. These nutrients yield the following number of calories:

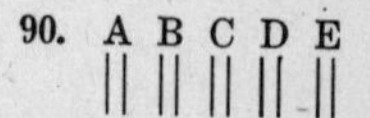

NUTRIENT	CALORIES
1 gram of carbohydrate	4
1 gram of fat	9
1 gram of protein	4

A sample of bread was analyzed to contain 53.3 grams of carbohydrate, 1.6 grams of fat, and 9.1 grams of protein. The total number of calories available in the bread were (A) 100.0 (B) 53.03 (C) 264.0 (D) 16.0 (E) 213.2

91. In the accompanying pedigree of a family, all the individuals of the first and second generations are brown-eyed. However, the eye color of John is not

*From "How Light Interacts with Living Matter" by S. B. Hendricks, Sept. 1968, by *Scientific American, Inc.* All rights reserved.

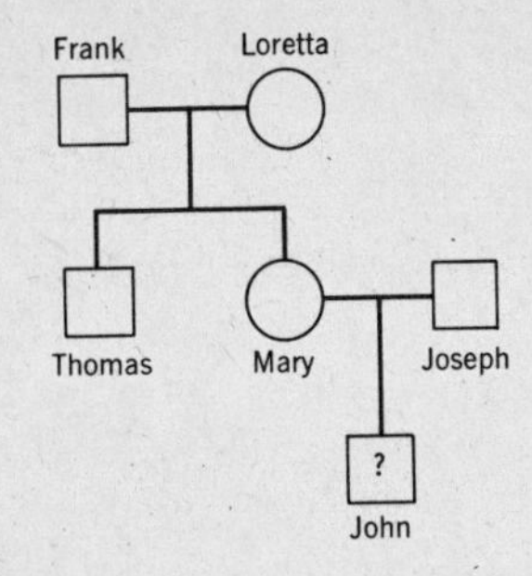

given. From this information, it can be determined that (A) John must be brown-eyed (B) John could be blue-eyed (C) if John were brown-eyed, Mary and Joseph would have to be homozygous for brown eyes (D) if John were blue-eyed, Mary would have to be heterozygous for brown eyes, and Joseph would have to be homozygous for brown eyes (E) if John were blue-eyed, Frank and Loretta would have to be homozygous for brown eyes

91. A B C D E

92. After incubation overnight, a Petri dish that had been exposed to the air for five minutes shows 8 large colonies and 3 small colonies of bacteria. The minimum number of bacteria that entered the Petri dish from the air was most likely (A) 3 (B) 8 (C) 11 (D) 33 (E) 88

92. A B C D E

93. A man who has hemophilia marries a normal homozygous woman. The chances of their daughters being normal carriers are (A) 25–75% (B) 25–50–25% (C) 50–50% (D) 100% (E) zero

93. A B C D E

94. It has been determined that the volume of a cell varies as the cube of its diameter and the surface as the square. As a cell grows in size, therefore the ratio of surface area to volume (A) increases (B) decreases (C) remains the same (D) depends on mitosis (E) is determined by its metabolic rate

94. A B C D E

95. The daily Calorie requirements of a young man 18 years old have been estimated to be 23–25 calories per pound. At this age, the average weight for someone 5′11″ tall is 154 pounds. His total requirements of Calories is about (A) 2,950–3,250 (B) 3,250–3,550 (C) 3,550–3,850 (D) 3,850–4,150 (E) 4,150–4,450

95. A B C D E

96. Inability to roll the tongue is a recessive trait. If 36 per cent of the people in a population cannot roll their tongues what is the percentage of recessive genes in the gene pool: (A) 3% (B) 6% (C) 36% (D) 60% (E) 72%

96. A B C D E

TYPE C

One way of measuring the rate at which plants consume oxygen is to use the following apparatus in which air is admitted in accurately measured amounts. All connections are airtight. There are three hoses leading into the container. One hose (1) is attached to a U-shaped glass tube partially filled with colored water; it serves as a manometer and measures the difference in pressure between the container and the outside. The second hose (2) ends in a hypodermic syringe,

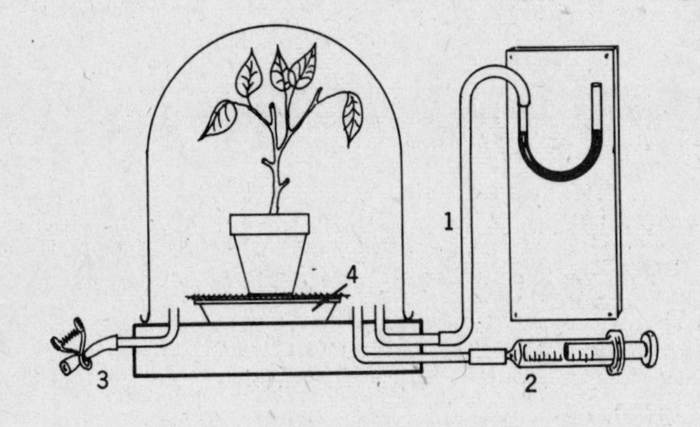

by which air is admitted into the container by measured amounts. The third hose (3) serves as a vent; it is opened until the columns of colored water in the manometer are at the same height, and then it is closed with a pinchcock. The plant is enclosed in a large opaque container. It is placed over a container of calcium chloride (4), which absorbs carbon dioxide.

After an interval of time, the plant will take up enough oxygen to cause a drop of pressure in the manometer. The piston of the syringe is pushed in until the air pressure is balanced, as indicated by the equal levels of the liquid in the manometer. A record is kept of the elapsed time and the volume of air used. The rate of consumption is computed by dividing the volume of air admitted from the syringe by the elapsed time in minutes.

97. When the plant takes in oxygen (A) the level of the right side of the manometer tube is higher (B) the level of the right side of the manometer tube is lower (C) the level of the left side of the manometer is lower (D) the level of the left side of the manometer is unchanged (E) the level of both sides of the manometer is equal

97. A B C D E

98. Since the plant is enclosed in an opaque container, an activity that is not carried on is (A) respiration (B) transpiration (C) photosynthesis (D) assimilation (E) growth

98. A B C D E

99. Once the experiment has started, the only hose(s) in use is (are) (A) 1 (B) 1 and 2 (C) 1, 2, and 3 (D) 2 (E) 2 and 3

99. A B C D E

100. If the plant were enclosed by a clear glass container, the amount of oxygen contained within it would (A) increase (B) decrease (C) remain the same (D) serve as an index of the respiratory quotient (E) keep both levels of the manometer equal

100. A B C D E

101. The purpose of using calcium chloride is (A) to serve as a control (B) to prevent photosynthesis from taking place (C) to speed up the rate of respiration (D) to eliminate the effects of oxygen on the manometer (E) to eliminate the effects of carbon dioxide on the manometer

101. A B C D E

102. In setting up and conducting the beginning of the experiment, the correct order of using the numbered parts is (A) 1–2–3–4 (B) 2–3–4–1 (C) 4–3–1–2 (D) 4–3–2–1 (E) 4–3–1–2

102. A B C D E

103. If a plant is consuming oxygen rapidly (A) the piston of the syringe will have to be pushed in a small distance over a long period of time (B) the piston of the syringe will have to be pushed in a small distance over a short period of time (C) the piston of the syringe will have to be pushed in a long distance over a long period of time (D) the piston of the syringe will have to be pushed in a long distance over a short period of time (E) the piston of the syringe will have to be pushed in a long distance while the pinchcock is open

103. A B C D E

104. If the experiment extends over several weeks, it will be necessary to take into account (A) the rate of water intake (B) the rate of water excretion (C) both temperature and barometric pressure (D) the intensity of light (E) the effect of natural plant hormones

104. A B C D E

PART V

Directions: **Each of the following situations is followed by five suggested answers. Select the one answer that is best in each case.**

105. Although corn is one of the main grain crops of the world, it is essentially a poor source of protein. The protein of corn is low in quantity, most of it being zein, which cannot be digested by man. Corn's most serious nutritional deficiency is its low content of lysine, one of the essential amino acids for man. Plant breeders in Colombia, South America recently backcrossed a mutant type of corn having a high-lysine content with a local strain of corn to produce a new variety of corn with the high-lysine factor. When this was

105. A B C D E

fed to pigs, they gained 73.2 pounds in 130 days, as compared with the controls living on ordinary corn, which gained only 6.6 pounds. Malnourished children showed improvement in health and weight when fed high-lysine corn as the sole source of protein. Scientists are now trying to increase the yield, as well as the protein content of the high-lysine corn. They believe that (A) corn can be developed with a higher yield of zein, to overcome protein shortages in many parts of the world (B) by feeding pigs on high-lysine corn, they can reserve selected corn with a high zein content solely for human use (C) corn can be used in the future to meet the protein needs of people in many parts of the world (D) coffee produced in Colombia for sale throughout the world can be improved in its amino acid content of zein (E) they should avoid inbreeding high-lysine corn because of the danger of lowering the production of zein in future generations

106. There are three chief ways of controlling virus diseases in man. The major method, up to now, has been to prepare a vaccine of killed or weakened virus that stimulates the body to form antibodies against the virus; such a vaccine is specific and is useful only against one particular virus. Chemotherapy is another potential method, but no safe chemical agents have yet been found that act against viruses the way the sulfa drugs or antibiotics act against large groups of bacteria. The third possibility is to rely on the cell's own defense against virus attack, through the production of the recently discovered substance, interferon, that also protects the body against other types of virus infections. Many types of research in protection against viruses are now taking place with the most promising being to (A) discover a vaccine that is effective against interferon (B) discover a sulfa drug that inhibits the action of bacteria against viruses (C) discover an antibiotic that inhibits the action of bacteria against viruses (D) find a method of artificially stimulating the production of interferon by the cells (E) try interferon on a number of diseases caused by bacteria

106. A B C D E

PART VI

Directions: **Each of the following questions has five suggested answers. Select the one answer that is best in each case.**

107. The following selection is taken from the writing of an early scientist: "A sprig of a nettle plant was put in a jar full of air fouled by breathing so as to extinguish a candle; it was placed in a room during the whole night; the next morning the air was found to be as bad as before. At 9 o'clock in the morning the jar was put in the sunshine and in the space of two hours the air was so much corrected that it was found to be nearly as good as common air." The best explanation of his results is: (A) The "jar full of air fouled by breathing" probably contained an excess of the gas carbon monoxide. (B) The fact that the "air was found to be as bad as before" was due to the process of photosynthesis taking place in the plant. (C) The plant performed the process of photosynthesis to produce air nearly as good as "common air." (D) The gas produced by the plant in the process which improved the air in the jar is called carbon dioxide. (E) The gas which was produced by the plant in the dark is called oxygen.

107. A B C D E

108. A pupil performed the following experiment: He tested a dry piece of white bread with iodine solution; the bread turned blue-black. Then he chewed another piece of white bread and tested it with Benedict's solution; the chewed bread turned red. He therefore concluded that, when a piece of bread is chewed, starch is changed into sugar. One error in the pupil's procedure was that he did not (A) test the chewed piece of bread for

108. A B C D E

starch (B) test saliva for starch (C) use nitric acid (D) test the dry piece of bread for sugar (E) consider the age of the bread

109. It has been found quite conclusively that treatment of sperm with relatively large doses of X-rays induces the occurrence of "true" gene mutations in a high proportion of the treated germ cells. Several hundred mutants have been obtained in this way in a short time. They are nearly all stable in their inheritance, and most of them behave in the manner typical of the Mendelian inheritance in organisms. The best organism for conducting this investigation is (A) dogs (B) pea plants (C) white rats (D) Drosophila flies (E) white mice

109. A B C D E

ANSWER KEY: TEST NO. 8

1. (A)	**12.** (B)	**23.** (E)	**34.** (B)	**45.** (D)	**56.** (B)	**67.** (E)	**78.** (B)	**89.** (C)	**100.** (A)
2. (C)	**13.** (D)	**24.** (A)	**35.** (D)	**46.** (D)	**57.** (E)	**68.** (B)	**79.** (D)	**90.** (C)	**101.** (E)
3. (B)	**14.** (A)	**25.** (B)	**36.** (B)	**47.** (C)	**58.** (A)	**69.** (C)	**80.** (C)	**91.** (B)	**102.** (E)
4. (A)	**15.** (A)	**26.** (A)	**37.** (B)1	**48.** (A)	**59.** (A)	**70.** (E)	**81.** (D)	**92.** (C)	**103.** (D)
5. (D)	**16.** (D)	**27.** (D)	**38.** (A)4	**49.** (D)	**60.** (C)	**71.** (B)	**82.** (D)	**93.** (D)	**104.** (C)
6. (C)	**17.** (C)	**28.** (C)	**39.** (C)4	**50.** (E)	**61.** (B)	**72.** (E)	**83.** (B)	**94.** (B)	**105.** (C)
7. (A)	**18.** (E)	**29.** (A)	**40.** (A)1	**51.** (B)	**62.** (D)	**73.** (B)	**84.** (A)	**95.** (C)	**106.** (D)
8. (B)	**19.** (E)	**30.** (D)	**41.** (B)2	**52.** (B)	**63.** (D)	**74.** (C)	**85.** (C)	**96.** (D)	**107.** (C)
9. (B)	**20.** (D)	**31.** (C)	**42.** (A)	**53.** (E)	**64.** (E)	**75.** (A)	**86.** (E)	**97.** (B)	**108.** (D)
10. (A)	**21.** (B)	**32.** (C)	**43.** (E)	**54.** (B)	**65.** (B)	**76.** (A)	**87.** (C)	**98.** (C)	**109.** (D)
11. (A)	**22.** (C)	**33.** (A)	**44.** (D)	**55.** (D)	**66.** (B)	**77.** (B)	**88.** (A)	**99.** (B)	

ANSWERS EXPLAINED: TEST NO. 8

1. (A) red blood corpuscle
2. (C) white blood corpuscle
3. (B) plasma
4. (A) red blood corpuscle
5. (D) platelets
6. (C) white blood corpuscle
7. (A) red blood corpuscle
8. (B) plasma
9. (B) plasma
10. (A) red blood corpuscle
11. (A) red blood corpuscle

12. (B) Atoms of carbon are held together in chemical bonds (C—C). The atoms share electrons, which bond them together. This type of linkage is called a covalent bond, as contrasted with the ionic bond in which electrons are transferred from one atom to another.

13. (D) Villi are the tiny finger-like projections that line the inside of the small intestines. Their walls are one-celled thick, permitting the diffusion of nutrients through them into the capillaries and lacteals.

14. (A) A population is defined as the members of a species that breed together and that live together in an area. Examples: leopard frogs in a pond; dandelions in a lawn.

15. (A) The bison was almost wiped out because of the spread of civilization and unrestricted hunting. It is now protected by law.

16. (D) The lichen is a good example of mutualism; the alga makes food by photosynthesis, while the fungus furnishes it with water and minerals.

17. (C) All of these animals have jointed legs and an exoskeleton except the snail, which is a mollusk.

18. (E) The adrenal gland produces the hormone adrenin which stimulates the liver to convert stored glycogen to glucose; this is circulated quickly to the muscles by the stronger beat of the heart, supplying them with a greater source of energy.

19. (E) This chemical substance is produced along the length of the neurons, and at the synapses.

20. (D) The anther is the pollen case of the stamen, the male reproductive organ of flowers.

21. (B) Linnaeus classified organisms according to their structural characteristics, and gave them two names, a genus and a species name, i.e., *Homo sapiens*, man.

22. (C) As a tadpole goes through metamorphosis, it replaces its gills with lungs, and breathes air.

23. (E) All of these plants form flowers and are examples of spermatophytes except the fern, which is a member of the filicineae and reproduces by spores.

24. (A) The gene for color-blindness is located on the X chromosomes which help determine the sex of an individual.

25. (B) As a result of geographic separation and isolation, the finches on the various Galapagos Islands were prevented from interbreeding. Changes that took place in one island were not transmitted to another island. Consequently, over a long period of time, the genetic differences became accentuated to make the various finches different from one another.

26. (A) In sickle-cell anemia, the red blood cells take on a sickle shape when the oxygen supply is low. The hemoglobin molecule in such cells differs from normal hemoglobin by only a single amino acid out of a total of over 300. The sickle type of hemoglobin has the amino acid valine instead of the normal amino acid known as glutamic acid. It is thought that this change occurred as a mutation, changing one base in the genetic code for glutamic acid (UAG) into the code (UUG) for valine.

27. (D) A tropical forest occurs in regions of high temperatures and ample rainfall. Grassland has considerably less rainfall and may be marked by drought. Taiga includes the northernmost forests. Tundra is the treeless region in the far north.

28. (C) All animals ingest, or take in, food.

29. (A) One-celled animals such as paramecium reproduce asexually by dividing in half.

30. (D) Only green plants carry on photosynthesis.

31. (C) All living things make more protoplasm from nutrients.

32. (C) The protoplasm of living things has the ability to respond to the environment.

33. (A) Paramecium may reproduce sexually by conjugation, during which two cells come to lie next to each other and exchange parts of their micronuclei.

34. (B) Fertilization occurs when two unlike gametes, the sperm and the ovum, unite.

35. (D) Transpiration is carried on by plants which give off water vapor through their leaves.

36. (B) The gametes formed during meiosis have the monoploid number of chromosomes.

37. (B)1 These are all parts of a leaf concerned with the process of photosynthesis except nitrogen-fixing bacteria, which live in the nodules of the roots of legumes, and convert nitrogen into nitrates.

38. (A)4 These are all concerned with the activity of the small intestine except gastric juice, which is produced by the stomach.

39. (C)4 They are all vitamins except aureomycin, which is an antibiotic.

40. (A)1 They are all concerned with the reproductive organs, or gonads, except synapse, which is an area in the nervous system where an impulse crosses over from one neuron to another.

41. (B)2 All these methods are used in plant breeding except salivary chromosome, which is a giant chromosome found in the salivary gland of Drosophila larvae.

42. (A) All organisms vary; this may be due to environment, or heredity, or a combination of both.

43. (E) A gene change will produce a new trait.

44. (D) This is evidence that an ancestor of the whale had hind legs. Through the years, these vanished, but there are still some genes present for the formation of hip bones.

45. (D) The ancestor of the horse used these splints when they were toes. With the passage of time, they grew smaller until all that remains are these small bones.

46. (D) A distant vertebrate ancestor had a functional third eyelid. Among the other changes that took place since that time, this became very much smaller, and remains as a vestige in the inside corner of the eye.

47. (C) Dissolved nutrients are absorbed into the capillaries and lacteals through the thin wall of the villi of the small intestine.

48. (A) The wavelike contractions and expansions of the alimentary canal squeeze the food along.

49. (D) After it has been absorbed into the bloodstream, dissolved food is transported throughout the body in the circulatory system.

50. (E) Bile acts on fat particles in the small

intestine, breaking them down to smaller size, or emulsifying them.

51. (B) Food is made soluble through two phases of digestion: mechanical, in which it is broken down to smaller size; chemical, in which enzymes convert it to soluble end-products.

52. (B) The time decreased from 200 seconds to 40 seconds, an improvement of 160 seconds.

53. (E) The time for the ninth trial was 11 seconds, the same as the time for the eighth trial.

54. (B) Learning does not take place all at once. There are periods of progress followed by slight setbacks. As learning continues, it becomes more and more perfect.

55. (D) After a number of trials and errors, an individual can learn to solve a problem perfectly.

56. (B) He took less time for the first trial, and did it perfectly at the sixth trial.

57. (E) It would be necessary to try it. It is not correct to draw conclusions from only one or two cases.

58. (A) If a sperm carrying an X chromosome fertilized an egg, the offspring will have XX chromosomes, and will be female; if it carries a Y chromosome, the offspring will have XY chromosomes and will be male.

59. (A) He rang a bell each time there was a salivary reflex; i.e., a dog was fed, and there was a flow of saliva. He changed this reflex by having saliva flow only at the sound of the bell.

60. (C) It replaced an oak and pine forest and will remain dominant as long as the conditions are unchanged.

61. (B) It originated in the Americas and was exported to other parts of the world.

62. (D) They become part of sedimentary rock when the sediment with which they are covered is slowly changed into rock. When sedimentary rock under pressure becomes formed into metamorphic rock, any fossils in it generally become distorted and destroyed.

63. (D) They may have any of the four blood types.

64. (E) They died out long before man appeared on the scene about a million years ago. They lived during the Mesozoic Era which began about 185 million years ago and ended about 60 million years ago.

65. (B) It gives birth to its young alive; they develop in its uterus; it feeds its young on milk.

66. (B) It consists of a chain of ganglia outside of the central nervous system, and controls such activities as heartbeat, and secretion.

67. (E) Underproduction of thyroxin causes a child to become a cretin. ACTH is produced by the pituitary gland and stimulates the production of cortisone by the adrenal cortex.

68. (B) The evaporation of sweat has a cooling effect on the body. In hot weather, the blood vessels in the skin are dilated, permitting more blood to flow by the sweat glands, and the excretion of more sweat.

69. (C) The medulla controls the rate at which the diaphragm and chest wall muscles contract and expand in breathing. When the concentration of carbon dioxide increases in the blood passing through the medula, it stimulates the breathing rate.

70. (E) There is a greater concentration of auxin on the shaded side of the stem. This causes the cells on that side to elongate, making the stem bend toward the light.

71. (B) The part of the cerebrum that controls memory is highly developed in man. Specific areas of the cerebrum have been mapped that are the centers of its other activities, including vision, hearing, speech, voluntary movement, etc.

72. (E) Insects that have arisen as resistant mutations have survived; the nonresistant types were killed by the DDT. After a while, only the resistant type survived.

73. (B) When water is boiled, it loses its content of dissolved oxygen.

74. (C) Mold spores are present in dust and in the air. Under favorable conditions of moisture, they germinate into a mold.

75. (A) The gene pool consists of the sum total of all the genes for all the traits in a given population. According to the Hardy-Weinberg principle, the gene pool remains stable from generation to generation if mating is at random and no new factors are introduced, such as mutation, selection or migration. Under such theoretical conditions, the species would remain practically the same.

76. (A) Homeostasis refers to the ability of an organism to maintain a stable, internal environment. Although there may be changes in the external environment, and differences in the concentrations of various substances, the semi-permeable membrane controls the passage of water and salts into or out of the cell. Marine fish are not able to survive in fresh water because they are not able to maintain homeostatic water balance. They are adapted to a water environment in which there is a relatively high concentration of salts.

77. (B) When ovulation in humans takes place, one egg ordinarily leaves the ovary and enters the oviduct. If it is fertilized there, an embryo will develop. If more than one egg is produced by the ovary, multiple ovulation takes place. The eggs enter the oviduct at about the same time. If fertilization takes place, more than one embryo will form and multiple births will occur.

78. (B) During cross-pollination, a breeder transfers pollen from one flower to another. The stamens are first removed from this latter flower, and then it is covered with a paper bag to keep it from being pollinated by any other pollen.

79. (D) DNA, or deoxyribonucleic acid, has been identified as the complex nucleic acid occurring in the chromosome which makes up the genes.

80. (C) Eohippus, which lived about 60 million years ago, has been linked with the horse of today through a series of intermediate forms which showed gradual changes up to the present day.

81. (D) These structures were used in the distant ancestor, but through changes, they became useless; there are still genes present for them, and so they continue to appear.

82. (D) This is one of the effects of radiation sickness; others are—skin eruptions, nausea, diarrhea, bleeding, fever, and general weakness.

83. (B) The sudden increase in speed from a stationary position to about 17,500 miles an hour within a few minutes, subjects the astronaut to considerable gravitational stress. If the blood is prevented from reaching the brain, he blacks out, or becomes unconscious.

84. (A) The mammalian ovum has little or no food. It is completely dependent on the mother's bloodstream for receiving its food and oxygen, and for getting rid of its wastes.

85. (C) The housefly may carry the germs of typhoid fever on its feet. The Aedes mosquito may carry the virus of yellow fever.

86. (E) Both plants and animals have evolved mechanisms to respond to the changing daily cycle of light and dark. This is known as photoperiodism.

87. (C) Photosynthesis takes place in the light. Photoperiodism refers to the effect of the changing daily cycle of light and dark.

88. (A) In all the three processes named, light acts through absorption of a small colored molecule known as a chromophore.

89. (C) Red light with an absorption maximum at a wavelength of 660 nanometers is most effective in preventing flower formation.

90. (C) The number of grams of each nutrient is multiplied by its Calorie value, and then the Calorie values are added together; i.e., $53.3 \times 4 = 213.2$; $1.6 \times 9 = 14.4$; $9.1 \times 4 = 36.4$.

91. (B) John could receive a gene for blue eyes from both Mary and Joseph, if they were heterozygous for brown eyes; in this case, he could be blue-eyed.

92. (C) Each bacterium reproduced by fission continuously until it resulted in so many bacteria that the mass of them appeared visible to the naked eye as a colony. Each colony developed from one bacterium.

93. (D) The possible results are shown below:

	$\underline{X}$	Y
X	$X\underline{X}$	XY
X	$X\underline{X}$	XY

Female: 100% normal carrier
Male: 100% normal

94. (B) The surface increases only as the square of its diameter, while the volume increases as the cube. As the cell grows, the increase in surface lags behind the increase in volume.

95. (C) The average number of Calories per pound is 24; for 154 pounds, this would be 24×154 or 3,696 Calories.

96. (D) The percentage of people who cannot roll their tongues is 36 percent, or 0.36 of the population. They may be represented as *tt*, since inability to roll the tongue is a recessive trait (Key: *T*—ability to roll the tongue; *t*—inability to roll the tongue). To find the percentage of *t*, take the square root of 0.36, which is 0.6. Thus, 6 out of every 10, or 60% of the alleles for the tongue rolling trait are recessive.

97. (B) When the plant takes in oxygen, there is a drop in pressure under the bell jar. This causes the liquid on the left side of the manometer to rise, with a consequent lowering on the right.

98. (C) Photosynthesis can take place only in the presence of light.

99. (B) Tube 1 is connected to the manometer, and transmits the change in air pressure within the bell jar to it. After some oxygen has been taken in by the plant, the piston of the syringe connected with tube 2 is pushed in to balance the air pressure and to give an indication of the volume of air used.

100. (A) Photosynthesis would be carried on in the presence of light, and oxygen would be given off by the plant.

101. (E) The plant takes in oxygen and gives off carbon dioxide during the process of respiration. If the carbon dioxide were not absorbed by calcium chloride, it would exert pressure on the manometer and interfere with the effects of oxygen consumption.

102. (E) The plant is enclosed in an opaque container containing calcium chloride (4); then the vent (3) is adjusted so that the level of the manometer tubes is equalized (1); when oxygen has been taken in by the plant, and the manometer levels have changed, air is admitted into the container in measured amounts through the syringe (2).

103. (D) It will be used to admit a large amount of air to take the place of the oxygen that was used up.

104. (C) Air pressure within the container is increased by rising temperature of the air, and decreased by falling temperature; an increase in barometric pressure causes an increase in the air pressure within the container and decreased barometric pressure causes a reduction.

105. (C) In view of the good results obtained with high-lysine corn, it is apparent that it holds great promise as a source of protein in poverty areas of the world.

106. (D) Interferon is useful against many types of viruses. If its production by the cells can be stimulated, it may serve as an effective broad means of protection against virus diseases.

107. (C) In the light, the plant takes in carbon dioxide (and water) to make glucose by the process of photosynthesis. Oxygen is given off as a by-product. The removal of carbon dioxide and the addition of oxygen improved the air.

108. (D) He should have performed a control to see whether the bread had sugar to begin with. If he had tested the dry piece of bread for sugar and found it lacking, he might have been correct in assuming that chewing a piece of bread digests its starch into sugar. Another needed control is to test the saliva for the presence of sugar.

109. (D) One of the reasons that Drosophila flies, or fruitflies, are used in genetic research is that they reproduce in large numbers every two weeks. This permits a geneticist to study about 25 generations a year. At this rate, there would be thousands of individuals, increasing the possibility of discovering a mutant.

PRACTICE COLLEGE BOARD ACHIEVEMENT TEST IN BIOLOGY

NO. 9

PART I—TYPE A

Directions: **The following diagram shows parts of a green plant cell. Certain of the parts are labeled with letters. Next to each statement from 1–11, write the letter of the specific structure most closely related to it. A label may be used once, more than once, or not at all.**

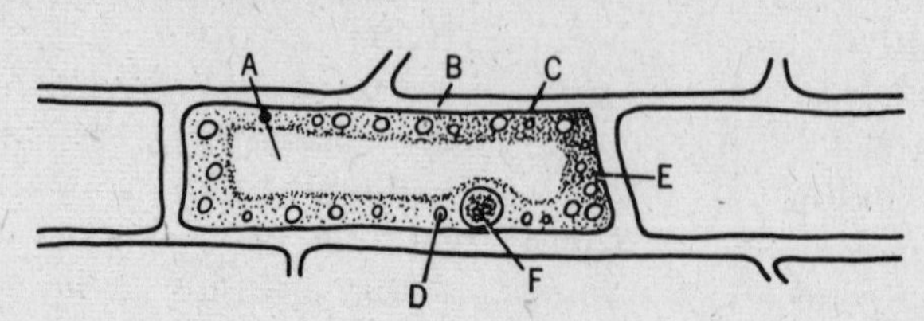

1. Serves to give support to the cell 1. ....
2. Stores water and minerals 2. ....
3. Carries on carbohydrate synthesis 3. ....
4. Controls the reproduction of the cell 4. ....
5. Controls the diffusion of gases in and out of the cell 5. ....
6. Contains cellulose 6. ....
7. Contains hereditary materials 7. ....
8. Releases oxygen as a by-product 8. ....
9. Serves to keep the cell turgid 9. ....
10. Streams throughout the cell 10. ....
11. Contains mitochondria 11. ....

TYPE B

Directions: **Each of the following statements is followed by five suggested answers. Select the one answer which is best in each case.**

12. Niacin will most likely be found in foods containing vitamin (A) A (B) B_1 (C) C (D) D (E) E — 12. A B C D E
13. The blood type known as the universal donor is (A) A (B) B (C) AB (D) O (D) Rh factor — 13. A B C D E
14. All of the following are functions of the skin except (A) protection (B) sensation (C) excretion (D) manufacture of vitamin D (E) exhalation — 14. A B C D E
15. Which process is represented by the following equation: — 15. A B C D E

$$6CO_2 + 12H_2O \xrightarrow[\text{energy}]{\text{enzymes}} C_6H_{12}O_6 + 6O_2 + 6H_2O$$

(A) fermentation (B) lactation (C) photosynthesis (D) aerobic respiration (E) anaerobic respiration

16. A nerve cell that transmits impulses from a sense organ to the nervous system is known as a(n) (A) sensory neuron (B) motor neuron (C) associative neuron (D) plexus (E) ganglion — 16. A B C D E
17. The weakness in Darwin's theory of how evolution occurs was his inability to explain the (A) mechanisms which produce variations (B) reasons for overproduction (C) role played by natural selection (D) adaptations of living organisms for survival (E) inheritance of acquired characteristics — 17. A B C D E
18. A fatty substance that is suspected of playing a role in heart disease is (A) chloromycetin (B) choroid coat (C) chitin (D) cholesterol (E) chlorotone — 18. A B C D E
19. The fine branching extensions from the cell body of a nerve cell are known as (A) dendrites (B) end brushes (C) terminal branches (D) ciliated epithelia (E) synapses — 19. A B C D E
20. ACTH is an example of (A) an antibiotic (B) an antitoxin (C) an antidote (D) a hormone (E) a sulfa drug — 20. A B C D E
21. An increase in the diameter of an oak tree is caused chiefly by the activity of the (A) vascular ducts (B) bark (C) lenticels (D) cambium (E) annual rings — 21. A B C D E
22. Primitive man first appeared on earth (A) at the time of the dinosaurs (B) at the beginning of the Mesozoic era (C) at the time of archaeopteryx (D) about sixty million years ago (E) about a million years ago — 22. A B C D E
23. According to Lamarck's theory (A) genes are inherited in linked groups (B) acquired characteristics are inherited (C) there is a struggle for — 23. A B C D E

existence (D) overproduction takes place (E) mutations are caused by radioactivity

24. After a child is born, it does not form any new (A) nerve cells (B) bone cells (C) muscle cells (D) epithelial cells (E) blood cells

24. A B C D E

25. What effect does the hydrolytic action of enzymes have on organic molecules? (A) They are converted to more complex form (B) Their hydrogen ion concentration is increased (C) Their hydrogen ion concentration is decreased (D) They become chemically inactive (E) They become smaller

25. A B C D E

26. During respiration, chemical bond energy is transferred from C—C and C—H bonds of carbohydrates to chemical bonds in the formation of molecules of (A) DNA (B) RNA (C) protein (D) ATP (E) oxygen

26. A B C D E

27. A chain of bonded amino acid molecules forms an organic compound called a (A) polynucleotide (B) polypeptide (C) polysaccharide (D) polyploid (E) polyamine

27. A B C D E

28. In order to grow bacteria, the minimum material required, in addition to a sterile Petri dish of nutrient agar, would be (A) steam sterilizer and polluted water (B) incubator and transfer needle (C) autoclave and spores (D) methylene blue stain (E) centrifuge and Bunsen burner

28. A B C D E

TYPE C

Directions: **For each of the terms listed from 29–40, write the letter (A), (B), (C) or (D), depending on whether it**

(A) applies only to A
(B) applies only to B
(C) applies to both A and B
(D) applies to neither A nor B

(A) Sexual reproduction (C) Both
(B) Asexual reproduction (D) Neither

29. egg 29. A B C D

30. aorta 30. A B C D

31. cutting 31. A B C D

32. nucleus 32. A B C D

33. blastula 33. A B C D

34. runner 34. A B C D

35. polar bodies 35. A B C D

36. zygote 36. A B C D

37. bud 37. A B C D

38. chromosomes 38. A B C D

39. spore 39. A B C D

40. ATP 40. A B C D

TYPE D

Directions: **The following questions require two answers. One of the items in the left column** (A), (B), **or** (C) **is related to four of the items in the right column** (numbered 1–5). **Select the letter of the item which is so related; then select the number of the remaining item which does not belong.**

EXAMPLE:

(A) paramecium
(B) ameba
(C) spirogyra

1. cilia
2. macronucleus
3. trichocyst
4. pyrenoid
5. micronucleus

41. (A) microscopic cell
(B) female gamete
(C) protozoa

1. coccus
2. sperm
3. ameba
4. sparrow egg
5. elephant ovum

41. A B C 1 2 3 4 5

42. (A) arthropod
(B) echinoderm
(C) mollusk

1. praying mantis
2. leech
3. sand crab
4. centipede
5. spider

42. A B C 1 2 3 4 5

43. (A) plasmolysis
(B) osmosis
(C) mitosis

1. chromosome
2. spindle
3. centrosome
4. chloroplast
5. aster

43. A B C 1 2 3 4 5

44. (A) protein
(B) lipid
(C) carbohydrate

1. albumen
2. glucose
3. starch
4. sucrose
5. fructose

44. A B C 1 2 3 4 5

45. (A) pistil
(B) stamen
(C) calyx

1. style
2. stigma
3. anther
4. ovary
5. ovule

45. A B C 1 2 3 4 5

46. (A) parthenogenesis
(B) alternation of generations
(C) vegetative propagation

1. Christmas fern
2. scouring rush
3. liverwort
4. plant lice
5. moss

46. A B C 1 2 3 4 5

47. (A) budding
(B) fission
(C) spore formation

1. pleurococcus
2. yeast
3. ameba
4. bacillus
5. diatom

47. A B C 1 2 3 4 5

48. (A) three-chambered heart
(B) four-chambered heart
(C) two-chambered heart

1. salamander
2. frog
3. toad
4. trout
5. newt

48. A B C 1 2 3 4 5

TYPE E

Directions: **The five lettered headings below are followed by a list of numbered statements. For each numbered statement, choose the one lettered heading (A–E) which is most closely related to it.**

(A) Segregation (C) Mutation
(B) Sex-linkage (D) Unit characters
(E) Incomplete dominance

49. Brown-eyed parents have a blue-eyed child

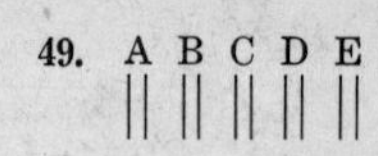

50. A trait seems to skip a generation

51. When black Andalusian roosters are mated with white Andalusian hens, all the offspring appear blue

51. A B C D E

52. An albino frog was recently discovered

53. Hemophilia affects mostly males

53. A B C D E

54. A 3 : 1 ratio is obtained when hybrids are crossed

54. A B C D E

55. In humans, eye color and skin color are inherited separately

55. A B C D E

(A) Phototropism (C) Thigmotropism
(B) Geotropism (D) Hydrotropism
(E) Chemotaxis

56. How does a plant grow after it has been blown over by the wind?

56. A B C D E

57. How does paramecium react to salt?

57. A B C D E

58. How does a plant grow on a window sill?

58. A B C D E

59. How does euglena move in a partly shaded pool?

59. A B C D E

60. How does a morning glory plant grow along a string?

60. A B C D E

PART II

In a study of the rate of growth of *Bacillus coli*, some of the bacteria were introduced into a freshly sterilized test tube containing a nutrient liquid at 6 A.M. The test tube was incubated at 37°C. At hourly intervals, samples of the liquid were taken and the number of bacteria per cubic centimeter determined. The accompanying graph shows the results obtained from 6 A.M. to 11 P.M.

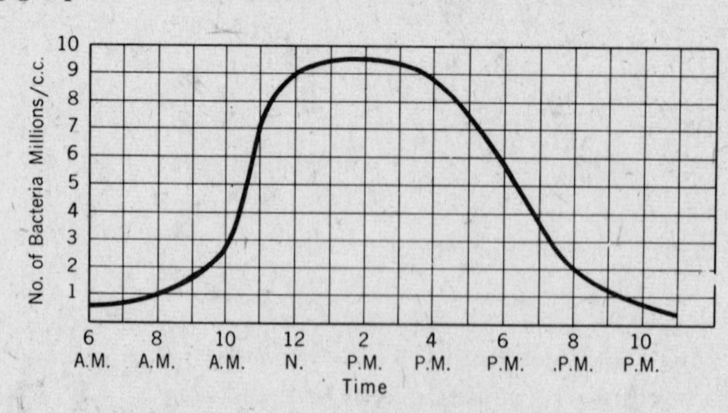

61. The greatest increase in growth took place (A) 8–9 A.M. (B) 9–10 A.M. (C) 10–11 A.M. (D) 12 noon–1 P.M. (E) 1–2 P.M.

62. There was no perceptible increase in the rate of multiplication of the bacteria at (A) 6–7 A.M. (B) 7–8 A.M. (C) 11 A.M.–12 noon (D) 4–5 P.M. (E) at no time shown on graph

63. All the bacteria had died at (A) 6–7 P.M. (B) 7–8 P.M. (C) 9–10 P.M. (D) 10–11 P.M. (E) at no time shown on graph

63. A B C D E

64. The greatest number of bacteria per cc. were present at (A) 12 noon (B) 1 P.M. (C) 2 P.M. (D) 3 P.M. (E) 4 P.M.

64. A B C D E

65. Since bacteria multiply by binary fission, (A) their rate of growth is determined by proper food and temperature conditions (B) their rate of growth would be doubled by doubling both the amount of food and the temperature (C) their rate of growth would be reduced in half by dividing the time of the experiment in half (D) the graph was in error, because the bacteria should have continued to increase their growth indefinitely (E) the graph was in error, because it should have shown twice the rate of growth each hour

65. A B C D E

66. The smallest number of bacteria per cc. were present at (A) 6 A.M. (B) 7 A.M. (C) 8 A.M. (D) 10 P.M. (E) 11 P.M.

66. A B C D E

PART III–TYPE A

***Directions:* Each item consists of an *assertion* (statement) in the left-hand column, and a *reason* on the right-hand side. Select**

(A) if both assertion and reason are true, and *are related* as cause and effect
(B) if both assertion and reason are true, but *are not related* as cause and effect
(C) if the assertion is true, but the reason is a false statement
(D) if the assertion is false, but the reason is a true statement
(E) if both assertion and reason are false statements

DIRECTIONS SUMMARIZED:

	(A)	(B)	(C)	(D)	(E)
Assertion:	True	True	True	False	False
Reason:	True	True	False	True	False
	Cause and effect	*Not* cause and effect			

	Assertion:		Reason:	
67.	Testosterone, or the male hormone, determines the development of male secondary sexual characters	BECAUSE	it is produced in the testes	67. A B C D E
68.	The largest part of the human brain is the cerebrum	BECAUSE	it is the seat of memory and thought	68. A B C D E
69.	Cancer can easily be cured in the late stages	BECAUSE	radioactive cobalt is now being used to treat cancerous growths	69. A B C D E
70.	Petrified trees in the Petrified Forest occasionally sprout twigs and leaves	BECAUSE	the climate has changed since the time when they first became petrified	70. A B C D E
71.	*Eohippus*, the ancestor of the horse, helped destroy dinosaur eggs	BECAUSE	it was an early warm-blooded mammal	71. A B C D E
72.	De Vries' mutation theory disproved Darwin's theory of evolution	BECAUSE	it established that the giraffe developed its long neck through use and disuse	72. A B C D E
73.	Plant breeders are now treating plants with radiations	BECAUSE	useful mutations may result from this treatment	73. A B C D E
74.	Microscopic organisms, called plankton, are the basis of life in the ocean	BECAUSE	they provide fish with immunity from their natural enemies	74. A B C D E
75.	The peas in a pod are identical	BECAUSE	they contain identical genes	75. A B C D E

TYPE B

Directions: **Each of the following statements is followed by five suggested answers. Select the one answer which is best in each case.**

76. Identical twins are alike because they (A) develop in the same uterus (B) develop from one egg and two sperm cells (C) have the same gene make-up (D) receive their nourishment from the same bloodstream during development (D) have similar traits

76. A B C D E

77. Cosmic rays are a threat to space travel because (A) they are generated by the third rocket stage (B) they can penetrate through the walls of a space vehicle (C) man is more sensitive to them than monkeys (D) they increase the effect of acceleration on the bloodstream (E) they are more dangerous to man in the weightless condition

77. A B C D E

78. Terracing on a hillside prevents erosion because (A) the crop is selected for its ability to bind the soil (B) it enriches the soil (C) it acts as a wind-break (D) organic matter is returned to the soil (E) it slows down the run-off of water

78. A B C D E

79. Carnivorous animals depend on green plants for their food (A) in order to achieve a balanced diet (B) because herbivorous animals have grinding teeth (C) because only green plants can make food (D) to be protected against vitamin deficiency diseases (E) because only omnivores are both carnivores and herbivores

80. Radioactive iodine is used to treat the thyroid gland because (A) it is absorbed almost exclusively by the thyroid gland (B) the thyroid gland is located in the neck, close to the surface of the body (C) it is too dangerous to be used on other parts of the body (D) its use as an antiseptic has been definitely proven (E) the parathyroid gland next to it is also sensitive to it

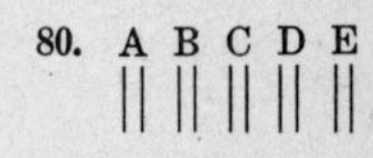

81. The whale has remains of useless hip bones because (A) all vertebrates have such bones (B) it was descended from an ancestor that used these bones (C) it is descended from the coelacanth (D) its comparative anatomy is like that of any other water animal (E) vertebrates have four limbs

81. A B C D E

82. The flower of a rose plant is its reproductive organ because (A) its thorns protect it against animals (B) its parts are arranged in groups of five (C) its petals attract bees for cross-pollination (D) it has been bred for its attractive appearance (E) it forms the seeds

82. A B C D E

83. Although it consists of only one cell, the ovum of a frog is visible to the naked eye because (A) the frog is a cold-blooded animal (B) it is larger than the sperm (C) it contains stored food (D) it undergoes oögenesis (E) it goes through cleavage

83. A B C D E

84. A student who dissected a frog knew that it was a female because (A) its uterus contained developing embryos (B) it had a brighter skin color than the other frogs (C) he did not see any Y chromosomes in its sperms (D) the pituitary gland was enlarged (E) the abdominal cavity contained eggs

84. A B C D E

85. Although the island of Madagascar is separated from Africa only by a narrow strait, many plants and animals common on the mainland are unknown on the island. This fact illustrates the principle of (A) incomplete dominance (B) independent assortment (C) evolutionary equilibrium (D) evolution in isolated populations (E) ecological succession

85. A B C D E

TYPE C

Directions: **In each of the following questions, there is a statement followed by four choices of which *one or more* are correct. Indicate the correct combination by choosing the following letters:**

(A) if only 1, 2, and 3 are correct

(B) if only 1 and 3 are correct
(C) if only 2 and 4 are correct
(D) if only 4 is correct
(E) if some other combination is correct

DIRECTIONS SUMMARIZED:

(A)	(B)	(C)	(D)	(E)
1, 2, 3 only	1, 3 only	2, 4 only	4 only	some other combination

86. Nitrogen-fixing bacteria
1. are parasitic organisms
2. are found on the roots of legume plants
3. give off toxins that destroy harmful soil bacteria
4. convert free nitrogen into nitrates

86. A B C D E

87. When clotting of blood takes place
1. fibrinogen is changed to fibrin
2. the white blood cells trap the red blood cells
3. the blood platelets play a role
4. hemoglobin is changed to oxyhemoglobin

87. A B C D E

88. Lymph
1. surrounds the cells of the body as intercellular fluid
2. flows in special vessels called lymphatics
3. is similar to blood, but has no red blood cells
4. originates in blood plasma

88. A B C D E

89. Vegetative propagation by bulbs
1. is an example of asexual reproduction
2. produces flowers that are true to type
3. takes place in tulips
4. requires the use of a scion in grafting

89. A B C D E

90. The bat is classified as a mammal because it
1. has hair
2. suckles its young offspring
3. has internal development of its young
4. forms eggs with the haploid number of chromosomes

90. A B C D E

91. The plant hormone auxin
1. stimulates the rate of photosynthesis
2. influences some plant tropisms
3. causes an increase in cell growth
4. speeds up cross-pollination

91. A B C D E

92. The left ventricle of the heart
1. sends blood into the aorta
2. receives blood from the right atrium
3. contains oxygenated blood
4. is connected with the right ventricle by a valve

92. A B C D E

93. The trachea
1. is lined with ciliated epithelial tissue
2. leads to the bronchi
3. is reinforced with rings of cartilage
4. serves as a passageway to the diaphragm

93. A B C D E

94. The small intestine
1. produces hydrochloric acid
2. receives digestive fluids from the liver and pancreas
3. contains the enzyme ptyalin
4. completes the digestion of food

94. A B C D E

PART IV–TYPE A

Directions: **Referring to the following selection, choose the best answer for each of the questions that follows:**

*"Students of behavior generally agree that the early experiences of animals (including man) have a profound effect on their adult behavior. D. O. Hebb of the University of Montreal goes so far as to state that the effect of early experience upon adult behavior is inversely correlated with age. This may be an oversimplification, but in general it appears to hold true. Thus the problem of the investigator is not so much to find out *whether* early experience determines adult behavior, but rather to discover *how* it determines adult behavior.

"Three statements are usually made about the effects of early experience. The first is that early habits are very persistent and may prevent the formation of new ones. This, of course, refers not only to the study of experimental animals but also to the rearing of children. The second statement is that early perceptions deeply affect all future learning. This concept leads to the difficult question whether basic perceptions – the way we have of seeing the world around us – are inherited or acquired. The third statement is simply that early social contacts determine adult social behavior. This, of course, is imprinting.

"Although imprinting has been studied mainly in birds, it also occurs in other animals. It has been observed in insects, in fishes and in some mammals."

95. The title that best befits this selection is: (A) Imprinting in Animals (B) Basic Perception in Animals (C) Habit Formation (D) The Effect of Early Experience (E) The Inheritance of Behavior

95. A B C D E

96. According to the selection, the chief problem of an investigator is to (A) discover whether early experience determines adult behavior (B) find out how early experience determines adult behavior (C) determine early experience from adult behavior (D) investigate the age at which early experience starts (E) investigate the age at which adult behavior starts

96. A B C D E

97. It is generally true that (A) there is no correlation between early experience and adult behavior (B) the effect of adult behavior on early experience is definitely predictable (C) the effect of early experience on adult behavior is directly correlated with age (D) the effect of adult behavior on early experience is directly correlated with age (E) the effect of early experience on adult behavior is inversely correlated with age

97. A B C D E

98. The meaning of imprinting can be inferred from the statement that (A) it has been studied in insects (B) it has been observed in some mammals (C) new habits may be prevented by old habits (D) basic perception is either inherited or acquired (E) early social contacts determine adult social behavior

98. A B C D E

TYPE B

Directions: **Each of the following problems is followed by five suggested answers. Select the one answer which is best in each case.**

99. A teaspoonful of sugar weighing four grams was burned in a bomb calorimeter surrounded by four kilograms of water. The temperature of the water was raised from 14.5°C to 18.5°C. The number of Calories in the sugar was (A) 4 (B) 8 (C) 12 (D) 16 (E) 32

99. A B C D E

100. In comparison to the number of offspring produced by an organism, the amount of parental care provided is usually (A) directly proportional (B) inversely proportional (C) twice as much (D) dependent on the nutritional factors in the environment (E) related to web of life

100. A B C D E

101. A man who is normal for color vision marries a normal heterozygous woman. The chances of their daughters being normal homozygous for color blindness is (A) zero (B) 50–50 (C) 25–75 (D) 25–50–25 (E) 100

101. A B C D E

*From "Imprinting in Animals" by Eckhard H. Hess, March 1958, by *Scientific American, Inc.*

102. An 18-year-old college freshman weighing 150 pounds requires the number of Calories shown in the following list of his activities for a single day. This is in addition to his basal metabolism rate which is estimated to be 14 Calories per pound.

102. A B C D E

ACTIVITY	HOURS	CALORIES USED PER POUND
Studying	5	0.2
Attending lectures	3	0.2
Walking	3	0.9
Playing ping pong	1	2.0
Swimming	½	3.6
Eating	1½	0.2
Dressing and undressing	½	0.3
Running	½	3.3
Typewriting	1	0.5

His Calorie requirements are (A) 3,400–3,600 (B) 3,600–3,800 (C) 3,800–4,000 (D) 4,000–4,200 (E) 4,200–4,400

103. When Mendel crossed pea plants that were hybrid for smooth seed form (smooth-dominant; wrinkled-recessive), he obtained 1,850 wrinkled seeds out of a total of 7,324. The number of smooth seeds was most likely (A) 1,850 (B) 3,700 (C) 5,474 (D) 7,324 (E) 9,174

103. A B C D E

104. When a piece of newsprint containing the letter "d" is examined with the microscope, the lenses produce an image that looks like (A) d (B) b (C) q (D) p (E) ᓇ

104. A B C D E

105. In demonstrating the action of saliva in starch digestion, a necessary control is to (A) test the saliva with iodine (B) test the saliva with Benedict's solution (C) test the saliva with nitric acid (D) test the Benedict's solution for starch (E) test the iodine solution for starch

105. A B C D E

PART V

106. One of the practical outcomes of space exploration has been the development of infra-red photography as an aid to agriculture. Flying at altitudes as high as 60,000 feet, airplanes of the U.S. Air Force's Weather Service have taken aerial photographs that are interpreted by a computer to distinguish cornfields from other vegetation. On the basis of a rating scale, cornfields may show up on the computer with the fungus corn blight disease, to a severe, heavy or mild degree, or as being healthy. On the special film, healthy growth appears red, unhealthy growth shows in different shades of green. With this information, the best step for the farmer to take in order to protect the summer's crop is to: (A) cross-hybridize with a resistant variety of corn (B) spray fungicide to destroy blighted portions of the crop to prevent the disease from spreading (C) spray red vegetable coloring over the field to assist the infra-red film in identifying the blighted areas (D) expose the corn plants to ultra-violet light so that the infra-red rays will not harm the plants (E) harvest the crop early and take advantage of early-season corn prices

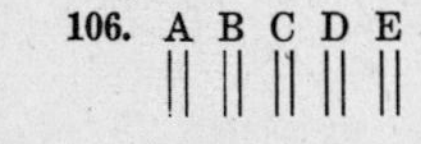

107. Animals fed vitamin B_{12} show increased growth. Pure vitamin B_{12} is extracted from waste materials left in vats in which antibiotics were made. Animals fed on a diet that includes wastes from the antibiotic vats grow faster than those fed only pure vitamin B_{12}. It is probable that (A) waste from the antibiotic vats contains a growth promoter other than vitamin B_{12} (B) pure vitamin B_{12} is not a growth promoter at all (C) vitamin B_{12} is a good growth promoter if it is in impure form (D) the waste material in the

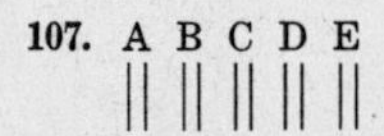

vats contains vitamin B_{12} which the process does not extract (E) antibiotics are better growth promoters than vitamin B_{12}

PART VI

108. A population originally inhabiting the entire area shown on the accompanying diagram has become separated into two separate populations by a barrier (water).

108. A B C D E

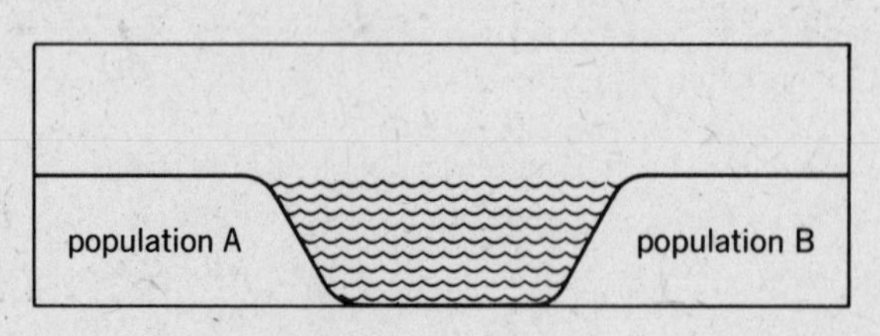

If the environment inhabited by population *A* undergoes severe changes and the environment of population *B* does not, the rate of evolution of population *A* will probably be (A) consistently slower than population *B* (B) consistently faster than population *B* (C) the same as population *B* (D) slower at first then faster than population *B* (E) dependent on the rate of evolution of population *B*

109. If the heart of a freshly killed frog is suspended in a beaker of Ringer's solution (which includes 0.45% common salt, NaCl), the heart will continue to beat at the same rate as that of a live frog. If the heart is then immersed in a beaker containing a 0.6% water solution of common salt, it will beat more slowly and weakly. If it is then replaced in Ringer's solution, it will recover its normal rate. The best interpretation of this demonstration is that (A) salt is harmful to the frog's tissues (B) there is not enough salt in the solution used in this experiment (C) there is too much salt in the solution used in this experiment (D) Ringer's solution is chemically identical to frog plasma (E) Ringer's solution provides a chemical environment similar to that of a normal heart

109. A B C D E

ANSWER KEY: TEST NO. 9

1. (B)	**12.** (B)	**23.** (B)	**34.** (B)	**45.** (A)3	**56.** (B)	**67.** (B)	**78.** (E)	**89.** (A)	**100.** (B)
2. (A)	**13.** (D)	**24.** (A)	**35.** (A)	**46.** (B)4	**57.** (E)	**68.** (B)	**79.** (C)	**90.** (A)	**101.** (B)
3. (D)	**14.** (E)	**25.** (E)	**36.** (A)	**47.** (B)2	**58.** (A)	**69.** (D)	**80.** (A)	**91.** (E)	**102.** (B)
4. (F)	**15.** (C)	**26.** (D)	**37.** (B)	**48.** (A)4	**59.** (A)	**70.** (D)	**81.** (B)	**92.** (B)	**103.** (C)
5. (C)	**16.** (A)	**27.** (B)	**38.** (C)	**49.** (A)	**60.** (C)	**71.** (D)	**82.** (E)	**93.** (A)	**104.** (D)
6. (B)	**17.** (A)	**28.** (B)	**39.** (B)	**50.** (A)	**61.** (C)	**72.** (E)	**83.** (C)	**94.** (C)	**105.** (B)
7. (F)	**18.** (D)	**29.** (A)	**40.** (D)	**51.** (E)	**62.** (A)	**73.** (A)	**84.** (E)	**95.** (D)	**106.** (B)
8. (D)	**19.** (A)	**30.** (D)	**41.** (A)4	**52.** (C)	**63.** (E)	**74.** (C)	**85.** (D)	**96.** (B)	**107.** (A)
9. (A)	**20.** (D)	**31.** (B)	**42.** (A)2	**53.** (B)	**64.** (C)	**75.** (E)	**86.** (C)	**97.** (E)	**108.** (B)
10. (E)	**21.** (D)	**32.** (C)	**43.** (C)4	**54.** (A)	**65.** (A)	**76.** (C)	**87.** (B)	**98.** (E)	**109.** (E)
11. (E)	**22.** (E)	**33.** (A)	**44.** (C)1	**55.** (D)	**66.** (E)	**77.** (B)	**88.** (E)	**99.** (D)	

ANSWERS EXPLAINED: TEST NO. 9

1. (B) cell wall
2. (A) vacuole
3. (D) chloroplast
4. (F) nucleus
5. (C) cell membrane
6. (B) cell wall
7. (F) nucleus
8. (D) chloroplast
9. (A) vacuole
10. (E) cytoplasm
11. (E) cytoplasm

12. (B) Niacin is a member of the vitamin B complex, which also includes vitamin B_1, or thiamin.

13. (D) Blood containing this type may be safely given in a transfusion to people who have other blood groups.

14. (E) The skin does not exhale air; this is done only through the lungs.

15. (C) Photosynthesis takes place within the chloroplasts of a green plant. Carbon dioxide (CO_2) and water (H_2O) are taken in. Chlorophyll becomes activated by light energy and decomposes the water molecules, releasing oxygen (O_2) as a gas; this occurs during the light phase. It is followed by a dark reaction during which carbon fixation takes place. A number of enzymes combine carbon dioxide and hydrogen atoms to form the six-carbon sugar, glucose, $C_6H_{12}O_6$.

16. (A) A sensory neuron is also known as an afferent neuron.

17. (A) Darwin's theory of natural selection was published after many years of careful observation and study. He came to his conclusions after he had accumulated much evidence to support them, mostly based on the observations he himself had made. He was frank to admit that there were aspects of evolution he could not explain, because he did not have enough data for a full understanding, i.e., how variations were caused, and how life originated on earth.

18. (D) Cholesterol deposits are found lining the inner surfaces of arteries in the condition known as atherosclerosis. This may lead to high blood pressure and possibly coronary thrombosis.

19. (A) One part of the cytoplasm surrounding the nucleus extends outward into these branching extensions; the other end of the cell body leads to the axon.

20. (D) ACTH is one of the hormones produced by the pituitary gland.

21. (D) Cambium consists of actively dividing cells which differentiate into xylem, phloem, and additional cambium cells, adding to the diameter of the trunk.

22. (E) The discoveries of Dr. Leakey and his wife at the Olduvai Gorge in Tanzania, Africa, indicate that a type of primitive man, Zinjanthropus, lived there about a million years ago. Some of their other discoveries point to types that may actually have lived 20–30 million years ago.

23. (B) Lamarck believed that organs either become developed or undeveloped as a result of use or disuse, and that the acquired characteristic is passed on to the offspring.

24. (A) The number of nerve cells does not increase after birth.

25. (E) A complex chain of organic molecules may be broken down during the process of hydrolysis. Water is added while enzymes break apart the bond between the molecules, and give rise to individual smaller molecules. Example: The disaccharide, maltose, is hydrolyzed by the enzyme maltase into the simpler molecules of glucose.

26. (D) During cellular respiration, chemical bond energy is released from glucose by a series of complex steps, and then stored in ATP (adenosine triphosphate), a storehouse for energy. ATP is derived from ADP (adenosine diphosphate) by the addition of a high energy phosphate molecule. When glucose is broken down, some of its hydrogen is removed, releasing energy. This energy is transferred to ADP, with the addition of a high energy phosphate molecule, to convert it to ATP. ATP is used by the cell in its metabolism including the synthesis of complex molecules.

27. (B) In dehydration synthesis, large molecules, such as those of proteins, are synthesized from the chemical combination of amino acids. In the bonding of amino acids, a bond forms between the carboxyl group of one amino acid and the amino group of the next amino acid. This bond is known as a peptide bond. Water molecules are released in the process. A chain of such amino acid molecules linked by many peptide bonds becomes a polypeptide.

28. (B) An incubator is used in providing the best temperature for bacterial growth. The transfer needle is needed to place a sample of bacteria on the surface of the agar.

29. (A) An egg is a female gamete that unites with a male gamete, the sperm, during fertilization.

30. (D) The aorta is the largest artery of the body, and is not directly related to reproduction.

31. (B) A cutting is a form of vegetative propagation, utilizing a stem or a leaf to produce a new plant.

32. (C) The nucleus contains the genes which determine heredity in all types of reproduction.

33. (A) The blastula is the hollow-ball stage that is formed during cleavage of a fertilized egg.
34. (B) A strawberry plant may form runners that serve to produce additional plants, as an example of vegetative propagation.
35. (A) Polar bodies are produced along with an ovum during oögenesis.
36. (A) The zygote is the cell resulting from the union of two gametes.
37. (B) A bud is an extension of a yeast cell or a hydra, that will develop into another organism.
38. (C) Chromosomes are present in the nucleus and are passed on to the offspring, where they determine heredity.
39. (B) A spore is produced by a mold plant and will germinate to form another mold.
40. (D) ATP (adenosine triphosphate) is a chemical that plays an important role in energy release during the process of respiration.
41. (A)4 All are microscopic except a sparrow egg, which has so much stored food that it can be seen with the naked eye.
42. (A)2 All are arthropods except leech, which is an annelid worm.
43. (C)4 All are structures that appear during mitosis except chloroplast, which contains chlorophyll and is active in photosynthesis.
44. (C)1 All are examples of carbohydrates except albumen, which is a protein.
45. (A)3 All are parts of a pistil except anther, which is part of a stamen.
46. (B)4 All are plants that have a life history that includes both a sporophyte and a gametophyte stage except plant lice, which are insects that reproduce either by fertilization or by parthenogenesis.
47. (B)2 All reproduce simply by splitting to form two equal cells, except yeast, which reproduces by budding, in which the daughter cells are unequal in size.
48. (A)4 All are amphibia, with a three-chambered heart except trout, a fish, which has a two-chambered heart.
49. (A) They each have a gene for blue eyes, which segregates into the gametes during reduction-division, and is combined with the other during fertilization; the child therefore receives two genes for blue eyes.
50. (A) When Mendel crossed pure tall and pure short pea plants, there were only tall hybrid plants in the next generation; when these were crossed, short plants reappeared.
51. (E) Neither black nor white is dominant. Both traits show up to give the blended blue appearance.
52. (C) A change occurred in the genes for pigmentation, to produce the characteristic of albinism, or lack of pigmentation.
53. (B) It is a recessive, sex-linked trait. A female needs two genes, one on each X chromosome, for hemophilia to appear. In a male, the Y chromosome has very few genes; only one gene for hemophilia on the X chromosome will be enough to make him a hemophiliac.
54. (A) The genes segregate into the gametes, and then recombine in a ratio of 25% homozygous dominant, 50% heterozygous dominant, and 25% pure recessive; the ratio of dominant to recessive is thus 3 : 1.
55. (D) The genes are on different chromosomes, and are inherited independently of each other.
56. (B) The stem grows upward; this is negative geotropism.
57. (E) The paramecium moves away from salt, illustrating negative chemotaxis.
58. (A) The plant turns toward the light; this is an example of positive phototropism.
59. (A) It moves toward the lighted side, illustrating positive phototropism.
60. (C) It grows around the string responding positively to contact.
61. (C) From 10 to 11 A.M., the number of bacteria per cubic centimeter increased from 2.8 million to 7 million.
62. (A) The number of bacteria remained at less than one million per cubic centimeter. The bacteria had not yet begun to multiply in large numbers during this first hour.
63. (E) There were still a little less than half a million per cubic centimeter left by 11 P.M.
64. (C) At that time, there were 9.5 million bacteria per cubic centimeter.
65. (A) They continue to multiply freely as long as the proper supply of food is present, under the correct temperature conditions. As their numbers increase, however, the food supply begins to be reduced; also their waste products change the content of the nutrient liquid, making conditions less favorable.
66. (E) At that time, the supply of bacteria had diminished to less than half a million per cubic centimeter.
67. (B) Testosterone brings about the development in man of such male secondary sexual characters as deep voice and growth of the beard.

68. (B) Man has the most highly developed cerebrum of all the animals, giving him the power of reasoning and memory.

69. (D) Cancer may be cured in the early stages, before it spreads to other parts of the body, by surgery, or by radiations from X-ray, radium, or radioactive isotopes such as those of cobalt and gold.

70. (D) The petrified trees turned to stone under water as their organic structure was gradually replaced by mineral deposits that covered them.

71. (D) Eohippus appeared after the dinosaurs had become extinct.

72. (E) The mutation theory added to Darwin's theory of natural selection by stating that favorable gene changes produce variations that are inherited, and result in new species. According to it, the giraffe gradually developed its long neck through a series of favorable mutations.

73. (A) Radiations may cause changes in the gene molecules, to produce organisms with new variations.

74. (C) Plankton consists of microscopic plant and animal life which serves as food for the small animals of the ocean, such as tiny crustaceans. These are used as food by small fish, which in turn are preyed on by larger fish.

75. (E) They are unlike because they develop from ovules that were fertilized by different pollen tubes. They not only have different sets of genes, but are also different because of environmental factors.

76. (C) They developed from one fertilized egg that split into two masses of cells during cleavage. Their gene make-up is therefore the same.

77. (B) Primary cosmic rays consist mostly of protons that travel with enormous energy and can penetrate a space vehicle and destroy tissue cells of a space traveler.

78. (E) Terraces are built up on the side of a slope in steplike fashion, and are planted with soil-binding plants. This holds the water back and prevents the loss of soil.

79. (C) Carnivorous animals eat herbivorous animals that depend on green plants for their food. Without green plants there would be no food.

80. (A) The thyroid gland concentrates the iodine within itself. The radiations given off by the iodine are effective in treating a cancerous growth of the gland. In case of an overactive thyroid gland, they may also reduce the production of thyroxin.

81. (B) A distant ancestor had hind legs attached to hip bones. With the passage of time, the legs disappeared, and the hip bones became useless. There are still some genes present for the development of these bones.

82. (E) The flower contains the male and female reproductive structures, the stamen and the pistil. Pollen from the stamen lands on the stigma of the pistil, after which a pollen tube grows down to the ovule and fertilizes it. Ovules mature into seeds.

83. (C) If all the stored food were removed, only a microscopic bit of living cytoplasm and the nucleus would be left.

84. (E) Eggs are produced in the ovaries, which become enlarged and fill the abdominal cavity.

85. (D) As a result of their being separated by the barrier of water, the plants and animals on the mainland were prevented from migrating to the islands and interbreeding. Therefore, changes that took place in one area were not transmitted to the other area. Consequently, over a long period of time, these changes became accentuated making the species quite different from each other. This illustrates the principle that evolution in isolated populations proceeds in different directions.

86. (C) These bacteria live in a symbiotic relationship with plants such as clover and alfalfa.

87. (B) In clotting, the blood platelets start a chain of reactions involving a number of substances including prothrombin, thrombin, and fibrinogen, and ending with the formation of a clot made of fibrin threads.

88. (E) All of the statements are true.

89. (A) No new genes are introduced in reproduction by bulbs, so that the new flowers retain the identical characteristics of the original plant.

90. (A) The bat is a flying mammal.

91. (E) Items 2 and 3 are correct. Auxin plays a role in phototropism, and geotropism.

92. (B) When the left ventricle contracts, it sends blood to the entire body, except the lungs, through the aorta.

93. (A) The cilia help keep bacteria and dust from entering the lungs; the rings of cartilage help keep the trachea open all the time.

94. (C) Digestion is started in the mouth and stomach, and is completed in the small intestine; there are three digestive fluids in the small intestine, namely, bile, pancreatic juice, and intestinal juice.

95. (D) Throughout the selection, there is reference to the effects of the early experiences of animals (including man) on their adult behavior.

96. (B) The problem of the investigator is stated as being this: not so much to find out *whether* early experience determines adult behavior, but rather to discover *how* it determines adult behavior.

97. (E) The work of D. O. Hebb of the University of Montreal is cited in support of the statement that the effect of early experience upon adult behavior is inversely correlated with age.

98. (E) The third statement about the effects of early experience, indicates that early social contacts determine social behavior. This process is referred to as imprinting.

99. (D) The temperature of 4 kg. of water was raised 4 degrees, giving a total of 16 Calories.

100. (B) The more offspring produced, the less care given, i.e., frog, fish; the fewer the offspring, the greater the care, i.e., humans and other mammals.

101. (B) The possible results are:

(Key: X = chromosome with gene for normal vision;
$\underline{X}$ = chromosome with gene for color-blindness;
XX = female; XY = male)

	X	Y
X	XX	XY
$\underline{X}$	X$\underline{X}$	$\underline{X}$Y

Female: 50% normal homozygous
50% normal heterozygous
Male: 50% normal
50% color-blind

102. (B) His basic energy requirement at 14 Calories per pound is 14 × 150, or 2,100 Calories. This is added to the Calories needed for the various activities, i.e.,

ACTIVITY	HOURS	CALORIES/ POUND	CALORIES FOR 150 POUNDS
Studying	5	0.2	× 150 = 150.0
Lectures	3	0.2	× 150 = 90.0
Walking	3	0.9	× 150 = 405.0
Ping pong, etc.	1, etc.	2.0, etc.	× 150 = 300, etc.
		Total for all activities –	1,605
			2,100
			3,705

103. (C) When hybrids are crossed, the results are in a ratio of 3 dominant : 1 recessive. There would therefore be about three times as many smooth seeds as wrinkled seeds.

104. (D) An object seen under the microscope appears to be reversed and inverted.

105. (B) This is to see whether the sugar came from the action of saliva on starch, or whether it was present in the saliva to begin with.

106. (B) The value of learning that part of a cornfield is suffering from blight is to make use of this information quickly in order to protect the rest of the crop. By spraying the blighted area, the farmer may be able to destroy the fungus parasite before it can spread to the healthy corn plants.

107. (A) Waste from antibiotic vats contains many substances. One of them, vitamin B_{12}, increases the growth of animals. The remainder of the wastes also promotes growth. Evidently there must be an additional growth promoter other than vitamin B_{12} in the waste.

108. (B) If there is a drastic change in the environment of population *A*, many organisms that are not adapted to the new conditions will die out. Those few that are adapted will survive and will pass their genes on to the next generation. By natural selection over a period of time, organisms that are fit for the new environment will continue to survive, increasing the gene frequency for the favorable characteristics. This may lead to evolutionary changes in population *A*.

109. (E) When the heart of a freshly-killed frog is removed from the body and placed in Ringer's solution, it will continue to beat. This solution provides a chemical environment similar to that of a normal heart. In addition to common salt, sodium chloride, it also contains potassium chloride, calcium chloride, sodium bicarbonate and water. Varying the composition of the solution alters the chemical environment and affects the heart.

PRACTICE COLLEGE BOARD ACHIEVEMENT TEST IN BIOLOGY

NO. 10

PART I—TYPE A

Directions: **The following diagram shows the structure of a leaf. Certain of the parts are labeled with letters. Next to each statement from 1–10, write the letter of the specific structure most closely related to it. A label may be used once, more than once, or not at all.**

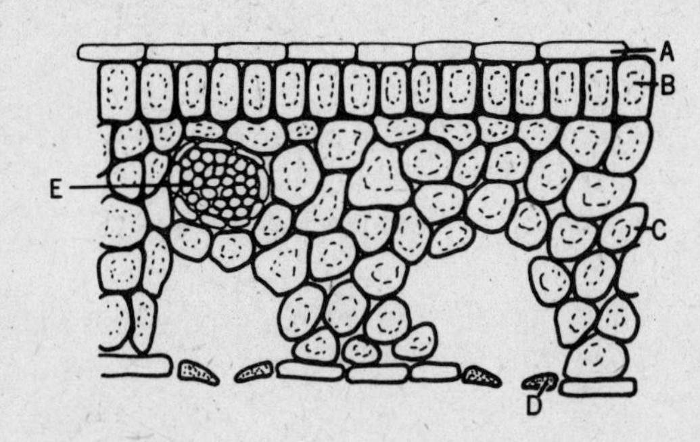

1. Regulates size of stomate — 1. A B C D E

2. Carries on major amount of photosynthesis — 2. A B C D E

3. Supports structure of the leaf — 3. A B C D E

4. Brings water to the cells of the leaf — 4. A B C D E

5. Regulates the entrance of oxygen for respiration — 5. A B C D E

6. Regulates the entrance of carbon dioxide for photosynthesis — 6. A B C D E

7. Contains the highest concentration of chloroplasts — 7. A B C D E

8. Conducts carbohydrates to other parts of the plant — 8. A B C D E

9. Controls the rate of transpiration — 9. A B C D E

10. Serves for protection — 10. A B C D E

TYPE B

Directions: **Each of the following statements is followed by five suggested answers. Select the one answer which is best in each case.**

11. A drop of anti-A blood typing serum was placed on the left side of a slide and a drop of anti-B serum on the right side. Then a drop of blood was mixed into the serum on the left and another drop into the serum on the right. If the blood was type A, the red blood cells would be clumped on (A) the right side only (B) the left side only (C) both sides (D) neither side (E) type AB side

12. Enzyme synthesis in a living cell is most likely to occur in the (A) centriole (B) centrosome (C) chromosome (D) ribosome (E) food vacuole

13. Referring to the accompanying structural formula,

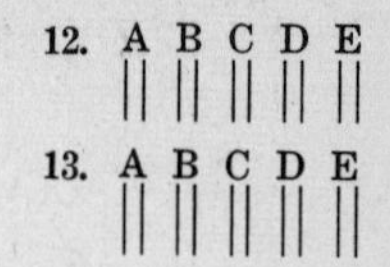

```
            CH2OH                          CH2OH
              |                              |
              C--------O                     C--------O
     H      / |         \       H H        / |         \      H
      \   /   H          \         \     /   H          \   /
        C                  C         C                    C
      /     OH          H   \   O  /  \    OH        H   /  \
    HO        \ |       | /    \ /     \   |         |  /     OH
                C-------C               C-----------C
                |       |               |           |
                H       OH              H           OH
```

the organic molecule depicted (A) is hydrolyzed to form glycerol (B) represents the building block of a type of protein (C) represents a building block utilized in the synthesis of a DNA molecule (D) represents a polysaccharide (E) is acted upon by the enzyme maltase

14. In a nephron of the kidney, reabsorption of glucose generally takes place by the process of (A) osmosis (B) diffusion (C) active transport (D) passive transport (E) cyclosis

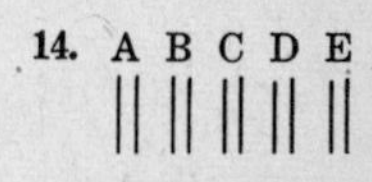

15. Which could be the sequence of a DNA code? (A) CAUGG (B) AUGGA (C) TUGGA (D) ACGTA (E) UCGGA

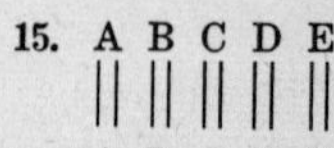

16. When a large breeding population is broken up into many small groups as a direct result of geographic isolation, it is most likely that (A) the rate of evolution of the original population will be speeded up (B) natural selection will not occur (C) natural selection will occur in isolated fashion (D) fewer mutations will occur within the populations than would have occurred if the population had not been broken up (E) gene frequencies for the population will remain constant

16. A B C D E

17. A plant which is a member of the legume family is (A) Queen Anne's lace (B) clover (C) wheat (D) buckwheat (E) corn

17. A B C D E

18. Two plants are classified in the same species if they (A) can be grafted together (B) have leaves of similar shape (C) have leaves of similar color (D) have flowers of similar color (E) can be crossed to produce fertile offspring

18. A B C D E

19. The expression Rh negative refers to (A) immunity to rheumatic fever (B) susceptibility to rheumatic fever (C) a blood type (D) a tropism (E) a rootlike structure of a minus strain of mold

19. A B C D E

20. Genes do not occur in pairs in the (A) zygote (B) zygospore (C) gamete (D) fertilized egg (E) cells of the gastrula

20. A B C D E

21. Thiamin is (A) a gene molecule (B) a metabolic waste (C) a sex-linked trait (D) an antibody (E) a vitamin

21. A B C D E

22. A plant which stores food in a tuber is the (A) potato (B) carrot (C) radish (D) sugar cane (E) tulip

22. A B C D E

23. Yeast is economically important because it carries on the process of (A) fermentation (B) oxidation (C) respiration (D) photosynthesis (E) carbohydrate synthesis

23. A B C D E

24. A measurement of 0.6 mm is equal to (A) 6 microns (B) 60 microns (C) 600 microns (D) 6,000 microns (E) 60,000 microns

24. A B C D E

25. Starfish have been found harmful to the industry involving (A) lobsters (B) oysters (C) codfish (D) crabs (E) flounder

26. Which of the following conditions is *not* essential in a self-sustained ecosystem? (A) a means of permitting the cycling of carbon between living organisms and their environment (B) a means of permitting the cycling of

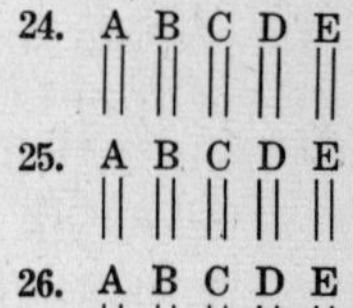

water between living organisms and their environment (C) a constant supply of energy (D) a living system capable of incorporating energy into organic compounds (E) equal numbers of plants and animals

TYPE C

Directions: **For each of the terms listed from 27–36, select the letter** (A), (B), (C), **or** (D), **depending on whether it**

(A) applies only to A
(B) applies only to B
(C) applies to both A and B
(D) applies to neither A nor B

(A) Conjugation	(C) Both
(B) Fertilization	(D) Neither

27. zygote — 27. A B C D

28. sperm — 28. A B C D

29. spore — 29. A B C D

30. isogamete — 30. A B C D

31. nucleus — 31. A B C D

32. zygospore — 32. A B C D

33. gastrula — 33. A B C D

34. bulb — 34. A B C D

35. heterogamete — 35. A B C D

36. cambium — 36. A B C D

TYPE D

Directions: **The following questions require two answers. One of the items in the left column** (A, B, or C) **is related to four of the items in the right column** (numbered 1–5). **Select the letter of the item which is so related; then select the number of the remaining item which does not belong.**

EXAMPLE:
(A) paramecium
(B) ameba
(C) spirogyra

1. cilia
2. macronucleus
3. trichocyst
4. pyrenoid
5. micronucleus

A B C 1 2 3 4 5 (A and 4 marked)

37. (A) evergreen plant
(B) deciduous plant
(C) monocot

1. fir
2. spruce
3. oak
4. pine
5. hemlock

37. A B C 1 2 3 4 5

38. (A) only plant cells
(B) only animal cells
(C) plant and animal cells

1. mitochondria
2. cytoplasm
3. chromatin
4. chlorophyll
5. chromosome

38. A B C 1 2 3 4 5

39. (A) skeletal system
(B) excretory system
(C) nervous system

1. ganglion
2. acetylcholine
3. cyton
4. coccyx
5. synapse

39. A B C 1 2 3 4 5

40. (A) vitamin disease
(B) endocrine disease
(C) contagious disease

1. diabetes
2. scurvy
3. pellagra
4. xerophthalmia
5. rickets

40. A B C 1 2 3 4 5

41. (A) antitoxin
(B) antibiotic
(C) antibody

1. secretin
2. penicillin
3. aureomycin
4. streptomycin
5. chloromycetin

41. A B C 1 2 3 4 5

42. (A) DNA
(B) ATP
(C) Rh

1. deoxyribose sugar
2. glucose
3. phosphate
4. purine
5. pyrimidine

42. A B C 1 2 3 4 5

TYPE E

Directions: **The five lettered headings below are followed by a list of numbered statements. For each numbered statement, choose the one lettered heading (A–E) which is most closely related to it.**

(A) Selection (C) Hybridization
(B) Inbreeding (D) Mutation
(E) Vegetative propagation

43. A nursery sells an apple tree bearing five varieties of apples

43. A B C D E

44. The seedless orange was discovered on a tree bearing oranges with seeds

44. A B C D E

45. Texas cattle have been crossed with Brahman cattle from India

45. A B C D E

46. A farmer saves his best ears of corn for planting

46. A B C D E

47. Burbank developed the Shasta daisy by crossing the American, English, and Japanese daisies

47. A B C D E

48. Prize animals are mated with members of the same litter

48. A B C D E

(A) Adrenal (C) Pancreas
(B) Thyroid (D) Pituitary
(E) Small intestine

49. A man is very thin, very active, and has bulging eyes

49. A B C D E

50. A boy who is being chased by a bully never ran so fast

50. A B C D E

51. The fat lady of the circus weighs 300 pounds

51. A B C D E

52. One of the clowns at the circus is a midget

52. A B C D E

53. A doctor notices that his patient's urine shows a positive test for sugar

53. A B C D E

PART II

In a study of variation in the number of petals contained in a daisy, a student plotted the following graph based on daisies he had collected in a sunny field.

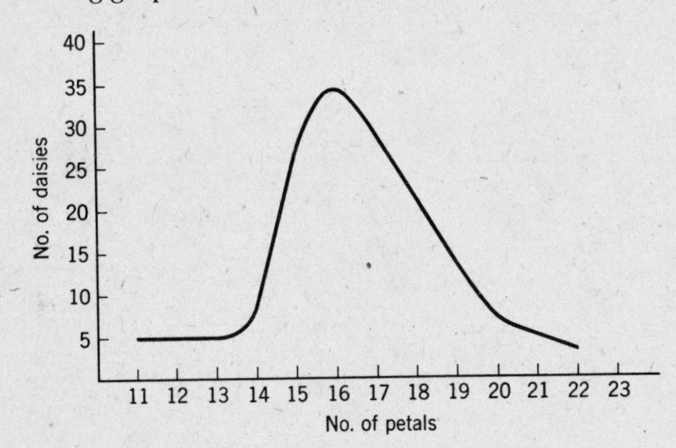

54. The most common number of petals is (A) 12 (B) 13 (C) 14 (D) 15 (E) 16

55. The least common number of petals is (A) 11 (B) 12 (C) 15 (D) 20 (E) 22

56. The curve is not symmetrical because (A) there were too few small daisies (B) there were too few large daisies (C) the total number of daisies was too large (D) the number of petals is determined by genes (E) the number of petals is a factor of the environment

57. Suppose he had collected the daisies in a shaded area: (A) the graph would still look the same (B) he would need a larger number of daisies to get the same graph (C) plants collected in shady and sunny areas cannot be compared (D) the flowers might be smaller, but the number of petals would be larger (E) none of the above answers would apply

58. From this study, the conclusion can be reached that (A) most daisies have an odd number of petals (B) the number of petals on a daisy is variable (C) there are not enough daisies to draw any conclusions (D) daisies are unsatisfactory flowers to study because they have so many petals (E) the mineral content of the soil must be determined before further studies are made

59. Daisies are members of the Compositae family; so are asters. If an equal number of asters were studied in the same way, the results would be (A) identical (B) different (C) unpredictable (D) easy to predict (E) none of the above answers would apply

54. A B C D E

55. A B C D E

56. A B C D E

57. A B C D E

58. A B C D E

59. A B C D E

PART V—TYPE A

Directions: **Each item consists of an *assertion* (statement) in the left-hand column, and a *reason* on the right-hand side. Select**

(A) if both assertion and reason are true, and *are related* as cause and effect
(B) if both assertion and reason are true, but *are not related* as cause and effect
(C) if the assertion is true, but the reason is a false statement
(D) if the assertion is false, but the reason is a true statement
(E) if both assertion and reason are false statements

DIRECTIONS SUMMARIZED:

	(A)	(B)	(C)	(D)	(E)
Assertion:	True	True	True	False	False
Reason:	True	True	False	True	False
	Cause and effect	*Not* cause and effect			

	Assertion:		Reason:
60.	The termite is classified as an insect	BECAUSE	it has six legs
61.	A green plant normally grows upward	BECAUSE	plant stems show positive geotropism
62.	ACTH is necessary for the production of cortisone	BECAUSE	it is produced in the pituitary gland
63.	A fruit hinders seed dispersal	BECAUSE	it keeps the seeds from being spread
64.	The ovum of an elephant is microscopic	BECAUSE	it contains practically no stored food
65.	Dinosaur footprints have been preserved in rock	BECAUSE	the heavy weight of the dinosaurs impressed their footprints into the rock
66.	Cosmic rays will not bother travelers in space	BECAUSE	the earth's atmosphere serves as a partial shield against cosmic rays
67.	Clover plants help enrich the soil	BECAUSE	their roots contain nitrogen-fixing bacteria

60. A B C D E
61. A B C D E
62. A B C D E
63. A B C D E
64. A B C D E
65. A B C D E
66. A B C D E
67. A B C D E

TYPE B

Directions: **Each of the following statements is followed by five suggested answers. Select the one answer which is best in each case.**

68. The Hardy-Weinberg principle is a mathematical formulation which explains why gene frequencies can remain constant from generation to generation, under certain conditions. Assuming that gene frequencies in a given population remain constant, it would be an indication that (A) evolution within the population would not take place (B) evolution would take place, but at a very slow rate (C) evolution would take place but at a very rapid rate (D) dominant characteristics would increase in the population and would eventually replace the recessive ones (E) recessive characteristics would increase in the population and would eventually replace the dominant ones

68. A B C D E

69. It is generally believed by scientists that chlorophyll and hemoglobin have a common origin. What is the best reason for believing this to be true? (A) All organisms contain either chlorophyll or hemoglobin (B) Heterotroph aggregates have the same basic structure as chlorophyll and hemoglobin (C) Chlorophyll can be changed to hemoglobin (D) Both substances perform the same biological function (E) Both substances are similar in chemical structure

69. A B C D E

70. The concentration of carbon dioxide in the atmosphere remains relatively constant at about 0.04% as a result of established equilibrium between the process of (A) assimilation and excretion (B) oxidation and photosynthesis (C) photosynthesis and assimilation (D) photosynthesis and reproduction (E) respiration and reproduction

70. A B C D E

71. Competition among the chipmunk population in a certain area could be expected to increase as a result of (A) a temporary increase in the chipmunk reproduction rate (B) an epidemic of rabies among chipmunks (C) an increase in the number of chipmunks killed on the highways (D) an increase in the number of hawks that prey on chipmunks (E) an increase in secondary consumers

71. A B C D E

72. The four-chambered heart represents an advance over the three-chambered heart because the four-chambered heart makes it possible to (A) deliver blood to the brain more rapidly (B) separate completely oxygenated and deoxygenated blood (C) vary blood temperatures to correspond with the

72. A B C D E

temperatures of the environment (D) circulate blood through the gill system more rapidly (E) prevent the blood in the left ventricle from backflowing into the left atrium

73. The methods of agriculture used by humans have created serious insect problems chiefly because these methods (A) increase soil erosion (B) provide concentrated areas of food for insects (C) increase the effectiveness of insecticides over a long period of time (D) grow crops in former desert areas (E) encourage insect resistance to their natural enemies

74. Under natural conditions large quantities of organic matter decay after each year's plant growth has been completed. As a result of such conditions (A) many plants are deprived of adequate food supplies (B) many animals are deprived of adequate food supplies (C) soils soon become exhausted if not fertilized (D) soils maintain their fertility (E) decomposers are placed at a disadvantage in the food chain they are part of

75. A plant breeder covered his flowers with paper bags to (A) keep stray pollen away (B) shade them from the sun's rays (C) protect the delicate organs from the wind (D) prevent self-pollination (E) keep them warm at night

76. Hawks are useful to the farmer because they (A) eat only sickly chickens (B) scare crows away from cornfields (C) destroy mice (D) eat harmful insects (E) keep other birds from eating his useful crops

77. The embryo of a man and the embryo of a rabbit are similar because (A) mutations occurred (B) they had a common ancestor (C) of survival of the fittest (D) they both carry on regeneration (E) they both have a backbone

78. There are no pure races of man in Europe because of (A) the existence of natural barriers (B) natural selection (C) artificial selection (D) inbreeding (E) extensive migration

79. The Drosophila fly is used in genetic research because (A) it is immune to typhoid fever (B) it is also known as the fruit fly (C) its salivary glands contain only two large chromosomes (D) the tradition of using it was established by Gregor Mendel (E) it breeds rapidly

80. The body of the woolly mammoth was preserved for thousands of years because (A) it was trapped in the La Brea tar pits (B) it was frozen in mud and ice (C) it was embedded in amber (D) it became petrified (E) its tough hide protected it against decay

81. A spider spins a web because (A) it is a habit (B) it is a tropism (C) it is an instinct (D) it is a conditioned reflex (E) it is a stimulus

73. A B C D E

74. A B C D E

75. A B C D E

76. A B C D E

77. A B C D E

78. A B C D E

79. A B C D E

80. A B C D E

81. A B C D E

TYPE C

Directions: **In each of the following questions, there is a statement followed by four choices, of which *one or more* are correct. Indicate the correct combination by choosing the following letters:**

(A) if only 1, 2, and 3 are correct
(B) if only 1 and 3 are correct
(C) if only 2 and 4 are correct
(D) if only 4 is correct
(E) if some other combination is correct

DIRECTIONS SUMMARIZED:

(A)	(B)	(C)	(D)	(E)
1, 2, 3 only	1, 3 only	2, 4 only	4 only	some other combination

82. Erosion on a hillside may be prevented by
1. strip cropping
2. terracing
3. contour plowing
4. reforestation

82. A B C D E

83. Herman J. Muller contributed
1. a theory to explain the action of auxin
2. the most recent fossil of primitive man
3. the use of the electron microscope for virus studies
4. to our knowledge of mutations

83. A B C D E

84. Cancer is difficult to control because
1. its cause is unknown
2. the symptoms are difficult to determine
3. it may spread through the body
4. it produces deadly antitoxins

84. A B C D E

85. The annual rings of a tree
1. may be used to determine the age of a tree
2. can be seen only in monocots
3. are added during the growing season
4. are composed of pith cells

85. A B C D E

86. Drugs such as morphine and codeine
1. relieve pain
2. are dangerous because they are habit-forming
3. are commonly taken in the form of special cigarettes
4. may be used with a doctor's prescription

86. A B C D E

87. Paramecium obtains food by
1. paralyzing its prey with its trichocysts
2. seizing it with its cilia
3. engulfing it
4. creating ciliary currents in the region of its oral groove

87. A B C D E

88. In order to build bones properly, the body needs
1. calcium and phosphorus
2. iodine and iron
3. vitamin D
4. vitamin B complex

88. A B C D E

89. In hemophilia
1. the blood does not clot properly
2. the gene for this condition is recessive
3. more males than females are affected
4. the marrow of the bones does not produce enough red blood cells

89. A B C D E

90. The temperature of the body is maintained by the
1. rate of breathing
2. storage of glycogen
3. rate of digestion
4. evaporation of sweat

90. A B C D E

PART IV—TYPE A

Directions: **Referring to the following selection, choose the best answer for each of the questions that follows:**

"Let me say one or two more things about the nature of species. When I was a young taxonomist, indeed even when I published my *Systematics and the Origin of Species* in 1942, the evolutionist had to fight the then still widespread *typological* concept of the species. There was a tendency in those days to treat each species *as if* it were a separate act of creation, monotypic and uniform. Now the populational approach to the species is so broadly adopted that there is a danger of falling into the opposite extreme, the danger of stressing too much the polytypic nature of species and overemphasizing the fact that species are composed of populations, of isolates, of subspecies.

"To be sure, a considerable percentage of species are polytypic, perhaps even the majority, but the differences between the various subspecies are far less important than the unifying ties, which derive from the same gene pool. In spite of superficial differences, the populations of a species seem to share the same homeostatic systems and the same physiological constants as a consequence of the same basic genotype. This ties the totality of local gene pools together into a single genetic system with great stability and strong cohesion."

91. According to the selection, all of the following statements about species express the modern point of view *except* (A) species are composed of isolates (B) species are polytypic (C) species are composed of subspecies (D) species are a separate act of creation (E) species show variation

91. A B C D E

92. In 1942, (A) evolution of species was not accepted by taxonomists (B) *Systematics and the Origin of Species* dealt with species and evolution (C) evolutionists and taxonomists were in exact agreement about species (D) the population approach to species was generally accepted (E) the concept of species evolution did not accept the idea of local gene pools

92. A B C D E

93. The differences between subspecies (A) are the cause of their having the same homeostatic systems (B) cause the typological concept of species (C) are less important than the unifying ties which derive from the same gene pool (D) lead to the same physiological constants (E) are due to their having a different basic genotype

93. A B C D E

94. The present concept of species is described as being (A) a populational approach (B) one of special creation (C) monotypic (D) uniform (E) typological

94. A B C D E

TYPE B

Directions: **Each of the following problems is followed by five suggested answers. Select the one answer which is best in each case.**

95. The Calorie value of 1 gram of each of the following nutrients has been determined:

NUTRIENT	CALORIES
carbohydrate	4
fat	9
protein	4

A sample of cottage cheese weighing 100 grams was analyzed to contain 19.5 proteins, 0.5 grams of fat, and 2.0 grams of carbohydrate. The total number of Calories available is (A) 4.5 (B) 78.0 (C) 90.5 (D) 100.0 (E) 195.0

95. A B C D E

96. The magnification of a microscope whose eyepiece is marked 7x and whose objective is marked 43x, is (A) 6 (B) 36 (C) 43 (D) 50 (E) 301

96. A B C D E

97. A hemophiliac man marries a normal heterozygous woman. The chances of their two sons having hemophilia are (A) 50–50 (B) 50–25–25 (C) 25–75 (D) zero (E) 100%

97. A B C D E

98. The basal metabolism of a normal adult man has been found to be such as to liberate 39 Calories per hour per square meter of skin surface. A person suffering from severe hyperthyroidism was tested, and the basal metabolism rate was determined as being (A) 39 Calories (B) 39% below normal (C) 100% above normal (D) 100% below normal (E) proportional to the production of adrenin

98. A B C D E

99. 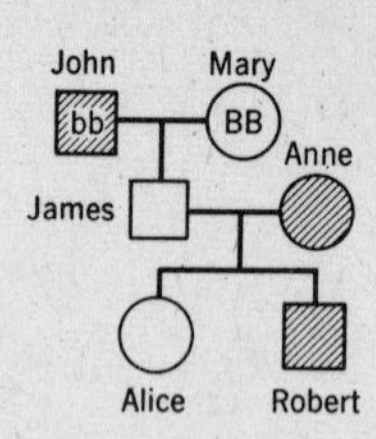

In the accompanying pedigree of a family, brown (B) eye color is indicated as ○ , and blue eye color (b) as ◍ . From this chart it can be determined that (A) Robert has the same genotype as his grandmother Mary (B) Robert has the same genotype as his father (C) Alice has the same genotype as her grandfather John (D) Alice has the same genotype as her father James (E) Alice has the same genotype as her grandmother Mary

99. A B C D E

100. When pink four o'clock flowers are crossed, the greatest number of flowers in the next generation would be (A) homozygous white (B) homozygous red (C) homozygous pink (D) heterozygous pink (E) heterozygous red

100. A B C D E

TYPE C

A goldfish was removed from an aquarium and placed on a glass plate. The front part of its body was covered with absorbent cotton that was soaking wet. Its tail was examined with a microscope, and the circulation of blood was studied.

Directions: **Select the one answer that is best in each of the following statements.**

101. The goldfish remains alive outside the aquarium because (A) its gills are surrounded by water (B) it is kept wet (C) as a lower animal, it has few reflexes (D) its breathing center is stimulated by the room temperature (E) it uses the absorbent cotton for artificial respiration

101. A B C D E

102. Some blood vessels were seen to be so small that blood cells passed through them in single file; these small blood vessels were (A) arteries (B) arterioles (C) capillaries (D) lacteals (E) venules

102. A B C D E

103. Small arteries could be recognized by (A) their valves (B) the spurting action of the blood in them (C) their attachment to the ventricle of the heart (D) their high content of oxygen (E) the absence of wastes in them

103. A B C D E

104. The heartbeat of the fish could be stimulated by the use of (A) a good quality fish food (B) massage (C) bubbling carbon dioxide through the water (D) replacing the wet absorbent cotton with dry absorbent cotton (E) adrenalin

104. A B C D E

PART V

105. In order to relieve the food shortages facing the ever-increasing world population, scientists are working on ways to farm the sea for food. Among their experiments, they are trying to learn how to grow prawns, shrimp, oysters, mussels and fish in special tanks and artificial ponds. In their research, they are faced with all of the following problems, except: (A) Not too much is known about fish diseases; an outbreak of disease can wipe out a sea-farmer's crop in a short time. (B) Some fish cannot live in ponds that are overcrowded. (C) If too much food is added to a pond, the fish may not survive. (D) Farming the sea for food would be unpopular with the farmers of our midwest because corn and fish are not nutritionally healthy when eaten with prawn, shrimp, mussles, oysters or fish. (E) Fish can live only in water that has the right temperature and amount of oxygen.

105. A B C D E

106. A farmer collected 203 pellets of indigestible remains dropped by barn owls on his farm. He took the pellets to a nearby museum for analysis of the type of food eaten by the owl. The following results were obtained:

106. A B C D E

429 meadow mice	18 jumping mice	1 squirrel
4 lemming mice	21 star-nosed moles	5 cottontail rabbits
1 pine mouse	1 Brewer's mole	23 unidentified mice
12 white-footed deer mice	96 short-tailed shrews	5 small birds

On the basis of these findings, the best course of action for the farmer to take would be to (A) kill all owls as a menace to his chickens (B) feed poisoned pellets to owls to get rid of them (C) offer special protection to barn owls (D) scare owls away with white mice (E) inform his neighbors that he now has proof that owls are a great menace to birds.

PART VI

107. Assume that a population of both white rabbits and gray rabbits is transferred to a white sandy desert from the field in which it normally lives. If the population survives, it is probably that the gene (A) frequency for white coat will increase (B) frequency for white coat will decrease (C) frequency for gray coat will increase (D) frequency for gray coat will remain unchanged (E) frequency for white coat will remain unchanged.

107. A B C D E

108. A substance named calvacin has been used to prevent the growth of certain types of tumors. It is obtained by extraction from a species of giant puffball mushroom. The most probably *immediate* use of this substance would be as a (A) supplement to the diet of cancer patients (B) supplement to the use of radioactive isotopes in treating cancer patients (C) basis for new cancer experiments (D) new method of treating cancer (E) new method of destroying unwanted growths in both plants and animals.

108. A B C D E

109. Lower forms of life such as protozoa, sponges and jellyfish have continued to exist since earliest geologic times. Which of the following statements would Darwin have considered as the best explanation for their continued existence to modern times?
(A) These organisms are still in the process of becoming extinct
(B) There is a fairly direct connection in evolution between the number of new forms that appear and the number of old ones that disappear
(C) Some of the simpler forms of life will survive no matter what environmental changes occur
(D) Some of the simpler organisms have a broad adaptation to the conditions of life
(E) Simpler organisms continue to exist because they are able to inherit acquired characteristics

109. A B C D E

ANSWER KEY: TEST NO. 10

1. (D)	**12.** (D)	**23.** (A)	**34.** (D)	**45.** (C)	**56.** (A)	**67.** (A)	**78.** (E)	**89.** (A)	**100.** (D)
2. (B)	**13.** (E)	**24.** (C)	**35.** (B)	**46.** (A)	**57.** (E)	**68.** (A)	**79.** (E)	**90.** (D)	**101.** (A)
3. (E)	**14.** (C)	**25.** (B)	**36.** (D)	**47.** (C)	**58.** (B)	**69.** (E)	**80.** (B)	**91.** (D)	**102.** (C)
4. (E)	**15.** (D)	**26.** (E)	**37.** (A)3	**48.** (B)	**59.** (C)	**70.** (B)	**81.** (C)	**92.** (B)	**103.** (B)
5. (D)	**16.** (A)	**27.** (C)	**38.** (C)4	**49.** (B)	**60.** (A)	**71.** (A)	**82.** (E)	**93.** (C)	**104.** (E)
6. (D)	**17.** (B)	**28.** (B)	**39.** (C)4	**50.** (A)	**61.** (C)	**72.** (B)	**83.** (D)	**94.** (A)	**105.** (D)
7. (B)	**18.** (E)	**29.** (D)	**40.** (A)1	**51.** (B)	**62.** (B)	**73.** (B)	**84.** (A)	**95.** (C)	**106.** (C)
8. (E)	**19.** (C)	**30.** (A)	**41.** (B)1	**52.** (D)	**63.** (E)	**74.** (D)	**85.** (B)	**96.** (E)	**107.** (A)
9. (D)	**20.** (C)	**31.** (C)	**42.** (A)2	**53.** (C)	**64.** (A)	**75.** (A)	**86.** (E)	**97.** (A)	**108.** (C)
10. (A)	**21.** (E)	**32.** (A)	**43.** (E)	**54.** (E)	**65.** (C)	**76.** (C)	**87.** (D)	**98.** (C)	**109.** (D)
11. (B)	**22.** (A)	**33.** (B)	**44.** (D)	**55.** (E)	**66.** (D)	**77.** (B)	**88.** (B)	**99.** (D)	

ANSWERS EXPLAINED: TEST NO. 10

1. (D) guard cells
2. (B) palisade layer
3. (E) vein
4. (E) vein
5. (D) guard cells
6. (D) guard cells
7. (B) palisade layer
8. (E) vein
9. (D) guard cells
10. (A) epidermis

11. (B) Anti-A blood typing serum contains an agglutinin that makes the red blood cells of type A blood clump together, or agglutinate. Anti-B serum contains an agglutinin that makes the red blood cells of type B blood clump. The blood mixed with each serum in this case was type A, since it clumped the red blood cells on the left side of the slide containing the anti-A serum.

12. (D) The ribosome is a tiny granule found in the endoplasmic reticulum of the cell. It contains most of the RNA of the cell, and is the site of protein synthesis, including enzyme formation.

13. (E) The structural formula of the disaccharide, maltose, is shown. It is a polymer of two glucose units. Under the influence of the enzyme maltase, it is hydrolyzed and converted into glucose.

14. (C) The reabsorption of glucose and other materials, takes place by active transport. Energy is released by the mitochondria of the nephron to move the glucose molecules through the membrane back into the bloodstream. This movement is opposite to that which would normally occur through diffusion.

15. (D) DNA is made of four kinds of nucleotides, in which the nitrogen bases may be arranged as follows: A (adenine) may be linked with T (thymine), and G (guanine) may be linked with C (cytosine). Combinations of these can give rise to various amino acids. RNA has U (uracil) in place of T. Since the sequence indicates the present of T, the code depicted is DNA.

16. (A) When members of a population are isolated into many small groups, they are prevented from breeding freely with each other. In such small populations, changes in the gene pool take place more readily. Random breeding would establish a new gene balance in each group that would be different from the others. New genetic characteristics would appear in each population. As the new types were prevented from breeding with the others, distinct species would occur.

17. (B) Other members of the clover family are alfalfa, peanut, bean and pea. These plants have nodules on their roots containing nitrogen-fixing bacteria. Their fruit is usually in the form of a pod.

18. (E) By definition, a species is a group of individuals that are so alike in structure that they can breed with one another and produce fertile offspring. They generally cannot breed with individuals of another species in nature.

19. (C) About 15% of the population has the Rh negative blood type.

20. (C) When reduction-division occurs, the chromosomes of a pair separate into the gametes to give the monoploid number.

21. (E) Thiamin, or vitamin B_1, is a member of the vitamin B complex.

22. (A) A tuber is an underground stem. The "eyes" are buds from which new plants will sprout when a piece of a potato is planted.

23. (A) When it ferments sugar, it gives off carbon dioxide and alcohol. The carbon dioxide causes dough to rise in the baking of bread and cake. The production of alcohol is important in the making of beer.

24. (C) One millimeter contains 1,000 microns. Therefore 0.6 mm. contain 600 microns ($1{,}000 \times 0.6$).

25. (B) Starfish feed on oysters by forcing the shells open and ingesting the soft material.

26. (E) An ecosystem is a self-sustaining community of plants and animals in relation to their physical environment. In order for it to be self-sustaining, it requires autotrophic organisms as its basis. These make their own food, using the constant source of energy available in the sun. The various cycles, such as carbon, water and nitrogen help to make the important elements available for the use of the organisms. As a result, matter and energy are constantly being exchanged between the community and the physical environment. Equal numbers of plants and animals are not significant in these relationships.

27. (C) A zygote is formed by the union of two gametes.

28. (B) A sperm is a male gamete that unites with a female gamete, the ovum, when fertilization occurs.

29. (D) A spore is an asexual cell, as in bread mold, that can reproduce a new plant.

30. (A) In conjugation, the gametes are alike.

31. (C) The nucleus of each gamete unites when a zygote is formed.

32. (A) After a zygote is formed, as in spirogyra and bread mold, a heavy wall forms around it.

33. (B) The gastrula stage is formed during cleavage of the fertilized egg.
34. (D) A bulb is an asexual structure of a plant such as a tulip.
35. (B) In fertilization, the gametes are different in size, shape, and activity.
36. (D) The cambium of a tree consists of a layer of actively dividing cells that differentiate to form xylem and phloem tissue.
37. (A)3 All are examples of evergreens except oak, which is a deciduous plant, losing its leaves in the winter.
38. (C)4 These are all present in both plant and animal cells except chlorophyll, which is found only in green plant cells.
39. (C)4 These are all related to the nervous system except coccyx, which consists of the small tailbones at the base of the spine.
40. (A)1 All are caused by lack of a vitamin (scurvy—vitamin C; pellagra—niacin; xerophthalmia—vitamin A; rickets—vitamin D) except diabetes, which is due to lack of the hormone insulin.
41. (B)1 All are examples of antibiotics except secretin, which is a hormone produced by the lining of the small intestine, and which stimulates the pancreas to secrete pancreatic juice into the small intestine.
42. (A)2 The DNA molecule is composed of smaller units called nucleotides, each of which contains a 5-carbon sugar (deoxyribose), a phosphate group and a nitrogen base which may be a purine (either adenine or guanine) or a pyrimidine (either cytosine or thymine). Glucose is an example of a monosaccharide.
43. (E) Branches (scions) of four other different apple trees were grafted on to an apple tree. Their genes were unaffected, and they bred true to the original type.
44. (D) A change took place in the genes, resulting in a mutation.
45. (C) Texas cattle had good beef qualities, but they were susceptible to Texas fever; the Brahman cattle were imported because they were immune to Texas fever, although their beef quality was inferior. When the two types were crossed, there were offspring that combined the desirable qualities of both.
46. (A) He selects the best ears each time, thus eliminating undesirable genes. Eventually, he will have corn seeds with the best genes.
47. (C) He combined the desirable qualities of each in the Shasta daisy; hardiness of the American daisy, large petals of the English daisy, and dazzling whiteness of the Japanese daisy.
48. (B) By inbreeding them no new genes are introduced, and they retain the desirable characteristics.
49. (B) He undoubtedly is suffering from an overactive thyroid gland. As a result of the extra amount of thyroxin, his metabolism is carried on at a faster rate.
50. (A) During times of emotional stress, a greater amount of adrenin is secreted by the adrenal glands; this causes an increased amount of glucose and oxygen to be brought to the muscles, supplying them with additional energy.
51. (B) Her thyroid gland is underactive, resulting in a lower metabolism rate. Her food is not entirely oxidized, and the excess is stored as fat.
52. (D) His pituitary gland did not secrete enough of the hormone that controls growth.
53. (C) The patient's pancreas may not be producing enough insulin, and he may have a case of diabetes. Insulin permits the cells to use glucose. When it is not present, glucose is excreted as a waste.
54. (E) A total of 34 daisies were found with 16 petals on each.
55. (E) Only 3 daisies were found with 22 petals on them.
56. (A) If there had been more small daisies with fewer petals, the left side of the curve would have risen more gradually.
57. (E) He could not draw any conclusions from his first study. He would have to try it.
58. (B) The number of petals varies from 11 per daisy to 22 per daisy, with most daisies having 16 petals.
59. (C) It is not possible to make accurate predictions of this sort, since living things vary. A similar study would have to be made of asters before any conclusions could be reached.
60. (A) The termite also has three body parts; a head, a thorax, and an abdomen. It is covered with an exoskeleton of chitin.
61. (C) Plant stems grow upward away from the force of gravity; this is an example of negative geotropism.
62. (B) ACTH stimulates the production of cortisone by the cortex of the adrenal gland.
63. (E) It is adapted for seed dispersal in many

ways: a cherry seed is discarded after the fruit has been eaten, far from the cherry tree; a cocklebur has hooks by which it attaches itself to a passing animal; dandelion seeds have a feathery parachute.

64. (A) Mammal embryos are nourished internally on nutrients provided by the mother's bloodstream.

65. (C) The footprints were embedded in mud which dried and hardened; then it was covered with sediment and water, and gradually turned to rock.

66. (D) Travelers in space are far above the atmosphere and are not shielded against cosmic rays, which can penetrate the body and damage tissue.

67. (A) These bacteria change nitrogen into nitrate compounds which are needed by plants.

68. (A) According to the Hardy-Weinberg principle, gene frequencies remain constant from generation to generation if mating is at random, and no new factors are introduced, such as mutation, selection or migration. Under such theoretical conditions, the species would remain predictably the same.

69. (E) Hemoglobin is closely related, chemically, to chlorophyll. However, the hemoglobin molecule contains iron, while the chlorophyll molecule contains magnesium. Both substances belong to a class of pigments known as porphyrins. According to the heterotroph hypothesis, primitive protein molecules may have developed, leading to porphyrin molecules that served as a common origin for chlorophyll and hemoglobin.

70. (B) During oxidation, living things take in oxygen and give off carbon dioxide and water as wastes. During photosynthesis, carbon dioxide and water are taken in by green plants and oxygen is given off as a waste. As a result, the concentrations of carbon dioxide and oxygen in the atmosphere remain relatively constant.

71. (A) If there is a temporary increase in the chipmunk reproduction rate, there would be an increase in the number of chipmunks that would be seeking food in the given area. This would result in increased competition among them for the available food and other necessary conditions for life.

72. (B) In the four-chambered heart of birds and mammals, deoxygenated blood entering from the body goes into the right atrium and then into the right ventricle. From here, it goes to the lungs and returns as oxygenated blood to the left atrium and then the left ventricle. In the three-chambered heart of amphibia, there is only one ventricle; oxygenated blood from the lungs returns to the ventricle where it mixes partially with the deoxygenated blood that entered from the body.

73. (B) Insects are man's greatest enemies. Although other forms of life have had difficulty maintaining themselves with the spread of civilization, insects have continued to multiply and spread. Man has assisted this growth of insects by planting crops in large fields providing them with concentrated areas of food, i.e., the potato beetle, and the corn borer do not have far to go for their food, once they settle in the right field.

74. (D) The organic material in plants is attacked by bacteria of decay, and fungi. The protein compounds are digested by them and broken down into amino acids, and then into simpler nitrates and other compounds. Thus, the fertility of soils is maintained.

75. (A) A plant breeder pollinates a flower with selected pollen, and then covers it with a paper bag to prevent the wrong type of pollen from being introduced by insects or by the wind.

76. (C) The hawk is a carnivorous bird that is useful because it destroys more rodents than birds. Mice are harmful because they eat grain.

77. (B) Mammals had a distant common ancestor. They changed since that time, but still retain enough of the same genes to appear similar in the embryo stages.

78. (E) Because of migrating and interbreeding, there has been considerable mixing of the genes.

79. (E) It produces a new generation every two weeks, permitting the study of many generations in one year.

80. (B) By being frozen, it was protected from bacterial decay.

81. (C) This is an inborn, unlearned, and automatic act, consisting of a series of reflexes.

82. (E) All of the items are correct.

83. (D) He received the Nobel Prize for his work on the production of mutations in drosophila flies that were exposed to X-rays.

84. (A) Cancerous growth may be detected when it is too late to be cured.

85. (B) Each spring, cambium is stimulated to produce an active growth of xylem cells, which form a concentric layer or annual ring within the trunk.

86. (E) All of the items are correct except number 3, which refers to marijuana.

87. (D) As cilia beat back and forth, they create currents which sweep water and particles of food into the oral groove.

88. (B) Vitamin D makes it possible for the body to utilize calcium and phosphorus in building bones and teeth.

89. (A) Hemophilia is a sex-linked, recessive trait. Since the Y chromosome has practically no genes, a male will have hemophilia if he has only one gene for this trait on the X chromosome. A female needs two genes, one on each of the two X chromosomes, to have the condition.

90. (D) As sweat evaporates, it cools the skin. In summer, we sweat more in order to cool off.

91. (D) In 1942, the evolutionist had to fight a point of view that was still widespread in those days. There was a tendency to treat each species as if it were a separate act of creation.

92. (B) The author of this selection refers to the fact that he published the *Systematics and the Origin of Species* in 1942. In it, he referred to the process of evolution at the species level.

93. (C) In the second paragraph, the author states that the differences between the various species are not as important as their common characteristics. These are unifying ties, arising from their sharing the same gene pool.

94. (A) In contrast to the old approach to the concept of species, the populational approach to species is broadly accepted. In the former approach, there had been a tendency to treat each species as if it were a separate act of creation, monotypic and uniform.

95. (C) The total number of Calories can be estimated by obtaining the energy value of all the nutrients, and adding them, i.e. 19.5 gr. proteins × 4 Calories per gram = 78.0; 0.5 gr. fats × 9 Calories per gram = 4.5; 2.0 gr. carbohydrates × 4 Calories per gram = 8.0.

96. (E) The eyepiece magnification is multiplied by the objective magnification, i.e., $7 \times 43 = 301$.

97. (A) The possible results are:
(Key: X = chromosome with gene for normal clotting; $\underline{X}$ = chromosome with gene for hemophilia; XX = female; XY = male)

	$\underline{X}$	Y
X	X$\underline{X}$	XY
$\underline{X}$	$\underline{X}\underline{X}$	$\underline{X}$Y

Male: 50% normal
50% hemophiliac
Female: 50% normal carrier
50% hemophiliac

98. (C) In hyperthyroidism, there is an excessive production of thyroxin, leading to a considerable increase in the basal metabolism rate. This results in an increased rate of oxidation, and an increase in the number of calories liberated.

99. (D) James is heterozygous for brown eyes, since he received a gene for brown eyes from his father, John, and a gene for blue eyes from his mother, Mary. His daughter Alice is also heterozygous for brown eyes, receiving a gene for brown eyes from him, and a gene for blue eyes from Anne.

100. (D) When hybrids are crossed, the results are in a 1:2:1 ratio. In this example of incomplete dominance, the homozygous types are red (RR) and white (WW). Pink is heterozygous (RW). The possible results are:

	R	W
R	RR	RW
W	RW	WW

25% red
50% pink
25% white

101. (A) The gills absorb the oxygen in the water.

102. (C) Capillaries are the smallest blood vessels that reach into every part of the body.

103. (B) The blood spurts in arteries because of the contractions and expansions of the ventricle of the heart.

104. (E) Adrenalin is a hormone that makes the heart beat more rapidly and more strongly. It is frequently administered to humans who have had a heart attack.

105. (D) All of the problems stated in the question are practical challenges to scientists seeking to expand the world's sources of food. The type of sea animals mentioned are now used as a source of food in many parts of the world. They are nutritious and help to sustain life. If corn is added to the diet, it could only improve the basic nutrition.

106. (C) The owl is a very useful part of his farm environment, since it destroys animals that eat his grain and his crops. The evidence indicates that owls rarely go after birds, but live mainly on rodents and other small mammals.

107. (A) By natural selection, the protective coloration of the white rabbits will make them more fit to survive in the white sandy desert. They will blend with the background more readily, and escape the notice of predators to a greater extent than the gray rabbits. More of the white rabbits will survive and reproduce, thus increasing the gene frequency for white coat.

108. (C) Tumors are growths of cells, and are types of cancer. If calvacin prevents the growth of certain types of tumors, it may be useful in controlling cancer. It would therefore be used in new cancer experiments to determine whether it can be useful in this way.

109. (D) Darwin stated that the organism that possessed favorable variations would be most fit, in the struggle for existence, and would survive. The lower forms of life are broadly adapted to the conditions surrounding them, and have continued to survive.

Index